Neuro-Urology Research

Neuro-Urology Research

A Comprehensive Overview

Edited by

Anne M.J. Verstegen, PhD
Department of Medicine, Beth Israel Deaconess Medical Center/
Harvard Medical School, Boston, MA, United States

ELSEVIER
ACADEMIC PRESS
An imprint of Elsevier

Academic Press is an imprint of Elsevier
125 London Wall, London EC2Y 5AS, United Kingdom
525 B Street, Suite 1650, San Diego, CA 92101, United States
50 Hampshire Street, 5th Floor, Cambridge, MA 02139, United States
The Boulevard, Langford Lane, Kidlington, Oxford OX5 1GB, United Kingdom

ISBN: 978-0-12-822455-7

For information on all Academic Press publications visit our website at https://www.elsevier.com/books-and-journals

Publisher: Stacy Masucci
Acquisitions Editor: Stacy Masucci
Editorial Project Manager: Sara Pianavilla
Production Project Manager: Omer Mukthar
Cover Designer: Miles Hitchen

Typeset by TNQ Technologies

Contents

Section II
Fundamental and translational neuro-urology research

5. How treatment of lower urinary tract symptoms can benefit from basic research

Jason P. Van Batavia

6. "Translational effects of neuro-urology research on clinical practice"; Patient population—specific lower urinary tract symptoms

Ingrid Hoeritzauer

7. Effect of androgens and estrogens on bladder/lower urinary tract function

Anne E. Turco and Chad M. Vezina

Section IV
Research directions and research opportunities

11. Impact of spinal neuromodulation on spinal neural networks controlling lower urinary tract function

Parag Gad and Evgeniy Kreydin

12. Neural control of continence

Danielle J. Gordon, Abdo E. Kabarriti and Jeffrey P. Weiss

Contributors

Michael B. Chancellor, Department of Urology, Oakland University William Beaumont School of Medicine, Royal Oak, MI, United States

William C. de Groat, Department of Pharmacology and Chemical Biology, University of Pittsburgh School of Medicine, Pittsburgh, PA, United States

Parag Gad, SpineX Inc, Los Angeles, CA, United States; Rancho Research Institute, Rancho Los Amigos National Rehabilitation Center, Downey, CA, United States

Joel C. Geerling, University of Iowa, Department of Neurology, Iowa City, IA, United States

Danielle J. Gordon, Department of Urology, SUNY Downstate Health Sciences University, Brooklyn, NY, United States

Alexander L. Green, ALG Nuffield Department of Surgical Sciences, Oxford & Department of Neurosurgery, John Radcliffe Hospital, Oxford, United Kingdom

Ingrid Hoeritzauer, Department of Clinical Neurosciences, Royal Infirmary Edinburgh, NHS Lothian and Centre for Clinical Brain Sciences, University of Edinburgh, Edinburgh, United Kingdom

William F. Jackson, Department of Pharmacology and Toxicology, Michigan State University, East Lansing, MI, United States

Abdo E. Kabarriti, Department of Urology, SUNY Downstate Health Sciences University, Brooklyn, NY, United States

Takeya Kitta, Department of Renal and Urologic Surgery, Asahikawa Medical University, Asahikawa, Hokkaido, Japan; Department of Urology, University of Pittsburgh School of Medicine, Pittsburgh, PA, United States

Evgeniy Kreydin, SpineX Inc, Los Angeles, CA, United States; Rancho Research Institute, Rancho Los Amigos National Rehabilitation Center, Downey, CA, United States; Institute of Urology, Keck School of Medicine of University of Southern California, Los Angeles, CA, United States

Teruyuki Ogawa, Department of Urology, University of Pittsburgh School of Medicine, Pittsburgh, PA, United States

Holly A. Roy, University of Plymouth and Department of Neurosurgery, Derriford Hospital, Plymouth, United Kingdom

Douglas Strand, University of Texas Southwestern Medical Center, Department of Urology, Dallas, TX, United States

Margaret M. Tish, University of Iowa, Department of Neurology, Iowa City, IA, United States

Anne E. Turco, University of Wisconsin-Madison, Molecular and Environmental Toxicology, Madison, WI, United States

Nathan R. Tykocki, Department of Pharmacology and Toxicology, Michigan State University, East Lansing, MI, United States

Jason P. Van Batavia, Division of Pediatric Urology, Children's Hospital of Philadelphia, University of Pennsylvania Perelman School of Medicine, Philadelphia, PA, United States

Anne M.J. Verstegen, Department of Medicine, Beth Israel Deaconess Medical Center, Harvard Medical School, Boston, MA, United States

Chad M. Vezina, University of Wisconsin-Madison, Department of Comparative Biosciences, George M. O'Brien Center for Benign Urology Research, Madison, WI, United States

Jeffrey P. Weiss, Department of Urology, SUNY Downstate Health Sciences University, Brooklyn, NY, United States

Naoki Yoshimura, Department of Urology, University of Pittsburgh School of Medicine, Pittsburgh, PA, United States; Department of Pharmacology and Chemical Biology, University of Pittsburgh School of Medicine, Pittsburgh, PA, United States

About the editor

Anne M.J. Verstegen, PhD

Dr. Verstegen is a neuroscientist researcher specializing in the investigation of brain circuits and in linking the activity of defined subpopulations of neurons with specific neurobehavioral and physiological processes.

She graduated from the master's program "Experimental and Clinical Neuroscience" at Utrecht University in the Netherlands. She was awarded her PhD degree from the Neuroscience and Brain Technologies Department at the Italian Institute of Technology. She then moved to Boston, Massachusetts, to start a postdoctoral researcher position at Beth Israel Deaconess Medical Center. Since 2017, her research has focused on mapping bladder-control pathways in the brain and spinal cord, applying neuroscience (methodology) to the underexplored field of neuro-urology. Her research interests include the identification of the etiology and the unraveling of neural mechanisms underlying lost bladder-control as this often accompanies neurological disorders.

Beth Israel Deaconess Medical Center, Department of Medicine, Boston, MA, United States; Harvard Medical School, Boston, MA, United States.

Preface

Why read this book? To gain a comprehensive overview of neuro-urology research.

The lower urinary tract (LUT) is controlled by complex mechanisms. Chapters in this book review the neural pathways, cell types, and neurotransmitters involved in micturition (urination) and review how neural control is ultimately important for the bladder to function. This book combines pieces of the puzzle to clearly outline what is known and what is still not known about how the brain and bladder communicate with each other.

Because of the heterogeneity of symptoms associated with abnormal urine storage and voiding, many mechanisms of LUT disorders have remained unclear. These knowledge gaps complicate the translation of our current understanding to clinical conditions. Sections in this book also focus on the applicability of basic research to clinical practice.

What conditions are discussed? Examples of 'conditions' that frequently affect normal bladder function and that are detailed in this book include the following:

- Aging
- Benign prostatic hyperplasia
- Emotional trauma
- Fluctuating sex hormone levels
- Fowler's syndrome
- Neurological diseases (including dementia, multiple sclerosis, multiple system atrophy, and stroke)
- Spinal cord injury
- Traumatic brain injury

Who is the audience? This book is intended for urology researchers and neuro-urologists, of course, because of the emphasis on and synthesis of urology and neuroscience, but also healthcare practitioners in patient-facing settings as most chapters elaborate on how research may relate to human patients. Additionally, clinical urologists may appreciate the detailed explanation of basic research techniques and discussions about how fundamental research can inform the clinic.

Finally, this book is for clinicians and researchers who are less familiar with the neuro-urology field and want to learn about the neural control of LUT function and the effects of functional and dysfunctional neural circuits on bladder function. Clinical presentations are explained and new areas for study presented. This book is written in an intentionally understandable way for nonresearcher-clinicians and nonclinician-researchers alike.

This book highlights the evolution of neuro-urology research, emphasizes how technology is advancing research toward the goal of a greater appreciation and better understanding of 'the neural component' of LUT function, and advocates for improved treatments for LUT dysfunction. Chapters in this book review the science that has served as the foundation of our current knowledge in the field and look forward to the research that will drive its future.

Anne M.J. Verstegen

Preface

Acknowledgments

No author has ever published without the generous support and assistance from many people. I am grateful for the time and effort that the following contributors put into making this book a reality: Michael B. Chancellor, William C. de Groat, Parag Gad, Joel C. Geerling, Danielle J. Gordon, Alexander L. Green, Ingrid Houritzauer, William F. Jackson, Abdo E. Kabarriti, Takeya Kitta, Evgeniy Kreydin, Teruyuki Ogawa, Holly A. Roy, Doug Strand, Margaret M. Tish, Anne E. Turco, Nathan R. Tykocki, Jason VanBatavia, Chad M. Vezina, Jeffrey P. Weiss, and Naoki Yoshimura.

In particular, I would like to thank Natalie Klymko for contributing the skillful drawings of several figures in Chapter 1.

I also wish to thank Cassandra Seifert (medical student at the University of Maryland Medical Center, Class of 2025) for excellent feedback and editorial support for the introductory chapter.

Finally, I owe special thanks to Patrick Fuller for his careful editing and constant encouragement; to my colleagues at Beth Israel Deaconess Medical Center for their kind collaboration; and to my dear friends and family for their continued love and support.

Chapter 1

Neuro-urology research: a comprehensive overview

Anne M.J. Verstegen
Department of Medicine, Beth Israel Deaconess Medical Center, Harvard Medical School, Boston, MA, United States

Neuro-urology studies how the brain and spinal cord control lower urinary tract (LUT) function. Neuro-urology research focuses on diseases and functional disorders of the LUT, including the bladder and urethra, and pelvic floor musculature that can derive from spinal injuries and neurological disorders.

Despite the fact that LUT dysfunction is very common, much is still unknown in the field of neuro-urology. Lower urinary tract symptoms (LUTS) include symptoms experienced during voiding or storage of urine. In addition to debilitating physical challenges, LUT dysfunction often causes adverse mental health outcomes including anxiety and depression. "Quality-of-life" surveys among study participants with spinal cord injuries report that bowel and bladder dysfunctions have a significant negative impact on their lives (Anderson, 2004; Hanson & Franklin, 1976; Mcgee & Grill, 2015). Furthermore, caregivers to individuals who have lost the ability to void normally rank LUTS as being among the greatest burdens to navigate (Stewart et al., 2003; Irwin et al., 2005; Coyne et al., 2009b; Gotoh et al., 2009).

Several large epidemiological studies evaluating the prevalence of LUTS (Irwin et al., 2011; Coyne et al., 2009a) conclude that the symptoms are highly prevalent and affect over 60% of men and women aged 40 or more years, with some variability depending on the study population (Przydacz et al., 2020). Meanwhile global population predictions include a significant percentage per population increase in older adults in the years to come (Department of Economic and Social Affairs, 2013). Thus, the worldwide population affected by LUTS is expected to grow.

LUT dysfunction can arise from disorders of the LUT or from disorders of the central nervous system (CNS) such as Parkinson's disease or traumatic brain injury (TBI). In individuals with neurological diseases, bladder problems are usually due to discoordination between the brain, spinal cord, and LUT, and the majority of neurological disorders exhibit LUT problems secondarily. Because investigations and management of bladder problems have focused mostly on dysfunctional outflow at the level of the bladder, the field will benefit from more fundamental research aimed at the nervous system control of LUT function in health and disease.

Research in neuro-urology is further relevant to public health because it is a route to understanding how bladder function is controlled and how urinary continence is maintained. This type of fundamental knowledge is needed to improve pharmacological approaches and interventional strategies and thereby reduce the burdens of human disability.

Outline

This "niche" book describes the current status of the neuro-urology field including the latest discoveries, explains in detail some of the neuroscience tools that can be used when studying the neural control of the LUT, and discusses potential future directions for research.

Chapter 1—Neuro-urology research: a comprehensive overview

The first chapter gives an introduction to the field of neuro-urology and neuro-urology research. In this field, fundamental research is needed for a better understanding of the central signaling pathways, as well as to identify the cells and specific genes that play critical roles in these pathways and that may be targets for new therapeutics to treat LUT dysfunction. After

Neuro-Urology Research. https://doi.org/10.1016/B978-0-12-822455-7.00013-1

a general introduction, we review approaches to research in the Past and in the Present. This is followed by a discussion of research topics, and research opportunities for the future. The final part of this chapter is a "Tools" section that provides an introductory guide to useful research tools and the "neuroscience toolbox."

Following the Introductory chapter, the book is divided into four sections:

- Neuroscience in Urology Research
- Fundamental and Translational Neuro-Urology Research
- Neurobiological Tools Applied to Neuro-Urology Research
- Research Directions and Research Opportunities.

Neuroscience in Urology Research (Chapters 2–4) reviews traditional and current studies, including the knowledge base for what neuron subtypes and specific neuroregulators are present in the brain micturition centers, voluntary versus reflex micturition control, and how bladder physiology readouts can be used for studying the effects of neuro-stimulations and the extent of functional recovery after manipulations.

Chapter 2—Barrington's nucleus: a century of progress identifying neurons that control micturition

Barrington's nucleus, also known as the "Pontine micturition center," is critical for micturition. Chapter 2 focuses on the anatomy of the brainstem region where this nucleus is located. It describes the structures and nuclei in the vicinity that may or may not play a role in micturition behavior themselves. It reviews what the neuron subpopulations typical for Barrington's nucleus are, the projection targets of neuron subtypes, and how activity in Barrington's nucleus is regulated by upstream, input-providing brain sites. This chapter details everything that is currently known about these hindbrain neurons and their role in micturition control and pelvic organ function.

Chapter 3—Voluntary versus reflex micturition control

The activity of smooth and striated muscles of the bladder and urethral outlet is coordinated by a complex neural control system that involves the brain, spinal cord, and peripheral autonomic ganglia. Chapter 3 summarizes the research that has led to our current understanding of the micturition reflex pathway and all its functional components. In addition, because in most socialized mammals the micturition reflex is under conscious inhibitory control from the forebrain, this chapter describes which connections and brain regions modulate the brainstem circuitry for voluntary micturition control.

Chapter 4—The bladder as a readout in neuroscience research

Understanding how the brain controls micturition—and how this control may be lost—is an important focus of neuroscience research. Bladder physiology can be used as a readout to generate a more complete view of important brain sites involved in bladder function. This readout can also be incorporated into studies that use bladder function as an index of damage resulting from a disease process, or of regeneration, and it is expected that studies like these will ultimately lead to more insight. Furthermore, there is growing interest in the role of the bladder in social and emotional disorders, and, accordingly, there is emerging research and literature that studies the links between bladder physiology and these conditions and behaviors. The authors of Chapter 4 consolidate these research studies.

Fundamental and Translational Neuro-Urology Research (Chapters 5–7) discusses the translational potential of basic research for patients and the impact of neuro-urology research on clinical practice. For example, through the reviewing of case studies or by describing how diseases can be modeled, collaborations between basic research and clinical medicine can advance both fields. Included here are the workings of bladder afferent signaling and the sensation of bladder stretch, the impact of a diversity of neurological diseases on bladder function, and the role of sex hormones in health and in the setting of LUT dysfunction. Chapters in this section aim to provide more insight into pathologies endemic to specific patient populations and potential areas for treatment development.

Chapter 5—How treatment of lower urinary tract symptoms can benefit from basic research

Afferent pathways arising from the LUT involve a delicate interplay between chemical and mechanical signaling mechanisms. Central to this chapter is understanding how neurotransmitters and receptors in the sensory limb of the system interact to produce bladder sensation, the promoting versus suppressing of bladder afferent activity during bladder filling, and the unique role of the urothelium as a mechanosensory organ. Conventional therapies for LUTS rely on receptors in the bladder. In this chapter, the focus is taken from the bladder and shifted to the bladder afferents. Here, LUTS are looked at as a consequence of sensory signaling dysfunction. For this, the role of neuronal afferents in specific clinical conditions of overactive and underactive bladder is explored. The chapter further addresses how basic science research into afferent neuronal pathways can lead to novel therapeutic targets for treatment of common LUTS.

Chapter 6— "Translational effects of neuro-urology research on clinical practice"; Patient population–specific lower urinary tract symptoms

Bladder problems disproportionally affect patients with neurological diseases. This chapter familiarizes the reader with patient populations in which LUTS most often occur and describes current treatments as well as how ongoing basic research may lead to new treatment modalities. The chapter covers LUTS presenting in dementia, stroke, Parkinson's disease, multiple system atrophy, multiple sclerosis, spinal cord injury (SCI), spina bifida, cauda equina syndrome, Fowler's syndrome, and functional neurological disorders. The chapter aims to generate greater acknowledgment of LUTS caused by neurological disease, and greater awareness of the challenges LUT dysfunction causes for people living with neurological disease.

Chapter 7—Effect of androgens and estrogens on bladder/lower urinary tract function

Sex hormone concentrations have an influence on male and female bladder anatomy and physiology and on bladder-related disease. This chapter focuses on research that studies the effects of androgens and estrogens, and of changing hormone levels, on bladder and LUT function. The authors also discuss the impact of hormones on healthy systems and in disease states, and the relationship between age-related hormone changes and urinary disease processes.

Neurobiological Tools Applied to Neuro-Urology Research (Chapters 8–9) supplements "A quick guide to the 'neuroscience toolbox'" found in this introductory Chapter 1. This section introduces research techniques. Transcriptome profiling using RNA sequencing and calcium imaging of neural activity are explained in great detail. The question of how these tools can be applied to neuro-urology research to elucidate and subsequently probe the underlying neuronal "wiring diagram" of LUT function is explored.

Chapter 8—Transcriptomic identification of cell types in the lower urinary tract

RNA sequencing can be used to understand the molecular differences between two samples by identifying "differentially expressed genes." Using this new technique developed in the mid-2000s, the different cell types in entire organs can be identified and gene expression profiles can be compared between healthy and pathological states. The transcriptome of individual cells from, for example, a specific part of the LUT, or from a region in the spinal cord or brain, can be sequenced and studied further. Several protocols for bulk RNA sequencing, tissue digestion, and single-cell RNA sequencing (scRNA-Seq), as well as an overview of bioinformatics and many reference papers in this chapter provide the reader with deeper insight into transcriptomics.

Chapter 9—Exploring urinary bladder neural circuitry through calcium imaging

This chapter describes a range of imaging tools that are being used to measure calcium signals. These methods can be applied to image live neurons and to study bladder-associated neural circuits. Pros and cons are addressed for the use of each of these state-of-the-art tools including their use in the recording of single-cell excitability and the examination of neuronal circuitry in both in vivo and ex vivo tissue sections. The imaging techniques are compared with non–imaging-based approaches for recording neural activity. Finally, promising and even newer tools that allow for visualization of membrane potentials and ion flux throughout a neuron are explored. The authors give useful resources to support the successful use of these tools in every laboratory.

Finally, **Research Directions and Research Opportunities** (Chapters 10–12) discusses research directions that remain underexplored or have high therapeutic potential. Chapters included in this section give more insight into the periaqueductal gray (PAG) (the bladder control command center, which because of its complexity and heterogeneity of both cell types and functions is still not completely fathomed), into neural networks that are indispensable for the regulation of micturition and micturition-related behaviors, and into brain regions that support the maintenance of continence.

Chapter 10—The periaqueductal gray and control of bladder function

The midbrain PAG is "caught in the middle" as a receiver of sensory information regarding bladder fullness and as a sender of integrated information to Barrington's nucleus. This chapter summarizes research focused on the PAG as a relay center and its functioning as a switch for the micturition reflex. This important brain region has been explored using brain imaging techniques, pharmacological assays, immunohistological analyses, and electrophysiological methods.

Chapter 11—Impact of spinal neuromodulation on spinal neural networks controlling lower urinary tract function

This chapter focuses on the role of the spinal cord in LUT control. It reviews and discusses spinal networks, SCI leading to severe LUT dysfunction, and a range of commonly used approaches and experimental techniques for stimulating spinal neural networks. One example of these techniques is transcutaneous electrical spinal cord neuromodulation, which may be used for improving LUT function after SCI.

Chapter 12—Neural control of continence

Functional brain imaging in patients with urge incontinence and Fowler's syndrome has provided important insight into brain areas that are involved in the abnormal sense of urgency. This chapter discusses how brain regions activate and deactivate during bladder filling and voiding to explain incontinence upon neurological insult and to better understand neural circuits in the control of continence. Furthermore, the chapter provides a comprehensive review of treatments for clinical conditions causing incontinence and focusses on how potential future treatments can be targeted to neural pathways. Conditions discussed include suprapontine lesions, bladder outlet obstruction, interstitial cystitis, diabetes mellitus and detrusor underactivity, and aging.

Introduction to neuro-urology

This Introduction is divided into three parts. First, past and present research in neuro-urology is reviewed. This is followed by a discussion of research questions and future research directions for the field. Finally, a **"Tools" section** provides an introductory guide to useful research tools and the "neuroscience toolbox."

Neuro-urology is focused on understanding how the brain regulates bladder function. A century ago the physiologist Dr. F.J. Barrington performed lesion experiments and found that a specific region in the brainstem was necessary for micturition. Cats with focal lesions in this specific region could not void their bladder, underscoring the importance of the implicated area in bladder physiology (Barrington, 1925, 1927). Since then, studies have narrowed down the exact anatomical location of cells participating in micturition behavior to a nucleus in the pons positioned inferior to the fourth ventricle and medial to the locus coeruleus (Fig. 1.1). The functionally important "Barrington's nucleus" is critical for normal bladder function in several species, including in humans.

The major supraspinal influence on the LUT arises from Barrington's nucleus. Barrington's nucleus neurons directly innervate bladder motor neurons and indirectly innervate external urethral sphincter (EUS) motor neurons through spinal interneurons (Blok et al., 1998). Bladder motor neurons in the intermediolateral (IML) cell column of the sacral spinal cord give rise to preganglionic parasympathetic fibers (Nadelhaft et al., 1992) (Figs. 1.1 and 1.2). Somatic innervation of the EUS originates from fibers of EUS motor neurons located in the ventral horn at lumbar and lumbosacral spinal cord levels (Karnup & de Groat, 2020; Nadelhaft & Vera, 1996). These EUS motor neuron fibers travel through the pudendal nerve, which is the major somatic nerve carrying motor information to the EUS. Excitation of the descending pathway results in simultaneous bladder contraction and EUS relaxation (Fig. 1.2). Loss of brain influence, as can occur, for example, as a result of spinal cord injury (SCI), leads to detrusor overactivity and bladder contractions against a closed sphincter. This precipitating detrusor sphincter dyssynergia (DSD) results in incomplete bladder emptying, loss of compliance because of high pressures, and possible reflux of urine to the kidneys (vesicoureteral reflux) (Birder & Drake, 2009; Fowler et al., 2008; Fowler & Griffiths, 2010). Barrington's nucleus neurons, through their projections to the spinal cord level where the sacral pelvic nerves arise, likely control motor neurons involved in other pelvic functions as well (Kawatani et al., 2021; Rouzade-Dominguez et al., 2003; Salas et al., 2008).

Sensory feedback from the visceral and somatic structures travels to the spinal cord by "hitchhiking" on the autonomic and somatic motor nerves, as they contain both afferent (sensory) and efferent (motor) axons (see Fig. 1.2) (Gomez-Amaya et al., 2015). Primary afferents travel to the lateral dorsal horn, sacral parasympathetic nucleus (SPN), and dorsal gray

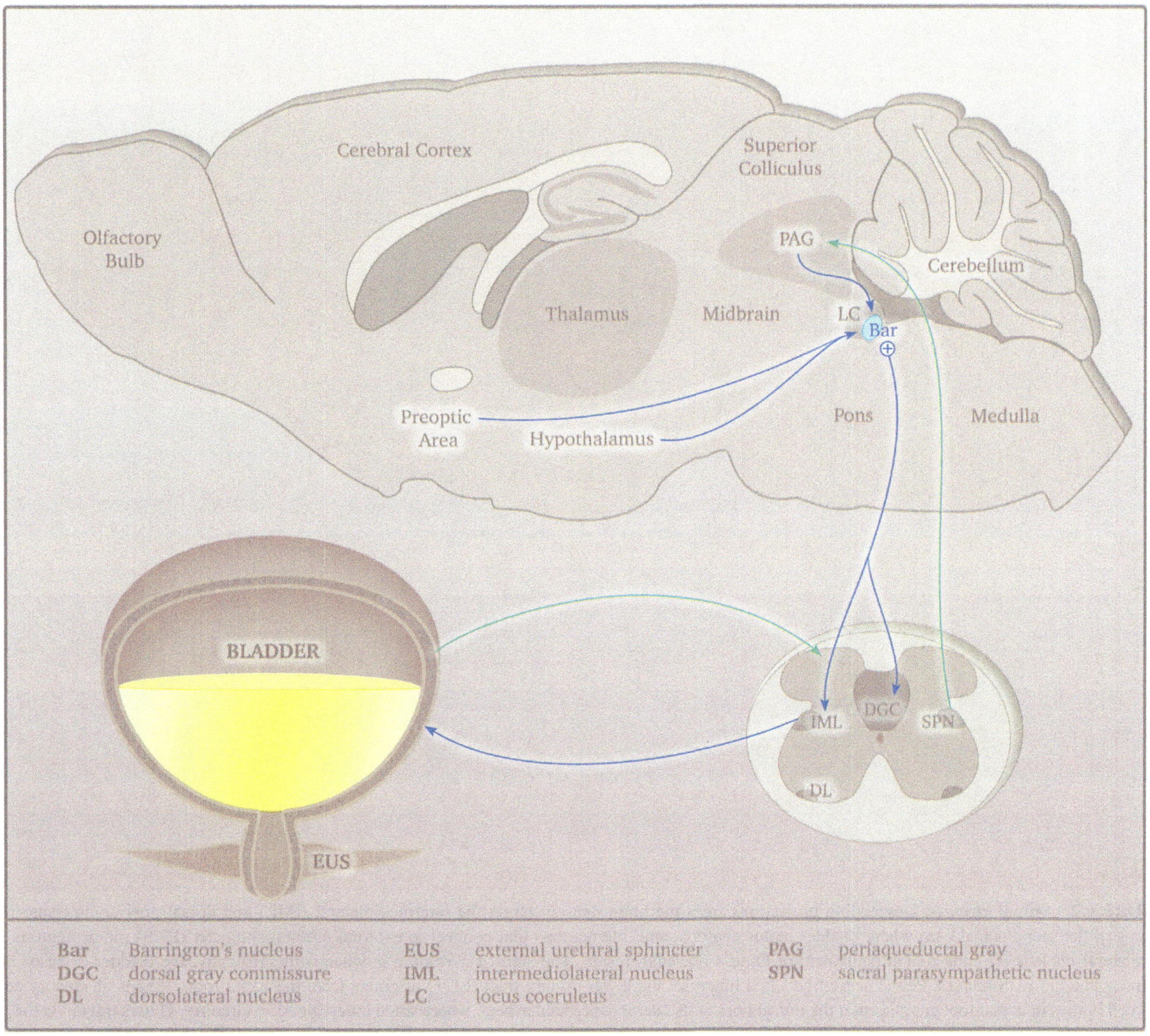

FIGURE 1.1 Schematic of mouse brain anatomy and micturition pathways. During bladder filling, afferent neuron firing increases and activates the spinobulbospinal (micturition) reflex pathways. The afferent activity signal arising from mechanoreceptors in the bladder wall passes by the spinal cord, to ascend and relay in the midbrain periaqueductal gray (PAG) before ultimately reaching Barrington's nucleus in the brainstem. In addition to axonal projections from PAG neurons, several other brain regions also innervate Barrington's nucleus, for example, specific nuclei within the hypothalamus area. Excitation of the descending pathway stimulates the parasympathetic outflow to the bladder and inhibits the pudendal outflow to the external urethral sphincter (EUS) (not shown), to initiate micturition. In green: lower urinary tract and spinal cord to brain afferent pathways. In blue: brain to spinal cord efferent pathways and inputs to Barrington's nucleus.

commissure (DGC), and because these are also sites of parasympathetic preganglionics, the interconnected network of neurons and interneurons must be coordinated to regulate bladder and sphincter activity.

Bladder filling increases afferent neuron firing and this activates the spinobulbospinal (SBS) micturition reflex pathway. In the SBS reflex, afferent pelvic nerve stimulation transmits an ascending signal that passes through relay centers before ultimately reaching Barrington's nucleus. The signal then descends to the spinal cord motor neurons contralateral to the side of the afferent nerve stimulation (Degroat, 1975; Noto et al., 1991).

When bladder distention exceeds a threshold, the afferent activity signal arising from mechanoreceptors in the bladder wall ascends and passes through the PAG center. Axonal projections from PAG neurons to Barrington's nucleus activate neurons in the latter nucleus (De Groat and Yoshimura, 2009; Verstegen et al., 2019). As afferent signals from spinal interneurons synapse in PAG, information is also relayed to several forebrain regions including the insula where the sensation of bladder fullness is interpreted. The insula and other cortical sites then send inhibitory signals back to PAG and

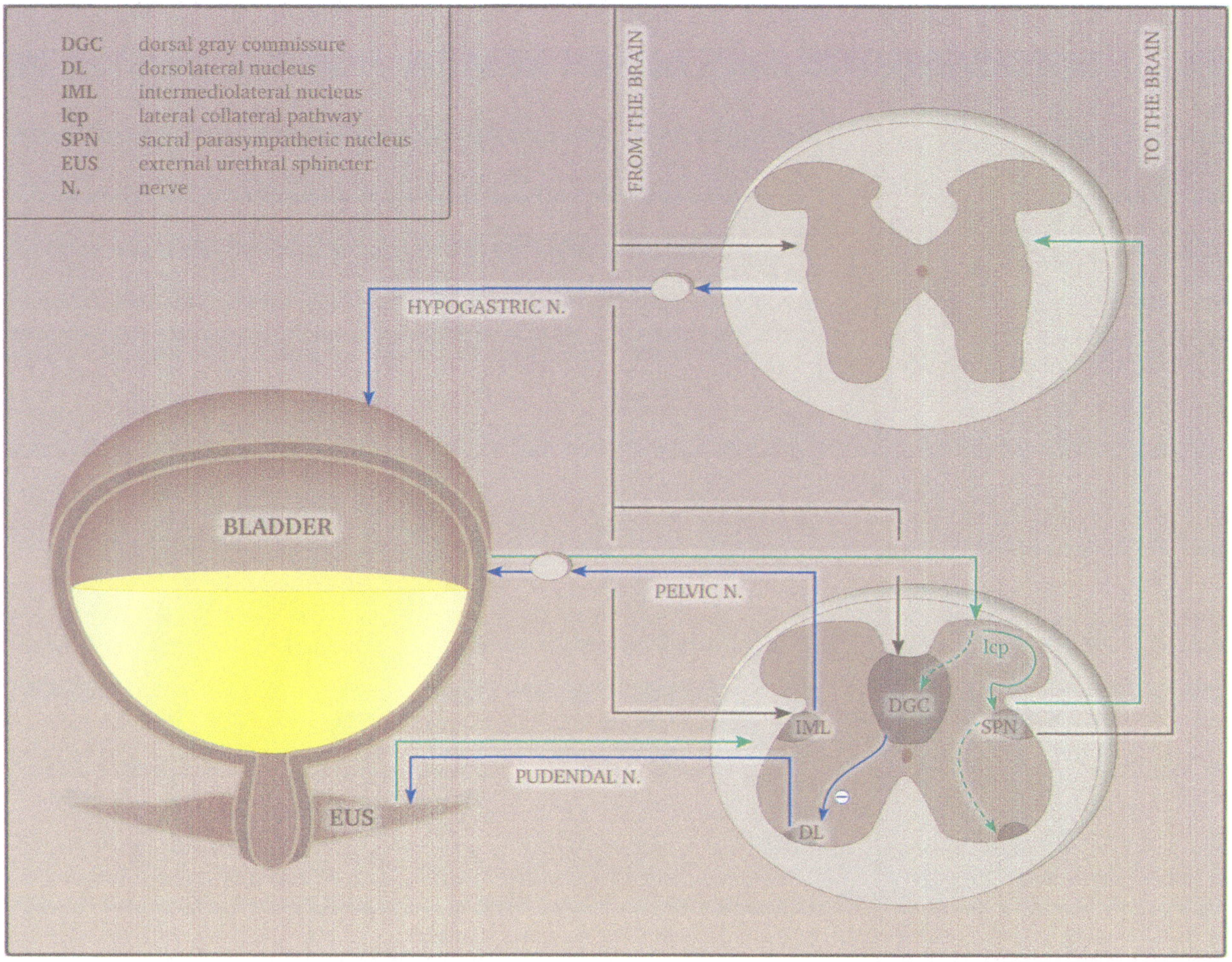

FIGURE 1.2 **Spinal circuits.** Descending projections from the brain densely target the intermediolateral (IML) and dorsal gray commissure (DGC) regions of the sacral spinal cord where bladder motor neurons, and interneurons that connect to external urethral sphincter (EUS) motor neurons in the dorsolateral nucleus (i.e., the rodent equivalent of Onuf's nucleus), reside. The start of a voiding is when, synergistically, the bladder detrusor muscle contracts and the EUS relaxes. Pelvic nerve fibers that innervate the lower urinary tract (LUT) originate from the sacral roots. Axons of sensory neurons with cell bodies in dorsal root ganglia enter the dorsal horn at its lateral superficial aspect, where some innervate local circuitry. Others travel via the lateral collateral pathway and terminate within the gray matter—sacral parasympathetic nucleus (SPN) (depicted down). During the storage of urine, distention of the bladder produces low-level afferent firing. Via the lumbosacral spinal cord, the signal reaches neurons at the thoracolumbar spinal cord level (depicted up) for stimulation of the sympathetic outflow in the hypogastric nerve to the bladder and outlet. This facilitates bladder relaxation and urethral and prostatic smooth muscle contraction. Additionally, afferent firing stimulates the pudendal outflow, which results in contraction of the EUS. These responses occur by spinal reflex pathways and represent guarding reflexes that accommodate bladder filling. When bladder distention reaches a threshold, projections to the supraspinal sites activate the excitatory descending pathway. In green: LUT afferent pathways. In blue: spinal cord efferent pathways for LUT innervation.

to other brainstem sites to suppress voiding until a voluntary decision about voiding is made by higher brain regions that are involved in organizing executive control, such as the prefrontal cortex (Fowler et al., 2008).

In addition to bladder emptying, bladder filling also activates the autonomic nervous system. Sympathetic outflow from the rostral lumbar spinal cord provides noradrenergic excitatory and inhibitory input to the bladder and urethra (Anderson, 1993). Sympathetic nerve activation via the hypogastric nerve relaxes the urinary bladder and contracts the urethral and prostatic smooth muscles, thereby accommodating bladder filling (Fig. 1.2). Storage of urine in the bladder is thought to be predominantly integrated at the level of the spinal cord. Increased firing in the somatic pudendal nerve together with high outlet resistance during filling is termed the "guarding reflex." This mechanism supports urinary continence when the bladder is full (De Groat, 2006). In health, urinary continence thus relies on a complex control system that includes the urethral sphincters and detrusor muscle, neurons at lumbosacral and thoracolumbar spinal cord levels, the PAG, Barrington's nucleus, and a higher brain network.

LUT dysfunction: LUT dysfunction can present in the form of incontinence or retention. This is, in part, dependent on the nature of the neurons involved and the location of lesion or damage.

- Bladder overactivity, including urinary frequency, is the result of loss of voluntary control over bladder function.
- Urinary retention is a condition in which the bladder does not empty completely or does not empty at all with urination. This can be caused by a pathomechanical outflow blockage preventing urine from leaving the body or by an interruption in communication between the brain and urinary system.

Incontinence symptoms range in severity from occasionally leaking urine when coughing or sneezing to a very sudden urge to urinate and not reaching a bathroom in time. It can also be that the sensation of bladder fullness is lost and therefore there is no attempted bathroom visit. Importantly, since the LUT is innervated differently in biological females and males, different types of LUT symptoms and dysfunction are likely to occur more, or exclusively, in a specific biological sex.

Fundamental research is desperately needed for a more complete understanding of the signaling pathways, to identify genes and cells that play critical roles in these pathways, and to find new or better targeted therapeutics for LUT dysfunction. Novel tools for studying neurons, circuits, and brain functions have revealed a high degree of complexity in the brain and in the "bladder circuits" in the brain. With newer approaches and technology available, we will soon be able to fill these knowledge voids.

The past and present of neuro-urology research

Research in the past century has advanced the understanding of the neural control of LUT function. In recent years especially, great strides have been made with the discovery and use of key tools in functional genomics (i.e., transgenic models) to generate models for human diseases. Here we present an overview of the evolution of research from old to newer approaches and techniques and arguments for behavioral assessments to be made in awake, freely moving animals of both sexes.

Electrical stimulation, pharmacological manipulation, and electrolytic lesions applied to specific brain areas were tolls of early studies of bladder function and control. These types of stimulations and lesions affect multiple neuron groups and present interpretive challenges as they indiscriminately impact cell bodies and axons of passage. With the identification of neuron subpopulations and the advent of Cre-lox technology (see "Tools" section in this chapter) to target them, findings are becoming increasingly more precise.

Barrington's nucleus is a key brain site for regulating micturition. Recent studies (Keller et al., 2018; Verstegen et al., 2019) have shown that cell type—specific stimulation of glutamatergic neurons prompts voiding, while selective ablation of glutamatergic Barrington's nucleus neurons causes retention. The PAG in the midbrain is another critical brain region for micturition regulation. It serves as a relay and a coordination center on the ascending limb of the micturition reflex pathway before signals reach Barrington's nucleus. Early studies in cats and rats reported that upon stimulation of pelvic nerve afferents, the elicited field potentials in PAG had a much shorter latency than potentials recorded in the hindbrain region (Degroat, 1975; Duong et al., 1999; Noto et al., 1989). Further support for the PAG being an important relay site in the micturition reflex pathway came from tract tracing studies in cats, which revealed a greater number of axonal inputs from sacral spinal cord neurons detected in the PAG compared to axons directly innervating hindbrain Barrington's nucleus neurons (Blok et al., 1995; Holstege & Mouton, 2003). "Circuit mapping" in the brain and spinal cord has revealed additional anatomical sites and specific cell types that are connected to each other and to the brain micturition centers. It is possible that these formerly undiscovered locations and cell types are involved in regulating LUT function.

Because voiding should not occur randomly when the bladder is full but only when and where appropriate (in humans), when the environment is safe, or with distinct social behaviors such as scent marking (in animals), the micturition reflex is under voluntary control. Consequentially, damage to cortical or other brain sites that modulate the activity of the brainstem centers involved in the micturition reflex may lead to loss of control of volitional LUT function. Furthermore, transection of the spinal cord often results in loss of the conscious sensation of bladder fullness.

In addition to circuit mapping, functional brain imaging (e.g., MRI) has identified brain regions that exhibit changes in neural activation during bladder filling specific to normal micturition and that differ from that seen in patients with incontinence. Imaging in human subjects is usually performed in the awake state, and sometimes in combination with urodynamic evaluation. Historically, bladder function studies in animal subjects were rarely performed in awake and unrestrained animals, despite the fact that most types of anesthesia alter the activity of neurons across the CNS (Pandita et al., 2000; Yoshiyama et al., 1994). While this limitation is largely explained by technical barriers, it also points to a general absence of behavioral context in the earlier reports. For a complete understanding of the central control of bladder filling, continence, and neural influence in a naturalistic setting, investigations should ideally employ awake, freely behaving animals (see "Tools" section in this chapter).

Cell type–specific transgenic laboratory mice (see "Tools" section in this chapter) are a popular model in fundamental research studies of bladder function and micturition behavior. Like humans, mice are continent; they void in specific places at specific times (Verstegen et al., 2019, 2020), and social stress produces retention (Wood et al., 2009). Male mice are further continent in that they mark their territory by controlled voiding (Keller et al., 2018). Voiding behavior is disturbed in animals in which specific neuron populations have been ablated (Verstegen et al., 2019), in urothelial-specific integrin beta-1 knockout animals (Kanasaki et al., 2013), and in animals as a result of the normal aging process (Kamei et al., 2018). Thus, mice can model aspects of normal and disturbed mammalian bladder function including urine storage and voiding. Cre-driver mice combined with injections of viral vectors that require the Cre-recombinase enzyme allow for targeted delivery of genetic tools to a particular type of neuron. This combination permits the interrogation of discrete cell populations with a high degree of spatial and temporal precision. With optogenetics technology, neurons are genetically modified to express a light-sensitive channel. Their activity can then be activated or inhibited using light stimulation (see "Tools" section in this chapter).

Groundbreaking recent technological advances enable studies of the transcriptome at the level of single cells. scRNA-Seq (see "Tools" section in this chapter) promises to provide valuable insight into cellular heterogeneity, which may significantly improve our understanding of biology and human disease (Angerer et al., 2017). Additionally, there is an important role for big data science in integrating the enormous volume and complexity of transcriptomic data from thousands of cells generated in a single experiment.

Finally, another aspect of this research that has come to the forefront in the past several years is the critical importance of including both biological sexes as subjects. The lack of inclusion of biological females in medical trials—because investigators worried that fluctuating hormones would add too many confounding variables to their studies—has resulted in lagging knowledge regarding how LUT function is organized in females, as well as how new drugs and treatments specifically affect women's LUT function and health.

Research questions and directions in the neuro-urology field

This section describes research directions in the field. First by examining four "Research Topics" that are current "hot items" and then by exploring potential directions for future studies. The section also links specific topics to the chapters in this book.

Research topic 1: neuroanatomical sites for micturition behavior

Barrington's nucleus and the PAG are the brainstem nuclei that give rise to the bulbospinal tracts and that coordinate urinary storage and voiding. Barrington's nucleus is a cytoarchitecturally defined population of neurons that controls pelvic motor functions. Progress has been made toward the identification of molecular markers, properties, and the functions of neurons in Barrington's nucleus. This is detailed in Chapter 2. A more thorough understanding of Barrington's nucleus neurons and their connectivity is critical to understanding how the brain communicates with the bladder.

The PAG is of interest to neuro-urology studies because, in addition to connecting to Barrington's nucleus, its neurons receive sensory information regarding bladder fill state (Blok & Holstege, 1994). The midbrain PAG and its neurons are implicated in many behaviors other than, and complementary to, micturition behavior. For example, glutamatergic PAG neurons play a role in pain modulation and aggressive behavior. Reports of studies on bladder function related to the "black box" PAG are central to Chapter 10, which reviews and summarizes pharmacological and imaging studies, immunohistochemistry for detecting selective antigens, and extracellular local field potential recordings. The chapter offers a clearer view of the complex circuitry within and involving the PAG including its columnar organization, its functional cell populations and their neuronal markers, and its ability to act as a "decision switch" for voluntary and reflex micturition.

Imaging studies using functional magnetic resonance imaging (fMRI) and positron emission tomography (PET) have revealed brain regions in addition to the two micturition centers described above that activate or inactivate during bladder filling and voiding. Examples of identified additional brain areas, which are directly or indirectly involved in bladder function, are described in Chapters 4, 6, and 12 and include the pedunculopontine tegmentum (PPTg), the subthalamic nucleus (STN), the hypothalamus, and other (sub)cortical sites. During bladder filling in healthy subjects, specific cortical sites such as the insula and prefrontal cortex activate. Conversely, during bladder voiding, the PAG and Barrington's nucleus activate. Imaging studies have also been performed in human patients with LUT dysfunction diagnoses ranging from urge incontinence to Fowler's syndrome retention. Chapter 4 focuses on specific brain sites implicated in bladder control and on potentially meaningful functions for them. Chapter 12 furthermore discusses whether there is a correlation between the absence of activation in certain regions in persons with urge incontinence and the importance of those brain sites for maintaining continence.

Both afferent and efferent bladder pathways converge at the level of the sacral spinal cord. This is therefore an important site for the regulation of LUT function. Axons from pelvic nerve afferents that innervate the LUT originate from the dorsal root ganglia (DRG) of roots S2-4. Central axons form Lissauer's tract at the lateral superficial aspect of the dorsal horn and terminate within the gray matter—SPN, as well as give off collaterals to the dorsal commissure. Moreover, spinally descending Barrington's nucleus neurons also reach the sacral segments where bladder and EUS motor neurons reside in the IML and Onuf's nucleus (or the rodent equivalent dorsolateral motor nucleus) (see Chapters 3 and 5 and Fig. 1.2).

Research topic 2: neural circuits involved in bladder function

The micturition control network spans all levels of the nervous system and can be divided into three major parts: the brainstem micturition centers, the spinal nuclei, and the higher brain regions that process and relay sensory information from the spinal centers to the brainstem centers. The micturition reflex pathway functions to produce an efficient void when the bladder is full by simultaneously contracting the detrusor muscle and relaxing the urethral sphincter. This requires coordination between different components of the LUT and spinal storage mechanisms and is directed by circuitry in the rostral brainstem. In the micturition reflex pathway, bladder distention increases afferent fiber activity. These afferent fibers activate PAG and Barrington's nucleus brain "nodes," which, in turn, activate downstream targets and the initiation of voiding. Tracing and functional studies are revealing with greater clarity the brain regions and neurons synaptically connected to PAG and Barrington's nucleus (Hou et al., 2016; Tish & Geerling, 2020; Verstegen et al., 2019). It remains unknown whether any of these upstream sites are also involved in maintaining continence during bladder filling.

The spinal cord centers and spinal circuits involved in micturition include the SPN, Onuf's nucleus, and the sympathetic chain (Fowler et al., 2008). Chapter 11 (Impact of spinal neuromodulation on spinal neural networks controlling lower urinary tract function) focuses in particular on these spinal networks and the current models of spinal control of LUT function. While sensory information via afferent neurons is critical during storage of urine and for alerting the CNS that it is time to void, efferent nerves coordinate the precise events that lead to bladder emptying and voiding. Via a descending pathway, Barrington's nucleus neurons reach sacral segments where bladder and EUS motor neurons, and interneurons connecting to them, reside. In addition to projections to supraspinal sites, neurons in the SPN also act locally to induce inhibition of the spinal motor neurons to the EUS. Furthermore, the sympathetic storage reflex, which is triggered by A-fiber activity in the pelvic nerve and leads to increased sympathetic efferent nerve activity to facilitate bladder storage and larger storage volumes, is inhibited when pressure increases above the threshold for micturition (see Chapters 3 and 5).

Higher brain regions integrate sensory information and conscious control of LUT function (see Chapters 3 and 12). While the bladder fills, preparations have to be made in anticipation of the moment of voiding. Ascending signals (such as mechanosensation) are thought to be relayed through the PAG to the hindbrain (Griffiths & Tadic, 2008). Via synapses in PAG, ascending signals are also sent to more rostral brain regions where bladder volume is interpreted (Fowler et al., 2008). Even when the bladder is full, voiding can be suppressed by inhibition from frontal brain sites until the person or animal find themselves in a safe and socially accepted environment. It remains to be determined whether this voluntary continence control impinges on Barrington's nucleus, on specific neuron subtypes in Barrington's nucleus, or on the reflex pathway in general. Central to Chapter 3 are anatomical, neurophysiological, and brain imaging studies that have provided insight into the neural circuitry and neurotransmitter systems controlling voluntary and reflex micturition.

Functional brain imaging techniques have been applied to elucidate normal neural control. fMRI and PET revealed fore- and midbrain centers that can potentially provide input to and modulate the brainstem circuitry to switch on or off to initiate micturition. Changes in neural activity during bladder filling potentially point to brain sites involved in maintaining continence. The strong overlap of brain regions involved in bladder regulation and brain sites for affective processing such as the PAG, cingulate cortex, and insular cortex furthermore suggests a link between neural circuits underlying behaviors and bladder function. The activation of a part of the anterior cingulate gyrus (ACG) during bladder filling implies the involvement of emotion in micturition. It is possible that emotional affective disorders that influence these networks may in turn have an impact on bladder function (see Chapters 4 and 12). Similarly, visceral pain often produces strong emotional responses relative to other types of pain.

Our knowledge of the neural control networks, the identity of neuron subpopulations that play a functional role in specific aspects of micturition behavior, the afferent and efferent connections, and the effect of variables such as age and sex hormone concentrations on LUT function is rapidly advancing. This characterization of cells, their molecular signature, and signaling pathways provides new paths for therapeutic research and avenues for future investigation.

Research topic 3: the functional brain-bladder connection

Bladder physiology can be used as a readout for enhancing our understanding of the neural control of LUT function and thereby informing at the CNS level on effective treatment of diseases that resulted in impaired bladder control. Chapter 4 provides an overview and discussion of studies that performed electrical stimulation at specific locations in the brain (i.e., in Bar, in thePPTg, parabrachial nuclei (PBN), basal ganglia, STN, lateral hypothalamic area, and the primary motor cortex) while taking quantitative and qualitative measurements from the LUT. Similarly, manipulations can also be performed at the bladder end while recording neural activity.

Quantification of cFos expression levels is another means of assessing how neuronal activity correlates with bladder filling and voiding. cFos is an immediate early gene (IEG) that functions as a transcription factor and regulates downstream target genes. It is widely used as a functional marker of activity in neurons after a stimulus or activation. Even though induction of cFos does not differentiate the temporal sequence in which neuronal activation occurred, Chapter 10 reviews how cFos expression studies support other studies of the PAG suggesting that there are distinct afferent and efferent LUT-related PAG pathways.

Bladder physiology readouts can also be used outside of research focused on LUT pathology. Micturition behavior studies can promote insight into how the autonomic nervous system participates in bladder control, or into how social behaviors are shaped in the brain. Furthermore, metastudies are being used to examine if bladder dysfunction may be a consequence of environmental exposures or of certain stressful occupations (see Chapters 4 and 7). All these ambitions rely on the ability to obtain consistent bladder physiology readouts.

Research topic 4: the "brain cause" of common lower urinary tract symptoms

What are the most common forms of brain—bladder "miscommunication" that may lead to LUTS? How does a complete understanding of the neural micturition circuits and identification of the neurons that regulate LUT function promote treatment options for patients?

LUT afferents monitor bladder volume during the storage phase as well as bladder contraction amplitude during voiding (Kanai & Andersson, 2010). Dysfunction in sensory signaling can result in various clinical LUTS such as urgency, idiopathic detrusor overactivity, and underactive bladder. LUTS such as these are observed in diabetes mellitus, where neuropathy causes impaired sensory signaling (see Chapter 5). A-fibers and C-fibers play a role in the normal and abnormal physiological functioning of the LUT. A-fibers respond to bladder distention and contraction. Under pathological conditions, C-fibers, which respond to noxious stimuli, are recruited and become dominant over A-fibers. Pathological sensory changes can affect bladder and LUT function (see Chapter 5), and result in the perception of pain. Vice versa, bladder pain can cause LUT dysfunction and may lead to pathological sensory changes in the urological system. Therefore, research that focuses on pain signaling and alterations occurring from painful LUTS has the potential to provide valuable insight to the development of therapeutics.

Human case studies show that damage to the frontal lobes or hypothalamus often results in urge incontinence (Andrew & Nathan, 1965; Sakakibara, 2015; Tish & Geerling, 2020). Chapter 6 focuses on neurological disorders that are frequently accompanied by LUTS and explains mechanisms for how neurological disorder or trauma can lead to urinary problems. Potential interruptions or damage to the brain pathways are also reviewed in Chapter 12, which discusses the significance of fMRI observations. For example, the lack of increased activation of the insula and PAG in patients with urge incontinence could be the result of changes in peripheral firing rather than impaired supraspinal processing.

Transection of the spinal cord also affects normal LUT reflexes. Neuromodulation of spinal circuitry is beneficial for improving LUT function after injury to the spinal cord. Chapter 11 proposes how activity in local spinal circuits can be "targeted" to rescue LUT function after SCI by using existing and experimental neuromodulation techniques.

Lastly, it is well known that sex hormones affect LUT function. The action of testosterone and estradiol at different levels of the brain—bladder axis on male and female physiology and benign bladder-related disease is discussed in Chapter 7. An example of a study reviewed in Chapter 7 is one that examines the impact of estradiol treatment on pudendal nerve damage in a rat model of stress incontinence (Kane et al., 2004); the pudendal nerve innervates the pelvic floor and the EUS and is often injured during childbirth. Different circuits innervating the LUT in females and males may account for different presentation of LUTS. Furthermore, aging-related changes in sex hormone concentrations are clear contributors to urinary disease processes.

Even though LUTS are likely caused or exacerbated by dysfunction of neural circuits controlling bladder function, significant knowledge gaps remain in our understanding of the cellular and synaptic circuits that control reflex and voluntary micturition. Deviations from normal neural control or insults to the neural connections can have consequences for urine storage ability. Fundamental research studies seek to address mechanisms of action and the underlying pathological mechanisms in neural circuit dysfunction leading to LUTS.

Opportunities for future research: The development of improved treatments for LUTS depends on the elucidation of the neural circuit basis for the control of bladder function. This will include understanding how essential cell populations in key brain sites within these circuits are connected, both anatomically and functionally. While the key players in the larger network controlling micturition have been identified, there is still significant potential for mapping connections that are not yet known. Barrington's nucleus, for instance, is richly innervated by input-providing neurons in the PAG, hypothalamus, and other afferent regions. The specific subtypes of neurons that provide synaptic input from these sites to the micturition center, as well as which subtypes of neurons they target, are yet to be revealed. Furthermore, distinct subcortical inputs and networks may direct particular aspects of micturition behavior (Hou et al., 2016; Keller et al., 2018). Neurons in the preoptic area (or other areas of the hypothalamus) likely mediate complex homeostatic or affective behavioral modulation of voiding. In addition, dedicated neural networks may exist for social behaviors such as territory marking (Desjardins et al., 1973) or the formation of ultrasonic vocalizations that occur simultaneously with scent marking (Chen et al., 2021). While the exact forebrain locations that are important for regulating LUT function and maintaining continence remain to be determined, lesion studies and case reports clearly establish that one or more sites within the frontal lobe exert a major modulatory influence over micturition (Tish & Geerling, 2020). How forebrain cells interface with hindbrain neurons for organizing micturition, and whether these pathways are direct or involve interneurons that relay some of this input to Barrington's nucleus, remains to be investigated.

The regulatory networks that are required for maintaining continence are incompletely resolved. In addition to being integrated at the spinal level, some components of EUS behavior are under brain control (Keller et al., 2018). It is not known whether Barrington's nucleus in general or if specific neurons within this nucleus are important for voluntary EUS control. The brain's control of bladder function via sympathetic motor neurons is also an interesting topic for further investigation. Urine storage in the bladder is supported by increased sympathetic activity via the hypogastric nerve. At the same time, loss of signaling from the brain results in detrusor overactivity (Birder & Drake, 2009; Fowler et al., 2008; Fowler & Griffiths, 2010). It will be essential to locate and understand the sites in the CNS that either directly interact with LUT-controlling neurons in the spinal cord or affect activity of the neurons in brain micturition centers. Additionally, further investigation will be required to understand how primary afferent neurons connect to motor neurons of the efferent limb of the system at the level of the spinal cord.

Endogenous and exogenous testosterone and estradiol have effects on bladder structure, function, and disease prevalence. The body's natural response to changes in levels of circulating testosterone and estradiol is mediated through hormone receptors that alter downstream neuron function. These hormonal effects differ depending on the type of neuron. The mechanism of action by which hormones in circulation have effects on the LUT system is far from completely understood in both health and disease states. In disease, it is possible that the pathophysiology of the specific condition which alters hormone levels may affect or dictate a specific LUT dysfunction. Furthermore, LUT dysfunction resulting from fluctuations in hormone levels attributed to natural variables such as age is yet to be fully explored. Studying detrusor and sphincter function with exogenous testosterone or estrogen supplementation can lead to novel insights and ultimately prove beneficial for patient populations. There is major scientific value in studying changes in density of estrogen and androgen receptors in the brain and whether different expression patterns correlate with overactive bladder or with aging.

As described in "Research Topic 3" above, bladder function can be used as an index for disease progression. Additionally, stimulation of specific brain or spinal cord regions and subsequent monitoring of the LUT response may reveal promising therapeutic targets (see Chapter 4). Such stimulation could be applied to neurons in models for conditions (e.g., multiple system atrophy, normal pressure hydrocephalus, Parkinson's disease, and spina bifida) to promote a greater understanding of the disease at the cellular level and potentially identify therapeutic targets. Targeted stimulation of the STN in Parkinson's disease patients provides one example of potential uses of bladder physiology readout research. Analogous to STN stimulation in Parkinson's disease, it would be expected that stimulation of a proposed "brain urine storage control center" would decrease detrusor activity or increase EUS activity.

Bladder function may also be evaluated when studying the extent of regeneration after a treatment. The incontinence phenotype secondary to TBI is used to study beneficial effects of cis P-tau antibodies on posttraumatic brain degeneration and cognitive dysfunction. Repetitive moderate TBI changes the urinary pattern. The phenotype is fully rescued by antibody therapy that reduces cis P-tau accumulation following injury (Albayram et al., 2019). In similar experimental settings, antibodies against nerve growth factor (NGF) (Seki et al., 2002, 2004) and antinerve fiber growth inhibitory protein Nogo-A (Schneider et al., 2019) have been used to counter disease progression and to study how SCI leads first to spinal shock and then to bladder-EUS dyssynergia. Deeper analysis of the functioning of neural circuits controlling bladder function after trauma is expected to lead to insights into how brain degeneration leads to bladder dysfunction, as well as to novel strategies for treating these disorders.

The use of existing "setups" such as deep brain stimulation (DBS), for improving motor function in patients with Parkinson's disease, could potentially be extended and utilized for the treatment of other neurodegenerative conditions such as TBI. DBS is hypothesized to work by overriding the pathological network activity and regulating axonal output (Sakakibara et al., 2018). The authors of Chapter 11 question whether LUT regulation can be retrained after a pathology and introduce a method for improving urine storage in the bladder using transcutaneous electrical spinal cord neuromodulation. Neuromodulation may also work to quiet nociceptive bladder afferent c-fibers or for suppressing the sensation of urge signaled from an overactive bladder. Thus, electrical stimulation approaches are useful for understanding neuron activity patterns in healthy physiological conditions as well as when functional control is disrupted.

Optogenetic stimulation (see "Tools" section in this chapter), in which neuron activity is manipulated with light, permits testing of the function of specific neurons and has the therapeutic potential to reduce symptoms in brain and bladder disorders. Furthermore, ArchT silencing technology, in which the optical neuronal silencer ArchT acts to hyperpolarize neurons, might be of great use in causal analysis of neural circuits, and, additionally, for therapeutic applications because of the powerful net suppression of activity (Han et al., 2011).

Finally, very little is known about the neuro-cognitive mechanics underlying the processing of bladder sensory information within the brain. More research is needed to understand interoceptive processes such as the perception of a full bladder and the effects of social and emotional disorders on bladder function. Animal models and human studies can provide insight into the links between LUT physiology and normal sensation or altered micturition behavior in response to physiological stress, pain, opioid use, or trauma.

While we are gaining a better understanding of the neural pathways themselves, research into how to target these pathways will be beneficial for further defining therapeutic targets for treatment of incontinence (see Chapter 12). Such research will also provide a more thorough understanding of the mechanisms of action that are still lacking for some treatments currently in use. In summary, in order to advance treatment of what are some of the most prevalent clinical disorders in humans, we must be able to identify the types of sensory neurons, interneurons and motor neurons, and the receptors on these neurons, at the level of the brain, the spinal cord and urothelium, in both females and males.

A quick guide to the "neuroscience toolbox"

How the brain orchestrates behavioral control is complex and far from completely understood. With the advent of novel scientific tools, the field of neuro-urology will advance and provide a greater mechanistic understanding of the neural control of LUT function. Below is a selection of transformative techniques that can be applied to advance neuro-urology research. These will be briefly introduced, and the reader is referred to reviews and helpful technical publications.

Transgenic animal models: knock-in and knockout (mice)

Genetically engineered animal models provide the opportunity to define essential gene functions (Gu et al., 1994; Rajewsky et al., 1996). Mice are the primary species utilized for genetically targeting cell types and circuit connections.

- A knockout mouse is a laboratory mouse in which a part of a DNA sequence has been deleted or an artificial piece of DNA replaces the original sequence. This disruption renders the gene nonfunctional and thereby provides insights into the function(s) of the gene.
- A conditional knockout has critical coding exons of a gene flanked by special loxP sequences (i.e., is commonly referred to as "floxed." Read more below under "Cre-lox system and genetic tools" section). The knockout can be induced at a specific time or restricted in expression to specific tissues. This is particularly useful for loss-of-function studies and for analyses of genes that have essential functions early in development.
- A knock-in mouse is created by replacing a portion of the gene sequence with an artificially generated version of the allele. The most common transgenic knock-in mice are mouse lines that have a protein (i.e., Cre recombinase or GFP reporter sequence) introduced into the DNA for which expression is under the transcriptional control of the endogenous locus. A knock-in mutation does not necessarily interfere with the function of the gene.
- A floxed-STOP reporter mouse (aka an indicator mouse) has a reporter sequence that is preceded by a "STOP" segment that is flanked by loxP sites. Expression of a recombinase allows for recombination of the two loxP sites and removal of the STOP codon. The reporter sequence in this mouse is a transgene that "indicates" a recombination event has occurred in a given cell at some point in time during development and provides a permanent record of this event (Dymecki et al., 2002).

Transgenic mice and CRISPR/Cas9 technology

Traditional strategies for genetic engineering of transgenic mice rely on homologous recombination (HR) in embryonic stem (ES) cells to deliver targeting cassettes flanked by long regions of homology (homology arms of ~3–10 kb) to the gene of interest (Skarnes et al., 2011). Correctly targeted ES cells are then introduced into mouse embryos and the resulting chimeric mice can be used to transfer the mutant allele to subsequent generations.

ES cell and bacterial artificial chromosome (BAC) transgene engineering have given way to editing genes in zygotes (for review see Lanigan et al., 2020). CRISPR/Cas9 (Clustered Regularly Interspaced Short Palindromic Repeats) technology is useful for making conditional knockout or recombinase/reporter knock-in mouse lines. This directed genome editing allows for more rapid and more efficient generation of floxed or knock-in alleles in any chosen genetic background. CRISPR/Cas9 gene-editing components are delivered directly to single-cell mouse zygotes of any strain. The Cas9 nuclease creates targeted double-strand breaks (DSB). Following the RNA-guided cleavage of a specific site, the target cell is flooded with large quantities of a donor template. This donor template contains the desired insertion or modification in between segments of DNA that are homologous to the blunt ends of the cleaved DNA. The homology-directed repair pathway, which is the cell's natural DNA-repair mechanism, inserts the desired genetic material at the location of the DSB.

Optimal guide RNA sequences (consisting of a ~20-nucleotide "target DNA sequence" and an 80-mer scaffold ["tracrRNA"] part) with high affinity and specificity can be designed in silico using programs such as "Benchling" and can be cloned either in the lab or using commercial services (e.g., "Synthego"). Lastly, for the knock-in cassette to be inserted in or after the endogenous coding sequence, either a short, self-cleaving viral 2A peptide or an internal ribosome entry site (IRES) is generally employed. Expression of the IRES-dependent second gene is typically less efficient than that of the first gene (Mizuguchi et al., 2000). Improved targeting efficiency has been demonstrated using the Easi-CRISPR (Efficient additions with ssDNA inserts-CRISPR) targeting strategy (Quadros et al., 2017) and Fig. 1.3).

NB: A limitation of the (Easi-)CRISPR technique is the possible generation of unwanted mutations in addition to the desired insertion in the specific location on the genome. After confirming germline transmission (i.e., that animals have inheritable alleles present in gametes), the founder animals should be bred to wildtype animals to eliminate any off-target allele variants.

- Useful references for CRISPR technology/genetic engineering: Lanigan et al. (2020), Leonova & Gainetdinov (2020), Lino et al. (2018) and the Zhang lab website (zlab.bio/guide-design-resources).

Cre-lox system and genetic tools

Site-specific recombinase systems have transformed genetics in mice by providing the means to delete, insert, invert, or exchange chromosomal DNA (Branda & Dymecki, 2004). Two of the most exciting and versatile genetic tools developed over the last 40 years are the Cre-lox and FLP-FRT technologies (The Jackson Laboratory website). These "tools" permit precise spatial, and often even temporal, regulation of gene expression. Cre-lox technology was introduced in the 1980s (Sauer & Henderson, 1988; Sternberg & Hamilton, 1981) and is based on the ability of the P1 bacteriophage cyclization recombinase gene (Cre) to catalyze recombination between pairs of "locus of crossover (x) in P1" (loxP) sites. A significant advantage of the Cre-lox system is that by modifying the orientation of the loxP sites the system can be leveraged to achieve several different outcomes. For example, if the loxP sites are in opposite orientation, and are flanking a piece of targeted DNA, the introduction of Cre recombinase results in the sequence being inverted. If however the loxP sites are in the same orientation, the introduction of Cre recombinase results in removal (deletion) of the targeted sequence (i.e., gene).

The FLP-FRT system is similar to the Cre-lox system and is becoming more frequently used in mouse-based research. Like Cre, FLP is a tyrosine recombinase, but it originates from the yeast *Saccharomyces cerevisiae* rather than from a bacteriophage (Sadowski, 1995). FLP recognizes a pair of FLP recombinase target (FRT) sequences that flank a genomic segment of interest. Although described around the same time as Cre-lox, FLP-FRT was not originally as efficient at recombining genes in mammals because the native FLP is labile at mammalian body temperatures (Buchholz et al., 1998). However, modifications to the FLP coding sequence produced a more thermostable "enhanced" FLP enzyme, FLPe. Further sequence changes produced FLPo, a mouse codon-optimized version whose recombination efficiency approaches that of Cre. The FLP-FRT system behaves in the same way as the Cre-Lox system with respect to orientation of the FRT sites.

A relative newcomer to the field is the D6 recombination system, or Dre-rox. Like Cre-lox, Dre-rox is a bacteriophage-derived recombinase in which the Dre recombinase cleaves DNA at one of two identical rox site sequences. The first rox site DNA sequence is then ligated to the cleaved DNA of the other rox site, leaving behind a single rox site after recombination.

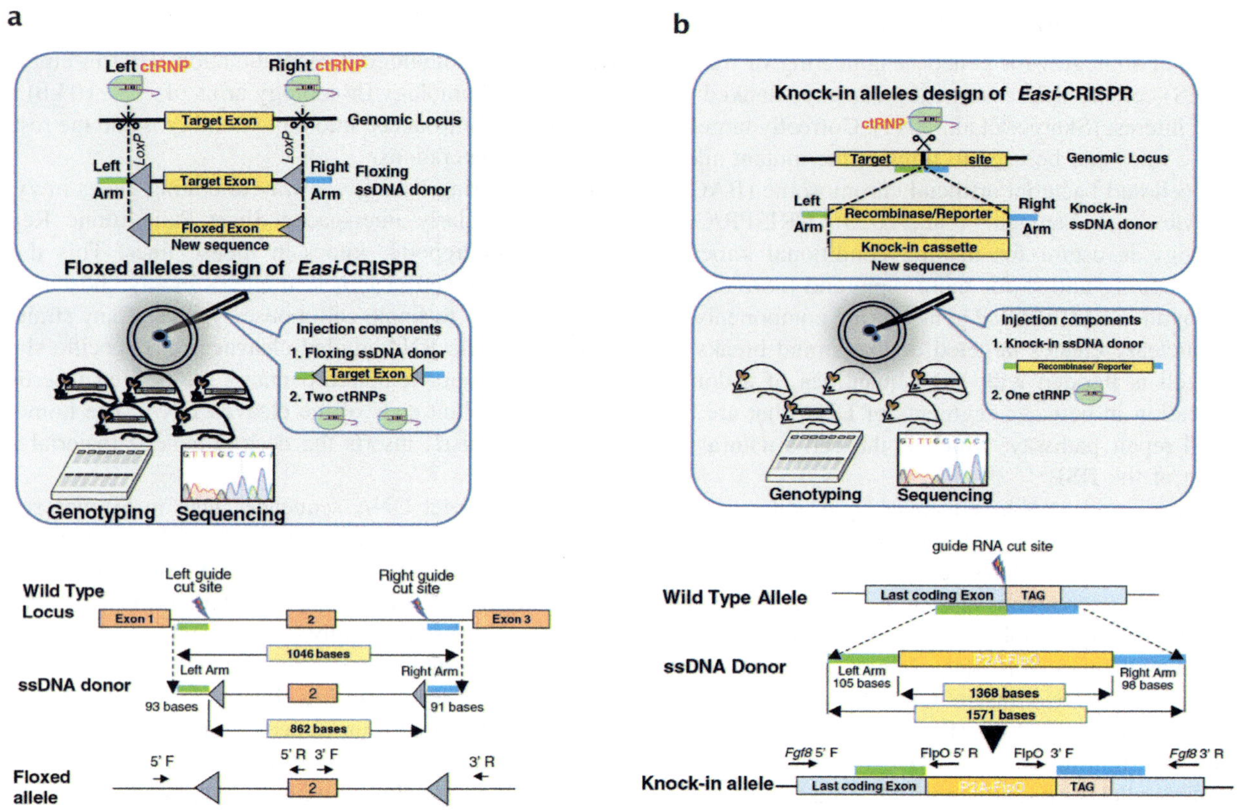

FIGURE 1.3 Homology directed repair with Easi-CRISPR targeting strategy. Typical scheme for insertion of loxP sites (left) and a knock-in targeting cassette (right). Single-stranded DNA is injected into mouse zygotes, together with preassembled crRNA, tracrRNA guide RNAs and Cas9 ribonucleoprotein (ctRNP) complexes. The floxed alleles design (A) shows a floxed exon with two loxP sequences flanking a functional exon of the gene. The "donor template" DNA has a left and right homology arm homologous to the endogenous DNA. Injection into zygotes, with two ctRNPs, results in the replacement of the target exon with the floxed mutant exon. Knock-in alleles design (B) shows a knock-in cassette with recombinase or reporter sequence. The donor DNA has a left and right homology arm homologous to endogenous DNA. Only a single ctRNP is required, and injection into zygotes results in insertion of the knock-in cassette at the cut site. *From Quadros, R. M., Miura, H., Harms, D. W., Akatsuka, H., Sato, T., Aida, T., Redder, R., Richardson, G. P., Inagaki, Y., Sakai, D., Buckley, S. M., Seshacharyulu, P., Batra, S. K., Behlke, M. A., Zeiner, S. A., Jacobi, A. M., Izu, Y., Thoreson, W. B., Urness, L. D., … Gurumurthy, C. B. (2017). Easi-CRISPR: A robust method for one-step generation of mice carrying conditional and insertion alleles using long ssDNA donors and CRISPR ribonucleoproteins.* Genome Biology, 18, 92.

Temporal control over recombinase expression can be obtained through use of a recombinase–steroid receptor fusion protein that can be activated by systemic administration of a hormone such as the estrogen analogue, 4-hydroxy tamoxifen (Metzger et al., 1995). This engineered recombinase allows for short periods of recombinase activation (Metzger & Chambon, 2001).

Neural circuit tracing

Tracing neuronal circuits is an essential component (and goal) of neuroscience research. Modern genetic manipulations, in particular using viral vectors, have revolutionized our ability to trace neuronal pathways. For example, genetically driven tracers can start from predetermined cells and label axons as they project to target sites where neurotransmitters or neuropeptides are released via their synaptic terminals. Different versions of tracers can be engineered to improve membrane localization and trafficking or for expression in specific cellular compartments such as the nucleus. It is helpful to become familiar with the terms anterograde and retrograde to understand neuronal tracing. Anterograde refers to movement toward the axonal projections whereas retrograde is movement toward the cell body.

Recombinant adeno-associated viruses (rAAVs) are commonly used vehicles for in vivo gene transfer. Viruses are easier and faster to produce than transgenic animals. rAAVs are particularly useful when coexpression of multiple transgenes is required (Luo et al., 2018). Transgenic mouse models may be used in conjunction with viral vectors. Microinjecting AAVs that require Cre for expression of a transgene in "Cre-driver" mice allow for targeted delivery of genetic tools. For example, AAVs can be used to deliver genetically encoded receptors, sensors, and effectors to a particular type of neuron or cell within the spatial area of AAV injection and Cre expression overlap. This powerful approach is used for dissecting brain circuits, for recording neural activity patterns, and for performing loss- and gain-of-function studies that elucidate mechanisms of action underlying specific behaviors.

AAVs can also be delivered systemically via the bloodstream using coat protein variants that efficiently pass through the blood–brain barrier. The Gradinaru lab developed AAV capsids to facilitate this less-invasive gene transfer to the central and peripheral nervous systems using an intravenous route of administration. The AAV variants AAV-PHP.B and AAV-PHP.S have been shown to efficiently transduce neurons and glia (Chan et al., 2017; Deverman et al., 2016) in C57Bl6/J mice (Hordeaux et al., 2018; Liguore et al., 2019). AAV-PHP.B has already been used to treat and model neurodegenerative diseases with widespread pathology (Wang et al., 2014).

Interestingly, specific AAV serotypes have been shown to preferentially transduce different types of neurons. For example, AAV5/6/8/9 and PHP.S exhibit enhanced tropism for DRG neurons compared to other serotypes (Mickle & Gereau, 2018). Furthermore, a modified capsid AAV2 has been created with the unique ability to transport in a retrograde manner, rAAV2-retro (Tervo et al., 2016).

AAVs for transsynaptic tracing: One approach for transsynaptic tracing uses existing AAV tracing tools. AAV-Cre and AAV-FLP, specifically those with serotypes 1 and/or 9, possess the ability to "tag" postsynaptic neural targets with the recombinase, hereby allowing recombinase-specific expression of transgenes in synaptically coupled second-order neurons (Zingg et al., 2017, 2020). This is useful for identifying neuron populations in downstream nuclei that receive direct input from a source nucleus. Furthermore, transsynaptic tracing allows for axonal tracing and functional manipulations of a postsynaptic, input-defined neuronal population (see also "Highlighted Approach 1 and 2" below in Box 1.1 and Box 1.2).

To overcome low efficiency transneuronal spread and to unlock transgene expression, it is suggested that an amplification step such as Cre is embedded. AAV1-Cre and AAV1-FLP transsynaptic tracers spread in the anterograde direction without exhibiting neurotoxicity or uncontrollable spread across multiple synapses. Additionally, there is no AAV "leaking" from fibers of passage. One downside of AAV1s is that they exhibit retrograde spread properties, and this must be controlled for.

Another transsynaptic tracer that is optimized for efficient and specifically monosynaptic anterograde tracing has been recently developed. This tool uses wheat germ agglutinin (WGA), which has natural anterograde properties. In this design, the WGA is fused to the red fluorophore mCherry to enable identification of synaptically coupled second-order neurons (Tsai et al., 2022).

Combinatorial, intersectional, and subtractive approaches: In addition to using a single recombinase system (i.e., Cre or FLP alone), it is also possible to exploit both FLP and Cre together (Farago et al., 2006). The use of a dual recombinase-mediated intersectional genetic approach has significantly enhanced the precision of labeling and gene manipulation through the identification of more restricted cell populations that are defined by combinations of expressed genes. Additionally, a suite of new Dre recombinase drivers has further expanded the ability to perform intersectional genetic targeting (Han et al., 2021). Forthcoming recombinase systems will increase capabilities even further.

The sophisticated INTRSECT (intronic recombinase sites enabling combinatorial targeting) allows for transgene expression within populations defined either positively or negatively by multiple genetic features (Fenno et al., 2014) (Fig. 1.4).

BOX 1.1 Highlighted Approach 1, one defined input

"Highlighted Approach 1" uses anterograde transport to specifically label neurons in the target brain region (site B) that receive inputs from a defined brain region (site A): AAV1-Cre (injected in site A) >> AAV-(Cre-dependent)-XFP (injected in site B).

BOX 1.2 Highlighted Approach 2, one defined output

"Highlighted Approach 2" uses retrograde transport to specifically label neurons in the target brain region (site A) that send outputs to a defined brain region (site B): AAV-(Cre-dependent)-XFP (injected in site A) << AAV2 retro-Cre (injected in site B). The use of specifically "AAV2 retro-Cre" is recommended for optimal (retrograde transport) efficiency (Weinholtz & Castle, 2021).

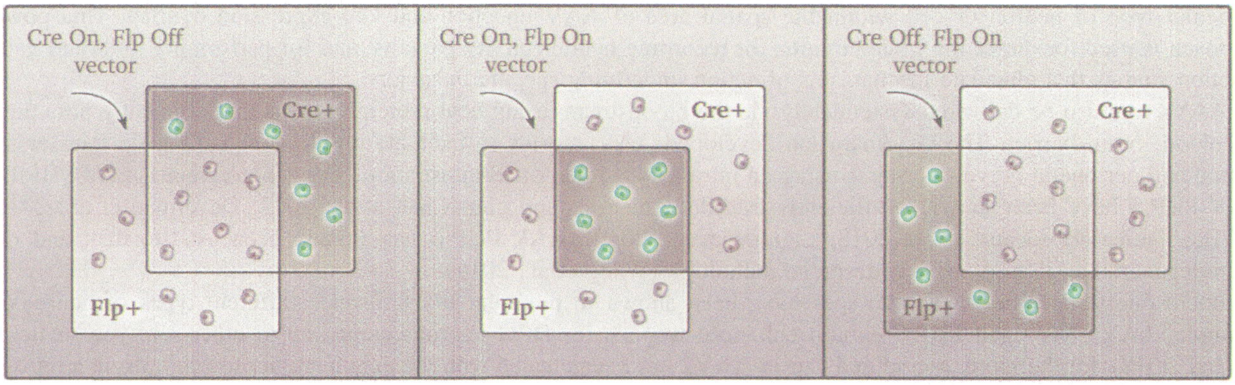

FIGURE 1.4 Dual recombinase-mediated intersectional genetic approach. Schematics representing target populations of cells expressing Cre and FLP. Transduced cells are shown in green. Left panel: using a subtractive approach, Cre_{on}/FLP_{off} AAV vectors direct gene expression to cells that are positive for Cre but not for FLP. Combinatorial approach in the middle: Cre_{on}/FLP_{on} vectors will express in cells that are both Cre and FLP positive. Right panel, exclusion approach: Cre_{off}/FLP_{on} vectors allow for expression of proteins within cells that are positive for FLP but not for Cre.

Retrograde tracing and labeling: Retrograde tracing with the Modified Rabies technique is used to identify direct monosynaptic inputs to the genetically targeted neurons ((Callaway & Luo, 2015) and Fig. 1.5). In this technique, the rabies virus has been modified in two ways: (1) It carries a fluorescent protein to label the cells it has infected, and (2) it is modified to lack the gene for the rabies glycoprotein (RG) and therefore cannot spread further without exogenously supplemented RG. Glycoproteins provide the RV access to the host cell by making spikes that interact with specific cell surface receptors. The modified RV is instead pseudotyped to carry avian sarcoma leukosis virus (ASLV) envelope A glycoprotein (EnvA). EnvA specifically interacts with the ASLV receptor TVA. "Helper"-AAVs with RG and TVA are introduced 3 weeks prior to injection with RV. The mCherry protein fused to TVA is useful for identifying starter cells (Watabe-Uchida et al., 2012).

- Additional useful references for mapping, neural circuits, tracing tools: Atasoy et al. (2008), Betley & Sternson (2011), Callaway & Luo (2015), Deverman et al. (2016), Dymecki et al. (2002), Gunaydin et al. (2014), Lowell (2019), Luo et al. (2018), Zingg et al. (2017).

Transcriptomics for identifying the gene expression profile of cells

"Transcriptome profiling," to reveal and study the transcriptome within individual neurons, is applicable to neuro-urology research. Analyses of genomic DNA sequences can generate a molecular understanding of cells. Studies of specific regions of the brain, the spinal cord, and the bladder are revealing cell type diversity and subgroups of neurons (e.g., interneurons, peptidergic [motor] neurons).

Because expression of protein-encoding messenger RNA (mRNA) correlates well with cellular traits and changes in cellular state, the transcriptome can be used as a proxy for studying the proteome (Haque et al., 2017). Since most scRNA-Seq techniques rely on primer extension from the polyadenylated tail of mRNAs, these scRNA-Seq techniques primarily quantify protein-coding transcripts. However, studies are starting to focus on other classes of transcripts and in particular microRNAs (for the regulation of gene activity and expression of mRNAs) by inferring miRNA activity from single-cell mRNA expression (Nielsen & Pedersen, 2021). Because the absolute quantity of RNA molecules is small in a single living cell, many cells are required for studying mRNAs or other non–protein-coding forms of cellular RNA transcripts.

RNA sequencing techniques provide the opportunity to identify cells in the LUT or CNS based on their RNA expression pattern. Specific cells can then be studied; their activity can be manipulated and activity patterns recorded. Functional assessment can be supplemented with circuit mapping for a more complete understanding of the neural circuitry of neurons involved. This technique has the potential to greatly advance treatment for LUTS by identifying specific, even unique, cellular and molecular features that can be selectively targeted by designer pharmacologics.

- Read more about the different RNA-Seq methods in Chapter 8—Transcriptomic identification of cell types in the lower urinary tract. Chapter 8 presents applications of this tool for researchers to start using it.

Spatially resolving gene expression profiles in intact biological samples

Single-cell transcriptome profiling provides a means to identify cell types. However, when cells are dissociated for sequencing analysis, their spatial context is lost (Zhuang, 2021). Image-based single-cell transcriptomic approaches

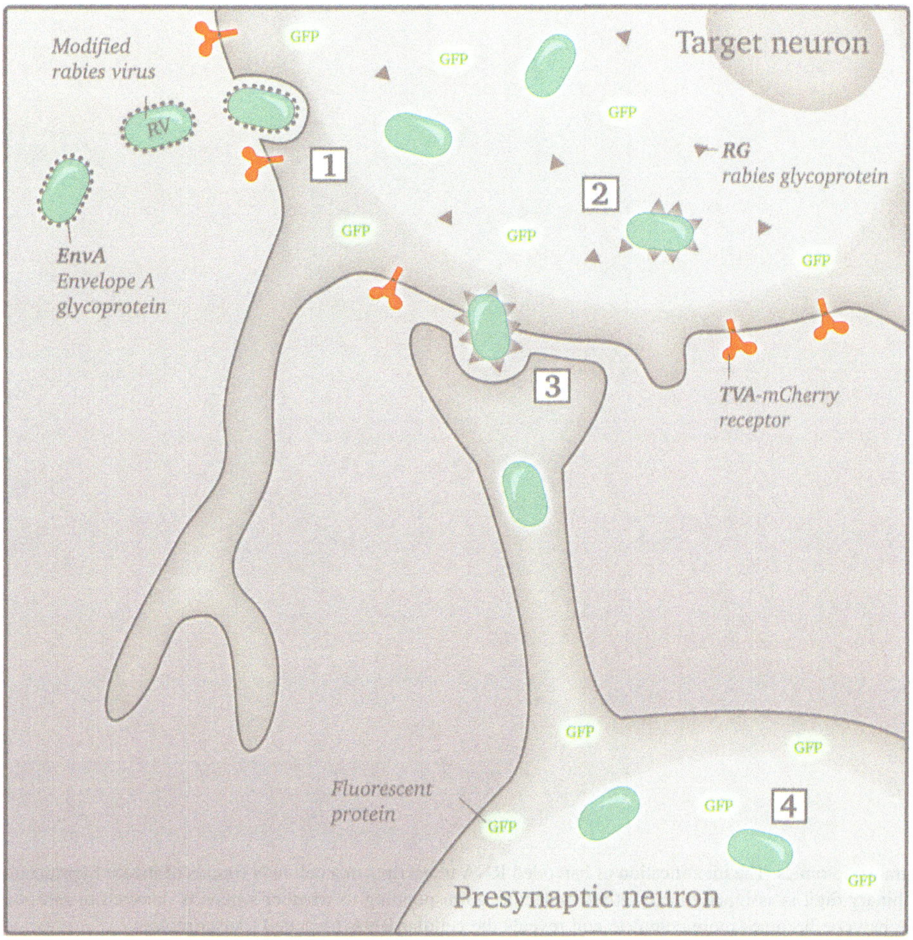

FIGURE 1.5 Retrograde tracing with modified glycoprotein (G)-deleted rabies virus (RV). Target (starter) neurons are supplemented with TVA receptor and rabies glycoprotein (RG) prior to microinjection with the modified RV. Modified RV is pseudotyped to carry avian sarcoma leukosis virus (ASLV) envelope A glycoprotein (EnvA), which interacts specifically with the ASLV receptor TVA (1). RV replication occurs and RG is incorporated into the viral envelope (2). RV is taken up by presynaptic nerve terminals (3), then replicates again in the presynaptic neuron and expresses fluorescent protein X (XFP) (4). The RV cannot spread further because RG is not present in the upstream neuron.

have been developed for spatially resolving gene expression profiles by directly imaging individual RNA molecules within intact biological samples with multiplexed fluorescence in situ hybridization (FISH) (Moffitt et al., 2019). Multiplexed error-robust fluorescence in situ hybridization (MERFISH) (Chen, Boettiger, et al., 2015) is one of these imaging methods. It is capable of simultaneously measuring the copy number and spatial distribution of hundreds to thousands of RNA molecules in a single cell. In addition, MERFISH can localize RNA transcripts at a subcellular level with $\leq$100 nm resolution. With MERFISH, each cellular RNA is targeted by short, complementary sequences that bind the RNA. Each set of targeting probes has four readout sequences distributed across them, which in turn bind fluorescently labeled readout probes. Multiple rounds of hybridization are run to detect the readout sequences in each round. A binary digit "1" or "0" is assigned corresponding to whether the gene is detected or not. Successive rounds of hybridization and imaging are used to identify the combination of readout sequences. The complete readout sequence profile comprises the N-bit barcode (Chen, Boettiger, et al., 2015) (see Fig. 1.6).

Thus, molecular marker genes within complex tissues can be imaged while they retain their spatial organization. Spatial transcriptome profiling can further be used to reveal subcellular RNA compartmentalization and cell cycle–dependent gene expression (Xia et al., 2019). Finally, imaging of activity markers such as the induction of IEGs simultaneously with genes of known functional importance may potentially uncover a functional role for distinct cells (Moffitt et al., 2019). This opens up new avenues for targeting neurons based on their projection patterns and activity, and for cell type–specific neurophysiology.

- Useful references for (spatial) transcriptomics technology: Bakken et al. (2018), Campbell et al. (2017), Chen, Boettiger, et al. (2015), Ecker et al. (2017), Macosko et al. (2015), Moffitt et al. (2016), Poulin et al. (2016).

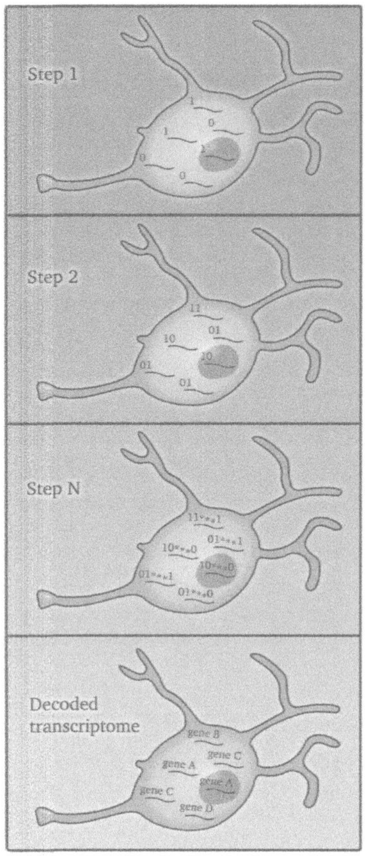

FIGURE 1.6 Spatial transcriptomics. The identification of barcoded RNA transcripts in a cell in N rounds of in situ hybridization and imaging. In every hybridization round, a binary digit is assigned to each RNA molecule, corresponding to whether a gene is detected or not (step 1). During subsequent rounds (step 2...N), the barcode becomes more complete and reveals the cellular RNA (decoded transcriptome).

Recording neural activity

Calcium imaging is used to measure activity of neurons and associated neural circuits in awake nonrestrained rodents. Recent advances have made it feasible to image calcium signals at high resolution. When neurons fire, calcium levels increase in the cytosol. Fluorescence signals that reflect cytosolic calcium levels in a cell population transduced with genetically encoded calcium indicators (GECIs) can then be monitored. The fiber photometry method detects changes in fluorescence—not only in cell bodies but also in axons which are considerably smaller—through an implanted optical fiber. Activity-dependent fluorescence emitted by cells in a tissue is collected simultaneously (as one signal). This signal is propagated back through the same patch cord that is used to deliver excitation light.

- Read more about measuring calcium signals, voltage imaging and other types of imaging in Chapter 9—Exploring urinary bladder neural circuitry through calcium imaging. Chapter 9 presents techniques for recording basal activity in distinct neurons involved in bladder control and for studying neural activity patterns in health and disease.

Fiber photometry during awake CMG

Urodynamic parameters such as bladder pressure are monitored by cystometrogram (CMG). For CMG recordings, saline is infused into the bladder through an indwelling catheter that is connected to a pressure transducer and coupled to a computerized recording system (Bjorling et al., 2015). Viewed via CMG, a voiding contraction starts with a steep rise in recorded intravesical pressure, which slows, then peaks, and declines rapidly as the bladder empties. The intercontractile interval (ICI) is the duration between two adjacent voiding contractions (Yao et al., 2019). Bladder physiology readouts can be combined with calcium imaging to study activity in specific cell populations with potential roles in LUT function and to investigate brain functions during urination (Verstegen et al., 2019).

Historically most urodynamic studies have been performed in anesthetized animals (De Groat and Wickens, 2013; Tanaka et al., 2003). However, because anesthesia itself changes the dynamics of calcium signals during bladder contractions, and may even induce dripping overflow incontinence (Yao et al., 2019), measurements recorded in awake mice are more ideal (Verstegen et al., 2019; Yao et al., 2019). Fiber photometry studies, when performed without CMG/continuous bladder filling, sometimes use a loop diuretic as an alternative approach for "collecting" enough voiding events during the neuronal activity recordings (Hou et al., 2016). Calcium imaging in conjunction with thermal imaging is described further below under micturition video thermography (MVT) and may be useful as a screening tool for LUTS and to link micturition behaviors with specific neuronal populations.

- Useful references for calcium imaging technology used in combination with (bladder) physiology readouts: Chen, Lin, et al. (2015), Foditsch et al. (2018), Verstegen et al. (2019), Yao et al. (2019).

Cell-specific manipulation of activity

Optogenetic stimulation or inhibition of neuron activity

Optogenetics uses microbial opsins. These light-gated ion channels or pumps absorb light at specific wavelengths. Photostimulation results in depolarization (activation), hyperpolarization (silencing), or signaling modulation of the illuminated neurons, depending on the type of opsin (Rogan & Roth, 2011). Through engineered point mutations in the opsin's channel design its kinetics can be modified. For example, mutating a cysteine residue in the 128th position or an aspartic acid residue in the 156th position of the opsin channel can delay closing of the channel. Alternatively, mutation of a glutamic acid residue in the 123rd position produces the E123-"ChETA" variant, which possesses accelerated channel closing and can be used for depolarizing neurons with frequencies of up to 200 Hz. Trafficking signals can also be added to the opsin for better localization to the cell membrane or cellular organelles.

Opsins used for interrogating discrete cell populations and neural circuitry underlying LUT control are expressed in specific cell populations via stereotaxic injection of an AAV that carries the opsin. Then, to turn the ion channel or ion pump–expressing neurons on or off, light is delivered into the tissue. The best-known opsin variants are Channelrhodopsin, Halorhodopsin, and Archaerhodopsin (Fig. 1.7). Channelrhodopsins are excited by blue light and allow for

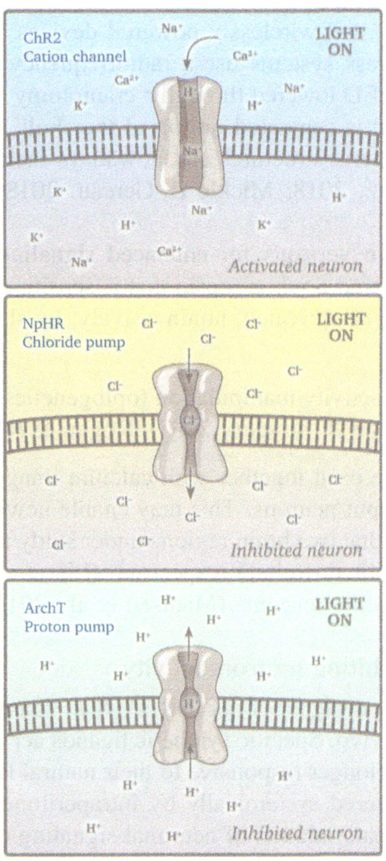

FIGURE 1.7 Common opsin variants used in optogenetics. The light-gated cation channel Channelrhodopsin-2 (ChR2) is excited by blue light (*upper panel*). The chloride pump Halorhodopsin (NpHR) (*middle*) and outward proton pump Archaerhodopsins (ArchT) (*lower panel*) are activated by yellow and green light, respectively.

nonspecific cation influx into the cell, resulting in depolarization. Channelrhodopsin-2 (ChR2) is the next generation and preferred over ChR1 because of its higher conductance at physiological pH (Lin et al., 2009).

Halorhodopsin (eNpHR3.0) and Archaerhodopsin are used for silencing cell populations (Chow et al., 2010; Han & Boyden, 2007; Zhang et al., 2011). Halorhodopsins are light-gated inward chloride pumps and cause hyperpolarization of the cell when triggered with yellow light, thus inhibiting action potential firing. Archaerhodopsin and its homolog ArchT, which is > threefold sensitive to light, are light-driven outward proton pumps. Expression of opsins in specific subsets of neurons allows for powerful, transient, and temporal control over neuronal activity by pulses of light. ChR2 and ArchT express well on neuronal membranes and can traffic long distances along neuronal axons.

Typical solutions for light delivery are tethered optical fiber—based systems in which the fiber optic is implanted into the brain using the stable nature of the brain—skull interface. Optical powers for LEDs are usually between 1 and 10 mW at the tip of the optic fiber, 0.25—2 mW have been reported for micro-LEDs (Zhang et al., 2019). The "brain tissue light transmission calculator" on the Deisseroth Lab website is a useful resource for calculating corresponding irradiances per mm^2.

An Arduino board and software (*Arduino cc, Sketch*) or a frequency generator can be used to turn the LED on or off at a specific frequency or pulse width duration. The optogenetics techniques can be combined with bladder physiology recordings to study specific cell populations with roles in LUT function (Hou et al., 2016; Keller et al., 2018; Verstegen et al., 2019; Yao et al., 2018).

Despite the ultrafast temporal resolution of optogenetics which allows for activating or silencing of neurons in the range of milliseconds, a disadvantage of the technique is its invasiveness. Other limitations include the relatively small targeting area as the light delivered from the optical fiber has limited spread. In addition, prolonged or too strong illumination can cause tissue damage. Finally, the patch cord connecting the optical fiber to the light source can prevent an animal from entering small enclosures or from engaging in normal interaction with others.

Optogenetics 2.0 and special tools

To enable sophisticated optogenetic manipulations with little disturbance of the animal's behavior, light delivery systems without fiber optic tethering are desirable (Montgomery et al., 2015). Researchers have developed battery-powered (Hashimoto et al., 2014; Iwai et al., 2011) and wirelessly powered devices (Kwon et al., 2015; Wentz et al., 2011) for delivering light to the mouse brain. Wireless systems use a radiofrequency power source and controller. Deeper brain regions can also be targeted with a micro-LED lowered through a craniotomy into the brain while the stimulator consisting of a power receiving coil and circuit board is cemented on top of the skull. Wireless systems can be adapted for stimulations to the brain, spinal cord, and peripheral structures and nowadays include subdermal implants (Ausra et al., 2021; Grajales-Reyes et al., 2021; Hibberd et al., 2018; Mickle & Gereau, 2018; Park et al., 2015; Shin et al., 2017) and injectable LEDs (Shin et al., 2017).

In the optogenetics field, various opsin versions for enhanced signaling and better targeting are being developed constantly. The Feng group at MIT engineered Soul, an opsin that is sensitive to even low-level light (Gong et al., 2020). Using "Soul," any mouse brain region can be activated, noninvasively, by shining light through the cranium or vertebral column (Gong et al., 2020).

- Additional useful references for neural activity manipulation (optogenetics) technology: Deisseroth (2015), Gong et al. (2020), Grajales-Reyes et al. (2021), Han & Boyden (2007), Rogan & Roth (2011), Verstegen et al. (2019).

NB: The optogenetics technique can be used together with calcium imaging and thus can link changes in activity in certain neurons to calcium signaling in output neurons. This may enable new hypothesis testing about neuron function in regulating a specific behavior. In the event that two brain regions under study are located in close proximity to one another, the use of opsins or GECIs with a (red-)shifted excitation spectrum (Lin et al., 2013), or of intersectional approaches by combining FLP and Cre recombinase—specific reagents (Madisen et al., 2015) can be leveraged.

Chemogenetics for stimulating or inhibiting neuron activity

Designer receptors exclusively activated by designer drugs (DREADDs)—based chemogenetics are research tools used for stimulating or inhibiting cells in vitro or in vivo. Specific synthetic ligands act on cells that express the modified muscarinic G protein—coupled receptors which are no longer responsive to their natural ligand acetylcholine. The DREADDs agonist *clozapine N-oxide (CNO)* can be administered systemically by intraperitoneal injection or in drinking water. Different events are evoked through chemogenetic manipulation of neuronal signaling depending on the type of G protein (Ciriachi et al., 2019; Rogan & Roth, 2011):

- Activation of Gq-mediated signaling induces burst firing,
- Activation of Gi-mediated signaling induces neuronal silencing,
- Activation of Gs-mediated signaling stimulates the cAMP–CREB pathway and Gs is, similarly to Gq, coupled to an excitatory signaling cascade.

At high concentrations (>10 mg/kg), pharmacokinetic analyses have shown that CNO back converts to clozapine, an antipsychotic drug that is able to cross the blood–brain barrier. Because this metabolite is active and can exert effects on cells, it is of critical importance to confirm doses of the ligand that activate DREADDs, specifically in DREADDs-expressing animals that do not elicit effects under control circumstances (Manvich et al., 2018). An alternative for DREADDs agonist CNO is Compound 21 (Jendryka et al., 2019), and different types of designer receptors also exist that interact with other small molecules. For example, Kappa-opioid receptors (KORD) are activated by the designer molecule salvinorin B (Vardy et al., 2015).

NB: Compared to optogenetics, chemogenetics is less invasive because it does not require the implantation of an optical fiber. Chemogenetic stimulation is also longer lasting as the receptor agonist is present in the tissue for several hours.

Diphtheria toxin–mediated ablation and tetanus toxin–induced neuronal silencing

For loss-of-function studies (i.e., to determine if a specific cell or neuron type is necessary for a certain behavior or physiological function), diphtheria toxin–mediated ablation and tetanus toxin–induced neuronal silencing can be used. Diphtheria toxin (DT) is the exotoxin secreted by the bacterium that causes diphtheria. DT gains entry to host cell cytosol via receptor-mediated endocytosis where it then inhibits protein synthesis and consequently causes cell death (Ciriachi et al., 2019). Ablation of neuronal cells can be achieved by infusing or injecting DT into the rodent brain. The expression of this transgene can be restricted to selective cell populations using a Cre recombinase–dependent system, resulting in cell type–specific ablation of cells.

Tetanus toxin (light chain: TeLC) allows for manipulation of neuronal signaling by eliminating synaptic vesicle exocytosis in the targeted neurons (Schiavo et al., 1992). TeLC has proteolytic activity on the presynaptic vesicle SNARE protein VAMP2. By cleaving VAMP2, TeLC completely and irreversibly abolishes neurotransmitter release from the "affected" cells.

- Useful references for neural activity manipulation (chemogenetics): Atasoy & Sternson (2018), Roth (2016), Smith et al. (2021), Sternson & Roth (2014), and (ablation and toxin-induced neuronal silencing) technology: Grégoire & Kmita (2014), Murray et al. (2011), Wiegert et al. (2017).

NB: A key advantage of these neuronal activity manipulation and recording techniques is that they allow experiments in awake, freely behaving animals without the need for anesthesia or mechanical restraint.

Channelrhodopsin-assisted circuit mapping

Channelrhodopsin-assisted circuit mapping (CRACM) is an approach for testing functional synaptic connectivity of neurons in specific and, if desired, in anatomically separate brain sites. ChR2 proteins can traffic along axonal membranes and light stimulation can trigger action potentials even when the ChR2-expressing axons are disconnected from their cell body, which is common with brain slice preparations. With CRACM, electrical currents, in response to photostimulation of the presynaptic neurons, are recorded from postsynaptic cells. Glutamate receptor antagonists (DNQX) or GABA-A receptor antagonists (bicuculline) may be used for "isolating" currents; when bath-applied, they block light-evoked excitatory or inhibitory postsynaptic currents (EPSCs or IPSCs, respectively). Tetrodotoxin (TTX) is a drug that blocks network firing and synchronous release but preserves synaptic transmission (Arendt et al., 2013). Photo-evoked currents that persist in the presence of TTX are considered to be evoked by monosynaptically connected cells.

Useful references for CRACM technology: Anastasiades et al. (2018), Hull et al. (2009), Kawatani et al. (2021), Krashes et al. (2014), Petreanu et al. (2009).

Bladder function readout that can be used with the described tools

Micturition video thermography void spot assay

Most types of anesthesia alter neuronal activity; therefore, it is important that studies are performed in awake models (Pandita et al., 2000; Yoshiyama et al., 1994). A simple, noninvasive assay for screening voiding behaviors in awake and freely behaving models in the laboratory is MVT (Fig. 1.8). During MVT, void spots and micturition behavior are recorded

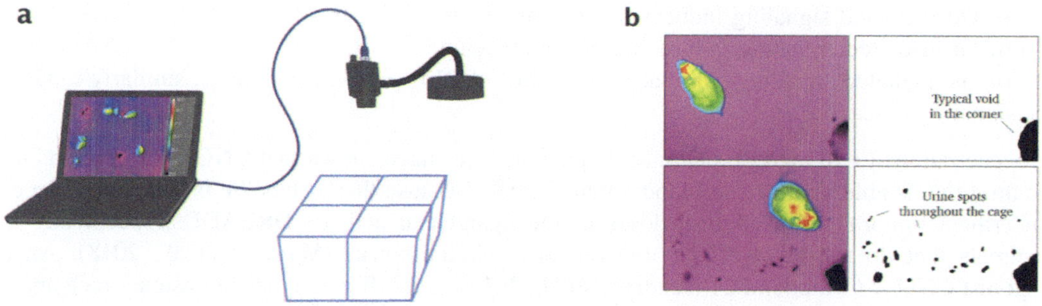

FIGURE 1.8 Micturition video thermography (MVT) void spot assay. Schematic representation of micturition video thermography (MVT) imaging setup; a thermal camera connected to a recording device, positioned above four enclosures on filter paper (A). Still images of MVT recordings illustrate specific voiding patterns (B). On the left, images from the thermal camera, and on the right, images with thresholding applied to the void spot(s) for void size calculation. MVT allows for micturition behavior analysis. *Modified from Verstegen, A. M. J., Tish, M., Szczepanik, L., Zeidel, M. L., & Geerling, J. C. (2020). Micturition video thermography in awake, behaving mice.* Journal of Neuroscience Methods, 331, *108449.*

from several mice simultaneously using a thermosensitive camera positioned above the behavior arena. Enclosures for MVT are assembled from laser-cut acrylic with tall walls that angle outward, allowing the view of inner cage corners at a low camera angle. The cages have no cage bottom and are open on top (Verstegen et al., 2020).

To provide optimal thermal contrast for identifying the border between the convective cooling of void spots relative to the dry filter paper on the cage floor, the lower display limit is set to 2°C below the ambient thermal emission of dry paper. The upper limit is set at 38°C, which is slightly higher than the core body temperature and thermal emission of mice and voided urine.

Video thermography recordings combine volume estimates and spatial information of void spot analysis and behavior with the temporal resolution of cystometry (CMG). Furthermore, due to variability of mouse behavior, the experimental design allows for repeat testing and within-animal controls. This repeatable technique is compatible with neuroscience techniques such as optogenetics and fiber photometry.

Video cystometry

Another method to reliably study bladder function without the need for suprapubic catheter implantation is X-ray video cystometry (Franken et al., 2021). This approach allows for monitoring bladder filling and bladder emptying in freely moving, nonanesthetized animals, over an extended period of time. By combining intravesical pressure measurements with X-ray—based fluoroscopy of the LUT this technique provides accurate and detailed information on urethral flow rate, intercontraction intervals, and residual bladder volume. The use of video cystometry and video thermography can potentially have broad applications ranging from the elucidation of molecular mechanisms of bladder control to drug development for LUT dysfunction.

Conclusions

Modern transgenic tools have enabled the mapping, monitoring, and manipulation of discrete cell populations in vivo. Successful application of these tools is aided by the identification of transcriptional markers to delineate individual cell groups within regions of high cellular heterogeneity. The ability to identify these cell types and subgroups using next-generation sequencing methods and to inform newer approaches that permit selective access to, and manipulation of, each subpopulation will enable a more complete understanding of neural control of bladder function in both health and disease in the future.

References

Albayram, O., Maciver, B., Mathai, J., Verstegen, A., Baxley, S., Qiu, C., Bell, C., Caldarone, B. J., Zhou, X. Z., Lu, K. P., & Zeidel, M. (2019). Traumatic Brain Injury-related voiding dysfunction in mice is caused by damage to rostral pathways, altering inputs to the reflex pathways. *Scientific Reports*.

Anastasiades, P. G., Marques-Smith, A., & Butt, S. J. B. (2018). Studies of cortical connectivity using optical circuit mapping methods. *The Journal of Physiology, 596*, 145—162.

Anderson, K. E. (1993). Pharmacology of lower urinary tract smooth muscles and penile erectile tissues. *Pharmacological Reviews, 45*, 253—308.

Anderson, K. D. (2004). Targeting recovery: Priorities of the spinal cord-injured population. *Journal of Neurotrauma, 21*, 1371–1383.

Andrew, J., & Nathan, P. W. (1965). The cerebral control of micturition. *Proceedings of the Royal Society of Medicine, 58*, 553–555.

Angerer, P., Simon, L., Tritschler, S., Wolf, A., Fischer, A., & Theis, F. J. (2017). Single cells make big data: New challenges and opportunities in transcriptomics. *Current Opinion in Systems Biology, 4*, 85–91.

Arendt, K. L., Sarti, F., & Chen, L. (2013). Chronic inactivation of a neural circuit enhances LTP by inducing silent synapse formation. *Journal of Neuroscience, 33*, 2087–2096.

Atasoy, D., Aponte, Y., Su, H. H., & Sternson, S. M. (2008). A FLEX switch targets Channelrhodopsin-2 to multiple cell types for imaging and long-range circuit mapping. *Journal of Neuroscience, 28*, 7025–7030.

Atasoy, D., & Sternson, S. M. (2018). Chemogenetic tools for causal cellular and neuronal biology. *Physiological Reviews, 98*, 391–418.

Ausra, J., Wu, M., Zhang, X., Vázquez-Guardado, A., Skelton, P., Peralta, R., Avila, R., Murickan, T., Haney, C. R., Huang, Y., Rogers, J. A., Kozorovitskiy, Y., & Gutruf, P. (2021). Wireless, battery-free, subdermally implantable platforms for transcranial and long-range optogenetics in freely moving animals. *Proceedings of the National Academy of Sciences, 118*. e2025775118.

Bakken, T. E., Hodge, R. D., Miller, J. A., Yao, Z., Nguyen, T. N., Aevermann, B., Barkan, E., Bertagnolli, D., Casper, T., Dee, N., Garren, E., Goldy, J., Graybuck, L. T., Kroll, M., Lasken, R. S., Lathia, K., Parry, S., Rimorin, C., Scheuermann, J. H., … Tasic, B. (2018). Single-nucleus and single-cell transcriptomes compared in matched cortical cell types. *PLoS One, 13*, e0209648.

Barrington, F. J. (1925). The effect of lesions of the hind- and mid-brain on micturition in the cat. *Quarterly Journal of Experimental Physiology, 15*, 81–102.

Barrington, F. J. (1927). Affections of micturition resulting from lesions of the nervous system. *Proceedings of the Royal Society of Medicine, 20*, 722–727.

Betley, J. N., & Sternson, S. M. (2011). Adeno-associated viral vectors for mapping, monitoring, and manipulating neural circuits. *Human Gene Therapy, 22*, 669–677.

Birder, L., & Drake, M. (2009). Neural control. In P. Abrams, L. Cardozo, S. Khoury, et al. (Eds.), *Incontinence* (pp. 167–254). Paris: Health Publications Ltd.

Bjorling, D. E., Wang, Z., Vezina, C. M., Ricke, W. A., Keil, K. P., Yu, W., Guo, L., Zeidel, M. L., & Hill, W. G. (2015). Evaluation of voiding assays in mice: Impact of genetic strains and sex. *American Journal of Physiology-Renal Physiology, 308*, F1369–F1378.

Blok, B. F. M., DE Weerd, H., & Holstege, G. (1995). Ultrastructural evidence for a paucity of projections from the lumbosacral cord to the pontine micturition center or M-region in the cat: A new concept for the organization of the micturition reflex with the periaqueductal gray as central relay. *The Journal of Comparative Neurology, 359*, 300–309.

Blok, B. F. M., & Holstege, G. (1994). Direct projections from the periaqueductal gray to the pontine micturition center (M-region). An anterograde and retrograde tracing study in the cat. *Neuroscience Letters, 166*, 93–96.

Blok, B. F. M., van Maarseveen, J. T., & Holstege, G. (1998). Electrical stimulation of the sacral dorsal gray commissure evokes relaxation of the external urethral sphincter in the cat. *Neuroscience Letters, 249*, 68–70.

Branda, C. S., & Dymecki, S. M. (2004). Talking about a revolution: Review the impact of site-specific recombinases on genetic analyses in mice. *Developmental Cell, 6*, 7–28.

Buchholz, F., Angrand, P.-O., & Stewart, A. F. (1998). Improved properties of FLP recombinase evolved by cycling mutagenesis. *Nature Biotechnology, 16*, 657–662.

Callaway, E. M., & Luo, L. (2015). Monosynaptic circuit tracing with glycoprotein-deleted rabies viruses. *Journal of Neuroscience, 35*, 8979–8985.

Campbell, J. N., Macosko, E. Z., Fenselau, H., Pers, T. H., Lyubetskaya, A., Tenen, D., Goldman, M., Verstegen, A. M. J., Resch, J. M., Mccarroll, S. A., Rosen, E. D., Lowell, B. B., & Tsai, L. T. (2017). A molecular census of arcuate hypothalamus and median eminence cell types. *Nature Neuroscience, 20*, 484–496.

Chan, K. Y., Jang, M. J., Yoo, B. B., Greenbaum, A., Ravi, N., Wu, W. L., Sánchez-Guardado, L., Lois, C., Mazmanian, S. K., Deverman, B. E., & Gradinaru, V. (2017). Engineered AAVs for efficient noninvasive gene delivery to the central and peripheral nervous systems. *Nature Neuroscience, 20*, 1172–1179.

Chen, K. H., Boettiger, A. N., Moffitt, J. R., Wang, S., & Zhuang, X. (2015). Spatially resolved, highly multiplexed RNA profiling in single cells. *Science, 348*.

Chen, Y., Lin, Y.-C., Kuo, T.-W., & Knight, Z. A. (2015). Sensory detection of food rapidly modulates arcuate feeding circuits. *Cell Reports, 160*, 829–841.

Chen, J., Markowitz, J. E., Lilascharoen, V., Taylor, S., Sheurpukdi, P., Keller, J. A., Jensen, J. R., Lim, M. K., Datta, S. R., & Stowers, L. (2021). Flexible scaling and persistence of social vocal communication. *Nature, 593*, 108–113.

Chow, B. Y., Han, X., Dobry, A. S., Qian, X., Chuong, A. S., Li, M., Henniger, M. A., Belfort, G. M., Lin, Y., Monahan, P. E., & Boyden, E. S. (2010). High-Performance genetically targetable optical neural silencing via light-driven proton pumps. *Nature, 463*, 98–102.

Ciriachi, C., Svane-Petersen, D., & Rickhag, M. (2019). Genetic tools to study complexity of striatal function. *Journal of Neuroscience Research, 97*, 1181–1193.

Coyne, K. S., Sexton, C. C., Thompson, C. L., Milsom, I., Irwin, D., Kopp, Z. S., Chapple, C. R., Kaplan, S., Tubaro, A., Aiyer, L. P., & Wein, A. J. (2009a). The prevalence of lower urinary tract symptoms (LUTS) in the USA, the UK and Sweden: Results from the epidemiology of LUTS (EpiLUTS) study. *BJU International, 104*, 352–360.

Coyne, K. S., Wein, A. J., Tubaro, A., Sexton, C. C., Thompson, C. L., Kopp, Z. S., & Aiyer, L. P. (2009b). The burden of lower urinary tract symptoms: evaluating the effect of LUTS on health-related quality of life, anxiety and depression: EpiLUTS. *BJU International,* 103 *Suppl, 3*, 4–11.

Degroat, W. C. (1975). Nervous control of the urinary bladder of the cat. *Brain Research, 87*, 201–211.

Deisseroth, K. (2015). Optogenetics: 10 years of microbial opsins in neuroscience. *Nature Neuroscience, 18*, 1213–1225.

Department of Economic and Social Affairs, P.D. (2013). World population ageing 2013. New York: United Nations. ST/ESA/SER.A/348.

Desjardins, C., Maruniak, J. A., & Bronson, F. H. (1973). Social rank in house mice: Differentiation revealed by ultraviolet visualization of urinary marking patterns. *Science, 182*, 9390941.

Deverman, B. E., Pravdo, P. L., Simpson, B. P., Kumar, S. R., Chan, K. Y., Banerjee, A., Wu, W. L., Yang, B., Huber, N., Pasca, S. P., & Gradinaru, V. (2016). Cre-dependent selection yields AAV variants for widespread gene transfer to the adult brain. *Nature Biotechnology, 34*, 204–209.

Duong, M., Downie, J. W., & Du, H.-J. (1999). Transmission of afferent information from urinary bladder, urethra and perineum to periaqueductal gray of cat. *Brain Research, 819*, 108–119.

Dymecki, S. M., Rodriguez, C. I., & Awatramani, R. B. (2002). Switching on lineage tracers using site-specific recombination. *Methods in Molecular Biology, 185*, 309–334.

Ecker, J. R., Geschwind, D. H., Kriegstein, A. R., et al. (2017). The BRAIN initiative cell census consortium: Lessons learned toward generating a comprehensive brain cell atlas. *Neuron, 96*, 542–557.

Farago, A. F., Awatramani, R. B., & Dymecki, S. M. (2006). Assembly of the brainstem cochlear nuclear complex is revealed by intersectional and subtractive genetic fate maps. *Neuron, 50*, 205–218.

Fenno, L. E., Mattis, J., Ramakrishnan, C., Hyun, M., Lee, S. Y., He, M., Tucciarone, J., Selimbeyoglu, A., Berndt, A., Grosenick, L., Zalocusky, K. A., Bernstein, H., Swanson, H., Perry, C., Diester, I., Boyce, F. M., Bass, C. E., Neve, R., Huang, Z. J., & Deisseroth, K. (2014). Targeting cells with single vectors using multiple-feature Boolean logic. *Nature Methods, 11*, 763–772.

Foditsch, E. E., Roider, K., Sartori, A. M., Kessler, T. M., Kayastha, S. R., Aigner, L., & Schneider, M. P. (2018). Cystometric and external urethral sphincter measurements in awake rats with implanted catheter and electrodes allowing for repeated measurements. *Jove, Journal of Visualized Experiments, 131*, e56506.

Fowler, C. J., & Griffiths, D. J. (2010). A decade of functional brain imaging applied to bladder control. *Neurourology and Urodynamics, 29*, 49–55.

Fowler, C. J., Griffiths, D., & Degroat, W. C. (2008). The neural control of micturition. *Nature Reviews Neuroscience, 9*, 453–466.

Franken, J., DE Bruyn, H., Rietjens, R., Segal, A., DE Ridder, D., Everaerts, W., Voets, T., & Vande Velde, G. (2021). X-ray videocystometry for high-speed monitoring of urinary tract function in mice. *Science Advances, 7*, eabi6821.

Gomez-Amaya, S. M., Barbe, M. F., de Groat, W. C., Brown, J. M., Tuite, G. F., Corcos, J., Fecho, S. B., Braverman, A. S., & Ruggieri Sr, M. R. (2015). Neural reconstruction methods of restoring bladder function. *Nature Reviews Urology, 12*, 100–118.

Gong, X., Mendoza-Halliday, D., Ting, J. T., Kaiser, T., Sun, X., Bastos, A. M., Wimmer, R. D., Guo, B., Chen, Q., Zhou, Y., Pruner, M., Wu, C. W.-H., Park, D., Deisseroth, K., Barak, B., Boyden, E. S., Miller, E. K., Halassa, M. M., Fu, Z., ... Feng, G. (2020). An ultra-sensitive step-function opsin for minimally invasive optogenetic stimulation in mice and macaques. *Neuron, 107*, 38–51.

Gotoh, M., Matsukawa, Y., Yoshikawa, Y., Funahashi, Y., Kato, M., & Hattori, R. (2009). Impact of urinary incontinence on the psychological burden of family caregivers. *Neurourology and Urodynamics, 28*, 492–496.

Grajales-Reyes, J. G., Copits, B. A., Lie, F., Yu, Y., Avila, R., Vogt, S. K., Huang, Y., Banks, A. R., Rogers, J. A., Gereau, R. W., & Golden, J. P. (2021). Surgical implantation of wireless, battery-free optoelectronic epidural implants for optogenetic manipulation of spinal cord circuits in mice. *Nature Protocols, 16*, 3072–3088.

Grégoire, D., & Kmita, M. (2014). Genetic cell ablation. *Methods in Molecular Biology, 1092*, 421–436.

Griffiths, D., & Tadic, S. D. (2008). Bladder control, urgency, and urge incontinence: Evidence from functional brain imaging. *Neurourology and Urodynamics, 27*, 466–474.

de Groat, W. C. (2006). Integrative control of the lower urinary tract: Preclinical perspective. *British Journal of Pharmacology, 147*, S25–S40.

de Groat, W. C., & Wickens, C. (2013). Organization of the neural switching circuitry underlying reflex micturition. *Acta Physiologica, 207*, 66–84.

de Groat, W. C., & Yoshimura, N. (2009). Afferent nerve regulation of bladder function in health and disease. *Handbook of Experimental Pharmacology, 194*, 91–138.

Gu, H., Marth, J. D., Orban, P. C., Mossmann, H., & Rajewsky, K. (1994). Deletion of a DNA polymerase beta gene segment in T cells using cell type-specific gene targeting. *Science, 265*, 103–106.

Gunaydin, L. A., Grosenick, L., Finkelstein, J. C., Kauvar, I. V., Fenno, L. E., Adhikari, A., Lammel, A., Mirzabekov, J. J., Airan, R. D., Zalocusky, K. A., Tye, K. M., Anikeeva, P., Malenka, R. C., & Deisseroth, K. (2014). Natural neural projection dynamics underlying social behavior. *Cell, 157*, 1535–1551.

Han, X., & Boyden, E. S. (2007). Multiple-color optical activation, silencing, and desynchronization of neural activity, with single-spike temporal resolution. *PLoS One, 2*, e299.

Han, X., Chow, B. Y., Zhou, H., Klapoetke, N. C., Chuong, A., Rajimehr, R., Yang, A., Baratta, M. V., Winkle, J., Desimone, R., & Boyden, E. S. (2011). A high-light sensitivity optical neural silencer: Development and application to optogenetic control of non-human primate cortex. *Frontiers in Systems Neuroscience, 5*, 18.

Hanson, R. W., & Franklin, M. R. (1976). Sexual loss in relation to other functional losses for spinal cord injured males. *Archives of Physical Medicine and Rehabilitation, 57*, 291–293.

Han, X., Zhang, Z., He, L., Zhu, H., Li, Y., Pu, W., Han, M., Zhao, H., Liu, K., Li, Y., Huang, X., Zhang, M., Jin, H., Lv, Z., Tang, J., Wang, J., Sun, R., Fei, J., Tian, X., ... Zhou, B. (2021). A suite of new Dre recombinase drivers markedly expands the ability to perform intersectional genetic targeting. *Cell Stem Cell, 28*, 1160–1176.e7.

Haque, A., Engel, J., Teichmann, S. A., & Lönnberg, T. (2017). A practical guide to single-cell RNA-sequencing for biomedical research and clinical applications. *Genome Medicine, 9*, 1−12.

Hashimoto, M., Hata, A., Miyata, T., & Hirase, H. (2014). Programmable wireless light-emitting diode stimulator for chronic stimulation of optogenetic molecules in freely moving mice. *Neurophotonics, 1*, 011002.

Hibberd, T. J., Feng, J., Luo, J., Yang, P., Samineni, V. K., Gereau, R. W. T., Kelley, N., Hu, H., & Spencer, N. J. (2018). Optogenetic induction of colonic motility in mice. *Gastroenterology, 155*, 514−528.e6.

Holstege, G., & Mouton, L. J. (2003). Central nervous system control of micturition. *International Review of Neurobiology, 56*, 123−145.

Hordeaux, J., Wang, Q., Katz, N., Buza, E. L., Bell, P., & Wilson, J. M. (2018). The neurotropic properties of AAV-PHP.B are limited to C57BL/6J mice. *Molecular Therapy, 26*, 664−668.

Hou, X. H., Hyun, M., Taranda, J., Huang, K. W., Todd, E., Feng, D., Atwater, E., Croney, D., Zeidel, M. L., Osten, P., & Sabatini, B. L. (2016). Central control circuit for context-dependent micturition. *Cell, 167*, 73−86.

Hull, C., Adesnik, H., & Scanziani, M. (2009). Neocortical disynaptic inhibition requires somatodendritic integration in interneurons. *Journal of Neuroscience, 29*, 8991−8995.

Irwin, D. E., Milson, I., Kopp, Z., Abrams, P., & Cardozo, L. (2005). Impact of overactive bladder symptoms on employment, social interactions and emotional well-being in six European countries. *BJU International, 97*, 96−100.

Irwin, D. E., Kopp, Z. S., Agatep, B., Milsom, I., & Abrams, P. (2011). Worldwide prevalence estimates of lower urinary tract symptoms, overactive bladder, urinary incontinence and bladder outlet obstruction. *BJU International, 108*, 1132−1138.

Iwai, Y., Honda, S., Ozeki, H., Hashimoto, M., & Hirase, H. (2011). A simple head-mountable LED device for chronic stimulation of optogenetic molecules in freely moving mice. *Neuroscience Research, 70*, 124−127.

Jendryka, M., Palchaudhuri, M., Ursu, D., van der Veen, B., Liss, B., Kätzel, D., Nissen, W., & Pekcec, A. (2019). Pharmacokinetic and pharmacodynamic actions of clozapine-N-oxide, clozapine, and compound 21 in DREADD-based chemogenetics in mice. *Scientific Reports, 9*, 4522.

Kamei, J., Ito, H., Aizawa, N., Hotta, H., Kojima, T., Fujita, Y., Ito, M., Homma, Y., & Igawa, Y. (2018). Age-related changes in function and gene expression of the male and female mouse bladder. *Scientific Reports, 8*, 2089.

Kanai, A., & Andersson, K. E. (2010). Bladder afferent signaling: Recent findings. *The Journal of Urology, 183*, 1288−1295.

Kanasaki, K., Yu, W., Bodungen Von, M., Larigakis, J. D., Kanasaki, M., Ayale de la Pena, F., Kalluri, R., & Hill, W. G. (2013). Loss of β1-integrin from urothelium results in overactive bladder and incontinence in mice: A mechanosensory rather than structural phenotype. *The FASEB Journal, 27*, 1950.

Kane, D. D., Kerns, J. M., Lin, D. L., & Damaser, M. S. (2004). Early structural effects of oestrogen on pudendal nerve regeneration in the rat. *BJU International, 93*, 870−878.

Karnup, S. V., & de Groat, W. C. (2020). Propriospinal neurons of L3-L4 segments involved in control of the rat external urethral sphincter. *Neuroscience, 425*, 12−28.

Kawatani, M., Degroat, W., Itoi, K., Uchida, K., Sakimura, K., Yamanaka, A., Yamashita, T., & Kawatani, M. (2021). Downstream projection of Barrington's nucleus to the spinal cord in mice. *Journal of Neurophysiology, 126*, 1959−1977.

Keller, J. A., Chen, J., Simpson, S., Wang, E. H.-J., Lilascharoen, V., George, O., Lim, B. K., & Stowers, L. (2018). Voluntary urination control by brainstem neurons that relax the urethral sphincter. *Nature Neuroscience, 21*, 1229−1238.

Krashes, M. J., Shah, B. P., Madara, J. C., Olson, D. P., Strochlic, D. E., Garfield, A. S., Vong, L., Pei, H., Watabe-Uchida, M., Uchida, N., Liberles, S. D., & Lowell, B. B. (2014). An excitatory paraventricular nucleus to AgRP neuron circuit that drives hunger. *Nature, 507*, 238−242.

Kwon, K. Y., Lee, H. M., Ghovanloo, M., Weber, A., & Li, W. (2015). Design, fabrication, and packaging of an integrated, wirelessly-powered optrode array for optogenetics application. *Frontiers in Systems Neuroscience, 9*, 69.

Lanigan, T. M., Kopera, H. C., & Saunders, T. L. (2020). Principles of genetic engineering. *Genes, 11*, 291.

Leonova, E. I., & Gainetdinov, R. R. (2020). CRISPR/Cas9 technology in translational biomedicine. *Cellular Physiology and Biochemistry, 54*, 354−370.

Liguore, W. A., Domire, J. S., Button, D., Wang, Y., Dufour, B. D., Srinivasan, S., & Mcbride, J. L. (2019). AAV-PHP.B administration results in a differential pattern of CNS biodistribution in non-human primates compared with mice. *Molecular Therapy, 27*, 2018−2037.

Lin, J. Y., Knutsen, P. M., Muller, A., Kleinfeld, D., & Tsien, R. Y. (2013). ReaChR: A red-shifted variant of channelrhodopsin enables deep transcranial optogenetic excitation. *Nature Neuroscience, 16*, 1499−1508.

Lin, J. Y., Lin, M. Z., Steinbach, P., & Tsien, R. Y. (2009). Characterization of engineered channelrhodopsin variants with improved properties and kinetics. *Biophysical Journal, 96*, 1803−1814.

Lino, C. A., Harper, J. C., Carney, J. P., & Timlin, J. A. (2018). Delivering CRISPR: A review of the challenges and approaches. *Drug Delivery, 25*, 1234−1257.

Lowell, B. B. (2019). New neuroscience of homeostasis and drives for food, water, and salt. *New England Journal of Medicine, 380*, 459−471.

Luo, L., Callaway, E. M., & Svoboda, K. (2018). Genetic dissection of neural circuits: A decade of progress. *Neuron, 98*, 256−281.

Macosko, E. Z., Basu, A., Satija, R., Nemesh, J., Shekhar, K., Goldman, M., Tirosh, I., Bialas, A. R., Kamitaki, N., Martersteck, E. M., Trombetta, J. J., Weitz, D. A., Sanes, J. R., Shalek, A. K., Regev, A., & Mccarroll, S. A. (2015). Highly parallel genome-wide expression profiling of individual cells using nanoliter droplets. *Cell, 161*, 1202−1214.

Madisen, L., Garner, A. R., Shimaoka, D., Chuong, A. S., Klapoetke, N. C., Li, L., van der Bourg, A., Niino, Y., Egolf, L., Monetti, C., Gu, H., Mills, M., Cheng, A., Tasic, B., Nguyen, T. N., Sunkin, S. M., Benucci, A., Nagy, A., Miyawaki, A., … Zeng, H. (2015). Transgenic mice for intersectional targeting of neural sensors and effectors with high specificity and performance. *Neuron, 85*, 942−958.

Manvich, D. F., Webster, K. A., Foster, S. L., Farrell, M. S., Ritchie, J. C., Porter, J. H., & Weinshenker, D. (2018). The DREADD agonist clozapine N-oxide (CNO) is reverse-metabolized to clozapine and produces clozapine-like interoceptive stimulus effects in rats and mice. *Scientific Reports, 8*, 3840.

Mcgee, M., & Grill, W. M. (2015). Methodologies for the restoration of bladder and bowel functions. *Encyclopedia of Computational Neuroscience*.

Metzger, D., & Chambon, P. (2001). Site- and time-specific gene targeting in the mouse. *Methods, 24*, 71–80.

Metzger, D., Clifford, J., Chiba, H., & Chambon, P. (1995). Conditional site-specific recombination in mammalian cells using a ligand-dependent chimeric Cre recombinase. *Proceedings of the National Academy of Sciences of the United States of America, 92*, 6991–6995.

Mickle, A. D., & Gereau, R. W. T. (2018). A bright future? Optogenetics in the periphery for pain research and therapy. *Pain, 159*(Suppl. 1), S65–s73.

Mizuguchi, H., Xu, Z., Akiko, I.-W., Uchida, E., & Hayakawa, T. (2000). IRES-dependent second gene expression is significantly lower than cap-dependent first gene expression in a bicistronic vector. *Molecular Therapy, 1*, 376–382.

Moffitt, J. R., Bambah-Mukkku, D., Eichhorn, S. W., Vaughn, E., Shekhar, K., Perez, J. D., Rubinstein, N. D., Hao, J., Regev, A., Dulac, C., & Zhuang, X. (2019). Molecular, spatial and functional single-cell profiling of the hypothalamic preoptic region. *Science, 362*, 6416.

Moffitt, J. R., Hao, J., Bambah-Mukku, D., Lu, T., Dulac, C., & Zhuang, X. (2016). High-performance multiplexed fluorescence in situ hybridization in culture and tissue with matrix imprinting and clearing. *Proceedings of the National Academy of Sciences, 113*, 14456–14461.

Montgomery, K. L., Yeh, A. J., Ho, J. S., Tsao, V., Iyer, S. M., Grosenick, L., Ferenczi, E. A., Tanabe, Y., Deisseroth, K., Delp, S. L., & Poon, A. S. Y. (2015). Wirelessly powered, fully internal optogenetics for brain, spinal and peripheral circuits in mice. *Nature Methods, 12*, 969–974.

Murray, A. J., Sauer, J. F., Riedel, G., Mcclure, C., Ansel, L., Cheyne, L., Bartos, M., Wisden, W., & Wulff, P. (2011). Parvalbumin-positive CA1 interneurons are required for spatial working but not for reference memory. *Nature Neuroscience, 14*, 297–299.

Nadelhaft, I., & Vera, P. L. (1996). Neurons in the rat brain and spinal cord labeled after pseudorabies virus injected into the external urethral sphincter. *Journal of Comparative Neurology, 275*, 502–517.

Nadelhaft, I., Vera, P. L., & Card, J. P. (1992). Central nervous system neurons labelled following the injection of pseudorabies virus into the rat urinary bladder. *Neuroscience Letters, 143*, 271–274.

Nielsen, M. M., & Pedersen, J. S. (2021). miRNA activity inferred from single cell mRNA expression. *Scientific Reports, 11*, 9170.

Noto, H., Roppolo, J. R., Steers, W. D., & Degroat, W. C. (1989). Excitatory and inhibitory influences on bladder activity elicited by electrical stimulation in the pontine micturition center in the rat. *Brain Research, 492*, 99–115.

Noto, H., Roppolo, J. R., Steers, W. D., & Degroat, W. C. (1991). Electrophysiological analysis of the ascending and descending components of the micturition reflex pathway in the rat. *Brain Research, 549*, 95–105.

Pandita, R. K., Fujiwara, M., Alm, P., & Andersson, K. E. (2000). Cystometric evaluation of bladder function in non-anesthetized mice with and without bladder outlet obstruction. *The Journal of Urology, 164*, 1385–1389.

Park, S. I., Brenner, D. S., Shin, G., Morgan, C. D., Copits, B. A., Chung, H. U., Pullen, M. Y., Noh, K. N., Davidson, S., Oh, S. J., Yoon, Y., Jang, K.-I., Samineni, V. K., Norman, M., Grajales-Reyes, J. G., Vogt, S. K., Sundaram, S. S., Wilson, K. M., Ha, J. S., … Rogers, J. A. (2015). Soft, stretchable, fully implantable miniaturized optoelectronic systems for wireless optogenetics. *Nature Biotechnology, 33*, 1280–1286.

Petreanu, L., Mao, T., Sternson, S. M., & Svoboda, K. (2009). The subcellular organization of neocortical excitatory connections. *Nature, 457*, 1142–1145.

Poulin, J. F., Tasic, B., Hjerling-Leffler, J., Trimarchi, J. M., & Awatramani, R. (2016). Disentangling neural cell diversity using single- cell transcriptomics. *Nature Neuroscience, 19*, 1131–1141.

Przydacz, M., Golabek, T., Dudek, P., Lipinski, M., & Chlosta, P. (2020). Prevalence and bother of lower urinary tract symptoms and overactive bladder in Poland, an Eastern European Study. *Scientific Reports, 10*, 19819.

Quadros, R. M., Miura, H., Harms, D. W., Akatsuka, H., Sato, T., Aida, T., Redder, R., Richardson, G. P., Inagaki, Y., Sakai, D., Buckley, S. M., Seshacharyulu, P., Batra, S. K., Behlke, M. A., Zeiner, S. A., Jacobi, A. M., Izu, Y., Thoreson, W. B., Urness, L. D., … Gurumurthy, C. B. (2017). Easi-CRISPR: A robust method for one-step generation of mice carrying conditional and insertion alleles using long ssDNA donors and CRISPR ribonucleoproteins. *Genome Biology, 18*, 92.

Rajewsky, K., Gu, H., Kühn, R., Betz, U. A., Müller, W., Roes, J., & Schwenk, F. (1996). Conditional gene targeting. *Journal of Clinical Investigation, 98*, 600–603.

Rogan, S. C., & Roth, B. L. (2011). Remote control of neuronal signaling. *Pharmacological Reviews, 63*, 291–315.

Roth, B. L. (2016). DREADDs for neuroscientists. *Neuron, 89*, 683–694.

Rouzade-Dominguez, M. L., Pernar, L., Beck, S., & Valentino, R. J. (2003). Convergent responses of Barrington's nucleus neurons to pelvic visceral stimuli in the rat: A juxtacellular labelling study. *European Journal of Neuroscience, 18*, 3325–3334.

Sadowski, P. D. (1995). The Flp recombinase of the 2-microns plasmid of Saccharomyces cerevisiae. *Progress in Nucleic Acid Research and Molecular Biology, 51*, 53–91.

Sakakibara, R. (2015). Lower urinary tract dysfunction in patients with brain lesions. *Handbook of Clinical Neurology, Autonomic Nervous System, Neurology of Sexual and Bladder Disorders, 130*, 269–287.

Sakakibara, R., Tateno, F., Yamamoto, T., Uchiyama, T., & Yamanishi, T. (2018). Urological dysfunction in synucleinopathies: Epidemiology, pathophysiology and management. *Clinical Autonomic Research, 28*, 83–101.

Salas, J. C., Iwasaki, H., Jodo, E., Schmidt, M. H., Kawauchi, A., Miki, T., Kayama, Y., Otsuki, M., & Koyama, Y. (2008). Penile erection and micturition events triggered by electrical stimulation of the mesopontine tegmental area. *American Journal of Physiology - Regulatory, Integrative and Comparative Physiology, 294*, R102–R111.

Sauer, B., & Henderson, N. (1988). Site-specific DNA recombination in mammalian cells by the Cre recombinase of bacteriophage P1. *Proceedings of the National Academy of Sciences of the United States of America, 85*, 5166–5170.

Schiavo, G. G., Benfenati, F., Poulain, B., Rossetto, O., de Laureto, P. P., Dasgupta, B. R., & Montecucco, C. (1992). Tetanus and botulinum-B neurotoxins block neurotransmitter release by proteolytic cleavage of synaptobrevin. *Nature, 359*, 832–835.

Schneider, M. P., Sartori, A. M., Ineichen, B. V., Moors, S., Engmann, A. K., Hofer, A.-S., Weinmann, O., Kessler, T. M., & Schwab, M. E. (2019). Anti-nogo-A antibodies as a potential causal therapy for lower urinary tract dysfunction after spinal cord injury. *Journal of Neuroscience, 39*, 4066–4076.

Seki, S., Sasaki, K., Fraser, M., Igawa, Y., Nishizawa, O., Chancellor, M. B., Degroat, W. C., & Yoshimura, N. (2002). Immunoneutralization of nerve growth factor in lumbosacral spinal cord reduces bladder hyperreflexia in spinal cord injured rats. *The Journal of Urology, 168*, 2269–2274.

Seki, S., Sasaki, K., Fraser, M., Igawa, Y., Nishizawa, O., Chancellor, M. B., Degroat, W. C., & Yoshimura, N. (2004). Suppression of detrusor-sphincter dyssynergia by immunoneutralization of nerve growth factor in lumbosacral spinal cord in spinal cord injured rats. *The Journal of Urology, 171*, 478–482.

Shin, G., Gomez, A. M., AL-Hasani, R., Jeong, Y. R., Kim, J., Xie, Z., Banks, A., Lee, S. M., Han, S. Y., Yoo, C. J., Lee, J. L., Lee, S. H., Kurniawan, J., Tureb, J., Guo, Z., Yoon, J., Park, S. I., Bang, S. Y., Nam, Y., … Rogers, J. A. (2017). Flexible near-field wireless optoelectronics as subdermal implants for broad applications in optogenetics. *Neuron, 93*, 509–521.e3.

Skarnes, W. C., Rosen, B., West, A. P., Koutsourakis, M., Bushell, W., Iyer, V., Mujica, A. O., Thomas, M., Harrow, J., Cox, T., Jackson, D., Severin, J., Biggs, P., Fu, J., Nefedov, M., DE Jong, P. J., Stewart, A. F., & Bradley, A. (2011). A conditional knockout resource for the genome-wide study of mouse gene function. *Nature, 474*, 337–342.

Smith, K. S., Bucci, D. J., Luikart, B. W., & Mahler, S. V. (2021). Dreadds: Use and application in behavioral neuroscience. *Behavioral Neuroscience, 135*, 89–107.

Sternberg, N., & Hamilton, D. (1981). Bacteriophage P1 site-specific recombination. I. Recombination between loxP sites. *Journal of Molecular Biology, 150*, 467–486.

Sternson, S. M., & Roth, B. L. (2014). Chemogenetic tools to interrogate brain functions. *Annual Review of Neuroscience, 37*, 387–407.

Stewart, W. F., VAN Rooyen, J. B., Cundiff, G. W., Abrams, P., Herzog, A. R., Corey, R., Hunt, T. L., & Wein, A. J. (2003). Prevalence and burden of overactive bladder in the United States. *World Journal of Urology, 20*, 327–336.

Tanaka, Y., Koyama, Y., Kayama, Y., Kawauchi, A., Ukimura, O., & Miki, T. (2003). Firing of micturition center neurons in the rat mesopontine tegmentum during urinary bladder contraction. *Brain Research, 965*, 146–154.

Tervo, D. G., Hwang, B. Y., Viswanathan, S., Gaj, T., Lavzin, M., Ritola, K. D., Lindo, S., Michael, S., Kuleshova, E., Ojala, D., Huang, C. C., Gerfen, C. R., Schiller, J., Dudman, J. T., Hantman, A. W., Looger, L. L., Schaffer, D. V., & Karpova, A. Y. (2016). A designer AAV variant permits efficient retrograde access to projection neurons. *Neuron, 92*, 372–382.

Tish, M., & Geerling, J. C. (2020). The brain and the bladder: Forebrain control of urinary (In)Continence. *Frontiers in Physiology, 11*, 658.

Tsai, N. Y., Wang, F., Toma, K., Yin, C., Takatoh, J., Pai, E. L., Wu, K., Matcham, A. C., Yin, L., Dang, E. J., Marciano, D. K., Rubenstein, J. L., Wang, F., Ullian, E. M., & Duan, X. (2022). Trans-Seq maps a selective mammalian retinotectal synapse instructed by nephronectin. *Nature Neuroscience, 25*, 659–674.

Vardy, E., Robinson, J. E., Li, C., Olsen, R. H. J., Diberto, J. F., Giguere, P. M., Sassano, F. M., Huang, X. P., Zhu, H., Urban, D. J., White, K. L., Rittiner, J. E., Crowley, N. A., Pleil, K. E., Mazzone, C. M., Mosier, P. D., Song, J., Kash, T. L., Malanga, C. J., Krashes, M. J., & Roth, B. L. (2015). A new DREADD facilitates the multiplexed chemogenetic interrogation of behavior. *Neuron, 86*, 936–946.

Verstegen, A. M. J., Klymko, N., Zhu, L., Mathai, J. C., Kobayashi, R., Venner, A., Ross, R. A., Vanderhorst, V. G. J. M., Arrigoni, E., Geerling, J. C., & Zeidel, M. L. (2019). Non-Crh glutamatergic neurons in Barrington's nucleus control micturition via glutamatergic afferents from the midbrain and hypothalamus. *Current Biology, 29*, 2775–2789.

Verstegen, A. M. J., Tish, M., Szczepanik, L., Zeidel, M. L., & Geerling, J. C. (2020). Micturition video thermography in awake, behaving mice. *Journal of Neuroscience Methods, 331*, 108449.

Wang, H., Yang, B., Qiu, L., Yang, C., Kramer, J., Su, Q., Guo, Y., Brown, R. H.,, Jr., Gao, G., & Xu, Z. (2014). Widespread spinal cord transduction by intrathecal injection of rAAV delivers efficacious RNAi therapy for amyotrophic lateral sclerosis. *Human Molecular Genetics, 23*, 668–681.

Watabe-Uchida, M., Zhu, L., Ogawa, S. K., Vamanrao, A., & Uchida, N. (2012). Whole-brain mapping of direct inputs to midbrain dopamine neurons. *Neuron, 74*, 858–873.

Weinholtz, C. A., & Castle, M. J. (2021). Intersectional targeting of defined neural circuits by adeno-associated virus vectors. *Journal of Neuroscience Research, 99*, 981–990.

Wentz, C. T., Bernstein, J. G., Monahan, P., Guerra, A., Rodriguez, A., & Boyden, E. S. (2011). A wirelessly powered and controlled device for optical neural control of freely-behaving animals. *Journal of Neural Engineering, 8*, 046021.

Wiegert, J. S., Mahn, M., Prigge, M., Printz, Y., & Yizhar, O. (2017). Silencing neurons: Tools, applications, and experimental constraints. *Neuron, 95*, 504–529.

Wood, S. K., Baez, M. A., Bhatnagar, S., & Valentino, R. J. (2009). Social stress-induced bladder dysfunction: Potential role of corticotropin-releasing factor. *American Journal of Physiology - Regulatory, Integrative and Comparative Physiology, 296*, R1671–R1678.

Xia, C., Fana, J., Emanuel, G., Hao, J., & Zhuang, X. (2019). Spatial transcriptome profiling by MERFISH reveals subcellular RNA compartmentalization and cell cycle-dependent gene expression. *Proceedings of the National Academy of Sciences, 116*, 19490–19499.

Yao, J., Li, Q., Li, X., Qin, H., Liang, S., Liao, X., Chen, X., Li, W., & Yan, J. (2019). Simultaneous measurement of neuronal activity in the pontine micturition center and cystometry in freely moving mice. *Frontiers in Neuroscience, Methods, 13*, 663.

Yao, J., Zhang, W., Liao, X., Li, Q., Liang, S., Li, X., Zhang, Y., Li, X., Wang, H., Qin, H., Wang, M., Li, J., Zhang, J., He, W., Zhang, W., Li, T., Xu, F., Gong, H., Jia, H., … Chen, X. (2018). A corticopontine circuit for initiation of urination. *Nature Neuroscience, 21*, 1541–1550.

Yoshiyama, M., Roppolo, J. R., & DE Groat, W. C. (1994). Alteration by urethane of glutamatergic control of micturition. *European Journal of Pharmacology, 264*, 417–425.

Zhang, Y., Castro, D. C., Han, Y., Wu, Y., Guo, H., Weng, Z., Xue, Y., Ausra, J., Wang, X., Li, R., Wu, G., Vázquez-Guardado, A., Xie, Y., Xie, Z., Ostojich, D., Peng, D., Sun, R., Wang, B., Yu, Y., … Rogers, J. A. (2019). Battery-free, lightweight, injectable microsystem for in vivo wireless pharmacology and optogenetics. *Proceedings of the National Academy of Sciences of the United States of America, 116*, 21427–21437.

Zhang, F., Vierock, J., Yizhar, O., Fenno, L. E., Tsunoda, S., Kianianmomeni, A., Prigge, M., Berndt, A., Cushman, J., Polle, J., Magnuson, J., Hegemann, P., & Deisseroth, K. (2011). The microbial opsin family of optogenetic tools. *Cell, 147*, 1446–1457.

Zhuang, X. (2021). Spatially resolved single-cell genomics and transcriptomics by imaging. *Nature Methods, 18*, 18–22.

Zingg, B., Chou, Z.-L., Zhang, Z.-G., Mesik, L., Liang, F., WIT Tao, H., & Zhang, L. I. (2017). AAV-mediated anterograde transsynaptic Tagging: Mapping Corticocollicular input-defined neural pathways for defense behaviors. *Neuron, 93*, 33–47.

Zingg, B., Peng, B., Huang, J., Tao, H. W., & Zhang, L. I. (2020). Synaptic specificity and application of anterograde transsynaptic AAV for probing neural circuitry. *Journal of Neuroscience, 40*, 3250–3267.

Section I

Neuroscience in urology research

Chapter 2

Barrington's nucleus: a century of progress identifying neurons that control micturition

Margaret M. Tish and Joel C. Geerling

University of Iowa, Department of Neurology, Iowa City, IA, United States

Introduction

Micturition, the act of urinating, is a complex behavior that many of us take for granted. However, a growing number of older individuals lose the ability to void (urinary retention) or to control when they void (urinary incontinence) (Johnson & Vaughan, 2019; Stewart et al., 2003; Verdejo et al., 2016). These patients report depression and loss of independence, and their caregivers rank these symptoms as their largest burdens (Irwin et al., 2006; Johnson & Vaughan, 2019; Mackay et al., 2018). Therefore, understanding how the brain controls micturition, and how this control may be lost, is an important focus of neuroscience research.

In this chapter, we focus on Barrington's nucleus (Bar), a small group of neurons that control pelvic motor functions, including micturition. Understanding these neurons and their connectivity is critical to understanding how the brain controls micturition. These neurons lie in a complex region, surrounded by many other populations of neurons that serve unrelated functions (Verstegen et al., 2017). Barrington's nucleus (Bar) was named after J.F. Barrington, who provided the first evidence that this region of the brainstem was necessary for micturition (Barrington, 1925, 1927). In the century since Barrington's work, we have learned a great deal more about the properties, connections, and functions of neurons in this region. In this chapter, after briefly reviewing the micturition reflex, we will describe the precise location and molecular markers that distinguish Bar neurons from surrounding neurons. Additionally, we will review what is known about regions synaptically connected to Bar neurons, which form a micturition control network spanning all levels of the nervous system.

The micturition reflex

The micturition reflex, in essence, triggers urination once the bladder fills to a pressure threshold (De Groat, 1975; De Groat et al., 2015). Most of the time, the bladder is in a storage phase, filling with urine delivered by the ureters. Under this predominant filling condition, the sympathetic nervous system, via the hypogastric nerves, helps inhibit bladder contraction while simultaneously contracting smooth muscle along the bladder neck, the internal urethral sphincter (Burnstock, 1990; Clemens, 2010; Elliott, 1907). The internal urethral sphincter is not under conscious control and relaxes or contracts in a reciprocal relationship with the bladder wall detrusor muscle (Denny-Brown & Robertson, 1933). As the bladder fills, its internal pressure increases, stretching the bladder wall. This activates the mechanosensory endings of thinly myelinated dorsal root ganglion neurons that project to the bladder via the pelvic nerves and send signals back to the spinal cord (Bahns et al., 1987; Clemens, 2010; Jänig & Koltzenburg, 1990; Mallory et al., 1989; Marshall et al., 2020).

This mechanosensory signal increases as the bladder continues to fill, ultimately reaching interneurons in the midbrain periaqueductal gray (PAG) matter (Vanderhorst et al., 1996), which are thought to form a sensory "switch" for reflex micturition (Stone et al., 2011). When activated, these PAG neurons send an excitatory signal to Bar (Blok & Holstege, 1994; Ding et al., 1998; Kuipers et al., 2006; Matsuura et al., 2000; Verstegen et al., 2019). Bar neurons, in turn, project their axons back down to the spinal cord. At lumbosacral spinal levels, these axons synapse in the intermediolateral (IML)

Neuro-Urology Research. https://doi.org/10.1016/B978-0-12-822455-7.00012-X

column and dorsal gray commissure (DGC) (Holstege et al., 1986; Hou et al., 2016; Keller et al., 2018; Loewy et al., 1979; Verstegen et al., 2017). The IML column contains autonomic preganglionic neurons that project axons through the pelvis to postganglionic neurons in the bladder wall, which in turn release acetylcholine to activate primarily M3 muscarinic receptors on the detrusor smooth muscle, causing bladder contraction (De Groat et al., 1982; Fetscher et al., 2002; Matsui et al., 2000). Inhibitory interneurons in the DGC directly inhibit motor neurons in the ventral horn, whose peripheral axons maintain tonic contraction of the external urethral sphincter (EUS). Excitatory input from Bar drives the DGC interneurons (Blok & Holstege, 1997) to inhibit these motor neurons, causing phasic relaxation of the EUS (Nadelhaft & Vera, 1996; Sato et al., 1978; Schrøder, 1981). Thus, upon bladder distention signaling via the PAG, Bar neurons send parallel excitatory signals to the spinal cord that simultaneously contract the bladder and relax the EUS, emptying the bladder.

Disruption of the micturition reflex can result from injuries to the spinal cord or lumbosacral nerve roots in the cauda equina (Denny-Brown & Robertson, 1933; Tai et al., 2006). Patients with such injuries progress from a transient loss of bladder tone (with tonic internal urethral sphincter contraction) to a condition known as "bladder sphincter dyssynergia" (Feloney & Leslie, 2020; Hou & Rabchevsky, 2014), where detrusor contraction (and reciprocal relaxation of the internal urethral sphincter) is no longer well-coordinated with relaxation of the EUS. In this state, the detrusor, internal urethral sphincter, and/or peripheral neurons develop spontaneous pressure waves with some reactivity to changes in bladder pressure, but without the discrete contractions and bladder–sphincter coordination required for complete bladder emptying (Denny-Brown & Robertson, 1933; Tai et al., 2006).

The British urologist F.J.F. Barrington made a fundamentally important contribution when, in the 1920s, he used progressively more proximal transections of the spinal cord and brainstem to reveal that the micturition reflex in cats is absent unless connectivity is maintained from the pons to the sacral spinal cord (Barrington, 1914, 1921). He then produced a series of lesions, localizing the region critical for reflex micturition to a focal zone in the dorsolateral pons–midbrain tegmentum. In cats with lesions that produced complete urinary retention, he observed:

> *... the common part of the lesions extends from just behind the level of the anterior end of the motor nucleus of the fifth nerve forwards for rather more than a millimeter to the extreme end of the hindbrain, stopping immediately behind the posterior end of the aqueduct: the decussation of the fourth nerves in the velum is present in the section containing it.*

Barrington, F. J. (1925). The effect of lesions of the hind- and mid-brain on micturition in the cat. *Experimental Physiology, 15*, 81–102.

Some investigators referred to this region broadly as the "Pontine micturition center" or PMC. However, this region of the hindbrain tegmentum contains many populations of neurons with diverse functions (see Neuroanatomical landscape surrounding Barrington's nucleus section below), and it was more than half a century before subsequent investigators determined the precise location of neurons critical for micturition (Loewy et al., 1979; Satoh, Shimizu, et al., 1978). Today, these neurons are referred to as Barrington's nucleus, abbreviated Bar (Ding et al., 2016; Paxinos & Franklin, 2013; Verstegen et al., 2017).

In addition to this core discovery, Barrington made another important observation (Morrison, 2008). He also noted that midbrain lesions just rostral to this zone did not cause urinary retention and instead produced a seeming lack of awareness that voiding was about to occur, reflected by an absence of behaviors cats typically exhibit before voiding. He claimed that lesions in or near what is now called the ventrolateral PAG produced "a permanent loss of consciousness of wanting to micturate or defaecate, but does not impair the performance of either of these functions" (Barrington, 1925). Thus, the micturition reflex circuit between the brainstem and spinal cord is necessary for the mechanics of voiding, but it is not sufficient for conscious control of voiding.

Conscious control is critical for urinary continence, but remains poorly understood. In early childhood, we gain the ability to suppress the micturition reflex, and this control grows over time. Continence begins with toilet training and proceeds through nighttime control to learning when it is socially appropriate to void and even withholding urination to the point of pain (Wu, 2010). However, in many individuals, conscious control of micturition is lost in adulthood. One-fifth of adults over the age of 65 will, to some degree, lose continent control (Landefeld et al., 2008). For some, this is due to a loss of central neural control, manifesting as urge incontinence, the sudden impulse to urinate with inability to overcome this urge (Drake, 2018). Urge incontinence results from damage to forebrain regions thought to modulate the activity of hindbrain regions that control the micturition reflex. Human case reports show that urge incontinence can result from damage to the frontal lobes or hypothalamus (Andrew & Nathan, 1964; Sakakibara, 2015; Tish & Geerling, 2020), and these regions of the brain have been identified in functional neuroimaging studies involving human micturition (Fowler & Griffiths, 2010; Khavari & Boone, 2019). These regions and their potential connectivity to the micturition reflex pathway are covered later in this chapter (see Afferent projections to Bar neurons section below).

Like humans, other mammals can control their micturition reflex. For example, while most rats and mice urinate intermittently and along their cage walls, dominant males leave many, small urine marks all over the cage when presented with female urine (Desjardins et al., 1973; Hou et al., 2016; Ralls, 1971). These observations offer an opportunity to study the neural control of micturition, beyond just reflex micturition, in experimental animals. Given the substantial and compelling evidence that Bar neurons represent a critical node, understanding these neurons will benefit investigators studying higher-level brain networks that control micturition.

Bar is located in the upper hindbrain (pontine) tegmentum, near the midbrain. A variety of other neurons in the pontine central gray (PCG) matter lie in close proximity to Bar (Verstegen et al., 2017). Each neuronal population in this region has a distinct set of molecular markers, circuit connections, and functions, most of which do not involve micturition. Before delving into neuronal subpopulations within Bar, we will first describe the location, markers, and connectivity of immediately surrounding populations, some of which were once thought to control micturition.

Neuroanatomical landscape surrounding Barrington's nucleus

The pontine tegmentum is a heterogeneous region, with many different cell types and functions. Many of these can be identified based on their distinct genetic markers and connectivity. In addition to populations bordering Bar, major nearby nuclei in the brainstem tegmentum include the parabrachial nucleus (laterally), the dorsal tegmental nucleus and dorsal raphe nucleus (medially), the pontine reticular formation (ventrally), and the vestibular nuclei (caudally). We will discuss the following populations here, due to their anatomical proximity and, in some cases, possible modulation of micturition: the locus coeruleus (LC), PCG, pre-LC, laterodorsal tegmental nucleus (LDT), and mesencephalic nucleus of the trigeminal nerve (MeV).

Locus coeruleus

The LC is directly adjacent to Bar. In rats, Bar neurons are ventromedial and rostral to LC neurons (Valentino et al., 1994). In mice, the LC directly opposes Bar on its lateral aspect (Verstegen et al., 2017). The LC is composed of noradrenergic neurons, which are the major source of norepinephrine in the central nervous system (Benarroch, 2009). With its widespread projections, the LC has been implicated in a variety of functions, including stress, arousal, cognition, and cerebral blood flow (Aston-Jones & Waterhouse, 2016; Carter et al., 2010; Chrousos, 2009; Gompf et al., 2010; Morin et al., 1997; Toussay et al., 2013). These neurons, and their connections to the rest of the brain, have been implicated in the development and progression of Parkinson's disease (Del Tredici & Braak, 2013; Wakamatsu et al., 2015; Wang et al., 2009), which often includes deficits in bladder control (Palma & Kaufmann, 2018; Sakakibara et al., 2017). LC neurons express tyrosine hydroxylase (TH) and other markers of noradrenergic neurons (Benarroch, 2018; Schmidt et al., 2019). Immunolabeling TH allows for clear differentiation between Bar and LC (Verstegen et al., 2017). Not surprisingly, given its axonal projections to the spinal cord and location within the region that Barrington identified as critical for micturition, the LC was studied for a possible role in bladder function.

Ventral LC neurons, like Bar neurons, project axons to the sacral spinal cord (Verstegen et al., 2017). Their axons branch diffusely, throughout the spinal gray matter (Bruinstroop et al., 2012). In contrast, Bar axons produce dense terminal fields primarily in the IML and DGC (Holstege et al., 1986; Hou et al., 2016; Keller et al., 2018; Loewy et al., 1979; Verstegen et al., 2017). Detailed electrophysiologic mapping within this region in rats did not identify micturition-associated changes in the LC (Tanaka et al., 2003). Likewise, focally stimulating the LC did not cause micturition (Yamao et al., 2001) and focal LC lesions that spared neurons medial and rostral to it caused little to no urinary retention in rats (Satoh, Shimizu, et al., 1978).

In contrast, others found that neurons in the LC do respond to bladder and colon distention, if pressures rise into the noxious range (Elam et al., 1986; Koyama et al., 1998; Page et al., 1992). Electrical stimulation in or near the LC was reported to cause bladder contraction in cats (Sasa & Yoshimura, 1994). Another study found that local field potentials in the LC and frontal cortex synchronized 10–30 s before micturition in rats (Manohar et al., 2017). Further, cell type–specific ablation of LC neurons that project axons to the lumbosacral spinal cord caused frequent voiding and bladder sphincter dyssynergia (Ma et al., 2022). These findings suggest involvement of the LC in micturition. Some Bar axons ramify locally within the brainstem, possibly targeting the LC (Valentino et al., 1996), and lesioning or inhibiting Bar was reported to prevent the activation of LC neurons during colon distention (Rouzade-Dominguez et al., 2001), supporting a hypothesized modulatory or arousal-related role for the LC in visceral functions, including micturition.

Pontine central gray

The PCG is a region medial to Bar that contains mostly GABAergic neurons, with many glutamatergic neurons scattered among them (Verstegen et al., 2017). Within the PCG in mice, GABAergic neurons that express the transcription factor FoxP2 border Bar along its ventromedial surface. Along the ventromedial and caudal borders of Bar, a separate population of glutamatergic neurons express neuropeptide S (Adori et al., 2015; Clark et al., 2011; Xu et al., 2004). The PCG also contains the dorsal tegmental nucleus, and directly in the midline, this region contains the caudal extent of the dorsal raphe nucleus; neither of these are implicated in micturition. Between these midline populations and Bar are many other PCG neurons. We know very little about most of these neurons, including whether any play a role in micturition. It is possible that GABAergic neurons in this region (possibly including the FoxP2-expressing GABAergic neurons that border Bar) provide inhibitory input that influences the spiking activity of Bar neurons. In cats, disinhibiting Bar and surrounding cells with bicuculline reduced the time between urinary events by over 50% (Mallory et al., 1991). In rats, previous investigators identified spiking activity outside Bar that was inversely proportional to bladder pressure and to neuronal firing within Bar (Koyama et al., 1999; Tanaka et al., 2003). Given this evidence that Bar neurons are modulated by on-going GABAergic input, neighboring neurons in the PCG represent a potentially important source of inhibitory input.

Pre-locus coeruleus

Another group of FoxP2-expressing neurons, known as the pre-locus coeruleus (pre-LC), lies along the dorsolateral aspect of Bar (Gasparini et al., 2019, 2021). Pre-LC neurons are glutamatergic (Verstegen et al., 2017) and a majority express prodynorphin (*Pdyn*) (Gasparini et al., 2021; Geerling et al., 2016; Lee et al., 2019). These neurons were named due to their location immediately rostral to the LC in rats (Geerling & Loewy, 2007; Geerling et al., 2011). Their distribution is more complicated in mice because they wrap around the LC and extend into the medial parabrachial nucleus (Gasparini et al., 2019, 2021; Jarvie & Palmiter, 2017). They have been implicated in sodium appetite because they are activated by dietary sodium deprivation and receive input from nucleus tractus solitarius (NTS) neurons that sense aldosterone and drive salt intake (Gasparini et al., 2021; Geerling & Loewy, 2006; Jarvie & Palmiter, 2017; Lee et al., 2019; Resch et al., 2017). Pre-LC neurons also receive input from hypothalamic neurons that control food intake (Garfield et al., 2015; Gasparini et al., 2019, 2021; Li et al., 2019). They project axons primarily to hypothalamic and other brain regions involved in homeostatic food intake, arousal, and reward (Shin et al., 2011). Besides their close proximity to Bar, there is currently no evidence linking pre-LC neurons to micturition or other pelvic functions.

Laterodorsal tegmental nucleus

The LDT is a prominent group of large cholinergic neurons in the PCG. These neurons are rostral and dorsomedial to Bar, and their distribution merges with the larger pedunculopontine tegmental (PPT) population of similarly large cholinergic neurons extending into the midbrain. These cholinergic neurons project axons to the forebrain, particularly the thalamus (Hallanger et al., 1987), and are thought to be involved in arousal and other behavioral functions (Kroeger et al., 2017; Saper et al., 2005; Van Dort et al., 2015). Other than their proximity to Bar, current information on the connectivity and functions of LDT neurons does not suggest a role in micturition.

Mesencephalic nucleus of the trigeminal nerve and fourth ventricle

Two other well-known structures in this region are the MeV and fourth ventricle. MeV neurons are much larger than surrounding neurons. At brainstem levels containing Bar, they are located between the LC and parabrachial nucleus. MeV neurons receive proprioceptive information from orofacial muscles, including muscle spindles in the jaw. These neurons are glutamatergic, but unlike surrounding glutamatergic populations, MeV neurons express the type 1 more than type 2 vesicular glutamate transporter (Vglut2) (Pang et al., 2006). In this region of the brainstem, the fourth ventricle is another important landmark. Where the cerebral aqueduct transitions caudally into the fourth ventricle, Bar lies just ventral to the ependymal lining of the ventricular floor. Dendrites of Bar neurons extend up through the PCG to the floor of the fourth ventricle (Verstegen et al., 2017), so Bar likely receives input from brain regions that send projections to this subventricular zone.

Discovery and characterization of the neurons in Barrington's nucleus

By the early 20th century, physicians had noted that injuries at all levels of the spinal cord led to urinary retention (Head & Riddoch, 1917), but the cerebral location of neurons responsible for the micturition reflex was unknown. In the 1920s,

Barrington provided experimental evidence that transecting the brainstem at levels up through the pons prevented reflex micturition and eliminated bladder tone (Barrington, 1921). He also discovered that transecting the upper pons or midbrain in cats did not reduce bladder tone or prevent reflex micturition, thus identifying the critical level as a narrow portion of the middle to upper pons (Fig. 2.1A). Next, to localize the precise region, Barrington used the first Horsley−Clarke stereotaxic

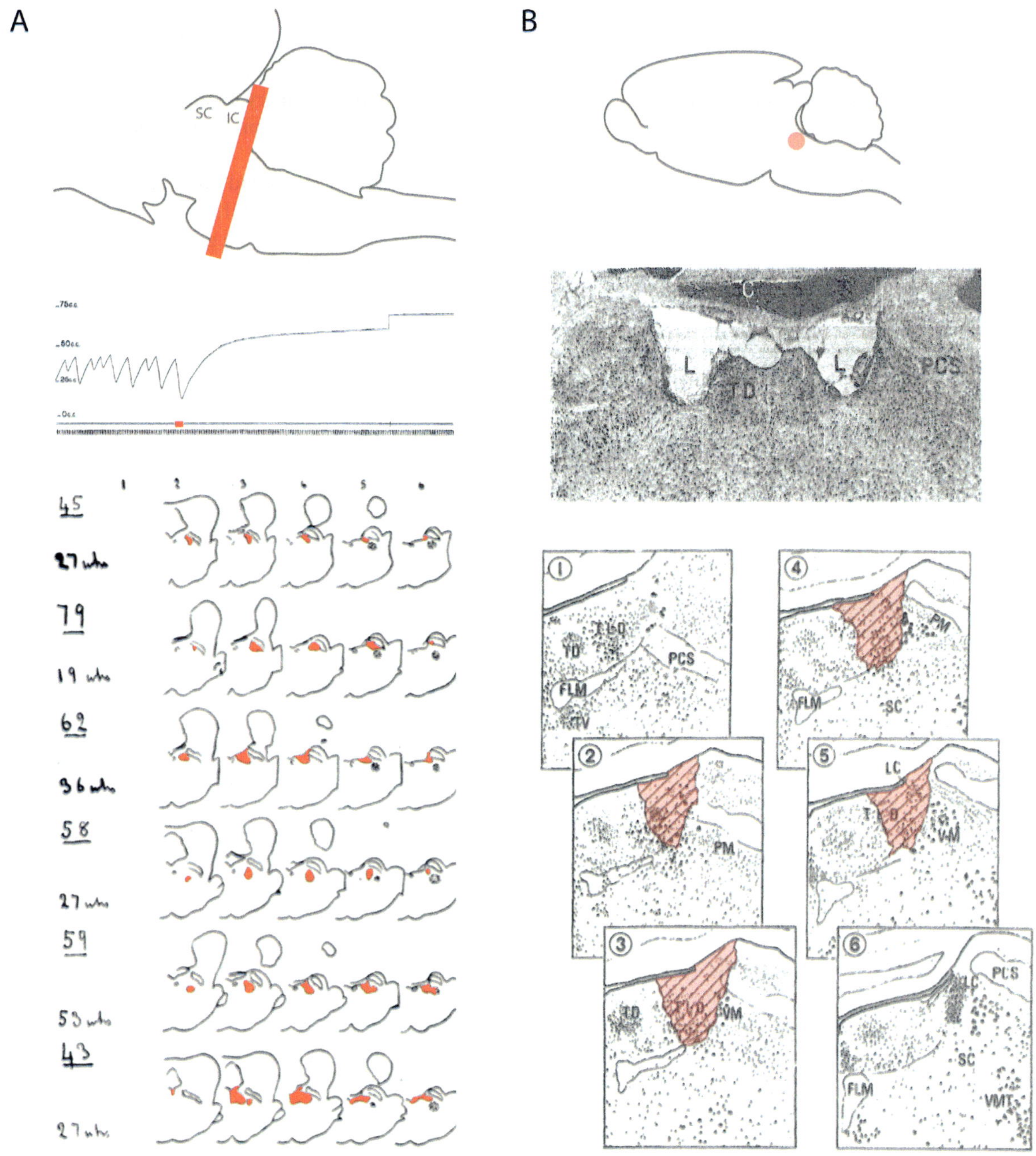

FIGURE 2.1 Transection and focal, stereotaxic lesions that produced urinary retention in cats and rats. (A) Barrington found that transecting the brainstem (*red line*) eliminated reflex micturition in cats. We added a *red* marker atop the original bladder tracing to highlight the time at which he transected the brainstem "through the posterior part of the inferior colliculi above and the middle of the pons below" (Fig. 5 from Barrington, 1921). Below are his drawings of stereotaxic brain lesions in cats with "residual urine permanently" (Fig. 5 from Barrington, 1925; we highlighted each lesion in *red*). (B) Smaller stereotaxic lesions helped further localize this region in rats. Photomicrograph shows bilateral electrolytic lesions "at the caudal part of the nucleus tegmentalis laterodorsalis" in a rat with urinary retention (Fig. 1 from Satoh, Shimizu, et al., 1978). Illustrations show lesion overlap in rats with urinary retention, across successive rostral-to-caudal levels of the brainstem tegmentum (Fig. 2 from Satoh, Shimizu, et al., 1978; *shaded red* for emphasis). *C*, cerebellum; *FLM*, medial longitudinal fasciculus; *IC/SC*, inferior/superior colliculi; *L*, lesion; *PCS*, superior cerebellar peduncle; *TD/TV*, nucleus tegmentalis dorsalis/ventralis; *VMT*, motor trigeminal nucleus.

apparatus (Schurr & Merrington, 1978) to produce bilateral electrolytic lesions in a variety of locations. His report on cats with focal brainstem lesions represents the first impactful use of stereotaxic brain surgery and remains the most important advance in understanding the neural control of micturition. Using this approach, Barrington identified a critical region of the pontine tegmentum where bilateral lesions caused complete urinary retention. This region was located ventral to the internal edge of the superior cerebellar peduncle, between the MeV and the anterior end of the hindbrain (Barrington, 1925). Cats with bilateral lesions in this location had urinary retention (they could no longer void, despite attempting to do so in their litter), while cats with lesions in other sites, or with a unilateral lesion in this site, exhibited normal voiding behavior. Barrington further observed that bilateral lesions in the midbrain, "on each side of the aqueduct in its ventral half" would result in "a permanent loss of consciousness of wanting to micturate or defecate, but does not impair the performance of either" and that more extensive lesions in this region produced urinary frequency as well (Barrington, 1925; Morrison, 2008). This was the first time the brainstem was implicated in micturition. Subsequent work over the past century more precisely localized and characterized the neurons involved.

Early on, Barrington's work went largely unappreciated, with few attempting to expand on his findings and determine how the pontine tegmentum influenced micturition. In 1933, Langworthy and Kolb found that intact cats completely void their bladder after infusing 60–90 mL of liquid through a catheter. The bladder pressure rose gradually during the infusion, from 5 up to 15 cm of H_2O, then quickly spiked before the bladder emptied (Langworthy & Kolb, 1933). However, after a brainstem transection at or below the level that Barrington had identified ("well below the acoustic colliculi"), the same cats would only release small amounts of liquid after as much as 125 mL had been infused and pressure had risen above 20 cm H_2O. After this, the bladder pressure would drop only slightly, with small releases of urine as more liquid was added. They noted that "this experiment could be repeated indefinitely without development of any adequate reflex micturition" (Langworthy & Kolb, 1933). A similar study in 1955 showed that ablating the dorsal half of the brainstem abolished the micturition reflex, while ablating the inferior colliculus had no effect (Tang, 1955). While these experiments were consistent with Barrington's finding, they did not improve his localization.

Complementing these early lesion studies, Ranson and colleagues revived a replica of the Horsley–Clarke stereotaxic apparatus to perform focal electrical stimulation. After testing 7500 total points in 65 cats, they observed bladder contraction after stimulating the ventromedial border of superior cerebellar peduncle as well as points in the periventricular gray matter and in the lateral reticular formation extending back to the spinal cord (Wang & Ranson, 1939). A year after his transection study, Tang also used cystometrograms combined with suction and electrolytic lesions to expand upon his findings and claimed that the "pontine micturition facilitatory area of Barrington" was located directly under the lateral angles of the periventricular gray at the isthmus region of the tegmentum (Ruch & Tang, 1956). A decade later, another group of investigators noted that stimulating the dorsolateral reticular formation of the rostral pons led to bladder contraction. They referred to this region as "Barrington's pontine detrusor nucleus." Its rostral half was medial to the superior cerebellar peduncle, and its caudal half was dorsomedial to the trigeminal motor nucleus (Kuru & Yamamoto, 1964). These studies narrowed the possible region containing neurons that may control the micturition reflex, but a lack of consensus allowed for continued debate and speculation.

While the above studies varied slightly on the precise location of neurons responsible for stimulating bladder contraction, all identified a region near the LC. Therefore, it is not surprising that many described this region as being part of the LC. Indeed, electrically stimulating "the region of the LC" evoked not only bladder contractions but also local field potentials in the sacral spinal cord. Additionally, electrically stimulating bladder afferents in the pelvic nerve evoked negative field potentials in the region of the LC (De Groat & Lalley, 1972; Lalley et al., 1972). Applying carbachol, which activates acetylcholine receptors, to the "optimal stimulus region" for electrically evoked bladder contraction in cats, evoked bladder contraction and sphincter relaxation resembling normal reflex micturition. These authors inferred from the injected region that catecholaminergic neurons were responsible, describing their optimal site as overlying a ventral part of the LC, known as LC_α (Sugaya et al., 1987). Another report in dogs found that electrical stimulation in the "LC or its subparts" produced micturition with bladder contraction (Nishizawa et al., 1988). Although these studies inferred involvement of LC neurons, several findings make this unlikely.

Neuroanatomic identification of Bar neurons

The neuroanatomy revolution of the 1970s provided techniques for directly tracing axonal projections, making it possible to determine precisely which neurons in the dorsal pontine tegmentum project axons to the sacral spinal cord. In rats, injecting tritiated amino acids into a tegmental region that includes what we now call "Bar" produced heavy axonal labeling in the sacral IML (Fig. 2.2A), while injections into the LC resulted in more diffuse labeling at all levels of the spinal cord (Loewy et al., 1979). Complementary injections of a retrograde tracer, horseradish peroxidase (HRP), into the lumbosacral spinal

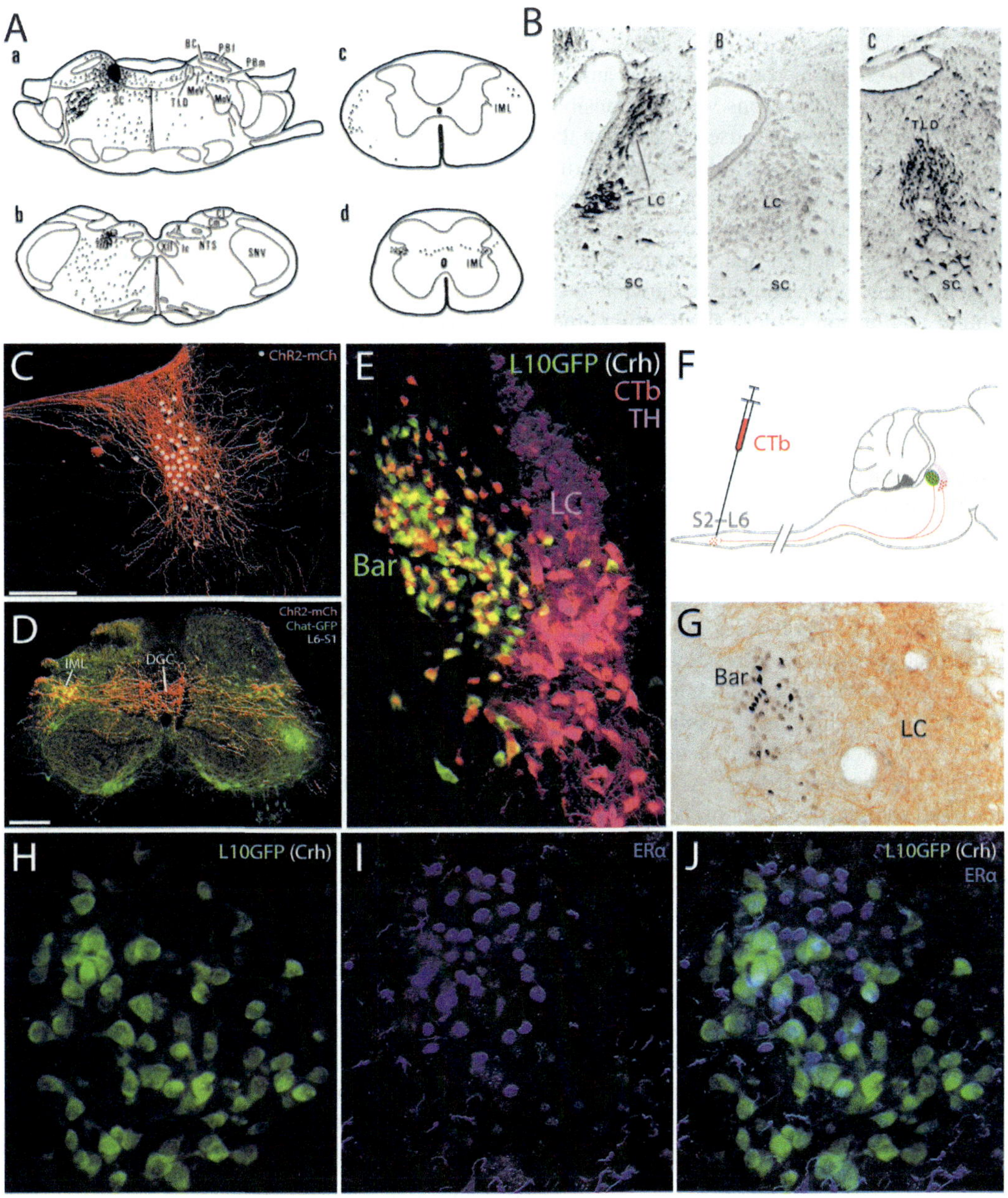

FIGURE 2.2 (A) Injecting [^{3}H]-labeled amino acids, marked by dots in (a), into the "nucleus tegmentalis laterodorsalis" (TLD) produced axonal transport through the medulla (b) and spinal cord (c, d) with extensive labeling in the sacral spinal cord (d). *BC*, brachium conjunctivum; *Cl*, lateral cuneate nucleus; *Cm*, medial cuneate nucleus; *Ic*, intercalated nucleus; *IML*, intermediolateral cell column of the spinal cord; *MeV*, mesencephalic nucleus of the trigeminal nerve; *MoV*, motor trigeminal nucleus; *NTS*, nucleus of the solitary tracts; *PBl*, lateral parabrachial nucleus; *PBm*, medial parabrachial nucleus; *SC*, subcoeruleus nucleus; *SNV*, spinal trigeminal nucleus; *X*, dorsal vagus nucleus; *XII*, hypoglossal nucleus (Fig. 1 from Loewy et al., 1979). (B) Retrogradely labeled neurons following injection of horseradish peroxidase (HRP) into the lumbosacral spinal cord (Fig. 2 from Loewy et al., 1979). Retrogradely labeled neurons in the locus coeruleus (LC, *left panel*) were not found after pretreatment with 6-hydroxydopamine (*middle panel*), while a prominent cluster of retrogradely labeled neurons was still present in the "TLD" (*right panel*). (C–F) Images excerpted from Figs. 9 and 10 of Verstegen et al. (2017) show anterograde and retrograde labeling of Bar^{Crh} neurons. (C) ChR2-mCherry (*red*) expression in Bar^{Crh} neurons following AAV injection in a *Crh-IRES-Cre;Chat-GFP* mouse (cell bodies indicated by *white dots*). (D) ChR2-mCherry in axons projecting to the spinal cord and ramifying at lumbosacral levels in a horizontal stripe that overlaps lateral cholinergic neurons (*green*) of both intermediolateral nuclei (IML) and the dorsal gray commissural nucleus (DGC) between them. Scale bars are 200 μm. (E, F) Retrogradely labeled neurons in Bar and LC following injection of CTb into the lumbosacral spinal cord (F) in a *Crh-IRES-Cre;R26-lsl-L10GFP* mouse. Many retrogradely labeled neurons in Bar express GFP and appear *yellow-orange*. All CTb-labeled neurons lateral to Bar were immunoreactive for tyrosine hydroxylase (TH) and were located in the ventral LC and subcoeruleus. (G) Bar contains a population of neurons with nuclear immunoreactivity for estrogen receptor alpha (*black*), just medial to the LC, whose neurons are immunoreactive for TH (*brown*; excerpted from panel C of Fig. 12 from Vanderhorst et al., 2005). (H–J) Combined labeling for Bar^{Crh} and Bar^{Esr1} neurons (unpublished example). (H) GFP fluorescence in a *Crh-IRES-Cre;R26-lsl-L10GFP* mouse identifies Bar^{Crh} neurons. (I) Estrogen receptor immunoreactivity (ERα, *blue*). (J) Combined labeling reveals complementary Bar^{Crh} and Bar^{Esr1} subpopulations, with colocalization in a smaller number of neurons.

cord labeled a dense cluster of neurons in both the LC and a neighboring region referred to at the time as "tegmentalis laterodorsalis," or TLD (Fig. 2.2B). Repeating these tracing experiments after LC neurons were destroyed with 6-hydroxydopamine (6-OHDA) produced similar anterograde labeling in the IML and similar retrograde labeling of TLD neurons, showing that non-LC neurons were responsible for the dense projections to the sacral IML (Loewy et al., 1979). Also in rats, electrolytic lesions were used to localize the region necessary for the micturition reflex. Urinary retention was observed after lesions that destroyed a large region of the pontine tegmentum that included the LC, "TLD," MeV, and medial PB. However, with smaller lesions, urinary retention was only observed after lesions that included the "TLD" (centered over Bar; Fig. 2.1B) and not those confined to the LC (Satoh, Tohyama, et al., 1978). Together, these axonal tracing and lesion results indicated that neurons responsible for the micturition reflex are located just outside the LC.

However, there was not yet a consensus on what to call this nucleus. By the early 1980s, this structure had been referred to as the "Pontine micturition center" (PMC) by some authors (Loewy et al., 1979; Nishizawa et al., 1988) and "Barrington's nucleus" by others (Sakumoto et al., 1978; Sastre et al., 1981; Tohyama et al., 1978). After injecting a fluorescent retrograde tracer into the sacral spinal cord, immunofluorescence labeling for corticotropin-releasing hormone (CRH) revealed that many of the spinally projecting "TLD" neurons contain CRH (Vincent & Satoh, 1984). Later studies confirmed that a cluster of neurons outside the LC were immunoreactive for CRH (Imaki et al., 1991; Valentino et al., 1995), and that many Bar neurons labeled by retrograde axonal transport after a tracer injection into the mouse spinal cord expressed a GFP Cre-reporter for *Crh* (Fig. 2.2E and F) (Verstegen et al., 2017). Beyond using this neuropeptide as a marker for Bar neurons, the role of CRH in stress physiology (Herman et al., 2016) led to a hypothesis that its expression in Bar neurons helps explain stress effects on micturition (Vincent & Satoh, 1984). However, stress did not alter the response of these CRH-immunoreactive neurons to bladder distention in rats (Rouzade-Dominguez, Pernar, et al., 2003), and eliminating expression of *Crh* in this region of the brainstem had no effect on micturition in mice (Verstegen et al., 2019). One study found that global overexpression of rat *Crh* in mice led to an increase in void number (Million et al., 2007), while another study found that overexpressing *Crh* in and around Bar in rats reduced voiding frequency (McFadden et al., 2012). Blocking CRH receptors in the spinal cord augmented bladder contractions, suggesting that CRH release may provide a negative feedback signal that restrains detrusor contraction or delays micturition (Ito et al., 2020; Kiddoo et al., 2006; Pavcovich & Valentino, 1995). A recent study also reported that subcutaneous injection of a CRH receptor antagonist prevented inhibitory effects of high-frequency (50 Hz) photostimulation of *Crh*-expressing neurons on bladder physiology (Van Batavia et al., 2021). The exact functional role of CRH remains unclear, but it is a useful marker for identifying a large and distinctive subpopulation of Bar neurons.

Identifying CRH as a marker for a subset of Bar neurons made it easier to use neuroanatomical techniques to study these neurons. In rats, a dense, oval-shaped group of CRH-immunoreactive neurons, referred to as the "core" of Bar, lies ventromedial to the rostral end of the LC. The LDT borders this oval core medially and rostrally (Valentino et al., 1994). The core of Bar can be easily identified via Nissl staining, but a looser collection of CRH-immunoreactive Bar neurons (retrogradely labeled from the spinal cord) extend outside this core (Valentino et al., 1995). Similar results were seen in Cre-reporter mice for *Crh*, which again revealed a dense core of Bar neurons medial to the LC (Fig. 2.3A; Hou et al., 2016; Verstegen et al., 2017). Along with CRH, NeuN can be used as a surrogate marker to separate Bar and LC neurons because it is expressed in Bar but not LC. The medial aspect of the LC cradles the lateral edge of Bar, except along the dorsolateral core, where a thin population of smaller neurons separates it from the dorsal LC (see Pre-locus coeruleus section above). LDT neurons in mice are dorsal, medial, and rostral to the core of Bar. As in rats, a loose collection of Bar^{Crh} neurons extend rostrally, ventrally, and caudally beyond the core (Verstegen et al., 2017). Of note, CRH immunoreactivity in rats and *Crh* expression in mice only mark approximately half the population of Bar neurons that project axons to the sacral spinal cord (Valentino et al., 1995; Verstegen et al., 2017).

Molecular characterization of Bar neurons

Neurons that elicit micturition were thought to be glutamatergic (Matsumoto et al., 1995; Yoshiyama et al., 1994, 1995), but this was not confirmed until the discovery of the Vglut2 (the sodium-linked cotransporter *Slc17a6*). Vglut2 expression was found in Bar but not in the LC (Stornetta et al., 2002). In mice, *Crh*-expressing neurons in this brainstem region also express Vglut2/*Slc17a6* (Hou et al., 2016). Further, neurons throughout the core of Bar express a Cre reporter for this glutamate vesicular transporter, but not for the GABA vesicular transporter (Vgat/*Slc32a1*), confirming that Bar neurons are excitatory and that throughout development, they are never GABAergic. Many surrounding neurons do express Vgat/*Slc32a1*, so the core of Bar and the LC form an easily distinguishable Vgat void in the pontine tegmentum (Verstegen et al., 2017).

This region contains multiple neuronal subtypes that may serve different functions. Some spinally projecting neurons within Bar increase their activity during the micturition cycle and are referred to as "ramp neurons," while others maintain

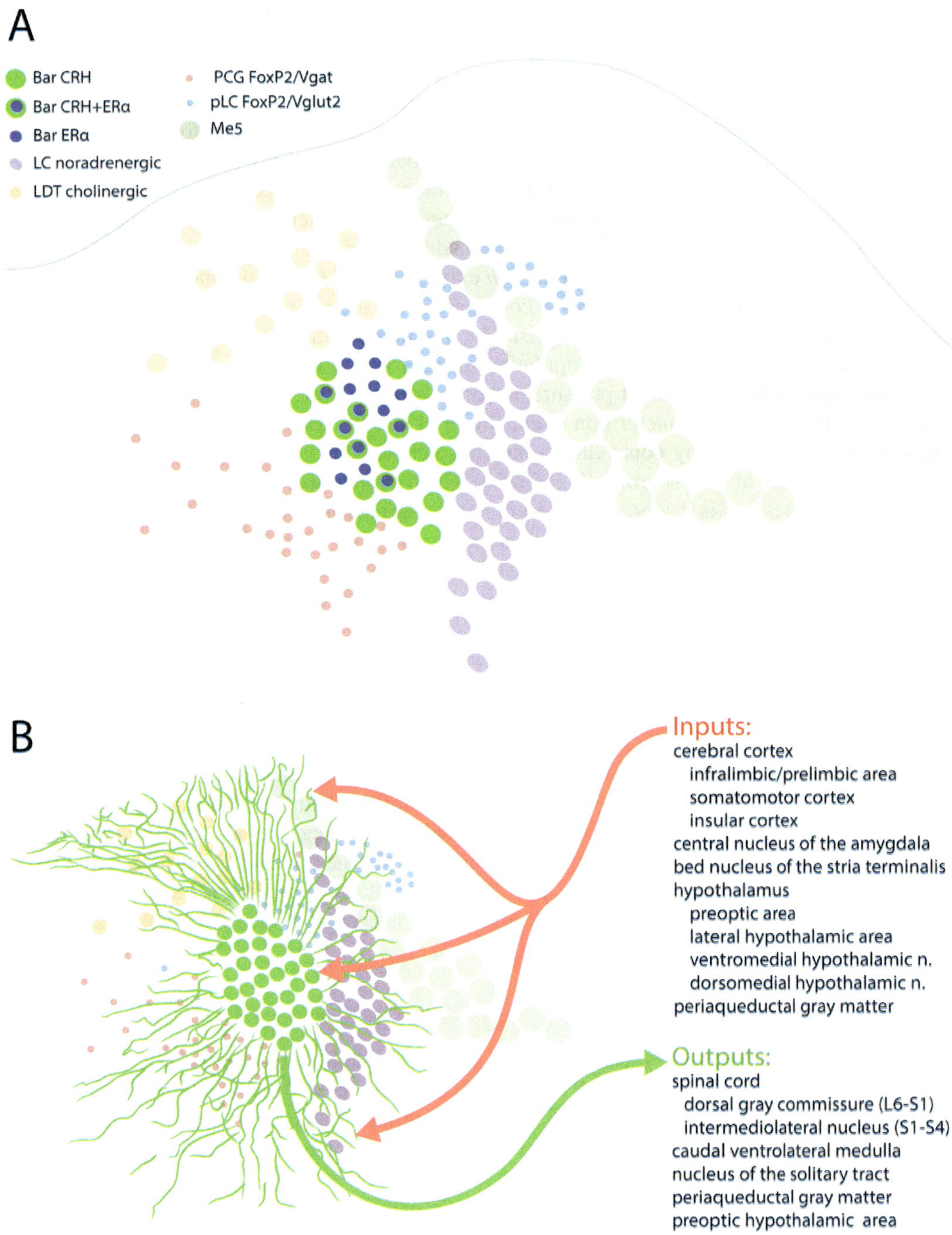

FIGURE 2.3 (A) Relative distributions of Bar neurons and surrounding populations in the mouse pontine tegmentum. (B) Overview of afferents to and efferent axonal projections from Bar neurons. *LDT*, laterodorsal tegmental nucleus; *Me5*, mesencephalic nucleus of the trigeminal nerve; *PCG*, pontine central gray; *pLC*, pre-locus coeruleus.

a more consistent firing rate and are referred to as "flat neurons" (Sasaki, 2005). As it became evident that the vast majority of Bar neurons are glutamatergic, while only half produce CRH, questions arose as to the role of the *Crh*-expressing subpopulation of glutamatergic neurons (Bar^{Crh}) relative to the larger population of glutamatergic neurons ($\mathrm{Bar}^{\mathrm{Vglut2}}$). When the overall population of glutamatergic neurons was stimulated in awake mice, immediate voiding occurred in 70% of trials (Verstegen et al., 2019). Under anesthesia, stimulating these glutamatergic neurons produced bladder contraction in 100% of trials and voiding in 91% of trials. However, stimulating the Bar^{Crh} subpopulation only provoked voiding in 6% of trials in awake mice, and the voids were not immediate or incontinent. Under anesthesia, voiding occurred in only

17% of trials, although bladder pressure typically increased. In addition to their different effects on micturition, activity in some Bar^{Vglut2} neurons began increasing before bladder pressure, while the Bar^{Crh} subpopulation activated in association with bladder pressure. Ablating Bar^{Vglut2} neurons led to urinary retention, often requiring euthanasia, while eliminating the Bar^{Crh} subpopulation resulted in milder changes, ranging from urinary frequency to retention.

Thus, Bar^{Crh} neurons are not necessary for voiding and may serve more of a modulatory role by augmenting detrusor contractions. In the absence of cholinergic bladder contraction, bladder morphology and function remain nearly normal in female, but not male mice (Matsui et al., 2000), so there may be a sex difference in the importance of Bar^{Crh} neurons. Additional studies found that activity of Bar^{Crh} neurons correlates with bladder pressure and that stimulating them drives detrusor contraction without triggering micturition until the bladder is sufficiently full (Hou et al., 2016; Ito et al., 2020; Keller et al., 2018). Together, these studies showed that *Crh* expression and CRH immunoreactivity are useful markers for a subpopulation of Bar neurons that augment detrusor contraction, but the neurons that produce this neuropeptide do not directly control the release of urine. Given the surprisingly minimal effects of eliminating Bar^{Crh} neurons, it will be important to explore additional roles for this subpopulation in other pelvic functions.

One study proposed that increasing bladder pressure and other effects of activating Bar^{Crh} neurons alters a spinal gating mechanism that ultimately controls the micturition event (Ito et al., 2020). However, a complementary subpopulation of Bar neurons may trigger micturition by controlling sphincter activity. Nuclear immunoreactivity for estrogen receptor alpha identifies a dorsal cluster of Bar neurons (Fig. 2.2G), which project axons to the spinal cord (Vanderhorst et al., 2005). Estrogen receptor expression is distinct from *Crh* in many Bar neurons, with coexpression in a minority of neurons in the core of Bar (Fig. 2.2H–J). Roughly 200 of these neurons (Bar^{Esr1}) distribute dorsally within the core, complementing a more ventral population of approximately 500 Bar^{Crh} neurons (Keller et al., 2018). Bar^{Esr1} neurons send a slightly different pattern of axonal projections to the lumbosacral spinal cord than Bar^{Crh} (see Efferent projections of Bar neurons section below).

Stimulating Bar^{Crh} neurons reliably increased bladder pressure, but rarely led to voiding. In contrast, stimulating Bar^{Esr1} neurons triggered voiding in 96% of trials in awake mice (Keller et al., 2018). Additionally, chemogenetic inhibition of Bar^{Esr1} neurons caused dominant male mice to void less profusely when presented with female urine. In contrast, Bar^{Crh} inhibition did not alter this behavior. Optogenetically silencing Bar^{Esr1} neurons during a spontaneous voiding event produced an immediate reversion in bladder pressure and sphincter activity (Keller et al., 2018). These results suggest that Bar^{Esr1} neurons have a more direct and immediate effect on micturition than Bar^{Crh} neurons. Due to the evidence that Bar^{Crh} neurons are neither necessary nor sufficient to trigger voiding (Verstegen et al., 2019), it will be important to test whether activating Bar^{Esr1} neurons is sufficient for micturition in the absence of Bar^{Crh} neurons. The function of estrogen receptor here is unknown. Ovariectomy and estrogen replacement had mild effects on micturition in rats (Fleischmann et al., 2002; Longhurst et al., 1992), but as with CRH, further work is needed to determine the functional role of estrogen receptors in Bar^{Esr1} neurons.

Additional features of Bar neurons

Relative to this information on subtypes of Bar neurons, we know less about their morphology. The somas of Bar neurons are mediumsized and oval or polygonal in shape. Typically, Bar neurons have 1–4 aspiny dendritic processes (Kawatani et al., 2020; Rouzade-Dominguez, Pernar, et al., 2003). Their dendritic arbors are extensive, spreading well beyond the conventional borders of Bar based on Nissl cytoarchitecture, CRH immunolabeling, or retrograde tracer labeling from the spinal cord (Verstegen et al., 2017).

These dendrites form three main bundles. The largest of these extends ventrally and rostrally, through the cholinergic neurons in the ventral LDT and into the sublaterodorsal region. Some of these dendrites run alongside the PPT nucleus in the midbrain reticular formation up to 500 μm from the core of Bar (Verstegen et al., 2017). The second bundle extends dorsally to cover a broad patch of the central gray matter beneath the rostral–lateral edge of the fourth ventricle. The third bundle extends behind Bar, overlapping the caudal dendritic arbors of LC neurons. Some of these dendrites also protrude laterally, punching through the center of the LC. These morphological data make it likely that additional brain regions provide synaptic input to Bar dendrites extending well outside the core, in addition to afferents identified in previous tracing studies that focused on the core of Bar. Of note, this information derives from neuroanatomical work in Bar^{Crh} neurons, and we do not yet know the dendritic architecture of or afferents to Bar^{Esr1} neurons.

Beyond micturition, Bar neurons may control other pelvic functions. Beyond a small number of studies (Rouzade-Dominguez et al., 2001; Rouzade-Dominguez, Miselis, et al., 2003; Rouzade-Dominguez, Pernar, et al., 2003; Salas et al., 2008), we have very little information about the role of Bar neurons in pelvic functions other than micturition. Bladder and bowel function are often affected together, with many people experiencing retention or incontinence in both systems (Panicker et al., 2019). In rats, most Bar neurons that were activated by bladder distention were also activated by colon distention (Rouzade-Dominguez, Pernar, et al., 2003). Bar neurons could also play a role in sexual activity due to inputs from hypothalamic nuclei associated with sexual behavior and output projections to spinal levels controlling pelvic function. One study in rats suggested that Bar neurons may not be involved in male sexual function, finding that stimulating sites in or near Bar induced bladder contraction without penile erection, while stimulating the subcoeruleus region ventral to Bar increased penile intracavernous pressure and reduced bladder pressure (Sugaya et al., 1998).

Human pontine tegmentum and micturition

In humans, the location and distribution of Bar neurons is not well understood. Similar to rodents, the brainstem tegmentum in humans contains neurons that express *CRH* and contain CRH immunoreactivity. These neurons form a cluster centered just ventromedial to the LC, whose neurons can be identified in human tissue sections due to their neuromelanin pigmentation (Austin et al., 1995). CRH-immunoreactive neurons here lack immunolabeling for TH and likely identify the location of Bar in the human brainstem (Blanco et al., 2013).

The location of Bar itself cannot be seen in magnetic resonance images of the human brain. However, the LC can be localized in adults, due to paramagnetic properties attributed to its dense neuromelanin content (Keren et al., 2015; Sasaki et al., 2006; Watanabe et al., 2019). This may make it possible to approximate the Bar location in adult patients, within several millimeters, relative to the LC. The spatial resolution of brainstem imaging is probably inadequate to definitively localize functional neuroimaging findings to Bar using current PET or fMRI techniques. Nonetheless, prior investigators identified a right-lateralized focus of activation associated with micturition, using PET (Blok et al., 1997), or with bladder filling in patients with urge incontinence, using fMRI (Griffiths et al., 2007).

Several reports of urinary retention after damage to the pontine tegmentum support the importance of this region for micturition (Sakakibara et al., 1996b; Tish & Geerling, 2020). Patients with pontine lesions resulting from multiple sclerosis experienced bladder hyporeflexia (Araki et al., 2003). Additionally, patients with ischemic or encephalitic brainstem lesions in the upper pons developed urinary retention (Sakakibara et al., 1998), including one with transient urinary retention and a transient inflammatory lesion involving the right tegmental region (Komiyama et al., 1998). In another patient, a hemorrhage in the right pontine tegmentum caused urinary retention with residual bladder volumes exceeding 1000 mL (Tish et al., 2022). In patients with Parkinson's disease gait symptoms, electrodes are sometimes implanted in or near this region in an attempt to target the PPT region rostral to Bar. However, when these electrodes deviate caudally, into the region containing Bar, micturition symptoms and urodynamic deficits can arise (Aviles-Olmos et al., 2011; Roy et al., 2018).

Despite a century of progress sparked by Barrington's initial discovery, much work lies ahead before we fully understand Bar neurons in the human brainstem. In experimental animals, it will be important to generate a comprehensive, molecular catalog of these neurons. Finally, while our ability to study the human brainstem is limited, replicating and extending at least the neuroanatomical findings from rodents in human tissue may provide future clinical insights.

Efferent projections of Bar neurons

Early tracing studies identified axonal projections from the region of Bar directly to the lumbosacral spinal cord in rats (Loewy et al., 1979; Satoh, Tohyama, et al., 1978). These axons leave Bar ventrally and through the caudal medulla, where some ramify and target a limited subset of regions, including the NTS, but most continue descending into the spinal cord. Within the spinal cord, most axons remain in the ipsilateral lateral funiculus (Loewy et al., 1979). These axons ramify within the central gray matter (lamina X) at every level of the spinal cord, but only the lumbosacral region contains a dense terminal field. The lumbosacral junction (L6-S1) contains a dense ramification across an expanded region of the central gray matter, known as the DGC. Coextensive with and just caudal to this level, midsacral levels contain dense, bilateral terminal fields in the IML nuclei, which contain autonomic preganglionic neurons, some of which contract the detrusor muscle (Holstege et al., 1986; Loewy et al., 1979). In addition to anterograde tracing from the pons and retrograde tracing from the spinal cord, viral tracer injections into the bladder and EUS consistently produced retrograde, transneuronal labeling of Bar neurons (Marson, 1997; Nadelhaft et al., 1992; Nadelhaft and Vera, 1995, 1996; Sugaya et al., 1997).

Just as Bar^{Crh} and Bar^{Esr1} neurons appear to play different roles in micturition, their axonal projection patterns to the lumbosacral spinal cord were reported to skew differently in the lumbosacral spinal cord. Selectively labeling axons of either Bar^{Crh} or Bar^{Esr1} neurons produces extensive axonal labeling in both the lumbosacral DGC and sacral IML (Fig. 2.2C and D). However, Bar^{Crh} axons were reported to project more to the sacral IML and Bar^{Esr1} axons more to the DGC (Keller et al., 2018). While neurons in the IML control detrusor contraction (De Groat et al., 2015), those in the DGC may inhibit the EUS (Blok et al., 1998; Grill et al., 1999). A stronger projection of Bar^{Crh} neurons to the IML supports a primary function in detrusor contraction, while a heavier projection of Bar^{Esr1} neurons to the DGC would help explain their rapid control over urine release. Subsequent work revealed that both Bar^{Crh} and Bar^{Esr1} axons extensively innervate the lumbosacral IML and that most autonomic preganglionic neurons in the IML receive monosynaptic glutamatergic input from both populations (Kawatani et al., 2021). While this information indicates an overlapping or possibly even redundant role of Bar^{Crh} and Bar^{Esr1} projections to the IML, whether Bar^{Esr1} neurons provide more synaptic input to DGC interneurons that control EUS relaxation remains unknown.

While it is often difficult to determine exactly which axonal tracts are damaged in human patients, urinary retention sometimes results from damage to regions of the lower brainstem through which Bar axons travel to reach the spinal cord. Voiding difficulty was the presenting symptom in a patient with an acute ischemic infarct in the right lateral medulla (Lee et al., 2008), and a retrospective case series identified 10 patients with urinary retention after lesions along a continuous column through the lateral tegmentum of the medulla (Cho et al., 2015). Just as Barrington found in cats, damage to almost any level of the human spinal cord can cause urinary retention (Head & Riddoch, 1917). As examples, patients with lesions ranging from midthoracic viral myelitis (Chern et al., 2017) to a thoracolumbar teratoma (Poeze et al., 1999) and sacral meningocele with syringomyelia (Kemp et al., 2014) all presented with urinary retention. Overall, the urinary retention resulting from lesions in this zone extending from the pontine tegmentum to the sacral spinal cord supports the results of neuroanatomical and basic scientific work in rodents and cats over the past century (Tish & Geerling, 2020).

In addition to the spinal cord, Bar neurons project axons to a limited number of sites in the brainstem and as far rostrally as the preoptic area. Conventional anterograde and retrograde axonal tracing revealed projections to the PAG matter and caudal medulla (Valentino et al., 1995). As discussed in the next section, some of these regions project reciprocally to Bar and modulate micturition, and they may receive information on the state of the bladder from Bar. At the time of this writing, we lack comprehensive information on the supraspinal axonal projections of Bar^{Crh} and Bar^{Esr1} neurons.

Afferent projections to Bar neurons

Mammals ranging from mice to humans are continent, retaining urine in the bladder until voiding is safe and socially appropriate. Bar seems to function as an on/off switch, in that its neurons are usually inactive, then phasically active during voids. It remains unclear what suppresses Bar neuronal activity between voids and the degree to which this inhibition is necessary for continence. Understanding urinary continence likely will require understanding the input signals that suppress and activate Bar neurons.

Bar neurons receive synaptic input from neurons in a variety of brain regions (Fig. 2.3B). Barrington and his close successors showed that transecting the brain above the level of Bar led to an increase in the frequency of voiding, which occurred at lower bladder pressures (Barrington, 1925; Langworthy & Kolb, 1933; Tang, 1955). Subsequent investigators activated and inhibited some of these regions to study their influence on Bar and bladder function (Green et al., 2012; Hou et al., 2016; Hyun et al., 2021; Matsuura et al., 2000; Stone et al., 2011; Verstegen et al., 2019). Additionally, human case reports suggest that brain lesions involving several of these regions alter bladder control (reviewed in Tish & Geerling, 2020). Relative to our growing understanding of Bar neurons, we know little about the afferent neurons that control them. Understanding this afferent network is critical to understanding continence, and the studies discussed in this section represent initial steps toward this goal.

Perhaps the heaviest source of direct synaptic input is the PAG, at levels of the midbrain just rostral to Bar. In rats and cats, conventional axon tracing techniques revealed a dense cluster of retrogradely labeled neurons in the lateral and ventrolateral PAG after retrograde tracer injections into Bar, as well as heavy axonal labeling in Bar after anterograde tracer injections into the PAG (Blok & Holstege, 1994; Ennis et al., 1991; Valentino et al., 1994). Selectively labeling the axons of glutamatergic neurons in the lateral PAG identified a similarly heavy projection in mice, with optogenetic confirmation that activating these axons evokes monosynaptic, glutamatergic excitatory responses in Bar neurons with and without *Crh* expression (Verstegen et al., 2019). Along with this excitatory projection from the lateral PAG, a small number of GABAergic neurons in the ventrolateral PAG also provide direct input to the core of Bar (unpublished observations). Given these neuroanatomical findings, it is not surprising that many studies have found an important role for the PAG in modulating micturition.

While there is ample evidence for a role of the PAG in micturition, separate studies report excitatory and inhibitory effects. Electrically stimulating the PAG in rats produced detrusor contractions, and continuous stimulation during bladder filling led to voiding at lower bladder volumes (Kruse et al., 1990). Different stimulation protocols and locations within the PAG were reported to produce either inhibition or stimulation of micturition (Green et al., 2012; Numata et al., 2008). Glutamate microinjections in the lateral and ventrolateral PAG stimulated voiding, and this could be prevented by inhibiting Bar, supporting the idea that Bar is an obligate output target for PAG-triggered voiding (Matsuura et al., 2000). When unit activity was recorded in the PAG during spontaneous micturition, four different activity patterns were identified. Forty-three percent of units were referred to as "tonic storage" (continually active across each storage/micturition cycle but more active during the storage phase). Twenty-nine percent were "tonic micturition" neurons, while 28% were only active during one phase of the cycle (15% storage, 13% micturition) (Liu et al., 2004).

Indeed, the PAG is a heterogenous region, consisting of both inhibitory and excitatory neurons. Injecting muscimol, a GABA receptor agonist, prolonged the delay between voids or completely suppressed micturition at some sites within the ventrolateral PAG. Conversely, disinhibiting PAG neurons with bicuculline, a GABA receptor antagonist, produced rapid detrusor contractions and a nearly continuous change in EUS activity with frequent voids (Stone et al., 2011). Within the PAG, the excitatory neurons that project to Bar are spread rostral-to-caudal across the lateral and ventrolateral PAG and heavily target the core of Bar, as well as its dorsal dendrites. Photostimulating these glutamatergic axons in Bar increased bladder pressure and produced immediate voids in 80% of trials (Verstegen et al., 2019). Clearly, the PAG plays an important role in controlling micturition, in large part by providing direct input to Bar. Much work lies ahead to better understand its complex modulatory role, particularly the roles of interneuronal circuitry within the PAG and of descending input from the forebrain.

Human case reports with lesions involving the PAG support a role in micturition. One patient presented with sudden voiding difficulty and urinary retention. The MRI revealed an isolated edematous lesion in the right PAG. Both the radiologic lesion and urinary retention resolved after steroid treatment (Yaguchi et al., 2004). Another patient with a small ischemic stroke involving the PAG had acute-onset urinary retention and remained catheter-dependent for more than 2 months (Liao et al., 2013). Primarily based on the more extensive evidence from experimental animals, the PAG is thought of as a control center for micturition, with complete control over Bar (Benarroch, 2010; Fowler et al., 2008; Liu et al., 2004). However, the extent to which its control over Bar is complete is unclear because Bar neurons receive direct input from several other regions.

Many forebrain regions project to Bar. Conventional axonal tracing studies in rats and cats identified afferents in the bed nucleus of the stria terminalis, preoptic area, lateral hypothalamic area (LHA), dorsomedial hypothalamic nucleus, ventromedial hypothalamic nucleus, posterior hypothalamic area, limbic association cortices (infralimbic, prelimbic, anterior cingulate, and insular), and somatomotor cortex (Canteras et al., 1994; Dong and Swanson, 2006a, 2006b; Hahn & Swanson, 2010; Holstege, 1987; Kuipers et al., 2006; Ribeiro-Barbosa et al., 1999; Rizvi et al., 1994; Valentino et al., 1994; Vertes & Crane, 1996). Retrograde, transneuronal tracing from Bar^{Crh} neurons labeled neurons in a largely similar set of forebrain regions (Hou et al., 2016; Hyun et al., 2021; Verstegen et al., 2019). In cats, electrical stimulation in several of these sites evoked detrusor contractions, while a minority of sites, including the lateral preoptic area, may inhibit bladder activity (Grossman & Wang, 1955; Kabat et al., 1936). Transection experiments in cats led to the hypothesis that a region of the posterior hypothalamus facilitates micturition, while neurons in one or more rostral forebrain regions exert an inhibitory effect (Ruch & Tang, 1956).

A mix of excitatory and inhibitory neurons in the LHA project axons to Bar (Venner et al., 2016; Verstegen et al., 2019). These axons target the core of Bar as well as Bar dendritic arbors extending up to the fourth ventricle. Unlike the PAG, optogenetically stimulating LHA afferents in Bar evoked voids less reliably and typically only after the mouse had moved into a corner. Bladder pressure increased less consistently in response to LHA afferent stimulation under anesthesia (Verstegen et al., 2019). This suggests the LHA exerts a less potent influence over Bar than the PAG. The preoptic area of the hypothalamus also contains populations of inhibitory and excitatory neurons that project to Bar and modulate micturition behavior. Inhibiting GABAergic neurons in this region reduced the void-marking behavior of dominant male mice (Hou et al., 2016). Along with the LHA and other subcortical afferents to Bar neurons, the preoptic area likely mediates complex homeostatic or affective modulation of voiding, further emphasizing the complexity of micturition behavior.

In addition to reports of florid urinary incontinence in patients with periventricular hypothalamic injuries as part of Wernicke's encephalopathy syndrome (Kuhn et al., 2012; Sakakibara et al., 1997), other, more focal hypothalamic lesions have been identified in patients with urinary incontinence. A 1965 case series included a 48-year-old woman with a cystic lesion compressing the anterior hypothalamus who developed urinary frequency and incontinence, with spastic and early detrusor contractions at low bladder volumes. When the cyst was evacuated, her urinary symptoms resolved (Andrew & Nathan, 1965). A more recent case series described three patients with pituitary adenomas compressing the hypothalamus presenting with urinary urgency, frequency, and incontinence (Yamamoto et al., 2005). These and other human cases indicate that the net effect of neurons in (or pathways through) the hypothalamus may be inhibitory, and that damage to this region disinhibits Bar, leading to urinary frequency and incontinence (Tish & Geerling, 2020).

Several regions of the cerebral cortex have been linked to Bar, but their role in micturition remains unclear. A cluster of pyramidal neurons in layer 5 of the mouse primary motor cortex, as well as parts of the somatosensory cortex, were infected following injections of a retrograde transneuronal tracer into the bladder wall. Optogenetic stimulation of neurons in the motor cortex variably increased bladder pressure, and voiding frequency increased during 4 min of continuous stimulation. Chemogenetic inhibition or ablation of neurons in this cortical area led to fewer voids and less total urine output (Yao et al., 2018). The exact pathway from motor cortex remains unclear, and axonal projections directly from the cortex to Bar appear minimal. Cortical control of micturition is likely complex and may involve coordination with many other brain regions.

While we are not aware of any human lesions studies with micturition deficits following focal damage to the primary motor cortex, there are many examples of urge incontinence following lesions to other cortical areas. An early case series presented 37 patients with various frontal lobe lesions and micturition deficits, in which 92% had hypertonic bladders that emptied at low volumes (Andrew & Nathan, 1964). A subsequent study of patients with intracranial tumors confirmed that only those in the frontal lobes (and only 14% of those) produced urgency and incontinence (Maurice-Williams, 1974). Another study of 33 patients with bladder symptoms after an ischemic stroke found that the majority of infarcts involved the frontal cortex or internal capsule (Khan et al., 1990). A study of 72 patients with acute stroke found that bladder symptoms were most common after a lesion in the frontal lobe, specifically its anterior and medial surfaces, the anterior periventricular white matter, or the genu of the corpus callosum (Sakakibara et al., 1996a). Other case reports have linked large anterior cerebral artery infarctions to urinary urgency and incontinence (Kumral et al., 2002; Lakhotia et al., 2016). While the precise forebrain region(s) that modulate micturition remain to be determined, these and other human lesion studies establish that one or more sites within the frontal lobe exert a major modulatory influence over micturition. Future work is needed to clarify which if any of these regions inhibit micturition via direct or indirect output to Bar neurons.

Conclusion

Since Barrington's discovery a century ago, our understanding of the neural control of micturition has expanded. Much of this work has focused on Bar neurons and their output to the spinal cord. While initial studies did not agree on a name or precise location for neurons in the "PMC," we now know that these neurons are located between the LC and LDT and include at least two glutamatergic subpopulations expressing specific genes (Fig. 2.3A). Further work is needed to clarify exactly how Bar^{Esr1} and Bar^{Crh} neurons work together to control micturition and possibly other pelvic functions. Beyond these partially overlapping subpopulations, we do not know how many additional Bar subtypes exist. Furthermore, the focus on long-range projections from Bar to the lumbosacral spinal cord has largely ignored its collateral axonal projections to other sites in the brainstem and spinal cord, the function of which remain obscure.

Gaining a more complete understanding of Bar neurons will add pieces to the puzzle of micturition control, but will not be enough to complete the picture. A full picture of the neural control of micturition will require understanding how other brain regions connect to and influence Bar neurons. Learning that Bar neurons extend dendrites well beyond its borders opens the possibility that a rich, distal set of synaptic inputs sculpt their activity. We do not yet know the specific subtypes of neurons that provide synaptic input to Bar from the PAG, hypothalamus, and other afferent regions, nor is it clear in all cases which Bar subtypes they target. Additionally, we know very little about the polysynaptic input pathways from the frontal lobe that impact micturition. We also know very little about local interneurons surrounding Bar, which may relay some of this input. A greater understanding of the relationship between Bar neurons and their synaptic network will provide a more complete understanding of micturition control.

References

Adori, C., Barde, S., Bogdanovic, N., Uhlén, M., Reinscheid, R. R., Kovacs, G. G., & Hökfelt, T. (2015). Neuropeptide S- and neuropeptide S receptor-expressing neuron populations in the human pons. *Frontiers in Neuroanatomy, 9*, 126.

Andrew, J., & Nathan, P. W. (1964). Lesions on the anterior frontal lobes and disturbances of micturition and defaecation. *Brain, 87*, 233–262.

Andrew, J., & Nathan, P. W. (1965). The cerebral control of micturition. *Proceedings of the Royal Society of Medicine, 58*, 553–555.

Araki, I., Matsui, M., Ozawa, K., Takeda, M., & Kuno, S. (2003). Relationship of bladder dysfunction to lesion site in multiple sclerosis. *The Journal of Urology, 169*, 1384–1387.

Aston-Jones, G., & Waterhouse, B. (2016). Locus coeruleus: From global projection system to adaptive regulation of behavior. *Brain Research, 1645*, 75–78.

Austin, M. C., Rice, P. M., Mann, J. J., & Arango, V. (1995). Localization of corticotropin-releasing hormone in the human locus coeruleus and pedunculopontine tegmental nucleus: An immunocytochemical and in situ hybridization study. *Neuroscience, 64*, 713–727.

Aviles-Olmos, I., Foltynie, T., Panicker, J., Cowie, D., Limousin, P., Hariz, M., Fowler, C. J., & Zrinzo, L. (2011). Urinary incontinence following deep brain stimulation of the pedunculopontine nucleus. *Acta Neurochirurgica, 153*, 2357–2360.

Bahns, E., Halsband, U., & Jänig, W. (1987). Responses of sacral visceral afferents from the lower urinary tract, colon and anus to mechanical stimulation. *Pflugers Archiv, 410*, 296–303.

Barrington, F. J. F. (1914). The nervous mechanism of micturition. *Experimental Physiology, 8*, 33–71.

Barrington, F. J. F. (1921). The relation of the hind-brain to micturition. *Brain, 44*, 23–53.

Barrington, F. J. (1925). The effect of lesions of the hind- and mid-brain on micturition in the cat. *Experimental Physiology, 15*, 81–102.

Barrington, F. J. (1927). Affections of micturition resulting from lesions of the nervous system. *Proceedings of the Royal Society of Medicine, 20*, 722–727.

Benarroch, E. E. (2009). The locus ceruleus norepinephrine system: Functional organization and potential clinical significance. *Neurology, 73*, 1699–1704.

Benarroch, E. E. (2010). Neural control of the bladder: Recent advances and neurologic implications. *Neurology, 75*, 1839–1846.

Benarroch, E. E. (2018). Locus coeruleus. *Cell and Tissue Research, 373*, 221–232.

Blanco, L., Yuste, J. E., Carrillo-De Sauvage, M. A., Gómez, A., Fernández-Villalba, E., Avilés-Olmos, I., Limousin, P., Zrinzo, L., & Herrero, M. T. (2013). Critical evaluation of the anatomical location of the Barrington nucleus: Relevance for deep brain stimulation surgery of pedunculopontine tegmental nucleus. *Neuroscience, 247*, 351–363.

Blok, B. F., & Holstege, G. (1994). Direct projections from the periaqueductal gray to the pontine micturition center (M-region). An anterograde and retrograde tracing study in the cat. *Neuroscience Letters, 166*, 93–96.

Blok, B. F., & Holstege, G. (1997). Ultrastructural evidence for a direct pathway from the pontine micturition center to the parasympathetic preganglionic motoneurons of the bladder of the cat. *Neuroscience Letters, 222*, 195–198.

Blok, B. F., Van Maarseveen, J. T., & Holstege, G. (1998). Electrical stimulation of the sacral dorsal gray commissure evokes relaxation of the external urethral sphincter in the cat. *Neuroscience Letters, 249*, 68–70.

Blok, B. F., Willemsen, A. T., & Holstege, G. (1997). A PET study on brain control of micturition in humans. *Brain, 120*(Pt 1), 111–121.

Bruinstroop, E., Cano, G., Vanderhorst, V. G., Cavalcante, J. C., Wirth, J., Sena-Esteves, M., & Saper, C. B. (2012). Spinal projections of the A5, A6 (locus coeruleus), and A7 noradrenergic cell groups in rats. *Journal of Comparative Neurology, 520*, 1985–2001.

Burnstock, G. (1990). Innervation of bladder and bowel. *Ciba Foundation Symposia, 151*, 2–18 (discussion 18–26).

Canteras, N. S., Simerly, R. B., & Swanson, L. W. (1994). Organization of projections from the ventromedial nucleus of the hypothalamus: A *Phaseolus vulgaris*-leucoagglutinin study in the rat. *Journal of Comparative Neurology, 348*, 41–79.

Carter, M. E., Yizhar, O., Chikahisa, S., Nguyen, H., Adamantidis, A., Nishino, S., Deisseroth, K., & De Lecea, L. (2010). Tuning arousal with optogenetic modulation of locus coeruleus neurons. *Nature Neuroscience, 13*, 1526–1533.

Chern, B. Z. H., Rajaraman, S., Verma, G., & Heng, K. W. J. (2017). Unusual case of acute urinary retention in a young female. *BMJ Case Reports, 2017*.

Cho, H. J., Kang, T. H., Chang, J. H., Choi, Y. R., Park, M. G., Choi, K. D., Sung, S. M., Park, K. P., & Jung, D. S. (2015). Neuroanatomical correlation of urinary retention in lateral medullary infarction. *Annals of Neurology, 77*, 726–733.

Chrousos, G. P. (2009). Stress and disorders of the stress system. *Nature Reviews Endocrinology, 5*, 374–381.

Clark, S. D., Duangdao, D. M., Schulz, S., Zhang, L., Liu, X., Xu, Y. L., & Reinscheid, R. K. (2011). Anatomical characterization of the neuropeptide S system in the mouse brain by in situ hybridization and immunohistochemistry. *Journal of Comparative Neurology, 519*, 1867–1893.

Clemens, J. Q. (2010). Basic bladder neurophysiology. *Urologic Clinics, 37*, 487–494.

De Groat, W. C. (1975). Nervous control of the urinary bladder of the cat. *Brain Research, 87*, 201–211.

De Groat, W. C., Booth, A. M., Milne, R. J., & Roppolo, J. R. (1982). Parasympathetic preganglionic neurons in the sacral spinal cord. *Journal of the Autonomic Nervous System, 5*, 23–43.

De Groat, W. C., Griffiths, D., & Yoshimura, N. (2015). Neural control of the lower urinary tract. *Comprehensive Physiology, 5*, 327–396.

De Groat, W. C., & Lalley, P. M. (1972). Reflex firing in the lumbar sympathetic outflow to activation of vesical afferent fibres. *The Journal of Physiology, 226*, 289–309.

Del Tredici, K., & Braak, H. (2013). Dysfunction of the locus coeruleus-norepinephrine system and related circuitry in Parkinson's disease-related dementia. *Journal of Neurology Neurosurgery and Psychiatry, 84*, 774–783.

Denny-Brown, D., & Robertson, E. G. (1933). *On the physiology of micturition* (Vol 56, pp. 149–190).

Desjardins, C., Maruniak, J. A., & Bronson, F. H. (1973). Social rank in house mice: Differentiation revealed by ultraviolet visualization of urinary marking patterns. *Science, 182*, 939–941.

Ding, S. L., Royall, J. J., Sunkin, S. M., Ng, L., Facer, B. A., Lesnar, P., Guillozet-Bongaarts, A., Mcmurray, B., Szafer, A., Dolbeare, T. A., Stevens, A., Tirrell, L., Benner, T., Caldejon, S., Dalley, R. A., Dee, N., Lau, C., Nyhus, J., Reding, M., ... Lein, E. S. (2016). Comprehensive cellular-resolution atlas of the adult human brain. *Journal of Comparative Neurology, 524*, 3127–3481.

Ding, Y. Q., Wang, D., Nie, H., Guan, Z. L., Lü, B. Z., & Li, J. S. (1998). Direct projections from the periaqueductal gray to pontine micturition center neurons projecting to the lumbosacral cord segments: An electron microscopic study in the rat. *Neuroscience Letters, 242*, 97–100.

Dong, H. W., & Swanson, L. W. (2006a). Projections from bed nuclei of the stria terminalis, dorsomedial nucleus: Implications for cerebral hemisphere integration of neuroendocrine, autonomic, and drinking responses. *Journal of Comparative Neurology, 494*, 75–107.

Dong, H. W., & Swanson, L. W. (2006b). Projections from bed nuclei of the stria terminalis, magnocellular nucleus: Implications for cerebral hemisphere regulation of micturition, defecation, and penile erection. *Journal of Comparative Neurology, 494*, 108–141.

Drake, M. J. (2018). Fundamentals of terminology in lower urinary tract function. *Neurourology and Urodynamics, 37*, S13–S19.

Elam, M., Thorén, P., & Svensson, T. H. (1986). Locus coeruleus neurons and sympathetic nerves: Activation by visceral afferents. *Brain Research, 375*, 117–125.

Elliott, T. R. (1907). The innervation of the bladder and urethra. *The Journal of Physiology, 35*, 367–445.

Ennis, M., Behbehani, M., Shipley, M. T., Van Bockstaele, E. J., & Aston-Jones, G. (1991). Projections from the periaqueductal gray to the rostromedial pericoerulear region and nucleus locus coeruleus: Anatomic and physiologic studies. *Journal of Comparative Neurology, 306*, 480–494.

Feloney, M. P., & Leslie, S. W. (2020). *Bladder sphincter dyssynergia.* StatPearls. Treasure Island (FL): StatPearls Publishing. Copyright © 2020, StatPearls Publishing LLC.

Fetscher, C., Fleichman, M., Schmidt, M., Krege, S., & Michel, M. C. (2002). M(3) muscarinic receptors mediate contraction of human urinary bladder. *British Journal of Pharmacology, 136*, 641–643.

Fleischmann, N., Christ, G., Sclafani, T., & Melman, A. (2002). The effect of ovariectomy and long-term estrogen replacement on bladder structure and function in the rat. *The Journal of Urology, 168*, 1265–1268.

Fowler, C. J., & Griffiths, D. J. (2010). A decade of functional brain imaging applied to bladder control. *Neurourology and Urodynamics, 29*, 49–55.

Fowler, C. J., Griffiths, D., & DE Groat, W. C. (2008). The neural control of micturition. *Nature Reviews Neuroscience, 9*, 453–466.

Garfield, A. S., Li, C., Madara, J. C., Shah, B. P., Webber, E., Steger, J. S., Campbell, J. N., Gavrilova, O., Lee, C. E., Olson, D. P., Elmquist, J. K., Tannous, B. A., Krashes, M. J., & Lowell, B. B. (2015). A neural basis for melanocortin-4 receptor-regulated appetite. *Nature Neuroscience, 18*, 863–871.

Gasparini, S., Resch, J. M., Gore, A. M., Peltekian, L., & Geerling, J. C. (2021). Pre-locus coeruleus neurons in rat and mouse. *American Journal of Physiology-Regulatory, Integrative and Comparative Physiology, 320*, R342–R361.

Gasparini, S., Resch, J. M., Narayan, S. V., Peltekian, L., Iverson, G. N., Karthik, S., & Geerling, J. C. (2019). Aldosterone-sensitive HSD2 neurons in mice. *Brain Structure and Function, 224*, 387–417.

Geerling, J. C., Kim, M., Mahoney, C. E., Abbott, S. B., Agostinelli, L. J., Garfield, A. S., Krashes, M. J., Lowell, B. B., & Scammell, T. E. (2016). Genetic identity of thermosensory relay neurons in the lateral parabrachial nucleus. *American Journal of Physiology-Regulatory, Integrative and Comparative Physiology, 310*, R41–R54.

Geerling, J. C., & Loewy, A. D. (2006). Aldosterone-sensitive neurons in the nucleus of the solitary tract: Efferent projections. *Journal of Comparative Neurology, 497*, 223–250.

Geerling, J. C., & Loewy, A. D. (2007). Sodium deprivation and salt intake activate separate neuronal subpopulations in the nucleus of the solitary tract and the parabrachial complex. *Journal of Comparative Neurology, 504*, 379–403.

Geerling, J. C., Stein, M. K., Miller, R. L., Shin, J. W., Gray, P. A., & Loewy, A. D. (2011). FoxP2 expression defines dorsolateral pontine neurons activated by sodium deprivation. *Brain Research, 1375*, 19–27.

Gompf, H. S., Mathai, C., Fuller, P. M., Wood, D. A., Pedersen, N. P., Saper, C. B., & Lu, J. (2010). Locus ceruleus and anterior cingulate cortex sustain wakefulness in a novel environment. *Journal of Neuroscience, 30*, 14543–14551.

Green, A. L., Stone, E., Sitsapesan, H., Turney, B. W., Coote, J. H., Aziz, T. Z., Hyam, J. A., & Lovick, T. A. (2012). Switching off micturition using deep brain stimulation at midbrain sites. *Annals of Neurology, 72*, 144–147.

Griffiths, D., Tadic, S. D., Schaefer, W., & Resnick, N. M. (2007). Cerebral control of the bladder in normal and urge-incontinent women. *NeuroImage, 37*, 1–7.

Grill, W. M., Bhadra, N., & Wang, B. (1999). Bladder and urethral pressures evoked by microstimulation of the sacral spinal cord in cats. *Brain Research, 836*, 19–30.

Grossman, R. G., & Wang, S. C. (1955). Diencephalic mechanism of control of the urinary bladder of the cat. *The Yale Journal of Biology and Medicine, 28*, 285–297.

Hahn, J. D., & Swanson, L. W. (2010). Distinct patterns of neuronal inputs and outputs of the juxtaparaventricular and suprafornical regions of the lateral hypothalamic area in the male rat. *Brain Research Reviews, 64*, 14–103.

Hallanger, A. E., Levey, A. I., Lee, H. J., Rye, D. B., & Wainer, B. H. (1987). The origins of cholinergic and other subcortical afferents to the thalamus in the rat. *Journal of Comparative Neurology, 262*, 105–124.

Head, H., & Riddoch, G. (1917). The automatic bladder, excessive sweating and some other reflex conditions, in gross injuries of the spinal cord. *Brain, 40*, 188–263.

Herman, J. P., Mcklveen, J. M., Ghosal, S., Kopp, B., Wulsin, A., Makinson, R., Scheimann, J., & Myers, B. (2016). Regulation of the hypothalamic-pituitary-adrenocortical stress response. *Comprehensive Physiology, 6*, 603–621.

Holstege, G. (1987). Some anatomical observations on the projections from the hypothalamus to brainstem and spinal cord: An HRP and autoradiographic tracing study in the cat. *Journal of Comparative Neurology, 260*, 98–126.

Holstege, G., Griffiths, D., DE Wall, H., & Dalm, E. (1986). Anatomical and physiological observations on supraspinal control of bladder and urethral sphincter muscles in the cat. *Journal of Comparative Neurology, 250*, 449–461.

Hou, X. H., Hyun, M., Taranda, J., Huang, K. W., Todd, E., Feng, D., Atwater, E., Croney, D., Zeidel, M. L., Osten, P., & Sabatini, B. L. (2016). Central control circuit for context-dependent micturition. *Cell, 167*, 73–86.e12.

Hou, S., & Rabchevsky, A. G. (2014). Autonomic consequences of spinal cord injury. *Comprehensive Physiology, 4*, 1419–1453.

Hyun, M., Taranda, J., Radeljic, G., Miner, L., Wang, W., Ochandarena, N., Huang, K. W., Osten, P., & Sabatini, B. L. (2021). Social isolation uncovers a circuit underlying context-dependent territory-covering micturition. *Proceedings of the National Academy of Sciences of the United States of America, 118*.

Imaki, T., Nahan, J. L., Rivier, C., Sawchenko, P. E., & Vale, W. (1991). Differential regulation of corticotropin-releasing factor mRNA in rat brain regions by glucocorticoids and stress. *Journal of Neuroscience, 11*, 585–599.

Irwin, D. E., Milsom, I., Kopp, Z., Abrams, P., & Cardozo, L. (2006). Impact of overactive bladder symptoms on employment, social interactions and emotional well-being in six European countries. *BJU International, 97*, 96–100.

Ito, H., Sales, A. C., Fry, C. H., Kanai, A. J., Drake, M. J., & Pickering, A. E. (2020). Probabilistic, spinally-gated control of bladder pressure and autonomous micturition by Barrington's nucleus Crh neurons. *Elife, 9*.

Jänig, W., & Koltzenburg, M. (1990). On the function of spinal primary afferent fibres supplying colon and urinary bladder. *Journal of the Autonomic Nervous System, 30*(Suppl. 1), S89–S96.

Jarvie, B. C., & Palmiter, R. D. (2017). HSD2 neurons in the hindbrain drive sodium appetite. *Nature Neuroscience, 20*, 167–169.

Johnson, T. M., 2nd, & Vaughan, C. P. (2019). Urological function and dysfunction in aging: Diagnosis and treatment. *Handbook of Clinical Neurology, 167*, 495–509.

Kabat, H., Magoun, H. W., & Ranson, S. W. (1936). Reaction of the bladder to stimulation of points in the forebrain and mid-brain. *Journal of Comparative Neurology, 63*, 211–239.

Kawatani, M., De Groat, W. C., Itoi, K., Uchida, K., Sakimura, K., Yamanaka, A., Yamashita, T., & Kawatani, M. (2021). Downstream projection of Barrington's nucleus to the spinal cord in mice. *Journal of Neurophysiology, 126*, 1959–1977.

Kawatani, M., Itoi, K., Talukder, A. H., Uchida, K., Sakimura, K., & Kawatani, M. (2020). Cholinergic modulation of CRH and non-CRH neurons in Barrington's nucleus of the mouse. *Journal of Neurophysiology, 124*, 443–457.

Keller, J. A., Chen, J., Simpson, S., Wang, E. H., Lilascharoen, V., George, O., Lim, B. K., & Stowers, L. (2018). Voluntary urination control by brainstem neurons that relax the urethral sphincter. *Nature Neuroscience, 21*, 1229–1238.

Kemp, J., Guzman, M. A., Fitzpatrick, C. M., & Elbabaa, S. K. (2014). Holocord syringomyelia secondary to tethered spinal cord associated with anterior sacral meningocele and tailgut cyst: Case report and review of literature. *Child's Nervous System, 30*, 1141–1146.

Keren, N. I., Taheri, S., Vazey, E. M., Morgan, P. S., Granholm, A. C., ASTON-Jones, G. S., & Eckert, M. A. (2015). Histologic validation of locus coeruleus MRI contrast in post-mortem tissue. *NeuroImage, 113*, 235–245.

Khan, Z., Starer, P., Yang, W. C., & Bhola, A. (1990). Analysis of voiding disorders in patients with cerebrovascular accidents. *Urology, 35*, 265–270.

Khavari, R., & Boone, T. B. (2019). Functional brain imaging in voiding dysfunction. *Current Bladder Dysfunction Reports, 14*, 24–30.

Kiddoo, D. A., Valentino, R. J., Zderic, S., Ganesh, A., Leiser, S. C., Hale, L., & Grigoriadis, D. E. (2006). Impact of state of arousal and stress neuropeptides on urodynamic function in freely moving rats. *American Journal of Physiology-Regulatory, Integrative and Comparative Physiology, 290*, R1697–R1706.

Komiyama, A., Kubota, A., & Hidai, H. (1998). Urinary retention associated with a unilateral lesion in the dorsolateral tegmentum of the rostral pons. *Journal of Neurology Neurosurgery and Psychiatry, 65*, 953–954.

Koyama, Y., Imada, N., Kawauchi, A., & Kayama, Y. (1999). Firing of putative cholinergic neurons and micturition center neurons in the rat laterodorsal tegmentum during distention and contraction of urinary bladder. *Brain Research, 840*, 45–55.

Koyama, Y., Imada, N., Kayama, Y., Kawauchi, A., & Watanabe, H. (1998). How does the distention of urinary bladder cause arousal? *Psychiatry and Clinical Neurosciences, 52*, 142–145.

Kroeger, D., Ferrari, L. L., Petit, G., Mahoney, C. E., Fuller, P. M., Arrigoni, E., & Scammell, T. E. (2017). Cholinergic, glutamatergic, and GABAergic neurons of the pedunculopontine tegmental nucleus have distinct effects on sleep/wake behavior in mice. *Journal of Neuroscience, 37*, 1352–1366.

Kruse, M. N., Noto, H., Roppolo, J. R., & De Groat, W. C. (1990). Pontine control of the urinary bladder and external urethral sphincter in the rat. *Brain Research, 532*, 182–190.

Kuhn, A. L., Hertel, F., Boulanger, T., & Diederich, N. J. (2012). Vitamin B1 in the treatment of Wernicke's encephalopathy due to hyperemesis after gastroplasty. *Journal of Clinical Neuroscience, 19*, 1303–1305.

Kuipers, R., Mouton, L. J., & Holstege, G. (2006). Afferent projections to the pontine micturition center in the cat. *Journal of Comparative Neurology, 494*, 36–53.

Kumral, E., Bayulkem, G., Evyapan, D., & Yunten, N. (2002). Spectrum of anterior cerebral artery territory infarction: Clinical and MRI findings. *European Journal of Neurology, 9*, 615–624.

Kuru, M., & Yamamoto, H. (1964). Fiber connections of the pontine detrusor nucleus (Barrington). *Journal of Comparative Neurology, 123*, 161–185.

Lakhotia, M., Pahadiya, H. R., Prajapati, G. R., Choudhary, A., Gandhi, R., & Jangid, H. (2016). A case of anterior cerebral artery A1 segment hypoplasia syndrome presenting with right lower limb monoplegia, abulia, and urinary incontinence. *Journal of Neurosciences in Rural Practice, 7*, 189–191.

Lalley, R. M., Degroat, W. C., & Mclain, P. L. (1972). *Activation of the sacral parasympathetic pathway to the urinary bladder by brainstem stimulation* (p. A386). Federation of American Societies for Experimental Biology.

Landefeld, C. S., Bowers, B. J., Feld, A. D., Hartmann, K. E., Hoffman, E., Ingber, M. J., King, J. T.,, Jr., Mcdougal, W. S., Nelson, H., Orav, E. J., Pignone, M., Richardson, L. H., Rohrbaugh, R. M., Siebens, H. C., & Trock, B. J. (2008). National Institutes of Health state-of-the-science conference statement: Prevention of fecal and urinary incontinence in adults. *Annals of Internal Medicine, 148*, 449–458.

Langworthy, O. R., & Kolb, L. C. (1933). The encephalic control of tone in the musculature of the urinary bladder. *Brain, 56*, 371–382.

Lee, S., Augustine, V., Zhao, Y., Ebisu, H., Ho, B., Kong, D., & Oka, Y. (2019). Chemosensory modulation of neural circuits for sodium appetite. *Nature, 568*, 93–97.

Lee, K. B., Jang, I. M., Roh, H., Ahn, M. Y., & Woo, H. Y. (2008). Transient urinary retention in acute right lateral medullary infarction. *The Neurologist, 14*, 312–315.

Liao, P.-W., Liu, S.-P., Yu, H.-J., Yeong-Shiau, P., & Ho, C.-H. (2013). Acute urinary retention after brainstem infarction. *Incont Pelvic Floor Dysfunct, 7*, 10–11.

Li, M. M., Madara, J. C., Steger, J. S., Krashes, M. J., Balthasar, N., Campbell, J. N., Resch, J. M., Conley, N. J., Garfield, A. S., & Lowell, B. B. (2019). The paraventricular hypothalamus regulates satiety and prevents obesity via two genetically distinct circuits. *Neuron, 102*, 653–667 e6.

Liu, Z., Sakakibara, R., Nakazawa, K., Uchiyama, T., Yamamoto, T., Ito, T., & Hattori, T. (2004). Micturition-related neuronal firing in the periaqueductal gray area in cats. *Neuroscience, 126*, 1075–1082.

Loewy, A. D., Saper, C. B., & Baker, R. P. (1979). Descending projections from the pontine micturition center. *Brain Research, 172*, 533–538.

Longhurst, P. A., Kauer, J., Leggett, R. E., & Levin, R. M. (1992). The influence of ovariectomy and estradiol replacement on urinary bladder function in rats. *The Journal of Urology, 148*, 915–919.

Mackay, W. G., Macintosh, T., Kydd, A., Fleming, A., O'kane, C., Shepherd, A., Hagen, S., Williams, C., Mundie, J., Russell, C., Rodgers, F., Maclachlan, M., Galbraith, R., Rankin, J., & Mciver, V. (2018). Living with an indwelling urethral catheter in a community setting: Exploring triggers for unscheduled community nurse "out-of-hours" visits. *Journal of Clinical Nursing, 27*, 866–875.

Mallory, B. S., Roppolo, J. R., & De Groat, W. C. (1991). Pharmacological modulation of the pontine micturition center. *Brain Research, 546*, 310–320.

Mallory, B., Steers, W. D., & De Groat, W. C. (1989). Electrophysiological study of micturition reflexes in rats. *American Journal of Physiology, 257*, R410–R421.

Manohar, A., Curtis, A. L., Zderic, S. A., & Valentino, R. J. (2017). Brainstem network dynamics underlying the encoding of bladder information. *Elife, 6*.

Marshall, K. L., Saade, D., Ghitani, N., Coombs, A. M., Szczot, M., Keller, J., Ogata, T., Daou, I., Stowers, L. T., Bönnemann, C. G., Chesler, A. T., & Patapoutian, A. (2020). PIEZO2 in sensory neurons and urothelial cells coordinates urination. *Nature, 588*(7837), 290–295.

Marson, L. (1997). Identification of central nervous system neurons that innervate the bladder body, bladder base, or external urethral sphincter of female rats: A transneuronal tracing study using pseudorabies virus. *Journal of Comparative Neurology, 389*, 584–602.

Ma, L., Tang, J. Y., Zhou, J. Y., Zhou, P., Zhu, C., Zhang, X., Wang, Y., Ding, J. Q., Jia, H. Q., & Gu, X. J. (2022). The role of lumbosacral innervating noradrenergic neurons in micturition control. *Brain Research, 1777*, 147754.

Matsui, M., Motomura, D., Karasawa, H., Fujikawa, T., Jiang, J., Komiya, Y., Takahashi, S., & Taketo, M. M. (2000). Multiple functional defects in peripheral autonomic organs in mice lacking muscarinic acetylcholine receptor gene for the M3 subtype. *Proceedings of the National Academy of Sciences of the United States of America, 97*, 9579–9584.

Matsumoto, G., Hisamitsu, T., & De Groat, W. C. (1995). Role of glutamate and NMDA receptors in the descending limb of the spinobulbospinal micturition reflex pathway of the rat. *Neuroscience Letters, 183*, 58–61.

Matsuura, S., Downie, J. W., & Allen, G. V. (2000). Micturition evoked by glutamate microinjection in the ventrolateral periaqueductal gray is mediated through Barrington's nucleus in the rat. *Neuroscience, 101*, 1053–1061.

Maurice-Williams, R. S. (1974). Micturition symptoms in frontal tumours. *Journal of Neurology Neurosurgery and Psychiatry, 37*, 431–436.

Mcfadden, K., Griffin, T. A., Levy, V., Wolfe, J. H., & Valentino, R. J. (2012). Overexpression of corticotropin-releasing factor in Barrington's nucleus neurons by adeno-associated viral transduction: Effects on bladder function and behavior. *European Journal of Neuroscience, 36*, 3356–3364.

Million, M., Wang, L., Stenzel-Poore, M. P., Coste, S. C., Yuan, P. Q., Lamy, C., Rivier, J., Buffington, T., & Taché, Y. (2007). Enhanced pelvic responses to stressors in female CRF-overexpressing mice. *American Journal of Physiology-Regulatory, Integrative and Comparative Physiology, 292*, R1429–R1438.

Morin, X., Cremer, H., Hirsch, M. R., Kapur, R. P., Goridis, C., & Brunet, J. F. (1997). Defects in sensory and autonomic ganglia and absence of locus coeruleus in mice deficient for the homeobox gene Phox2a. *Neuron, 18*, 411–423.

Morrison, J. F. (2008). The discovery of the pontine micturition centre by F. J. F. Barrington. *Experimental Physiology, 93*, 742–745.

Nadelhaft, I., & Vera, P. L. (1995). Central nervous system neurons infected by pseudorabies virus injected into the rat urinary bladder following unilateral transection of the pelvic nerve. *Journal of Comparative Neurology, 359*, 443–456.

Nadelhaft, I., & Vera, P. L. (1996). Neurons in the rat brain and spinal cord labeled after pseudorabies virus injected into the external urethral sphincter. *Journal of Comparative Neurology, 375*, 502–517.

Nadelhaft, I., Vera, P. L., Card, J. P., & Miselis, R. R. (1992). Central nervous system neurons labelled following the injection of pseudorabies virus into the rat urinary bladder. *Neuroscience Letters, 143*, 271–274.

Nishizawa, O., Sugaya, K., Noto, H., Harada, T., & Tsuchida, S. (1988). Pontine micturition center in the dog. *The Journal of Urology, 140*, 872–874.

Numata, A., Iwata, T., Iuchi, H., Taniguchi, N., Kita, M., Wada, N., Kato, Y., & Kakizaki, H. (2008). Micturition-suppressing region in the periaqueductal gray of the mesencephalon of the cat. *American Journal of Physiology-Regulatory, Integrative and Comparative Physiology, 294*, R1996–R2000.

Page, M. E., Akaoka, H., Aston-Jones, G., & Valentino, R. J. (1992). Bladder distention activates noradrenergic locus coeruleus neurons by an excitatory amino acid mechanism. *Neuroscience, 51*, 555–563.

Palma, J. A., & Kaufmann, H. (2018). Treatment of autonomic dysfunction in Parkinson disease and other synucleinopathies. *Movement Disorders, 33*, 372–390.

Pang, Y. W., Li, J. L., Nakamura, K., Wu, S., Kaneko, T., & Mizuno, N. (2006). Expression of vesicular glutamate transporter 1 immunoreactivity in peripheral and central endings of trigeminal mesencephalic nucleus neurons in the rat. *Journal of Comparative Neurology, 498*, 129–141.

Panicker, J. N., Marcelissen, T., Von Gontard, A., Vrijens, D., Abrams, P., & Wyndaele, M. (2019). Bladder-bowel interactions: Do we understand pelvic organ cross-sensitization? International Consultation on Incontinence Research Society (ICI-RS) 2018. *Neurourology and Urodynamics, 38*(Suppl. 5), S25–s34.

Pavcovich, L. A., & Valentino, R. J. (1995). Central regulation of micturition in the rat the corticotropin-releasing hormone from Barrington's nucleus. *Neuroscience Letters, 196*, 185–188.

Paxinos, G., & Franklin, K. B. J. (2013). *Paxinos and Franklin's the mouse brain in stereotaxic coordinates*. Amsterdam: Academic Press, an imprint of Elsevier.

Poeze, M., Herpers, M. J., Tjandra, B., Freling, G., & Beuls, E. A. (1999). Intramedullary spinal teratoma presenting with urinary retention: Case report and review of the literature. *Neurosurgery, 45*, 379–385.

Ralls, K. (1971). Mammalian scent marking. *Science, 171*, 443–449.

Resch, J. M., Fenselau, H., Madara, J. C., Wu, C., Campbell, J. N., Lyubetskaya, A., Dawes, B. A., Tsai, L. T., Li, M. M., Livneh, Y., Ke, Q., Kang, P. M., Fejes-Toth, G., Naray-Fejes-Toth, A., Geerling, J. C., & Lowell, B. B. (2017). Aldosterone-sensing neurons in the NTS exhibit state-dependent pacemaker activity and drive sodium appetite via synergy with angiotensin II signaling. *Neuron, 96*, 190–206 e7.

Ribeiro-Barbosa, E. R., Skorupa, A. L., Cipolla-Neto, J., & Canteras, N. S. (1999). Projections of the basal retrochiasmatic area: A neural site involved in the photic control of pineal metabolism. *Brain Research, 839*, 35–40.

Rizvi, T. A., Ennis, M., Aston-Jones, G., Jiang, M., Liu, W. L., Behbehani, M. M., & Shipley, M. T. (1994). Preoptic projections to Barrington's nucleus and the pericoerulear region: Architecture and terminal organization. *Journal of Comparative Neurology, 347*, 1–24.

Rouzade-Dominguez, M. L., Curtis, A. L., & Valentino, R. J. (2001). Role of Barrington's nucleus in the activation of rat locus coeruleus neurons by colonic distension. *Brain Research, 917*, 206–218.

Rouzade-Dominguez, M. L., Miselis, R., & Valentino, R. J. (2003). Central representation of bladder and colon revealed by dual transsynaptic tracing in the rat: Substrates for pelvic visceral coordination. *European Journal of Neuroscience, 18*, 3311–3324.

Rouzade-Dominguez, M. L., Pernar, L., Beck, S., & Valentino, R. J. (2003). Convergent responses of Barrington's nucleus neurons to pelvic visceral stimuli in the rat: A juxtacellular labelling study. *European Journal of Neuroscience, 18*, 3325–3334.

Roy, H. A., Pond, D., Roy, C., Forrow, B., Foltynie, T., Zrinzo, L., Akram, H., Aziz, T. Z., Fitzgerald, J. J., & Green, A. L. (2018). Effects of pedunculopontine nucleus stimulation on human bladder function. *Neurourology and Urodynamics, 37*, 726–734.

Ruch, T. C., & Tang, P. C. (1956). Localization of brain stem and diencephalic areas controlling the micturation reflex. *Journal of Comparative Neurology, 106*, 213–245.

Sakakibara, R. (2015). Lower urinary tract dysfunction in patients with brain lesions. *Handbook of Clinical Neurology, 130*, 269–287.

Sakakibara, R., Hattori, T., Fukutake, T., Mori, M., Yamanishi, T., & Yasuda, K. (1998). Micturitional disturbance in herpetic brainstem encephalitis; contribution of the pontine micturition centre. *Journal of Neurology Neurosurgery and Psychiatry, 64*, 269–272.

Sakakibara, R., Hattori, T., Yasuda, K., & Yamanishi, T. (1996). Micturitional disturbance after acute hemispheric stroke: Analysis of the lesion site by CT and MRI. *Journal of the Neurological Sciences, 137*, 47–56.

Sakakibara, R., Hattori, T., Yasuda, K., & Yamanishi, T. (1996). Micturitional disturbance and the pontine tegmental lesion: Urodynamic and MRI analyses of vascular cases. *Journal of the Neurological Sciences, 141*, 105–110.

Sakakibara, R., Hattori, T., Yasuda, K., Yamanishi, T., Tojo, M., & Mori, M. (1997). Micturitional disturbance in Wernicke's encephalopathy. *Neurourology and Urodynamics, 16*, 111–115.

Sakakibara, R., Tateno, F., Yamamoto, T., & Uchiyama, T. (2017). Bowel and bladder disorders in Parkinson's disease. *Nihon Rinsho, 75*, 111–118.

Sakumoto, T., Tohyama, M., Satoh, K., Kimoto, Y., Kinugasa, T., Tanizawa, O., Kurachi, K., & Shimizu, N. (1978). Afferent fiber connections from lower brain stem to hypothalamus studied by the horseradish peroxidase method with special reference to noradrenaline innervation. *Experimental Brain Research, 31*, 81–94.

Salas, J. C., Iwasaki, H., Jodo, E., Schmidt, M. H., Kawauchi, A., Miki, T., Kayama, Y., Otsuki, M., & Koyama, Y. (2008). Penile erection and micturition events triggered by electrical stimulation of the mesopontine tegmental area. *American Journal of Physiology-Regulatory, Integrative and Comparative Physiology, 294*, R102–R111.

Saper, C. B., Scammell, T. E., & Lu, J. (2005). Hypothalamic regulation of sleep and circadian rhythms. *Nature, 437*, 1257–1263.

Sasaki, M. (2005). Role of Barrington's nucleus in micturition. *Journal of Comparative Neurology, 493*, 21–26.

Sasaki, M., Shibata, E., Tohyama, K., Takahashi, J., Otsuka, K., Tsuchiya, K., Takahashi, S., Ehara, S., Terayama, Y., & Sakai, A. (2006). Neuromelanin magnetic resonance imaging of locus ceruleus and substantia nigra in Parkinson's disease. *NeuroReport, 17*, 1215–1218.

Sasa, M., & Yoshimura, N. (1994). Locus coeruleus noradrenergic neurons as a micturition center. *Microscopy Research and Technique, 29*, 226–230.

Sastre, J. P., Sakai, K., & Jouvet, M. (1981). Are the gigantocellular tegmental field neurons responsible for paradoxical sleep? *Brain Research, 229*, 147–161.

Satoh, K., Shimizu, N., Tohyama, M., & Maeda, T. (1978). Localization of the micturition reflex center at dorsolateral pontine tegmentum of the rat. *Neuroscience Letters, 8*, 27–33.

Satoh, K., Tohyama, M., Sakumoto, T., Yamamoto, K., & Shimizu, N. (1978). Descending projection of the nucleus tegmentalis latero-dorsalis to the spinal cord; studied by the horseradish peroxidase method following 6-hydroxy-DOPA administration. *Neuroscience Letters, 8*, 9–15.

Sato, M., Mizuno, N., & Konishi, A. (1978). Localization of motoneurons innervating perineal muscles: A HRP study in cat. *Brain Research, 140*, 149–154.

Schmidt, K., Bari, B., Ralle, M., Washington-Hughes, C., Muchenditsi, A., Maxey, E., & Lutsenko, S. (2019). Localization of the Locus Coeruleus in the mouse brain. *Journal of Visualized Experiments*, (145), https://doi.org/10.3791/58652. doi: 10.3791/58652. Erratum in: Journal of Visualized Experiments, 2019 Apr 8;(146).

Schrøder, H. D. (1981). Onuf's nucleus X: A morphological study of a human spinal nucleus. *Anatomy and Embryology, 162*, 443–453.

Schurr, P. H., & Merrington, W. R. (1978). The Horsley-Clarke stereotaxic apparatus. *British Journal of Surgery, 65*, 33–36.

Shin, J. W., Geerling, J. C., Stein, M. K., Miller, R. L., & Loewy, A. D. (2011). FoxP2 brainstem neurons project to sodium appetite regulatory sites. *Journal of Chemical Neuroanatomy, 42*, 1–23.

Stewart, W. F., Van Rooyen, J. B., Cundiff, G. W., Abrams, P., Herzog, A. R., Corey, R., Hunt, T. L., & Wein, A. J. (2003). Prevalence and burden of overactive bladder in the United States. *World Journal of Urology, 20*, 327–336.

Stone, E., Coote, J. H., Allard, J., & Lovick, T. A. (2011). GABAergic control of micturition within the periaqueductal grey matter of the male rat. *The Journal of Physiology, 589*, 2065–2078.

Stornetta, R. L., Sevigny, C. P., & Guyenet, P. G. (2002). Vesicular glutamate transporter DNPI/VGLUT2 mRNA is present in C1 and several other groups of brainstem catecholaminergic neurons. *Journal of Comparative Neurology, 444*, 191–206.

Sugaya, K., Matsuyama, K., Takakusaki, K., & Mori, S. (1987). Electrical and chemical stimulations of the pontine micturition center. *Neuroscience Letters, 80*, 197–201.

Sugaya, K., Ogawa, Y., Hatano, T., Koyama, Y., Miyazato, T., & Oda, M. (1998). Evidence for involvement of the subcoeruleus nucleus and nucleus raphe magnus in urine storage and penile erection in decerebrate rats. *The Journal of Urology, 159*, 2172–2176.

Sugaya, K., Roppolo, J. R., Yoshimura, N., Card, J. P., & De Groat, W. C. (1997). The central neural pathways involved in micturition in the neonatal rat as revealed by the injection of pseudorabies virus into the urinary bladder. *Neuroscience Letters, 223*, 197–200.

Tai, C., Roppolo, J. R., & De Groat, W. C. (2006). Spinal reflex control of micturition after spinal cord injury. *Restorative Neurology and Neuroscience, 24*, 69–78.

Tanaka, Y., Koyama, Y., Kayama, Y., Kawauchi, A., Ukimura, O., & Miki, T. (2003). Firing of micturition center neurons in the rat mesopontine tegmentum during urinary bladder contraction. *Brain Research, 965*, 146–154.

Tang, P. C. (1955). Levels of brain stem and diencephalon controlling micturition reflex. *Journal of Neurophysiology, 18*, 583–595.

Tish, M. M., Boes, A. D., & Geerling, J. C. (2022). Right tegmental hemorrhage with urinary retention: A case report. *Case Reports in Neurology, 14*, 68–71.

Tish, M. M., & Geerling, J. C. (2020). The brain and the bladder: Forebrain control of urinary (in)continence. *Frontiers in Physiology, 11*, 658.

Tohyama, M., Satoh, K., Sakumoto, T., Kimoto, Y., Takahashi, Y., Yamamoto, K., & Itakura, T. (1978). Organization and projections of the neurons in the dorsal tegmental area of the rat. *Journal für Hirnforschung, 19*, 165–176.

Toussay, X., Basu, K., Lacoste, B., & Hamel, E. (2013). Locus coeruleus stimulation recruits a broad cortical neuronal network and increases cortical perfusion. *Journal of Neuroscience, 33*, 3390–3401.

Valentino, R. J., Chen, S., Zhu, Y., & Aston-Jones, G. (1996). Evidence for divergent projections to the brain noradrenergic system and the spinal parasympathetic system from Barrington's nucleus. *Brain Research, 732*, 1–15.

Valentino, R. J., Page, M. E., Luppi, P. H., Zhu, Y., Van Bockstaele, E., & Aston-Jones, G. (1994). Evidence for widespread afferents to Barrington's nucleus, a brainstem region rich in corticotropin-releasing hormone neurons. *Neuroscience, 62*, 125–143.

Valentino, R. J., Pavcovich, L. A., & Hirata, H. (1995). Evidence for corticotropin-releasing hormone projections from Barrington's nucleus to the periaqueductal gray and dorsal motor nucleus of the vagus in the rat. *Journal of Comparative Neurology, 363*, 402–422.

Van Batavia, J. P., Butler, S., Lewis, E., Fesi, J., Canning, D. A., Vicini, S., Valentino, R. J., & Zderic, S. A. (2021). Corticotropin-releasing hormone from the pontine micturition center plays an inhibitory role in micturition. *Journal of Neuroscience, 41*, 7314–7325.

Van Dort, C. J., Zachs, D. P., Kenny, J. D., Zheng, S., Goldblum, R. R., Gelwan, N. A., Ramos, D. M., Nolan, M. A., Wang, K., Weng, F. J., Lin, Y., Wilson, M. A., & Brown, E. N. (2015). Optogenetic activation of cholinergic neurons in the PPT or LDT induces REM sleep. *Proceedings of the National Academy of Sciences of the United States of America, 112*, 584–589.

Vanderhorst, V. G., Gustafsson, J. A., & Ulfhake, B. (2005). Estrogen receptor-alpha and -beta immunoreactive neurons in the brainstem and spinal cord of male and female mice: Relationships to monoaminergic, cholinergic, and spinal projection systems. *Journal of Comparative Neurology, 488*, 152–179.

Vanderhorst, V. G., Mouton, L. J., Blok, B. F., & Holstege, G. (1996). Distinct cell groups in the lumbosacral cord of the cat project to different areas in the periaqueductal gray. *Journal of Comparative Neurology, 376*, 361–385.

Venner, A., Anaclet, C., Broadhurst, R. Y., Saper, C. B., & Fuller, P. M. (2016). A novel population of wake-promoting GABAergic neurons in the ventral lateral hypothalamus. *Current Biology, 26*, 2137–2143.

Verdejo, C., Méndez, S., & Salinas, J. (2016). Urinary tract dysfunction in older patients. *Medical Clinics of North America, 147*, 455–460.

Verstegen, A. M. J., Klymko, N., Zhu, L., Mathai, J. C., Kobayashi, R., Venner, A., Ross, R. A., Vanderhorst, V. G., Arrigoni, E., Geerling, J. C., & Zeidel, M. L. (2019). Non-Crh glutamatergic neurons in Barrington's nucleus control micturition via glutamatergic afferents from the midbrain and hypothalamus. *Current Biology, 29*, 2775–2789.e7.

Verstegen, A. M. J., Vanderhorst, V., Gray, P. A., Zeidel, M. L., & Geerling, J. C. (2017). Barrington's nucleus: Neuroanatomic landscape of the mouse "pontine micturition center". *Journal of Comparative Neurology, 525*, 2287–2309.

Vertes, R. P., & Crane, A. M. (1996). Descending projections of the posterior nucleus of the hypothalamus: Phaseolus vulgaris leucoagglutinin analysis in the rat. *Journal of Comparative Neurology, 374*, 607–631.

Vincent, S. R., & Satoh, K. (1984). Corticotropin-releasing factor (CRF) immunoreactivity in the dorsolateral pontine tegmentum: Further studies on the micturition reflex system. *Brain Research, 308*, 387–391.

Wakamatsu, K., Tabuchi, K., Ojika, M., Zucca, F. A., Zecca, L., & Ito, S. (2015). Norepinephrine and its metabolites are involved in the synthesis of neuromelanin derived from the locus coeruleus. *Journal of Neurochemistry, 135*, 768–776.

Wang, S. C., & Ranson, S. W. (1939). Autonomic responses to electrical stimulation of the lower brain stem. *Journal of Comparative Neurology, 8*, 437–455.

Wang, T., Zhang, Q. J., Liu, J., Wu, Z. H., & Wang, S. (2009). Firing activity of locus coeruleus noradrenergic neurons increases in a rodent model of Parkinsonism. *Neuroscience Bulletin, 25*, 15–20.

Watanabe, T., Tan, Z., Wang, X., Martinez-Hernandez, A., & Frahm, J. (2019). Magnetic resonance imaging of noradrenergic neurons. *Brain Structure and Function, 224*, 1609–1625.

Wu, H. Y. (2010). Achieving urinary continence in children. *Nature Reviews Urology, 7*, 371–377.

Xu, Y. L., Reinscheid, R. K., Huitron-Resendiz, S., Clark, S. D., Wang, Z., Lin, S. H., Brucher, F. A., Zeng, J., Ly, N. K., Henriksen, S. J., De Lecea, L., & Civelli, O. (2004). Neuropeptide S: A neuropeptide promoting arousal and anxiolytic-like effects. *Neuron, 43*, 487–497.

Yaguchi, H., Soma, H., Miyazaki, Y., Tashiro, J., Yabe, I., Kikuchi, S., Sasaki, H., Kakizaki, H., Moriwaka, F., & Tashiro, K. (2004). A case of acute urinary retention caused by periaqueductal grey lesion. *Journal of Neurology Neurosurgery and Psychiatry, 75*(8), 1202–1203. https://doi.org/10.1136/jnnp.2003.027516

Yamamoto, T., Sakakibara, R., Uchiyama, T., Liu, Z., Ito, T., Yamanishi, T., & Hattori, T. (2005). Lower urinary tract function in patients with pituitary adenoma compressing hypothalamus. *Journal of Neurology Neurosurgery and Psychiatry, 76*, 390–394.

Yamao, Y., Koyama, Y., Akihiro, K., Yukihiko, K., & Tsuneharu, M. (2001). Discrete regions in the laterodorsal tegmental area of the rat regulating the urinary bladder and external urethral sphincter. *Brain Research, 912*, 162–170.

Yao, J., Zhang, Q., Liao, X., Li, Q., Liang, S., Li, X., Zhang, Y., Wang, H., Qin, H., Wang, M., Li, J., Zhang, J., He, W., Zhang, W., Li, T., Xu, F., Gong, H., Jia, H., Xu, X., Yan, J., & Chen, X. (2018). A corticopontine circuit for initiation of urination. *Nature Neuroscience, 21*, 1541–1550.

Yoshiyama, M., Roppolo, J. R., & De Groat, W. C. (1994). Alteration by urethane of glutamatergic control of micturition. *European Journal of Pharmacology, 264*, 417–425.

Yoshiyama, M., Roppolo, J. R., & De Groat, W. C. (1995). Interactions between NMDA and AMPA/kainate receptors in the control of micturition in the rat. *European Journal of Pharmacology, 287*, 73–78.

Chapter 3

Voluntary versus reflex micturition control

Naoki Yoshimura[1,2], Michael B. Chancellor[3], Takeya Kitta[1], Teruyuki Ogawa[1] and William C. de Groat[2]

[1]Department of Urology, University of Pittsburgh School of Medicine, Pittsburgh, PA, United States; [2]Department of Pharmacology and Chemical Biology, University of Pittsburgh School of Medicine, Pittsburgh, PA, United States; [3]Department of Urology, Oakland University William Beaumont School of Medicine, Royal Oak, MI, United States

Introduction

Micturition is controlled by dynamic reflex activity that alternates between phases of storage and voiding of urine stored in the urinary bladder. The storage phase typically extends for hours, while the voiding phase lasts for a few seconds several times a day. To achieve these two phases of different durations, the coordinated activity of two functional units in the lower urinary tract—(1) a reservoir (the urinary bladder) and (2) an outlet consisting of the bladder neck, the urethra, and the urethral sphincter (Fry et al., 2005; de Groat et al., 2015)—is required.

The lower urinary tract is unusual in its pattern of activity and organization of neural control mechanisms. For example, the urinary bladder has only two modes of operation: storage and elimination. Thus, many of the neural circuits have switch-like or phasic patterns of activity (de Groat et al., 2015). Micturition also requires the integration of autonomic and somatic efferent mechanisms to coordinate the activity of visceral organs (i.e., bladder and urethra) with that of urethral striated sphincter (external urethral sphincter [EUS]) muscles (Morrison et al., 2005).

In addition, micturition is under voluntary control and depends on learned behavior that develops during maturation of the nervous system, whereas many other visceral structures (e.g., the gastrointestinal tract and cardiovascular system) can maintain a certain level of function even after extrinsic neural input has been eliminated. Reflex micturition control is organized in neural circuits in the peripheral nervous system, the spinal cord, and the brainstem, in particular the periaqueductal gray (PAG) and the Pontine micturition center (PMC), which can switch on the periodic transformation of the lower urinary tract from the mode of bladder filling to voiding (de Groat et al., 2015). However, the voluntary micturition control over this dynamic process of micturition, which is lacking in infants, but gained through maturation and learning usually by the age of 3 to 5 in most individuals, is highly dependent on central nervous system (CNS) pathways rostral to the brainstem circuitry (Figs. 3.1 and 3.2).

Thus, due to the complexity of the neural mechanisms regulating the lower urinary tract, micturition is sensitive to a wide variety of injuries, diseases, and chemicals that affect the nervous system (Fig. 3.1). Thus, neurologic mechanisms of both voluntary and reflex micturition control are an important consideration in the diagnosis and treatment of lower urinary tract dysfunction. This chapter reviews (1) the innervation of the urinary bladder and urethra, (2) the organization of the reflex pathways controlling urine storage and voiding, (3) the cortical and subcortical pathways controlling voluntary voiding, and (4) the neurotransmitters involved in voluntary and involuntary micturition control.

Neural circuits involved in reflex micturition control

Peripheral nervous system

The lower urinary tract is innervated by three sets of the peripheral nerves involving the parasympathetic, sympathetic, and somatic nervous systems (Fig. 3.3). Pelvic parasympathetic nerves arise at the sacral level of the spinal cord, excite the bladder, and relax the urethra. Lumbar sympathetic nerves inhibit the bladder body and excite the bladder base and urethra

Neuro-Urology Research. https://doi.org/10.1016/B978-0-12-822455-7.00003-9

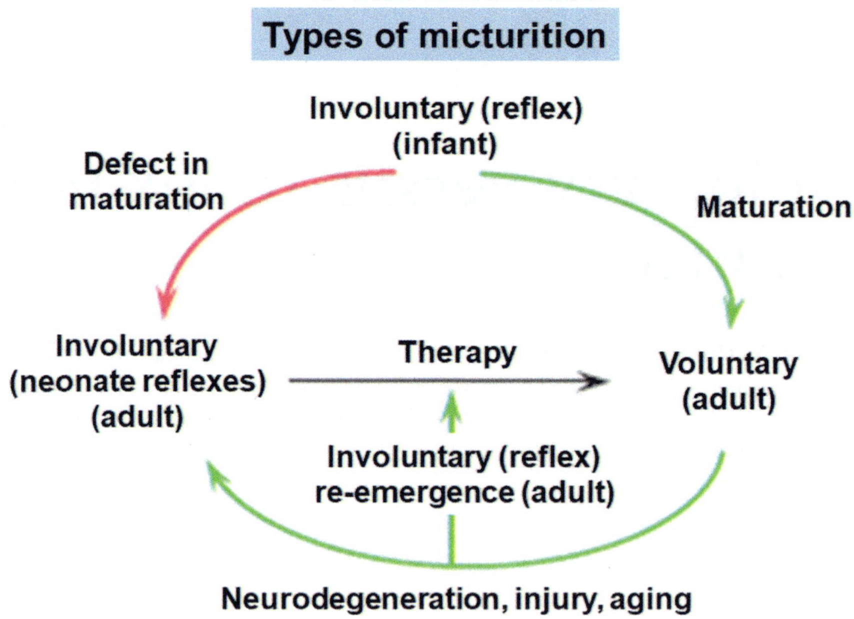

FIGURE 3.1 Bladder's "circle of life" influenced by maturation, pathologic processes, and aging. In infants, voiding is initiated and coordinated by reflex circuits. After maturation of central neural pathways, voiding is controlled voluntarily by neural circuitry in higher centers in the brain. A defect in neural maturation allows involuntary voiding to persist in adults. Aging, neural injury, or diseases can disrupt the central voluntary micturition neural pathways. In adults, pathologic processes can lead to the formation of new reflex circuitry by reemergence of primitive reflex mechanisms that were present in the infant or that appear as the result of synaptic remodeling in the spinal cord. The goal of therapy is to reverse the pathologic process and to reestablish normal voluntary control of voiding.

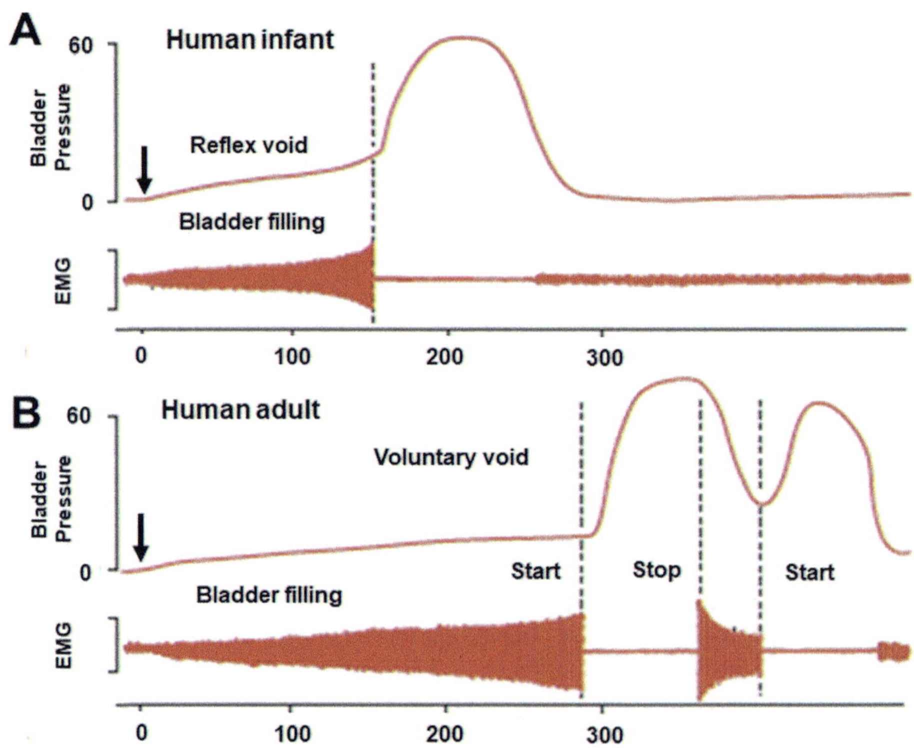

FIGURE 3.2 Combined cystometrogram and sphincter electromyogram (EMG) comparing reflex voiding responses in a human infant (A) and a voluntary voiding response in a human adult (B). The x-axis in all records represents bladder volume in milliliters, and the y-axis represents bladder pressure in centimeters of water and electrical activity of the electromyographic recording. On the left side of each trace, the arrows indicate the start of a slow infusion of fluid into the bladder (bladder filling). Vertical dashed lines indicate the start of sphincter relaxation that precedes by a few seconds the bladder contraction in (A) and (B). In (B), note that a voluntary cessation of voiding (stop) is associated with an initial increase in sphincter electromyographic activity followed by a reciprocal relaxation of the bladder. A resumption of voiding is again associated with sphincter relaxation and a delayed increase in bladder pressure. *From Yoshimura, N., & Chancellor, M. B. (2011). Physiology and pharmacology of the bladder and urethra. In A.J. Wein (Ed.), Campbell-Walsh Urology (10th edition). (Volume 3, Chapter 60, p. 1806). Saunders Elsevier.*

smooth muscles. Pudendal nerves excite the EUS striated muscles. These nerves contain both afferent (sensory) and efferent axons.

Parasympathetic pathways

The preganglionic parasympathetic fibers arise from neurons in the intermediolateral cell column of the sacral spinal cord (S2–S4), which is termed the sacral parasympathetic nucleus, and exit from the ventral roots to form the pelvic splanchnic

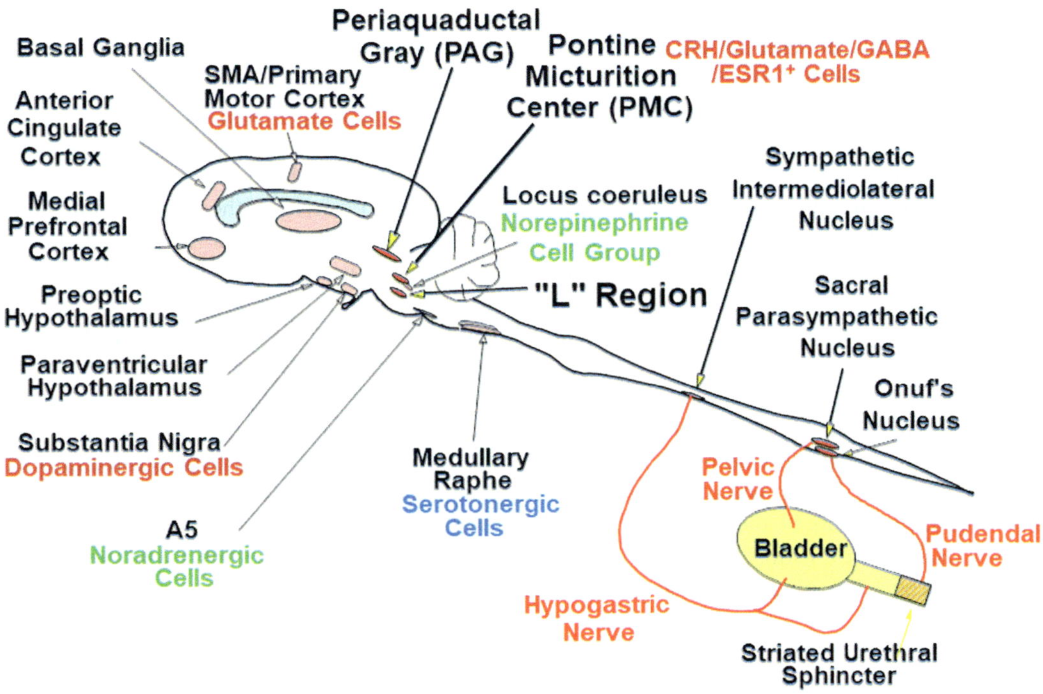

FIGURE 3.3 Micturition neuraxis and transmitters in cortical and subcortical regions. SMA, supplementary motor area.

nerves (de Groat et al., 2015). After passing through the pelvic plexus, the fibers synapse on terminal ganglia that innervate the detrusor smooth muscle and urethra. The parasympathetic postganglionic neurons in humans are located not only in the detrusor wall layer but also in the pelvic plexus (Yoshimura et al., 2010, chap. 23).

Sympathetic pathways

Sympathetic outflow from the rostral lumbar spinal cord provides a noradrenergic excitatory and inhibitory input to the bladder and urethra (de Groat et al., 2015). The peripheral sympathetic pathways follow a complex route that passes through the sympathetic chain ganglia (SCG) to the inferior mesenteric ganglia (IMG) and then through the hypogastric nerves to the pelvic ganglia (Kihara & de Groat, 1997). The activation of the sympathetic nerves induces relaxation of the bladder body and contraction of the bladder outlet and urethra, which contribute to the filling phase in the bladder. Urine storage is facilitated by bladder wall relaxation and accommodation promoted by action of noradrenaline at β-adrenergic receptors in bladder (Heymann, 2006) and activation of α-adrenergic receptors on the internal urethral sphincter smooth muscles, resulting in contraction of the bladder outlet.

Somatic pathways

The EUS motoneurons are located along the lateral border of the ventral horn of the sacral (S2−S4) spinal cord, in an area commonly referred to as Onuf's nucleus (de Groat et al., 2015). Propagation of EUS motoneuron activity through the pudendal nerve to the EUS elicits skeletal muscle contraction by activating nicotinic ACh receptors to provide both voluntary and reflex control over urinary continence.

In adults, the inhibition of voiding reflexes can be induced by activation of pudendal afferent inputs from visceral organs including the penis, vagina, rectum, perineum, urethral sphincter, and anal sphincter (Yoshimura et al., 2010, chap. 23). This is a part of the mechanisms of neuromodulation therapies, in which electrical stimulation of somatic afferent pathways projecting in the pudendal nerve to the caudal lumbosacral spinal cord can improve overactive bladder and urinary incontinence (Amend et al., 2011).

Afferent pathways

An intact afferent system is crucial to physiological bladder filling and voiding. During the filling phase, mechanoreceptors in bladder wall initiate visceral afferent activity that is carried by Aδ-fiber afferent axons in the pelvic nerve to interneurons in lumbosacral spinal cord. The pelvic nerve afferents consist of myelinated Aδ-fiber and unmyelinated C-fiber axons to monitor the volume of the bladder and the amplitude of the bladder contraction. The primary afferent neurons of the pelvic and pudendal nerves are contained in sacral dorsal root ganglia (DRG), whereas afferent innervation in the hypogastric nerves arises in the rostral lumbar DRG (de Groat et al., 2015).

Because capsaicin, the C-fiber afferent neurotoxin, does not block normal micturition reflexes in cats and rats, it is believed that C-fiber afferents are not essential for normal voiding (Cheng et al., 1993, 1999; de Groat et al., 1990; Maggi & Conte, 1990). However, during neuropathic conditions such as spinal cord lesions, there is recruitment of C-fibers as a new functional afferent pathway that can induce the spinal micturition reflex independent of the supraspinal control, resulting in detrusor overactivity (DO) and urinary incontinence (Cheng et al., 1995; de Groat & Yoshimura, 2006, 2009, 2010, 2012; Kanai, 2011).

Afferent fibers arising from the urethra are also important for modulating lower urinary tract function. Conduction velocities of cat pudendal nerve afferent fibers responding to electrical stimulation of the urethra are approximately twice as fast (45 m/s vs. 20 m/s) as pelvic nerve afferent fibers responding to the same stimulation (Bradley et al., 1973), suggesting the existence of an Aβ-fiber afferent population. Also, urethral afferents in the pudendal, pelvic, and hypogastric nerves serve as a flow sensor to facilitate the voiding. In the cat, pudendal nerve afferents responding to urine flow exhibit a slowly adapting firing pattern (Todd, 1964), while small myelinated or unmyelinated urethral afferents in the hypogastric nerves and myelinated urethral afferents in the pelvic nerves responding to urine flow or urethral distention exhibited rapidly adapting responses (Bahns et al., 1986, 1987). Stimulation of flow-sensing urethral afferents by intraurethral saline infusion enhances volume-induced reflex bladder contractions in rats (Jung et al., 1999) through the urethra-to-bladder facilitatory reflex.

Intraspinal pathways

Parasympathetic preganglionic neurons (PGNs) are located in the intermediolateral gray matter (laminae V—VII) in the sacral segments of the spinal cord (de Groat & Ryall, 1968; Nadelhaft & Booth, 1984; Nadelhaft et al., 1980, 1983), whereas sympathetic PGNs are located in medial (lamina X) and lateral sites (laminae V—VII) in the rostral lumbar spinal cord. EUS motoneurons are located in lamina IX in Onuf's nucleus in the cat and in the dorsolateral motor nucleus in the rat (de Groat et al., 2015; McKenna & Nadelhaft, 1986; Thor & de Groat, 2010).

The central axons of spinal afferent pathways carry the sensory information from the lower urinary tract through the dorsal roots to second-order neurons in the spinal cord (de Groat, 1986). Sacral afferent fibers of the pelvic and pudendal nerves enter the cord and travel rostrocaudally within Lissauer's tract at the apex of the dorsal horn and then give off collaterals that project around the dorsal horn to terminate in deeper laminae of the cord (de Groat et al., 2015).

Then, spinal interneurons that receive afferent inputs serve as relays to transmit information to efferent neurons in the cord or to supraspinal circuitry and thus these neurons can function as modulatory centers for the control of lower urinary tract function. Previous studies using retrograde transneuronal labeling after injection of pseudorabies virus (PRV) into the bladder, urethra, or EUS of the rat showed PRV-labeled spinal neurons located in the same general regions of the spinal cord that receive afferent input from the bladder including the dorsal commissure, laminae I and V, and lamina VII just dorsal and medial to the PGN (Marson, 1997; Nadelhaft et al., 1992; Nadelhaft and Vera, 1995, 1996, 2001; Sugaya et al., 1997; Vizzard et al., 1995). Some of these interneurons make excitatory and inhibitory synaptic connections with PGN (Araki & de Groat, 1996, 1997; de Groat et al., 1998; Miura et al., 2003) and participate in segmental spinal reflexes (de Groat et al., 1998), whereas others send long projections to supraspinal centers, such as the PAG (Fig. 3.3), PMC (aka Barrington's nucleus), hypothalamus, and thalamus that are involved in the supraspinal control of micturition (Birder et al., 1999; Blok et al., 1995; Blok & Holstege, 2000; Ding et al., 1997; Duong et al., 1999; Holstege & Mouton, 2003; McMahon & Morrison, 1982).

Organization of storage reflexes

Spinal reflex pathways

Bladder pressure remains low during the storage or bladder filling phase until the bladder volume exceeds the threshold for inducing voiding. Under normal conditions, the rise in intravesical pressure is linear with an increase in bladder volume until the volume reaches a high level (Sugaya et al., 1998). The accommodation of the bladder to increasing volumes of urine is primarily a passive phenomenon dependent on the viscoelastic characteristics of bladder wall and the quiescence of the parasympathetic efferent pathway (de Groat et al., 2015; Yoshimura et al., 2008).

During filling, bladder distension–induced afferent activity triggers sympathetic and somatic spinal reflexes at the spinal cord level, which contribute to the low intravesical pressure and high urethral pressure maintained by the tonic activity of smooth and striated muscle urethral sphincters, respectively (Griffiths & Fowler, 2013; Nishijima et al., 2005). Reflex activation of the sympathetic outflow inhibits neurally mediated contractions of the bladder during filling phase of micturition through β-adrenoceptor–mediated bladder smooth muscle relaxation and induces urethral smooth muscle contraction through α1-adrenoceptor activation (Fowler et al., 2008; Griffiths & Fowler, 2013; Kimura et al., 1995). The increased efferent firing in the pudendal nerve and the higher outlet resistance during filling termed the "guarding reflex" is responsible for the maintenance of urinary continence (Fig. 3.4A). Organization of these reflex responses in the lumbosacral spinal cord is confirmed by their persistence even after transection of the spinal cord at the thoracic levels (Fig. 3.4A) (Yoshimura et al., 2010, chap. 23).

Contraction of the EUS also induces firing in pudendal nerve afferent axons, which in turn activate inhibitory interneurons in the spinal cord that suppress reflex bladder activity (de Groat et al., 2001; McGuire et al., 1983) by inhibiting PGN and interneurons that facilitate the voiding reflex (de Groat et al., 1978, 1982). Thus, the bladder-to-EUS-to-bladder reflex pathway represents a second negative feedback mechanism in the spinal cord that promotes urinary continence. In addition, the EUS-to-EUS reflex through activation of afferents in the pudendal nerve also elicits reflex contractions of the EUS and contributes to continence (Thor & de Groat, 2010).

Taken together, the urine storage reflexes that contribute to the maintenance of urinary continence are organized mainly at the spinal cord level (Fig. 3.4A).

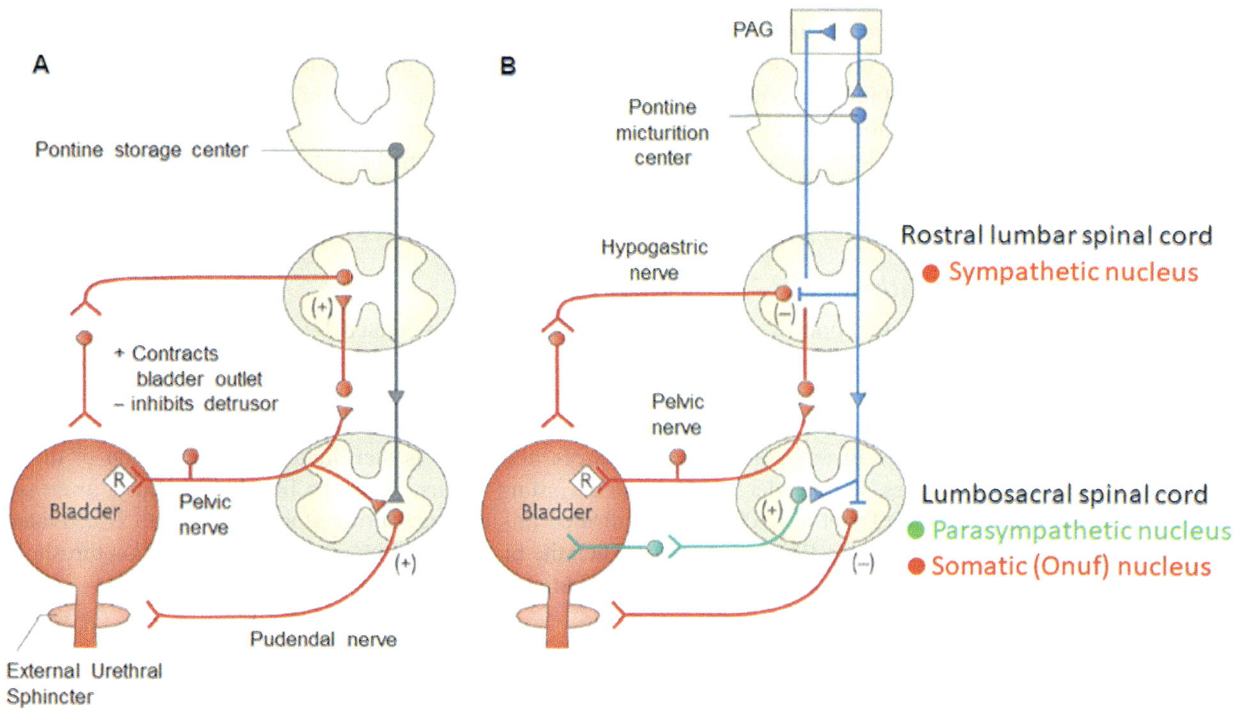

FIGURE 3.4 Neural circuits that control continence and micturition. A. Urine storage reflexes. During the storage of urine, distention of the bladder produces low-level vesical afferent firing. This in turn stimulates the sympathetic outflow in the hypogastric nerve to the bladder outlet (the bladder base and the urethra) and the pudendal outflow to the external urethral sphincter. These responses occur by spinal reflex pathways and represent guarding reflexes, which promote continence. Sympathetic firing also inhibits contraction of the detrusor muscle and modulates neurotransmission in bladder ganglia. A region in the rostral pons (the pontine storage center) might increase striated urethral sphincter activity. B. Voiding reflexes. During the elimination of urine, intense bladder afferent firing in the pelvic nerve activates spinobulbospinal reflex pathways (shown in blue) that pass through the Pontine micturition center. This stimulates the parasympathetic outflow to the bladder and to the urethral smooth muscle (shown in green) and inhibits the sympathetic and somatic outflow to the urethral outlet (shown in red). Note that the inhibition of sympathetic and somatic pathways during voiding reflexes could involve activation of interneurons such as GABA/glycinergic cells in the spinal cord. Ascending afferent input from the spinal cord might pass through relay neurons in the periaqueductal gray (PAG) before reaching the Pontine micturition center. Note that these diagrams do not address the generation of conscious bladder sensations, nor the mechanisms that underlie the switch from storage to voluntary voiding, both of which presumably involve cerebral circuits above the PAG. R represents receptors on afferent nerve terminals. *From de Groat, W.C., Yoshimura, N. (2015). Anatomy and physiology of the lower urinary tract. Handbook of Clinical Neurology. 130, 61–108.*

Supraspinal pathways

Some brainstem areas are reportedly involved in urine storage function. It has been demonstrated that electrical stimulation of the pontine urine storage center (PUSC), or L-region, located ventrolateral to the PMC not only excites the EUS but also inhibits reflex bladder activity, increases bladder capacity, and the inhibits the bladder excitatory effect of PMC stimulation (Holstege et al., 1986; Koyama et al., 1966; Kuru, 1965; Nishizawa et al., 1987; Sugaya et al., 2005) (Fig. 3.4A). Neurons in the region of the PUSC project to the nucleus raphe magnus (NRM) in the medulla, which contains neurons that in turn project to the lumbosacral spinal cord. Electrical (de Groat, 2002a; McMahon & Spillane, 1982; Morrison & Spillane, 1986; Sugaya et al., 1998) or chemical stimulation (Chen et al., 1993) in the NRM induces serotoninergic inhibition of reflex bladder activity (Fig. 3.3). Thus, neurons in the PUSC may activate descending inhibitory pathways to the sacral parasympathetic nucleus located at the intermediolateral cell column of the S2–S4 spinal cord (Sugaya et al., 2005).

In addition, electrical stimulation of the rostral pontine reticular formation (RPRF) ventral to the PMC in an area also known as the nucleus reticularis pontis oralis inhibits reflex bladder contractions in cats and rats (Kimura et al., 1995; Nishijima et al., 2005; Sugaya et al., 1987, 2005). Neurons in this region project to the spinal cord and also to nucleus reticularis gigantocellularis located in the rostral dorsal medulla. The RPRF projects to lumbosacral glycinergic inhibitory neurons that may mediate the inhibitory effects of RPRF stimulation (Sugaya et al., 2005).

However, it is not known how these urine storage areas in the brainstem are controlled via higher brain centers to achieve voluntary inhibition of the micturition (voiding) reflex.

Organization of voiding reflexes

The storage phase of the urinary bladder can be switched to the voiding phase either involuntarily or voluntarily (Fig. 3.4B). The former is readily demonstrated in the human infant when the volume of urine exceeds the micturition threshold. At this point, increased afferent firing from tension receptors in the bladder produces firing in the sacral parasympathetic pathways and inhibition of sympathetic and somatic pathways. The expulsion phase consists of an initial relaxation of the urethral sphincter followed by a contraction of the bladder, an increase in bladder pressure, and flow of urine. Relaxation of the urethral outlet is mediated by activation of a parasympathetic reflex pathway to the urethra that triggers the release of nitric oxide that induces urethral smooth muscle relaxation, as well as by removal of adrenergic and somatic excitatory inputs to the smooth and striated urethral sphincters, respectively.

Spinobulbospinal reflex pathways

The switch of lower urinary tract function between storage and voiding is mediated by a long-loop spino-bulbo-spinal voiding reflex, which is integrated through brainstem structures such as the PMC, also known as Barrington's nucleus, and the PAG (Figs. 3.3 and 3.4B) (Fowler et al., 2008; Griffiths & Fowler, 2013). During urine storage, as the bladder fills, sacral afferent signals from the bladder increase in strength until they exceed a certain threshold in the brainstem, specifically in the PAG. In the absence of any controlling influences, the reflex is then induced by activation of the PMC. Then, the urethral sphincter relaxes, the bladder contracts, and voiding occurs. After the bladder is empty, urine storage resumes. Brain imaging in the rat (Tai et al., 2009) has confirmed this picture: during storage, the PAG is activated (by afferent input from the bladder), while the PMC is inactive. When the bladder volume reaches the micturition threshold, the switch from storage to micturition is associated with PMC activation and enhanced PAG activity.

Brainstem circuitry

Role of PMC

Studies in cats using brain-lesioning and electrophysiological techniques revealed that reflex micturition is mediated by a spinobulbospinal pathway consisting of an ascending sensory limb that passes from the sacral spinal cord to circuitry in the rostral brainstem leading to activation of neurons in the PMC that send excitatory signals back to the sacral spinal cord to complete the reflex circuit (Fig. 3.4B).

In recordings of electrical activity in bladder efferent nerves, stimulation of bladder afferent nerves induces long latency discharges (120–150 ms) on bladder postganglionic nerves. Also, bladder afferent nerve stimulation evokes neuronal firing in the PMC at latencies ranging from 30 to 40 ms, and electrical stimulation in the PMC evokes bladder contractions and postganglionic nerve firing at latencies of 60–75 ms (de Groat, 1975; Noto et al., 1991). The sum of the latencies of the putative ascending (afferent-ponto-mesencephalic) and descending limbs (pontine-sacral efferent neuron) of the reflex approximates the latency of the entire reflex pathway (120 msec). The reflex firing elicited in cats and rats is not altered following supracollicular decerebration, but is eliminated by acute transection of neuraxis at any level caudal to the PMC (de Groat, 1975; de Groat & Ryall, 1969). These results indicate that the micturition reflex is mediated by a pathway passing through a switching center in the rostral pons including the PMC.

Previous histological and electrophysiological studies showed that PMC neurons project directly to bladder PGNs in the sacral spinal cord (Blok & Holstege, 1997) or polysynaptically via interneurons (Sasaki & Sato, 2013) in cats. The latter polysynaptic interneuronal pathway seems to be an important site where afferent feedback from the bladder during voiding can facilitate voiding bladder contractions (Kruse et al., 1991, 1992).

Also, there is the evidence that the PMC coordinates the functions of the colon and the lower urinary tract because the majority of bladder-responsive neurons (73%) in the PMC are also activated by colon distension (Rouzade-Dominguez et al., 2003a), and PRV tracing experiments showed that colon- and bladder-co-labeled neurons are present in the PMC (Rouzade-Dominguez et al., 2003b; Valentino et al., 2000).

Properties of neurons in the PMC

Single-unit recordings in the PMC of the cat (Fig. 3.5) (Bradley and Conway, 1966; de Groat et al., 1998; Koshino, 1970; Sasaki, 2002, 2004, 2005a, 2005b; Sugaya et al., 2003, 2005; Tanaka et al., 2003) and rat (Elam et al., 1986; Willette et al., 1988) with the bladder distended under isovolumetric conditions revealed several populations of neurons exhibiting firing correlated with reflex bladder contractions including (1) neurons that are silent in the absence of bladder activity but fire prior to and during reflex bladder contractions (direct neurons, 21%), (2) neurons that are active during the period between bladder contractions and are inhibited during contractions (inverse neurons, 51%), and (3) neurons that fire transiently at the beginning of bladder contractions (on-off neurons, 4%) (Fig. 3.5). Tonic firing that was not correlated with bladder activity was also identified in a large percentage (25%) of PMC neurons (termed independent neurons) (Fig. 3.5).

Also, a large portion of direct neurons project to the lumbosacral spinal cord (Sasaki, 2002, 2005b; Sugaya et al., 2003), whereas only a small percentage of inverse neurons send projections to the cord. Thus, it has been speculated that inverse neurons are involved in local inhibition inside the PMC and that both direct and inverse neurons exhibit excitatory synaptic responses to electrical stimulation of afferent axons in the pelvic nerve (de Groat et al., 1998).

More recently, subpopulation-specific functional properties have been characterized using transgenic mice. It has been demonstrated that a subpopulation of PMC neurons projecting the lumbosacral spinal cord expresses a neuropeptide transmitter; corticotropin-releasing hormone (CRH) in rodents (Vincent & Satoh, 1984) and that CRH-immunolabeled

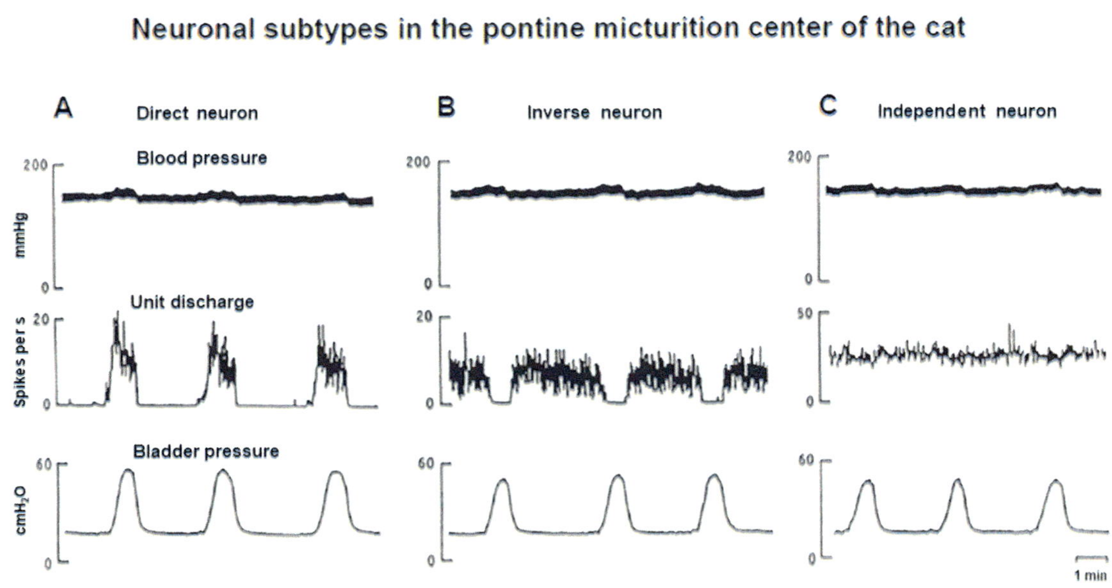

FIGURE 3.5 Relationship between single-unit activity in the PMC of a decerebrate, unanesthetized cat and reflex contractions of the urinary bladder. Top tracings are blood pressure, middle tracings are ratemeter recordings of unit activity in spikes per second, and the bottom tracings are bladder pressure in cm H_2O. Three types of neuronal activity are illustrated: (A) a direct neuron that only fired during a bladder contraction, (B) an inverse neuron that fired between bladder contractions and was inhibited during contractions, and (C) an independent neuron that exhibited continuous firing unrelated to bladder contractions. Small increases in blood pressure occurred during bladder contractions. The bladder was distended with saline and maintained under isovolumetric conditions. Horizontal calibration represents 1 min. The three neurons were studied at different times in the same animal. *From de Groat W.C., & Yoshimura, N. (2015). Anatomy and physiology of the lower urinary tract. Handbook of Clinical Neurology. 130, 61–108.*

neurons are also detected in the laterodorsal tegmental complex that contains the PMC in the human brain (Ruggiero et al., 1999). In rodents, these CRH-positive PMC cells comprise approximately half of all PMC neurons (Valentino et al., 1995; Verstegen et al., 2017). Also, the majority (94%) of mouse CRH-positive PMC neurons are reportedly glutamatergic excitatory cells (Hou et al., 2016) (Fig. 3.6). Recent studies using optogenetic and chemogenetic approaches have revealed that activation of CRH-positive PMC neurons can induce bladder contractions, but not efficient urine elimination from the bladder (Hou et al., 2016; Keller et al., 2018; Verstegen et al., 2019), and that genetic abatement of CRH-positive PMC neurons does not significantly alter normal, continent voiding behavior (Verstegen et al., 2019). However, the action of CRH-positive PMC neurons could be bladder volume dependent because voiding was induced after optogenetic activation of CRH-positive PMC neurons when the bladder is distended to more than half of its capacity (Ito et al., 2020). Thus, it is conceivable that CRH-positive, glutamatergic PMC neurons can augment bladder contractions, especially when the bladder is full, although they may play a minor role in reflex or voluntary micturition that depends on coordination between the bladder and the urethra.

In contrast, CRH-negative, glutamatergic PMC neurons seem to be essential for both reflex and voluntary micturition because elimination of all glutamatergic PMC neurons causes urinary retention, while only abating the CRH-positives subpopulation of glutamatergic PMC neurons does not (Verstegen et al., 2019). Furthermore, a proportion of CRH-negative, glutamatergic PMC neurons has estrogen receptor (ESR1) expression (Keller et al., 2018). Activation of these ESR1-positive glutamatergic PMC neurons, whose nerve terminals are primarily found in the interneuron-rich central region of the lumbosacral spinal cord, can elicit bursting activity of the EUS with intermittent EUS relaxation and efficient voiding (Keller et al., 2018). It is also shown that this ESR1-positive glutamatergic population is required for voluntary micturition induced by female odor stimulation in male mice (Keller et al., 2018). Thus, it is likely that ESR1-positive, CRH-negative glutamatergic neurons play an indispensable role in voluntary voiding with concomitant bladder contraction and urethral relaxation during urination events. In normal bladder emptying in humans, the start of sphincter relaxation usually precedes by a few seconds the bladder contraction (Fig. 3.2); therefore, ESR1-positive PMC neurons might particularly be important in the initiation of voiding, in which the EUS relaxation is the initial event.

Furthermore, it should be noted that EUS bursting activity during voiding is found in rats, mice (male > female) (Kadekawa et al., 2016; Keller et al., 2018), and dogs (Nishizawa et al., 1985), presumably related to territory marking behavior, but not in other species including humans and cats, which show the total cessation of EUS activity during voiding (Fig. 3.2). Previous studies using rats have reported that there is the lumbar spinal coordinating center (LSCC) consisting of interneurons in lamina X of the L3/L4 spinal cord that functions as a second component of spinal cord circuitry that contributes to EUS bursting and bladder–EUS coordination during voiding (Chang et al., 2007; Karnup & de Groat, 2020). Thus, it is possible to speculate that ESR1-positive, glutamatergic PMC neurons interact with LSCC interneurons to facilitate EUS bursting–mediated efficient voiding in rodents, whereas descending axons from ESR1-positive PMC neurons, which are also found in primates (Heymann, 2006), could terminate directly in the sacral spinal cord to activate GABAergic and/or glycinergic interneurons to directly suppress EUS motoneurons during voiding in other

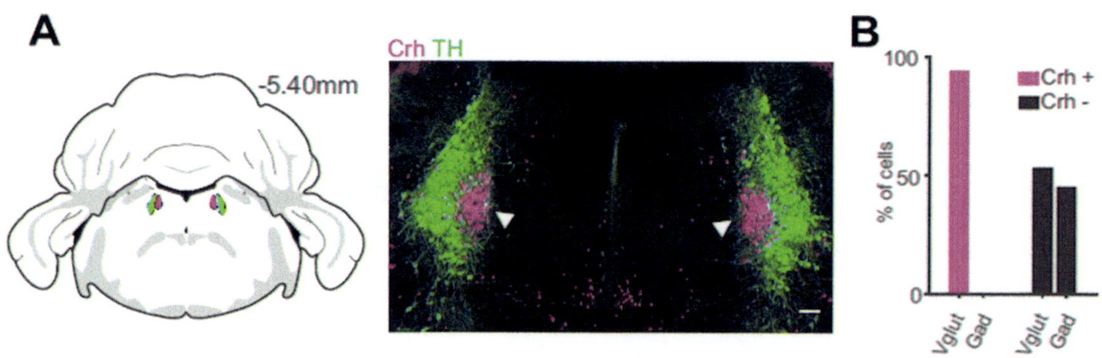

FIGURE 3.6 CRH-positive and negative neurons in the Pontine micturition center (PMC). (A) Schematic (left) and fluorescently labeled coronal section (right) of the pons illustrating the location of the PMC next to the locus coeruleus (LC). The PMC (arrowheads) is identified by tdTomato fluorescence (magenta) from a $CRH^{ires-Cre}$::$Rosa26^{lsl-tdTomato}$ mouse, whereas immunostaining against TH (green) highlights the LC. Scale bar: 100 μm. (B) Percentages of CRH-positive (magenta) and CRH-negative neurons (black) co-labeled by vesicular glutamate transporter (Vglut) or glutamic acid decarboxylase (GAD), which is a marker of glutamatergic or GABAergic neurons, respectively. This bar graph shows that there are three major subpopulations of PMC neurons, namely CRH-positive glutamatergic cells, CRH-negative glutamatergic cells, and CRH-negative GABAergic cells. *From Hou, X.H., Hyun, M., Taranda, J., Huang, K.W., Todd, E., Feng, D., Atwater, E., Croney, D., Zeidel, M.L., Osten, P., Sabatini, B.L. (2016). Central control circuit for context-dependent micturition. Cell. 167, 73–86.*

species such as cats (Blok, 2002) and possibly in humans. Future studies are warranted to further clarify the role of subpopulations of PMC neurons in micturition control.

In addition, there is another population of CRH-negative PMC neurons that are inhibitory GABAergic cells, evident as the expression of glutamic acid decarboxylases (GADs), the synthesizing enzyme of GABA (Hou et al., 2016) (Fig. 3.6). Although the functional role of these neurons or their projections to the spinal cord are not fully clarified, it can be speculated that inverse PMC neurons found in the cat, which are active during the period between bladder contractions, but inhibited during contractions (Fig. 3.5) (Sasaki, 2005a, 2005b; Verstegen et al., 2019), might be GABAergic to elicit local inhibition within the PMC to suppress PMC activity during the storage phase. Further studies are needed to clarify these points.

Role of the PAG

Early studies in cats (Gjone, 1966; Kabat et al., 1936; Koyama et al., 1962; Langworthy & Kolb, 1935; Skultety, 1959) revealed that stimulation at sites in the PAG could either excite or inhibit bladder activity, while the effects of stimulation were dependent on the degree of fullness in the bladder. Also, reflex bladder activity was enhanced by elimination of parts of the PAG by focal lesions or serial transections through the mesencephalon (Langworthy & Kolb, 1933; Ruch & Tang, 1956; Tang, 1955), suggesting that a mesencephalic bladder inhibitory center in the PAG tonically controls micturition.

An inhibitory region seems to be located in the dorsolateral margin of the rostral PAG (Numata et al., 2008) because chemical or electrical stimulation at this site inhibits reflex bladder contractions and the contractions induced by PMC stimulation. Injection of bicuculline, a $GABA_A$ receptor antagonist, into the PMC blocks the PAG-induced inhibition of PMC and provoked partial bladder contractions, indicating that GABA is the transmitter in the inhibitory pathway (Numata et al., 2008). Other sites in the PAG, especially the ventrolateral region of the PAG, seem to have a facilitatory role in micturition because electrical stimulation in the ventrolateral PAG evokes bladder contractions (Matsuura et al., 2000; Noto et al., 1989; Taniguchi et al., 2002) and firing on bladder postganglionic nerves (Noto et al., 1991). However, there is also the possibility that ventrolateral PAG neurons, some of which express vesicular GABA transporter and send afferent inputs to the PMC, may play an inhibitory role in micturition (Verstegen et al., 2019). Furthermore, it has been demonstrated that glutamatergic axons from the PAG densely innervate PMC neurons and that optogenetic stimulation of these PAG axon terminals in the PMC provokes voiding in mice (Verstegen et al., 2019).

Electrical recordings in the PAG indicate that it may serve as a relay and coordinating center on the ascending limb of the micturition reflex pathway. In the rat and cat, electrical stimulation of pelvic nerve afferents elicits field potentials in the PAG at the significantly shorter latency compared to the latency of field potentials in the PMC region (de Groat, 1975; Duong et al., 1999; Noto et al., 1989). Axonal tracing studies in the cat also showed more prominent direct axonal inputs from sacral spinal tract neurons to the PAG compared to those terminating in the PMC (Blok et al., 1995; Holstege and Mouton, 2003). Axonal tracing methods also identified projections from the PAG to the PMC (Blok & Holstege, 1994; Kuipers et al., 2006), indicating that ascending afferent information from the bladder is relayed through synapses in the PAG to the PMC. Thus, it has been proposed that the PAG has an essential role in the spinobulbospinal micturition reflex pathway (Holstege & Mouton, 2003; Noto et al., 1989; Tish & Geerling, 2020).

Brain imaging studies (Tai et al., 2009) in the rat also revealed that neuronal activity in the PAG increases during slow bladder filing, indicating that afferent activity from the bladder is received and processed in the PAG prior to micturition. On the other hand, during micturition, signals were detected in the PAG and the PMC. These results suggest that the PAG in the rat serves as a relay station for transmitting afferent information from the bladder to the PMC; however, the switching from urine storage to voiding requires activation of the PMC (de Groat & Wickens, 2013).

Pharmacological studies indicate that neuronal circuitry in the PMC and PAG allows the spinobulbospinal micturition reflex pathway to function as a switch that is either in a completely "off" mode (storage) or maximally "on" mode (voiding) (de Groat & Wickens, 2013). Injections of excitatory amino acids into the PMC (Mallory et al., 1991) or PAG (Taniguchi et al., 2002) evoke bladder contractions in cat and rat. On the other hand, microinjections of $GABA_A$ receptor agonists (muscimol) at the PMC or PAG increase the bladder volume threshold for inducing micturition without altering the voiding bladder contraction amplitude (Mallory et al., 1991; Matsumoto et al., 2004; Noto et al., 1991; Stone et al., 2011). However, injections of $GABA_A$ receptor (bicuculline) reduce the bladder volume threshold, indicating that tonic activation of inhibitory receptors in these centers can alter the set point of the micturition switch (Mallory et al., 1991; Noto et al., 1991; Stone et al., 2011). Because pharmacologic modulation of the PAG circuitry clearly alters the bladder volume threshold, it seems reasonable to conclude that PAG input to the PMC switching circuit also regulates the set point for the micturition switch. In addition, in vivo microdialysis studies using rats revealed that extracellular levels of glutamate and GABA in the PAG were significantly increased and decreased, respectively, during the micturition reflex with further enhancement of glutamate levels after acetic acid–induced bladder irritation, suggesting that glutamate and GABA are involved in the switching mechanisms in the PAG (Kitta et al., 2008, 2016a).

Spinal micturition pathways

Developmental changes in micturition pathways

The mechanisms involved in storage and periodic elimination of urine undergo marked changes during prenatal and postnatal development (Capek and Jelinek, 1956; de Groat et al., 1975, 1993a, 1998) (Fig. 3.1). In the neonatal stage, micturition is regulated by primitive reflex pathways organized in the spinal cord. As the CNS matures during the postnatal period, reflex voiding is eventually brought under voluntary control, which originates in the higher centers of the brain (Fig. 3.2).

In many species such as rats and cats, voiding in neonates is dependent on an exteroceptive somato-bladder reflex mechanism, which is organized in the sacral spinal cord and triggered when the mother licks the genital or perineal region of the young animal (de Groat et al., 1975; Maggi et al., 1986; Thor et al., 1986, 1989, 1990). Similar reflexes have been identified in human infants (Boehm & Haynes, 1966). The spinobulbospinal reflex through the PMC in the brainstem only emerges several weeks after birth in animals (de Groat et al., 1975; Thor et al., 1989) even though the brain is already wired to the spinal micturition pathways based on the PRV-based neurotracing (Sugaya et al., 1997) and neurophysiological studies (de Groat, 2002b; Sugaya & de Groat, 1994).

The mechanisms underlying the developmental suppression of the perineal-to-bladder reflex and its reemergence after spinal cord injury (SCI) were studied using patch clamp recordings in identified parasympathetic PGNs in spinal cord slice preparations from 6- to 28-day-old neonatal rats (Araki & de Groat, 1997). In slice preparations from 1- and 2-week-old rats, the dorsal interneuron–evoked excitatory postsynaptic currents (EPSCs) are large and of constant amplitude. However, in preparations from 3-week-old rats, an age when the spinobulbospinal micturition has emerged, the EPSCs are reduced in amplitude by 50%. Quantal analysis of unitary, glutamatergic EPSCs indicated that this reduction in amplitude was attributable to a decrease in the presynaptic release of glutamic acid and was not due to a change in the density or sensitivity of postsynaptic glutamatergic receptors (Araki and de Groat, 1997). Transection of the spinal cord between 1 and 2 weeks of age prevents the reduction in synaptic transmission that occurs at 3 weeks of age. Thus, synaptic remodeling in the sacral parasympathetic nucleus is likely to be an important process in the postnatal maturation of voiding reflexes.

The primitive reflex pathways organized in the spinal cord are also evident in human infants, as shown by the intravesical ice water test (Geirsson et al., 1999), which is known to induced C-fiber afferent-dependent bladder-cooling reflex (BCR) (Fall et al., 1990). BCR shown by reflex bladder contractions after instillation of ice water into the bladder is detected in most of normal infants and children younger than 4 years old, but becomes negative in those older than 6 years old (Geirsson et al., 1994, 1999). However, high positive rates of the BCR test were obtained even after the age of 6 in children, who had neurogenic lower urinary tract dysfunction due to spina cord lesions such as spina bifida with myelomeningocele (Gladh & Lindström, 1999). These data indicate that the C-fiber–mediated micturition reflex organized in the spinal cord, which contributes to involuntary voiding, disappears during postneonatal development at 4–6 years of age, but persists when the neuronal maturation in the spinal cord is impaired.

Reorganization of spinal micturition reflexes due to spinal cord legions

Although reflex voiding is brought under voluntary control which originates in the higher centers of the brain during postnatal development, sacral micturition reflexes reemerge in adults after neuronal damage in the spinal cord that induces the disruption of the spinobulbospinal micturition pathways and coordinated control via brain (Fig. 3.7).

Spinal cord lesions induced by traumatic SCI initially induce areflexic bladder and urinary retention, followed by the emergence of automatic micturition and eventually DO mediated by spinal reflex mechanisms. SCI also causes the emergence of a reflex similar to the neonatal exteroceptive micturition reflex that is activated by the mother licking the perineal region for the expelling of urine (de Groat, 2002b; Sugaya & de Groat, 1994). Although maximal voiding pressure is increased, voiding efficiency is reduced and bladder undergoes marked hypertrophy (Cheng & de Groat, 2004; Kruse et al., 1993, 1994). Bladder–sphincter coordination through the PMC is impaired, leading to detrusor sphincter dyssynergia (DSD) (de Groat et al., 1990, de Groat & Yoshimura, 2012; Yoshimura, 1999) (Fig. 3.7). Thus, the organization of the micturition reflex shows marked changes after SCI (Fig. 3.7). Any injury to the spinal cord such as blunt, degenerative, developmental, vascular, infectious, traumatic, and idiopathic injury can cause voiding dysfunction. In cats with chronic thoracic spinal cord transection, micturition is induced by C-fiber afferent pathways (Fig. 3.7C). It has been demonstrated that in chronic spinalized cats, desensitization of TRPV1-expressing C-fiber afferent pathways by

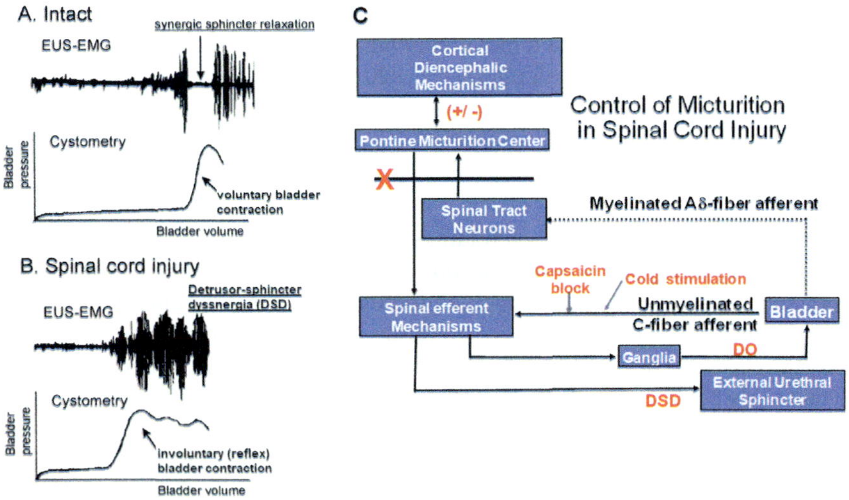

FIGURE 3.7 Combined cystometry and external urethral sphincter electromyography (EUS-EMG) recordings comparing voiding responses in a healthy adult (A, intact) and in a patient with spinal cord injury (B, SCI). The abscissa in all records represent bladder volume and the ordinates in cystometrograms represent bladder pressure. In panel A, a slow infusion of fluid into the bladder induces a gradual increase of EMG activity, but no apparent changes in bladder pressure. When a voluntary voiding starts, an increase of bladder pressure (voluntary bladder contraction) is associated with a cessation of EUS-EMG activity (synergic sphincter relaxation). On the other hand, in an SCI patient (B), the reciprocal relationship between bladder and sphincter controlled by the PMC is abolished. During bladder filling, involuntary (reflex) bladder contraction occurs in association with an increase in sphincter activity (detrusor sphincter dyssynergia; DSD). Loss of the reciprocal relationship between bladder and sphincter in SCI patients interferes with bladder emptying. (C) Schema showing organization of the parasympathetic excitatory reflex pathway to the detrusor muscle and the external urethral sphincter (EUS). Micturition is initiated by a supraspinal reflex pathway that passes through a center in the brainstem. The pathway is triggered by myelinated afferents (Aδ-fibers), which are connected to the tension receptors in the bladder wall. Injury to the spinal cord above the sacral segments (X) interrupts the connections between the brain and spinal autonomic centers and initially blocks micturition. However, following spinal cord injury, the Aδ-fiber afferent inputs are ineffective, thereby leading to detrusor areflexia in the acute phase; however, a spinal reflex mechanism later emerges that is triggered by unmyelinated vesical afferents (C-fibers). The C-fiber reflex pathway is usually weak or undetectable in animals with an intact nervous system. Stimulation of the C-fiber bladder afferents by instillation of ice water into the bladder (cold stimulation) activates voiding responses in patients with spinal cord injury, while ice water test is negative in healthy adults. Capsaicin-induced C-fiber desensitization reduces the C-fiber–mediated spinal micturition reflex in animals with spinal lesions. Also, the disruption of descending inhibition of sacral motor neurons innervating the EUS from the PMC results in DSD during voiding.

subcutaneously administered capsaicin, a C-fiber neurotoxin that binds to TRPV1 receptors, blocked voiding contractions and DO evident as nonvoiding bladder contractions during the storage phase, whereas capsaicin had no inhibitory effects on reflex bladder contractions in spinal intact cats (Cheng et al., 1999; de Groat et al., 1990). Thus, it is plausible that C-fiber bladder afferents that usually do not respond to bladder distension (de Groat & Yoshimura, 2009; Habler et al., 1990) become mechano-sensitive and initiate automatic micturition after SCI.

In a rat model of SCI induced by complete transection of the thoracic spinal cord, increased excitability of C-fiber bladder afferents after SCI also induces DO as evidenced by nonvoiding bladder contractions prior to micturition because desensitizing C-fiber afferents by systemic capsaicin administration completely suppressed these nonvoiding bladder contractions without affecting the voiding reflex (Cheng & de Groat, 2004; Cheng et al., 1995). Desensitization of C-fiber afferent pathways by capsaicin pretreatment also reduces DSD in chronic SCI rats (Seki et al., 2004). Furthermore, it has also been shown in the cat that C-fiber bladder afferents are responsible for cold-induced bladder reflexes via TRPM8 receptors (Birder, 2010; Fall et al., 1990). Chronic SCI in humans also causes the emergence of an unusual bladder reflex that is elicited by infusion of cold water into the bladder, which is blocked by intravesical capsaicin treatment (Geirsson et al., 1993, 1994, 1995). Thus, it is likely that cold- and capsaicin-sensitive C-fiber bladder afferents contribute to the reemergence of a spinally mediated involuntary micturition reflex to induce lower urinary tract dysfunction after SCI (Fig. 3.7C).

Neural circuits involved in voluntary micturition control

Lesioning and electrical stimulation studies indicate that voluntary control of micturition depends on connections between the frontal cortex and other forebrain structures, including the anterior cingulate cortex (ACC), insula, amygdala, bed nucleus of the stria terminalis (BnST), and septal nuclei, where electrical stimulation elicits excitatory bladder responses

(Andersson & Pehrson, 2003; de Groat et al., 1993a). Damage to the cerebral cortex, due to tumors, cerebrovascular disease, or neurodegenerative disease, appears to remove inhibitory control of the spinobulbospinal micturition pathways, resulting in lower urinary tract dysfunction (Yoshimura et al., 2010, chap. 23).

Cortical modulation of micturition

Human imaging studies

Evidence from human imaging studies has supported findings from preclinical studies and demonstrated the active control of micturition by the brain during filling and voiding (Nardos et al., 2014). Functional neuroimaging has been performed in healthy volunteers using single-photon emission computed tomography, positron emission tomography, functional magnetic resonance imaging, and near-infrared spectroscopy to observe activation in brain areas responsible for the perception of bladder fullness and the sensation of the desire to void during bladder filling, whereas other studies examined brain activity during micturition (DasGupta et al., 2007; Griffiths & Tadic, 2008). The brain is activated during bladder filling particularly in the PAG, the midline pons, the anterior and the midcingulate gyrus, the anterior insula, and bilaterally the frontal lobes, supplementary motor area (SMA), referred to as the "frontal micturition center" (Krhut et al., 2012, 2014) (Fig. 3.8). Participation of these brain regions such as thalamus, right insula, ACC, PAG, and brainstem L-region during bladder filling has also been identified in a more recent meta-analysis article of brain imaging studies in healthy adults (Arya et al., 2017).

The constellation of these cortical areas seems to "switch on and off" the spinobulbospinal micturition reflex. The results are consistent with the notion that the PAG receives information about bladder fullness and then relays this information to other brain areas involved in the control of bladder storage. In fMRI studies in men and women, activation during urinary storage occurs in the SMA, midcingulate cortex, insula, and right PFC, while the right anterior insula and midbrain PAG are more active at higher rather than at lower bladder volumes (Kuhtz-Buschbeck et al., 2005, 2009) (Fig. 3.8).

PET scan studies in healthy men and women revealed that during voiding, two cortical areas (the dorsolateral PFC and anterior cingulated gyrus) are active (i.e., exhibited increased blood flow). The hypothalamus including the preoptic area as well as the pons and the PAG also show activity in concert with voluntary micturition (Blok et al., 1997). Another PET study during voiding also confirmed that micturition is associated with increased activity in the pons, inferior frontal gyrus, hypothalamus, and PAG while also showing activity in several other cortical areas (postcentral gyrus, superior frontal

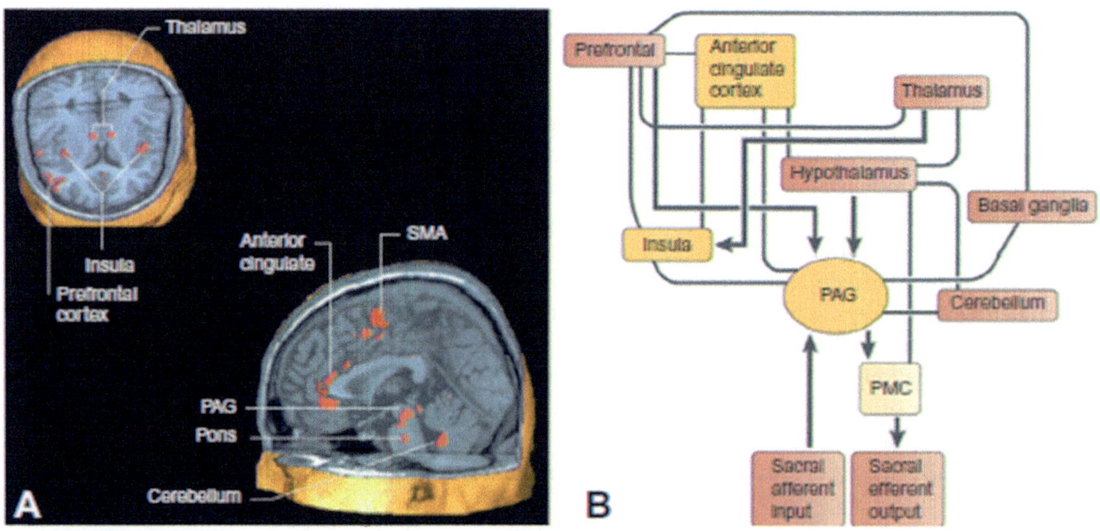

FIGURE 3.8 Brain areas involved in the regulation of urine storage. (A) A meta-analysis of positron emission tomography and functional magnetic resonance imaging (fMRI) studies that investigated which brain areas are involved in the regulation of micturition reveals that the thalamus, the insula, the prefrontal cortex, the anterior cingulate, the periaqueductal gray (PAG), the pons, the medulla, and the supplementary motor area (SMA) are activated during the urinary storage. (B) A conceptual framework, based on functional brain-imaging studies, suggesting a scheme for the connections between various forebrain and brainstem structures that are involved in the control of the bladder and the sphincter in humans. Arrows show probable directions of connectivity but do not preclude connections in the opposite direction. PMC, Pontine micturition center. *From de Groat, W.C., & Yoshimura, N. (2015). Anatomy and physiology of the lower urinary tract. Handbook of Clinical Neurology. 130, 61–108.*

gyrus, thalamus, insula, and globus pallidus) and the cerebellar vermis (Nour et al., 2000). Neuroimaging fMRI studies that evaluated the effect of voluntary contractions of the pelvic floor during urine withholding or during cold stimulation on brain activity revealed activation of the parietal cortex, cerebellum, putamen, and SMA (Seseke et al., 2006; Zhang et al., 2005) as well as activation of the frontal cortex, basal ganglia, and cerebellum (Seseke et al., 2006). It has also been reported using functional near-infrared spectroscopy (fNIRS) that similar PFC regions to those seen with fMRI are activated with bladder filling (Sakakibara et al., 2010). A recent article of meta-analysis of neuroimaging studies in healthy adults has confirmed that brain regions including the pons, cerebellum, insula, ACC, thalamus, and the frontal cortex are activated during the voiding phase of micturition (Harvie et al., 2019).

Based on the results, a simplified model of normal bladder filling is proposed, where the filling increases bladder and urethral afferent activity, which is received in the PAG, and mapped in the insula to form the basis of the desire to void, whereas the anterior cingulate gyrus (ACG) provides monitoring and control, and the PFC makes voiding decisions. The voiding reflex is continuously inhibited at the PMC as the bladder fills until the decision to void is made in the PFC. The increasing desire to void corresponds to a gradual increase in insular response (Griffiths & Fowler, 2013) (Fig. 3.8).

Animal studies

Similar to human imaging studies, in PRV-based tracing studies from the rat bladder, virus-labeled neurons are found in rostral regions in the hypothalamus (lateral medial preoptic and paraventricular nucleus), dorsal thalamus, the primary and secondary motor cortices, and entorhinal and piriform cortices. A recent study also showed that rabies-based retrograde transsynaptic labeling identified neurons occurred in prefrontal areas, including ACC and prelimbic cortex, as well as motor cortex and somatosensory cortex after rabies infection of CRH-positive PMC neurons in mice (Hou et al., 2016).

Although older studies demonstrated that electrical stimulation of the frontal lobe and the ACC elicited either inhibition or facilitation of bladder contractions in the cat (Gjone & Setekleiv, 1963), there has been limited information available regarding the cortical control of micturition. In the cat, neuronal activity recordings in the frontal brain structures such as the PFC and sensorimotor cortex showed that the majority (88%; 98 out of 112 cells) of PFC neurons were tonically activated during the bladder-relaxation phase, whereas the remaining neurons were tonically activated during the bladder-contraction phase (Yamamoto et al., 2010). Also, low-frequency stimulation of the frontal cortex induced frontal cortical inactivation, leading to bladder overactivity after a long delay in the cat (Pikov & McCreery, 2009). The results suggest that frontal cortical neurons modulate the micturition reflex, mostly in an inhibitory manner. It has also been reported that the ACC plays an inhibitory role in the control of micturition because neural activity in the ACC is significantly increased along with suppression of bladder overactivity after inhibition of adenosine A2A receptors, while electrical stimulation of the ACC inhibits the micturition reflex in a rat model of Parkinson's disease (PD) (Kitta et al., 2016b). However, a recent human case study revealed that electrical stimulation of the ACC at 50−150 Hz induces the urge to void followed by a micturition response (Patra et al., 2019), suggesting a dual role of the ACC in the control of micturition.

Furthermore, a recent study in mice showed that activities of layer 5 neurons in the primary motor cortex are tightly correlated with the onset of urination and that optogenetic stimulation of these neurons activates a direct projection to the PMC, which elicits a contraction of the bladder to initiate urination, clearly demonstrating the facilitatory role of this population of motor cortex neurons in micturition under a freely moving, conscious condition (Yao et al., 2018) (Fig. 3.3).

Subcortical modulation of micturition

The influence of the cortex on micturition could be mediated by a number of pathways, including not only direct cortical projections from the prefrontal cortex (PFC) and insular cortex to the PMC but also projections through the hypothalamus and the extrapyramidal system (de Groat et al., 2015). Studies in humans indicate that voluntary control of voiding is dependent on connections between the frontal cortex and the septal-preoptic region of the hypothalamus as well as on connections between the paracentral lobule and the brainstem. Lesions to these brain areas appear to directly increase bladder activity by removing cortical inhibitory control (de Groat et al., 2015). A recent rabies-based retrograde transsynaptic labeling study also revealed the connection of subcortical nuclei including the BnST, the medial preoptic area (MPOA), the lateral hypothalamic area (LHA), and the amygdala as well as midbrain regions such as the PAG, superior colliculus, and midbrain reticular formation with CRH-positive PMC neurons (Hou et al., 2016). Furthermore, another retrograde rabies tracing study also identified synaptic afferent connections to PMC from neurons in similar brain sites including the BnST, lateral preoptic area (LPOA), MPOA, median preoptic nucleus (MnPO), insular cortex, lateral hypothalamus area (LHA), the amygdala, and PAG (Verstegen et al., 2019).

Hypothalamus

Transneuronal virus tracing methods have identified virus-infected cells in several regions of the hypothalamus after injection of PRV into the lower urinary tract in animals (Nadelhaft et al., 1992; Sugaya et al., 1997; Vizzard et al., 1995) or directly to CRH-positive PMC neurons in mice (Hou et al., 2016). Histochemical studies using tracers also revealed that neurons in the PMC receive input from the caudal hypothalamus and that the paraventricular hypothalamic nucleus projects nonspecifically to all autonomic preganglionic motor neurons in the spinal cord, including the sacral parasympathetic and sphincter motor nuclei (Holstege & Mouton, 2003).

Although both excitatory and inhibitory effects on bladder activity are elicited by stimulation at different sites of the hypothalamus, the overall hypothalamic control seems to be facilitatory according to the studies of Tang and Ruch (Ruch & Tang, 1956; Tang, 1955), showing that after supracollicular decerebration, which eliminates the hypothalamic effects, the bladder volume threshold inducing micturition is markedly increased (de Groat et al., 1993a). A recent study directly showed the excitatory role of hypothalamus because glutamatergic axons from the LHA intensely innervate PMC neurons, and optogenetic stimulation of these axon terminals in the PMC induces voiding in mice (Verstegen et al., 2019).

The hypothalamic influence on bladder function may in turn be modulated by afferent inputs from the bladder because axonal tracing studies in rats have identified a spino-hypothalamic pathway arising from neurons in the region of sacral parasympathetic nucleus (Birder et al., 1999; Burstein et al., 1990), and 50% of these neurons projecting to the hypothalamus are activated by bladder afferents as evidenced by positive c-Fos staining after bladder irritation with dilute acetic acid (Birder et al., 1999). Brain imaging studies in human subjects also revealed that the caudal hypothalamus responds to changes in bladder volume (Athwal et al., 2001; Griffiths et al., 2007).

It has been reported that the MPOA, which is a part of the anterior hypothalamus, is densely labeled by modified rabies, after infection of CRH-positive PMC neurons in mice (Hou et al., 2016). These rabies-labeled MPOA neurons are largely GAD-positive, GABAergic cells, and chemogenetic inhibition of these neurons decreased the frequency of voids and increased voided volume per micturition, potentially due to disinhibition of inhibitory pathways to the PMC. Thus, it is likely that the MPOA modulates the voluntary micturition related with territorial marking behavior (Hou et al., 2016).

Basal ganglia, substantia nigra pars compacta, and ventral tegmental area

Single-unit recordings in the substantia nigra (SN)/ventral tegmental area (VTA) in ketamine-anesthetized cats during rhythmic reflex bladder contractions under isovolumetric conditions revealed firing patterns including (1) tonic (55%) and phasic (22%) neurons activated during the storage phase and (2) tonic (16%) and phasic (6%) neurons activated during the voiding phase (Sakakibara et al., 2002a). The large percentage of storage phase—activated neurons is consistent with other observations indicating that the dopaminergic neurons in the SN/VTA have a predominate inhibitory influence on micturition. For example, electrical stimulation in the SN terminates ongoing micturition (Yoshimura et al., 1992) and destruction of dopaminergic neurons using the toxin 1-methyl-4-phenyl-1,2,3,6-tetrahydropyridine (MPTP) or 6-hydroxydopamine facilitates the micturition reflex (Yoshimura et al., 1998a, 2003). Dopaminergic neurons in SN synapse with neostriatal GABAergic neurons that may be involved in micturition inhibitory mechanisms in the forebrain. In addition, D_1 dopaminergic receptors have been implicated in the control of GABAergic neurons in the PAG (Kitta et al., 2008), where GABAergic inhibition also plays an important role in the control of micturition (Stone et al., 2011).

Cerebellum

Brain imaging studies in humans have shown activation in the cerebellum in response to bladder distension (Griffiths et al., 2005; Kuhtz-Buschbeck et al., 2005; Seseke et al., 2006; Takao et al., 2008). This activation is consistent with studies in cats showing that electrical stimulation of bladder afferent nerves elicits neural activity in the cerebellum (Bradley and Teague, 1969a, 1969b). A possible inhibitory role of the cerebellum in the control of micturition is suggested by studies in cats showing that electrical stimulation of the cerebellar fastigial nucleus inhibits reflex bladder activity (Bradley and Teague, 1969a, 1969b; Martner, 1975), while ablation of the anterior vermis of the cerebellum results in bladder overactivity with increased reflex duration (Bradley and Teague, 1969a, 1969b). Cerebellectomy in dogs also induces bladder overactivity (Nishizawa et al., 1995) and cerebellar pathology in humans is accompanied by bladder overactivity (Zago et al., 2010). These data indicate that the cerebellum has a tonic inhibitory influence over the micturition reflex. Cerebellar projections to the nucleus subcoeruleus, nucleus locus coeruleus, and PAG have also been mentioned as possible pathways for cerebellar modulation of micturition (Dietrichs & Haines, 2002). Other evidence suggests that the cerebellum exerts a tonic inhibitory influence on micturition via a pathway passing through the mesencephalic reticular formation. Based on studies in decerebrate cats (Nishizawa et al., 1995), it was concluded that the cerebellum plays an inhibitory role during the storage phase of the micturition cycle and a facilitatory role during voiding.

Taken together, the neural control over micturition from subcortical sites as the hypothalamus, the basal ganglia, SN and VTA, and the cerebellum, which are located outside the spinobulbospinal reflex loop, has therefore been implicated to be under voluntary control in coordination with brain cortical areas.

Neurotransmitters in cortical and subcortical controls of micturition

Glutamate

Glutamic acid plays a major role in excitatory transmission at supraspinal sites including the PMC and PAG in the micturition reflex pathway. Exogenous L-glutamate or its analogue injected at sites (PMC or parabrachial nucleus) in the brainstem of supracollicular decerebrate or chloralose-anesthetized cats elicits voiding when the bladder is partially filled or increased frequency and amplitude of rhythmic bladder contractions when the bladder is filled above the micturition threshold volume and maintained under isovolumetric conditions (Chen et al., 1993; Mallory et al., 1991). On the other hand, injections of glutamic acid at some sites in the PMC elicits inhibition of isovolumetric contractions or initial excitation followed by inhibition (Mallory et al., 1991).

Administration of glutamatergic agonists into the region of the PMC in rats also elicits voiding or increases frequency and amplitude of bladder contractions (Matsuura et al., 2000; Rocha et al., 2001), whereas injection of agonists in the brain of rats and cats at other sites known to have inhibitory functions in micturition elicits inhibitory effects (Chen & Chai, 2002; Chen et al., 1993; Naka et al., 2009; Nishijima et al., 2012; Sugaya et al., 2014). Intracerebroventricular injection of AMPA or NMDA receptor antagonists blocks reflex bladder contractions in anesthetized rats, indicating that glutamatergic transmission in the brain is essential for voiding function (Yoshiyama & de Groat, 2005).

Recent studies using mice further demonstrated that glutamatergic PMC neurons including ESR1-positive cells are essential for the initiation of voluntary micturition and that glutamatergic axons from the motor cortex, hypothalamus, and the PAG can directly activate PMC neurons to induce micturition (Keller et al., 2018; Verstegen et al., 2019; Mukhopadhyay & Stowers, 2020; Yao et al., 2018).

Acetylcholine

Excitatory and inhibitory cholinergic influences on the micturition pathway have been identified at the supraspinal level using various techniques. In the rat brain, muscarinic receptor–mediated cholinergic mechanisms may be involved in both inhibitory and facilitatory modulation of the micturition reflex (Ishiura et al., 2001; Ishizuka et al., 2002; Yokoyama et al., 2001), and the muscarinic inhibitory mechanism seems to involve an activation of M1 muscarinic receptors (Yokoyama et al., 2001) and protein kinase C (Nakamura et al., 2003). One site of action can be localized to the midbrain–pons region because cholinergic agonists are effective after supracollicular decerebration in rats (Sillen et al., 1982). In the brainstem, microinjection of acetylcholine to the PMC in cats increased or decreased the threshold volume for inducing a reflex contraction of the bladder (Sugaya et al., 1987; Yoshimura & de Groat, 1997). These effects were blocked by atropine, indicating a role of muscarinic receptors. Nicotinic receptors are also involved in the control of voiding function since nicotinic receptor agonists, epibatidine, or nicotine injected into the lateral ventricle have an inhibitory effect on the micturition reflex in the rat (Lee et al., 2003; Masuda et al., 2006). A decreased volume threshold and increased micturition pressure were detected after administration of bethanechol, a muscarinic agonist, into the central circulation of the cross-perfused dog (O'Donnell, 1990).

GABA and glycine

Intrathecal injection of either $GABA_A$ or $GABA_B$ agonists increases bladder capacity and decreases voiding pressure and efficiency in normal rats (Igawa et al., 1993; Pehrson et al., 2002a) and also suppresses DO in rats with intravesical application of oxyhemoglobin, a nitric oxide (NO) scavenger (Pehrson et al., 2002a) or SCI (Miyazato et al., 2007). In addition, intravenous or intrathecal application of a GABA reuptake inhibitor (tiagabine) that increases endogenous GABA concentrations reportedly inhibits normal micturition in rats (Pehrson & Andersson, 2002).

Previous studies also showed that glycine, another inhibitory amino acid, acting on strychnine-sensitive receptors exerts an inhibitory effect on the micturition reflex pathway (de Groat, 1976; Miyazato et al., 2003; Yoshikawa et al., 2012) and is also involved in the inhibition of sphincter motoneurons during micturition (Shefchyk, 2001).

In addition, some studies have revealed that the level of glycine in the spinal cord is decreased by approximately 50% in rats with DO induced by chronic SCI, compared with spinal intact rats (Miyazato et al., 2003, 2005), and that dietary supplementation with glycine can restore bladder function along with an increase in the serum level of glycine in spinal cord injured rats (Miyazato et al., 2005). The level of GAD, the GABA synthetic enzyme, is also reduced in the spinal cord and lumbosacral DRG in spinal cord–injured rats with DO (Miyazato et al., 2008a) and DSD (Miyazato et al., 2008b), and both impaired functions are suppressed by intrathecal application of $GABA_A$ or $GABA_B$ receptor agonists (Miyazato et al., 2008a, 2008b). These results suggest that not only the disruption of coordinated neural control by PMC neurons including those expressing ESR1, which initiate the EUS relaxation during voiding (Keller et al., 2018), but also downregulation of spinal glycinergic and GABAergic mechanisms could contribute to the emergence of neurogenic DO and DSD after SCI.

GABA has also been implicated as an inhibitory transmitter at supraspinal sites where it can act on both $GABA_A$ and $GABA_B$ receptors (de Groat et al., 1993b, 1999; de Groat & Yoshimura, 2001; Kanie et al., 2000; Yoshimura & de Groat, 1997). Injection of $GABA_A$ receptor agonists into the PMC of decerebrate cats or into the PAG of rats suppresses reflex bladder activity and increases the volume threshold for inducing micturition (Mallory et al., 1991). These effects are reversed by bicuculline, a $GABA_A$ receptor antagonist, and bicuculline alone stimulates bladder activity and lowers the volume threshold for micturition, indicating that the micturition reflex pathway in the PMC and PAG is tonically inhibited by a GABAergic mechanism. Intracerebroventricular administration of melatonin increases bladder capacity in rats, and this effect is blocked by bicuculline, indicating that melatonin activates a GABAergic inhibitory mechanism in the brain (Matsuta et al., 2010). Intracerebroventricular injection of baclofen, a $GABA_B$ agonist, suppresses distention-evoked micturition in urethane-anesthetized rats, but unexpectedly this effect is not blocked by phaclofen, a $GABA_B$ receptor antagonist (de Groat et al., 1993c; de Groat & Yoshimura, 2001).

Dopamine

In the CNS, dopaminergic pathways exert inhibitory and facilitatory effects on the micturition reflex through D1-like (D1 or D5 subtypes) and D2-like (D2, D3, or D4 subtypes) dopaminergic receptors, respectively (Fig. 24) (Albanease et al., 1988; Hashimoto et al., 2003; Kontani et al., 1990; Seki et al., 2001; Yokoyama et al., 1999; Yoshimura et al., 1993, 1998b, 2003). In anesthetized cats, activation of dopaminergic neurons in the SN inhibits reflex bladder contractions via D1-like receptors (Yoshimura et al., 1992). In awake rats, a D1 dopaminergic antagonist (SCH 23390) facilitates the micturition reflex, whereas a D1 agonist (SKF 38393) does not alter reflex bladder contractions, suggesting that D1 receptor–mediated suppression of bladder activity is tonically active in the normal awake state (Seki et al., 2001). Conversely, activation of central D2-like dopaminergic receptors with quinpirole or bromocriptine facilitates the micturition reflex pathway in rats, cats, and monkeys (Kontani et al., 1990; Yokoyama et al., 1999; Yoshimura et al., 1993, 1998b, 2003). D2-like receptor–mediated facilitation of the micturition reflex may involve actions on spinal cord as well as actions on the brainstem because microinjection of dopamine to the PMC reduced bladder capacity and facilitated the micturition reflex in cats (de Groat et al., 1993c).

It is also known in cats that neurons in the SN pars compacta and the VTA, which are the major dopamine-containing nuclei in the midbrain, respond to the storage/micturition cycles during isovolumetric cystometry (Sakakibara et al., 2002b) and that dopamine levels in the striatum, where nigrostriatal dopaminergic nerves terminate (Fig. 3.3), increase during the storage phase of the micturition cycle (Yamamoto et al., 2005). Thus, central dopaminergic pathways appear to be involved in the control of the bladder function through actions on multiple receptors, such as D1 receptor–mediated inhibition and D2 receptor–mediated excitation, at different sites in the brain.

Therefore, imbalance of the brain dopaminergic mechanisms can induce an impairment of voluntary control of micturition (Ogawa et al., 2017). It has been documented that PD that is induced by dopamine depletion in the striatum or cerebrovascular diseases such as cerebral infarction often cause irritable bladder symptoms such as urinary urgency, frequency, and incontinence in association of bladder overactivity (Miyazato et al., 2017). Previous studies using animal models have demonstrated that PD-induced bladder overactivity is primarily due to a loss of D1 receptor–mediated inhibition of the micturition reflex (Araki et al., 2000; Araki & Kuno, 2000; Aranda & Cramer, 1993; Berger et al., 1987; Pavlakis et al., 1983), whereas D2 receptor–mediated facilitation of micturition is enhanced after cerebral infarction induced by middle cerebral artery occlusion (Yokoyama et al., 2002). Furthermore, a study using PET scanning showed that activation of the pons or the ACC seen in healthy volunteers was not observed in PD patients with bladder overactivity, suggesting the loss of ACC-mediated inhibitory control of micturition in PD (Kitta et al., 2006).

Serotonin (5-hydroxytryptamine)

Neurons containing 5-hydroxytryptamine (5-HT) in the raphe nucleus of the caudal brainstem send projections to the dorsal horn, as well as to the autonomic and sphincter motor nuclei in the lumbosacral spinal cord (Fig. 3.3). In cats, activation of raphe neurons or 5-HT receptors in the spinal cord inhibits reflex bladder contractions and firing of the sacral efferent pathways to the bladder (Chen et al., 1993; de Groat, 2002a; de Groat et al., 1993c; Ito et al., 2006; McMahon & Spillane, 1982) and also inhibits firing of spinal dorsal horn neurons elicited by stimulation of pelvic nerve afferents (Fukuda & Koga, 1991). Extracellular recordings of neuronal activity in the raphe nucleus correlating to storage/voiding cycles under the isovolumetric condition have revealed that the most common type of neuron ($\sim 50\%$) was a tonic storage type that exhibited increased firing at an interval between reflex bladder contractions in cats (Ito et al., 2006).

In rats, the administration of m-chlorophenylpiperazine (mCPP), which is an agonist for 5-$HT_{2A/C}$ receptors, suppressed efferent activity on bladder nerves and reflex bladder contractions (Steers & de Groat, 1989). These effects were blocked by mesulergine (a 5-HT_2 receptor antagonist) (Guarneri et al., 1996; Steers & de Groat, 1989). Intrathecal administration of methysergide, a 5-$HT_{1/2}$ antagonist, or zatosetron, a 5-HT_3 antagonist, decreased the micturition volume threshold in cats (Espey et al., 1998), implying that descending serotonergic pathways tonically depress the afferent limb of the micturition reflex through 5-HT_2 and/or 5-HT_3 receptors.

The role of 5-HT_1 receptors in bladder activity seems different in cats and rats. In cats, administration of 8-hydroxy-2-(di-n-propylamino)-tetralin (8-OH-DPAT), a 5-HT_{1A} receptor agonist, increased bladder capacity in chloralose-anesthetized cats, in which the bladder was irritated with acetic acid, but had only moderate effects on bladder activity in the absence of irritation (Thor et al., 2002). The drug also had a facilitatory effect on activity of the EUS. 8-OH-DPAT also inhibited reflex bladder activity in awake or chloralose-anesthetized, chronic spinal cord−injured cats but did not alter the somato-bladder excitatory reflex induced in spinal cats by tactile stimulation of the perigenital region (Gu et al., 2004). The effects of 8-OH-DPAT were blocked by WAY 100635, a 5-HT_{1A} receptor antagonist, which alone had no effect. These results indicate that 8-OH-DPAT acts in the spinal cord to inhibit the micturition reflex triggered by C-fiber bladder afferent axons and has much less effect on the spinobulbospinal reflex elicited by Aδ-afferents.

In contrast, 8-OH-DPAT administered intrathecally facilitated bladder activity in both normal and spinal cord−injured rats but not in rats in which bladder afferents were damaged by treatment with capsaicin at birth (Lecci et al., 1992). Conversely, administration of the 5-HT_{1A} receptor antagonist WAY 100635, which increases the firing rate of raphe neurons by blocking 5-HT_{1A} inhibitory autoreceptors, inhibits reflex bladder contractions in rats (Testa et al., 1999). The inhibition is antagonized by pretreatment with mesulergine, a 5-HT_2 receptor antagonist, indicating that 5-HT_2 receptors are involved in descending raphe/spinal inhibitory mechanisms (Testa et al., 1999). Similar inhibitory effects of another 5-HT_{1A} receptor antagonist, NAD-299, on the micturition reflex have been reported in rats (Pehrson et al., 2002b).

When the effects of intrathecal administration of WAY 100635 on the ascending and descending limbs of the micturition reflex pathway were examined in the anesthetized rat, WAY 100635 depressed bladder contractions evoked by electrical stimulation of the PMC but did not alter the evoked field potentials in the region of the PMC during electrical stimulation of afferent axons in the pelvic nerve, indicating that the drug suppresses the pathway from the brainstem to the spinal cord but does not alter the afferent pathway from the bladder to the PMC (de Groat, 2002a; Kakizaki et al., 2001). Thus, micturition in the rat is facilitated by stimulation of 5-HT_1 inhibitory autoreceptors, whereas in the cat 5-HT_1 receptor activation appears to act primarily through postsynaptic mechanisms to promote urine storage by enhancing sphincter activity and suppressing bladder activity (de Groat, 2006).

The sympathetic autonomic nuclei as well as the sphincter motor nuclei also receive a serotonergic input from the raphe nucleus (de Groat et al., 1979; Espey et al., 1998; Thor & Donatucci, 2004). Serotonergic activity mediated via 5-HT_2 and 5-HT_3 receptors enhances urine storage by facilitating sphincter reflexes in cats (Danuser & Thor, 1996; Espey et al., 1998). A previous study in rats also reported that activation of 5-HT_{2C} receptors at the spinal level enhances the urethral closure reflex induced by pudendal nerve−mediated urethral striated muscle contraction during sneezing, whereas 5-HT_{1A} receptors inhibit this response because intrathecally applied 8-OH-DPAT (a 5-HT_{1A} agonist) decreases urethral contractions during sneezing and mCPP (a 5-$HT_{2B/2C}$ agonist) increases them. Furthermore the effects of 8-OH-DPAT and mCPP are antagonized by intrathecal applications of WAY-100635, a selective 5-HT_{1A} antagonist, and RS-102221, a selective 5-HT_{2C} antagonist, respectively (Miyazato et al., 2009). More recently, it has been shown that serotonin depletion by systemic application of p-chlorophenylalanine (a serotonin synthesis inhibitor) induced a reduction in the urethral closure reflex during sneezing, which was then improved by CP-809101 (a 5-HT_{2C} agonist) or LP44 (a 5-HT_7 agonist), indicating the 5-HT_7 receptor also has facilitatory effects on the urethral sphincter function (Suzuki et al., 2018).

Recent studies have also demonstrated that direct administration of citalopram (a specific serotonin reuptake inhibitor; SSRI) into the PFC increased the extracellular level of 5-HT measured by microdialysis and that the micturition interval was increased after injection of MDL11939 (a 5-HT_{2A} antagonist) into the prelimbic cortex of the rat PFC, whereas SB269970 (a 5-HT_7 antagonist) injected into the PFC decreased it (Chiba et al., 2016, 2020). These results indicate that serotonergic mechanisms are involved in the PFC-mediated cortical control of micturition.

Stress-related peptides

Psychological stress reportedly plays a role in the exacerbation and development of lower urinary tract dysfunction, leading to irritative bladder symptoms and bladder pain. Corticotropin-releasing factor (CRF) has been implicated as an important neurohormone and neurotransmitter that mediates various stress responses in the CNS. Although recent studies showed that CRF-positive PMC neurons do not play an essential role in normal voiding behavior in mice (Verstegen et al., 2019), upregulation of CRF in PMC neurons reportedly contributes to stress-induced bladder dysfunction because increased expression of CRF in PMC neurons in male rats with resident-intruder stress was associated with the increases in micturition interval, micturition volume, and bladder capacity, which was blocked by a CRF_1 receptor antagonist (NBI-30775) (Wood et al., 2013). In contrast, female rats with water avoidance stress exhibited frequent urination and visceral hyperalgesia, along with increased cerebral blood flow in cortical regions including the posterior cingulate cortex, posterior insula, supplementary motor cortex, as well as in the thalamus, anterior hypothalamus, and the PMC, suggesting that activation of supraspinal micturition pathways is involved in stress-induced bladder dysfunction (Wang et al., 2017). In addition, the contribution of other stress-related neuropeptides such as bombesin (BB)-related peptides to bladder dysfunction has recently been studied in rats, in which centrally administered BB induced frequent urination through activation of brain BB1/BB2 receptors and 5-HT_7 serotonin receptors (Shimizu et al., 2016, 2017).

Conclusion

Micturition is a function accomplished by unique biomechanics of bladder and urethral muscles as well as by a complex neural control system located in the brain and the spinal cord. Storage of urine in the bladder until a socially acceptable moment arrives is predominantly integrated at the spinal cord level mediated through increased sympathetic activity, which relaxes the urinary bladder via activation of postsynaptic β3-receptors and contracts both urethral and prostatic smooth muscles via the α1-adrenoceptor. Reflex voiding is then activated in an all-or-none, switch-like manner by a spino-bulbospinal pathway that is relayed through PAG-PMC circuitry, which receives ascending afferent signals via spinal cord to PAG, and induces excitation of the sacral parasympathetic nucleus by descending PMC efferent signals via spinal cord to increase parasympathetic activity, leading to activation of postsynaptic muscarinic receptors to cause detrusor contraction and reciprocal relaxation of both urethral and prostatic smooth muscle by nitric oxide release. In addition, the rhabdosphincter is relaxed by inhibition of the pudendal motor nucleus in the sacral spinal cord by inputs from the PMC, especially through ESR1-positive glutamatergic PMC neurons.

During postnatal development, reflex voiding, which is dominant in infants, is replaced by voluntary control as afferent systems that relay information on the state of bladder fullness to higher brain centers are developed and as spinal micturition reflex mechanisms are downregulated in response to neuronal maturation and learned behavior. Afferent pathways that trigger storage and voiding reflexes as well as the sensations of bladder filling transmit activity from mechanoreceptors in the bladder through second-order neurons in the spinal cord to central processing areas in the PAG and forebrain (insula, anterior cingulate) before reaching the frontal cortex. An interaction between the PAG and the PMC generates efferent signals that descend from the PMC to the spinal cord to initiate the appropriate motor responses necessary for voiding (i.e., a bladder contraction and reciprocal urethral sphincter relaxation).

In the CNS, glutamic acid has also been identified as an excitatory transmitter in bladder and sphincter reflex pathways at spinal and supraspinal levels, while GABA and glycine have been identified as inhibitory transmitters. Monoamines (serotonin, norepinephrine, and dopamine) have facilitatory and/or inhibitory modulatory functions that can vary depending on the central pathway and the species.

Owing to the complexity of the neural mechanisms that regulate urine storage and voiding, these processes are sensitive to various neural injuries and diseases. SCI has been the focus of considerable attention because it eliminates voluntary and supraspinal control leading to DSD and inefficient voiding, as well as plasticity in central and peripheral neural pathways leading to DO, inefficient urine storage, and incontinence. Several types of peripheral and central neuroplasticity have been identified including (1) emergence of primitive neonatal micturition reflexes and (2) remodeling of spinal circuitry and sensitization of bladder silent C-fiber afferents leading to the emergence of a spinal micturition reflex. Also, dysfunction of various brain regions due to vascular or neurological diseases, tumors, or injuries can reduce cortical inhibitory mechanisms, resulting in the emergence of uninhibited activation of the spinobulbospinal reflex micturition pathway, leading to the overactive bladder condition.

Due to the recent advancement in neuro-urological research, the neural circuit involved in reflex micturition through spinal or spinobulbospinal pathways has increasingly been clarified. However, the neural circuit and neurotransmitter systems that directly or indirectly control voluntary micturition are still not well understood despite the recent advancement in human imaging techniques. Thus, genetic tools, including conventional and conditional mouse mutants, targeting different neural pathways or transmitter system components in the CNS, will be essential to further advance our understanding of the endogenous mechanisms underlying voluntary micturition control in normal or pathological conditions. The utilization of complementary methodologies, such as cell type–specific Cre-driver lines, and viral and optogenetic tools will help to further dissect the function of genetically defined neurocircuits in the context of reflex and voluntary control of micturition, leading to the development of target-specific, new therapeutic modalities for the treatment of lower urinary tract dysfunction.

References

Albanease, A., Jenner, P., Marsden, C. D., & Stephenson, J. D. (1988). Bladder hyperreflexia induced in marmosets by 1-methyl-4-phenyl-1,2,3,6-tetrahydropyridine. *Neuroscience Letters, 87*, 46–50.

Amend, B., Matzel, K. E., Abrams, P., de Groat, W. C., & Sievert, K. D. (2011). How does neuromodulation work. *Neurourology and Urodynamics, 30*(5), 762–765.

Andersson, K. E., & Pehrson, R. (2003). CNS involvement in overactive bladder: Pathophysiology and opportunities for pharmacological intervention. *Drugs, 63*(23), 2595–2611.

Araki, I., & de Groat, W. C. (1996). Unitary excitatory synaptic currents in preganglionic neurons mediated by two distinct groups of interneurons in neonatal rat sacral parasympathetic nucleus. *Journal of Neurophysiology, 76*(1), 215–226.

Araki, I., & de Groat, W. C. (1997). Developmental synaptic depression underlying reorganization of visceral reflex pathways in the spinal cord. *Journal of Neuroscience, 17*(21), 8402–8407.

Araki, I., Kitahara, M., Oida, T., & Kuno, S. (2000). Voiding dysfunction and Parkinson's disease: Urodynamic abnormalities and urinary symptoms. *The Journal of Urology, 164*(5), 1640–1643.

Araki, I., & Kuno, S. (2000). Assessment of voiding dysfunction in Parkinson's disease by the international prostate symptom score. *Journal of Neurology Neurosurgery and Psychiatry, 68*(4), 429–433.

Aranda, B., & Cramer, P. (1993). Effects of apomorphine and L-dopa on the parkinsonian bladder. *Neurourology and Urodynamics, 12*(3), 203–209.

Arya, N. G., Weissbart, S. J., Xu, S., & Rao, H. (2017). Brain activation in response to bladder filling in healthy adults: An activation likelihood estimation meta-analysis of neuroimaging studies. *Neurourology and Urodynamics, 36*(4), 960–965.

Athwal, B. S., Berkley, K. J., Hussain, I., Brennan, A., Craggs, M., Sakakibara, R., Frackowiak, R. S., & Fowler, C. J. (2001). Brain responses to changes in bladder volume and urge to void in healthy men. *Brain, 124*(Pt 2), 369–377.

Bahns, E., Ernsberger, U., Janig, W., & Nelke, A. (1986). Functional characteristics of lumbar visceral afferent fibres from the urinary bladder and the urethra in the cat. *Pflügers Archiv, 407*(5), 510–518.

Bahns, E., Halsband, U., & Janig, W. (1987). Responses of sacral visceral afferents from the lower urinary tract, colon and anus to mechanical stimulation. *Pflügers Archiv, 410*(3), 296–303.

Berger, Y., Blaivas, J. G., DeLaRocha, E. R., & Salinas, J. M. (1987). Urodynamic findings in Parkinson's disease. *The Journal of Urology, 138*(4), 836–838.

Birder, L. A. (2010). Urothelial signaling. *Autonomic Neuroscience, 153*(1–2), 33–40.

Birder, L. A., Roppolo, J. R., Erickson, V. L., & de Groat, W. C. (1999). Increased c-fos expression in spinal lumbosacral projection neurons and preganglionic neurons after irritation of the lower urinary tract in the rat. *Brain Research, 834*(1–2), 55–65.

Blok, B. F. (2002). Central pathways controlling micturition and urinary continence. *Urology, 59*(5 Suppl. 1), 13–17.

Blok, B. F. M., Deweerd, H., & Holstege, G. (1995). Ultrastructural evidence for a paucity of projections from the lumbosacral cord to the pontine micturition center or M-region in the cat–a new concept for the organization of the micturition reflex with the periaqueductal gray as central relay. *Journal of Comparative Neurology, 359*(2), 300–309.

Blok, B. F., & Holstege, G. (1994). Direct projections from the periaqueductal gray to the pontine micturition center (M-region). An anterograde and retrograde tracing study in the cat. *Neuroscience Letters, 166*(1), 93–96.

Blok, B. F., & Holstege, G. (1997). Ultrastructural evidence for a direct pathway from the pontine micturition center to the parasympathetic preganglionic motoneurons of the bladder of the cat. *Neuroscience Letters, 222*(3), 195–198.

Blok, B. F., & Holstege, G. (2000). The pontine micturition center in rat receives direct lumbosacral input. An ultrastructural study. *Neuroscience Letters, 282*(1–2), 29–32.

Blok, B. F., Sturms, L. M., & Holstege, G. (1997). A PET study on cortical and subcortical control of pelvic floor musculature in women. *Journal of Comparative Neurology, 389*(3), 535–544.

Boehm, J. J., & Haynes, J. L. (1966). Bacteriology of "midstream catch" urines. Studies in newborn infants. *American Journal of Diseases of Children, 111*(4), 366–369.

Bradley, W. E., & Conway, C. J. (1966). Bladder representation in the pontine-mesencephalic reticular formation. *Experimental Neurology, 16*(3), 237–249.

Bradley, W., Griffin, D., Teague, C., & Timm, G. (1973). Sensory innervation of the mammalian urethra. *Investigative Urology, 10*(4), 287–289.

Bradley, W. E., & Teague, C. T. (1969a). Cerebellar regulation of the micturition reflex. *The Journal of Urology, 101*(3), 396–399.

Bradley, W. E., & Teague, C. T. (1969b). Cerebellar influence on the micturition reflex. *Experimental Neurology, 23*(3), 399–411.

Burstein, R., Wang, J. L., Elde, R. P., & Giesler, G. J., Jr. (1990). Neurons in the sacral parasympathetic nucleus that project to the hypothalamus do not also project through the pelvic nerve—a double labeling study combining Fluoro-gold and cholera toxin B in the rat. *Brain Research, 506*(1), 159–165.

Capek, K., & Jelinek, J. (1956). The development of the control of water metabolism. I. The excretion of urine in young rats. *Physiologia Bohemoslovaca, 5*(1), 91–96.

Chang, H. Y., Cheng, C. L., Chen, J. J., & de Groat, W. C. (2007). Serotonergic drugs and spinal cord transections indicate that different spinal circuits are involved in external urethral sphincter activity in rats. *American Journal of Physiology - Renal Physiology, 292*(3), F1044–F1053.

Chen, S. Y., & Chai, C. Y. (2002). Coexistence of neurons integrating urinary bladder activity and pelvic nerve activity in the same cardiovascular areas of the pontomedulla in cats. *The Chinese Journal of Physiology, 45*(1), 41–50.

Cheng, C. L., & de Groat, W. C. (2004). The role of capsaicin-sensitive afferent fibers in the lower urinary tract dysfunction induced by chronic spinal cord injury in rats. *Experimental Neurology, 187*(2), 445–454.

Cheng, C. L., Liu, J. C., Chang, S. Y., Ma, C. P., & de Groat, W. C. (1999). Effect of capsaicin on the micturition reflex in normal and chronic spinal cord-injured cats. *American Journal of Physiology, 277*(3 Pt 2), R786–R794.

Cheng, C. L., Ma, C. P., & de Groat, W. C. (1993). Effects of capsaicin on micturition and associated reflexes in rats. *American Journal of Physiology, 265*(1 Pt 2), R132–R138.

Cheng, C. L., Ma, C. P., & de Groat, W. C. (1995). Effect of capsaicin on micturition and associated reflexes in chronic spinal rats. *Brain Research, 678*(1–2), 40–48.

Chen, S. Y., Wang, S. D., Cheng, C.-L., Kuo, J. S., de Groat, W. C., & Chai, C. Y. (1993). Glutamate activation of neurons in cardiovascular reactive areas of the cat brain stem affects urinary bladder motility. *American Journal of Physiology, 265*, F520–F529.

Chiba, H., Kitta, T., Ohmura, Y., Higuchi, M., Kon, M., Nakamura, M., Yoshioka, M., & Shinohara, N. (2020). Serotonin in the rat prefrontal cortex controls the micturition reflex through 5-hydroxytryptamine 2A and 5-hydroxytryptamine 7 receptors. *International Journal of Urology, 27*(8), 684–689.

Chiba, H., Mitsui, T., Kitta, T., Ohmura, Y., Moriya, K., Kanno, Y., Yoshioka, M., & Shinohara, N. (2016). The role of serotonergic mechanism in the rat prefrontal cortex for controlling the micturition reflex: An in vivo microdialysis study. *Neurourology and Urodynamics, 35*(8), 902–907.

Danuser, H., & Thor, K. B. (1996). Spinal 5-HT2 receptor-mediated facilitation of pudendal nerve reflexes in the anaesthetized cat. *British Journal of Pharmacology, 118*(1), 150–154.

DasGupta, R., Kavia, R. B., & Fowler, C. J. (2007). Cerebral mechanisms and voiding function. *BJU International, 99*(4), 731–734.

Dietrichs, E., & Haines, D. E. (2002). Possible pathways for cerebellar modulation of autonomic responses: Micturition. *Scandinavian Journal of Urology & Nephrology - Supplementum*, (210), 16–20.

Ding, Y. Q., Zheng, H. X., Gong, L. W., Lu, Y., Zhao, H., & Qin, B. Z. (1997). Direct projections from the lumbosacral spinal cord to Barrington's nucleus in the rat: A special reference to micturition reflex. *Journal of Comparative Neurology, 389*(1), 149–160.

Duong, M., Downie, J. W., & Du, H. J. (1999). Transmission of afferent information from urinary bladder, urethra and perineum to periaqueductal gray of cat. *Brain Research, 819*(1–2), 108–119.

Elam, M., Thoren, P., & Svensson, T. H. (1986). Locus coeruleus neurons and sympathetic nerves: Activation by visceral afferents. *Brain Research, 375*(1), 117–125.

Espey, M. J., Du, H. J., & Downie, J. W. (1998). Serotonergic modulation of spinal ascending activity and sacral reflex activity evoked by pelvic nerve stimulation in cats. *Brain Research, 798*(1–2), 101–108.

Fall, M., Lindstrom, S., & Mazieres, L. (1990). A bladder-to-bladder cooling reflex in the cat. *Journal of Physiology, 427*, 281–300.

Fowler, C. J., Griffiths, D., & de Groat, W. C. (2008). The neural control of micturition. *Nature Reviews Neuroscience, 9*(6), 453–466.

Fry, C., Brading, A. F., & Hussain, M. (2005). Cell Biology. In P. Abrams, L. Cardozo, S. Khoury, & A. Wein (Eds.), *Incontinence* (pp. 313–362). Plymouth, UK: Health Publication Ltd.

Fukuda, H., & Koga, T. (1991). Midbrain stimulation inhibits the micturition, defecation and rhythmic straining reflexes elicited by activation of sacral vesical and rectal afferents in the dog. *Experimental Brain Research, 83*, 303–316.

Geirsson, G., Fall, M., & Lindstrom, S. (1993). The ice-water test–a simple and valuable supplement to routine cystometry. *British Journal of Urology, 71*(6), 681–685.

Geirsson, G., Fall, M., & Sullivan, L. (1995). Clinical and urodynamic effects of intravesical capsaicin treatment in patients with chronic traumatic spinal detrusor hyperreflexia. *The Journal of Urology, 154*(5), 1825–1829.

Geirsson, G., Lindström, S., & Fall, M. (1999). The bladder cooling reflex and the use of cooling as stimulus to the lower urinary tract. *The Journal of Urology, 162*(6), 1890–1896.

Geirsson, G., Lindstrom, S., Fall, M., Gladh, G., Hermansson, G., & Hjalmas, K. (1994). Positive bladder cooling test in neurologically normal young children. *The Journal of Urology, 151*(2), 446–448.

Gjone, R. (1966). Excitatory and inhibitory bladder responses to stimulation of 'limbic', diencephalic and mesencephalic structures in the cat. *Acta Physiologica Scandinavica, 66*(1), 91–102.

Gjone, R., & Setekleiv, J. (1963). Excitatory and inhibitory bladder responses to stimulation of the cerebral cortex in the cat. *Acta Physiologica Scandinavica, 59*, 337–348.

Gladh, G., & Lindström, S. (1999). Outcome of the bladder cooling test in children with neurogenic bladder dysfunction. *The Journal of Urology, 161*(1), 254–258.

Griffiths, D., Derbyshire, S., Stenger, A., & Resnick, N. (2005). Brain control of normal and overactive bladder. *The Journal of Urology, 174*(5), 1862–1867.

Griffiths, D. J., & Fowler, C. J. (2013). The micturition switch and its forebrain influences. *Acta Physiologica, 207*(1), 93–109.

Griffiths, D., & Tadic, S. D. (2008). Bladder control, urgency, and urge incontinence: Evidence from functional brain imaging. *Neurourology and Urodynamics, 27*(6), 466–474.

Griffiths, D., Tadic, S. D., Schaefer, W., & Resnick, N. M. (2007). Cerebral control of the bladder in normal and urge-incontinent women. *NeuroImage, 37*(1), 1–7.

de Groat, W. C. (1975). Nervous control of the urinary bladder of the cat. *Brain Research, 87*(2–3), 201–211.

de Groat, W. C. (1976). Mechanisms underlying recurrent inhibition in the sacral parasympathetic outflow to the urinary bladder. *Journal of Physiology, 257*(2), 503–513.

de Groat, W. C. (1978). Inhibitory mechanisms in the sacral reflex pathways to the urinary bladder. In R. W. Ryall, & J. S. Kelly (Eds.), *Iontophoresis and transmitter mechanisms in the mammalian central nervous system* (pp. 366–368). Holland: Elsevier.

de Groat, W. C. (1986). Spinal cord projections and neuropeptides in visceral afferent neurons. *Progress in Brain Research, 67*, 165–187.

de Groat, W. C. (2002a). Influence of central serotonergic mechanisms on lower urinary tract function. *Urology, 59*(5 Suppl. 1), 30–36.

de Groat, W. C. (2002b). Plasticity of bladder reflex pathways during postnatal development. *Physiology & Behavior, 77*(4–5), 689–692.

de Groat, W. C. (2006). Integrative control of the lower urinary tract: Preclinical perspective. *British Journal of Pharmacology, 147*, S25–S40.

de Groat, W. C., Araki, I., Vizzard, M. A., Yoshiyama, M., Yoshimura, N., Sugaya, K., Tai, C., & Roppolo, J. R. (1998). Developmental and injury induced plasticity in the micturition reflex pathway. *Behavioural Brain Research, 92*(2), 127–140.

de Groat, W. C., & Booth, A. M. (1993). In C. A. Maggi (Ed.), *Synaptic transmission in pelvic ganglia* (pp. 291–347). London: Harwood Academic Publishers.

de Groat, W. C., Booth, A. M., & Krier, J. (1979). In C. M. Brooks, K. Koizumi, & A. Sato (Eds.), *Interaction between sacral parasympathetic and lumbar sympathetic inputs to pelvic ganglia* (pp. 234–247). Tokyo: University of Tokyo Press.

de Groat, W. C., Booth, A. M., Milne, R. J., & Roppolo, J. R. (1982). Parasympathetic preganglionic neurons in the sacral spinal cord. *Journal of the Autonomic Nervous System, 5*(1), 23–43.

de Groat, W. C., Booth, A. M., & Yoshimura, N. (1993). Neurophysiology of micturition and its modification in animal models of human disease. In C. Maggi (Ed.), *The autonomic nervous system* (pp. 227–290). London: Harwood Academic Publishers.

de Groat, W. C., Douglas, J. W., Glass, J., Simonds, W., Weimer, B., & Werner, P. (1975). Changes in somato-vesical reflexes during postnatal development in the kitten. *Brain Research, 94*(1), 150–154.

de Groat, W. C., Fraser, M. O., Yoshiyama, M., Smerin, S., Tai, C., Chancellor, M. B., Yoshimura, N., & Roppolo, J. R. (2001). Neural control of the urethra. *Scandinavian Journal of Urology & Nephrology - Supplementum*, (207), 35–43. discussion 106-25.

de Groat, W. C., Griffiths, D., & Yoshimura, N. (2015). Neural control of the lower urinary tract. *Comprehensive Physiology, 5*(1), 327–396.

de Groat, W. C., Kawatani, M., Hisamitsu, T., Cheng, C.-L., Ma, C.-P., Thor, K., Steers, W., & Roppolo, J. R. (1990). Mechanisms underlying the recovery of urinary bladder function following spinal cord injury. *Journal of the Autonomic Nervous System, 30*(Suppl. l), S71–S77.

de Groat, W. C., Nadelhaft, I., Milne, R. J., Booth, A. M., Morgan, C., & Thor, K. (1981). Organization of the sacral parasympathetic reflex pathways to the urinary bladder and large intestine. *Journal of the Autonomic Nervous System, 3*(2–4), 135–160.

de Groat, W. C., Roppolo, J. R., Yoshimura, N., & Sugaya, K. (1993). In Y. Taché, D. Wingate, & T. Burks (Eds.), *Neural control of the urinary bladder and colon* (pp. 167–190). Boca Raton, FL: CRC Press.

de Groat, W. C., & Ryall, R. W. (1968). The identification and characteristics of sacral parasympathetic preganglionic neurones. *Journal of Physiology, 196*(3), 563–577.

de Groat, W. C., & Ryall, R. W. (1969). Reflexes to sacral parasympathetic neurones concerned with micturition in the cat. *Journal of Physiology, 200*(1), 87–108.

de Groat, W. C., & Wickens, C. (2013). Organization of the neural switching circuitry underlying reflex micturition. *Acta Physiologica, 207*(1), 66–84.

de Groat, W. C., & Yoshimura, N. (2001). Pharmacology of the lower urinary tract. *Annual Review of Pharmacology and Toxicology, 41*, 691–721.

de Groat, W. C., & Yoshimura, N. (2006). Mechanisms underlying the recovery of lower urinary tract function following spinal cord injury. *Progress in Brain Research, 152*, 59–84.

de Groat, W. C., & Yoshimura, N. (2009). Afferent nerve regulation of bladder function in health and disease. *Handbook of Experimental Pharmacology*, (194), 91–138.

de Groat, W. C., & Yoshimura, N. (2010). Changes in afferent activity after spinal cord injury. *Neurourology and Urodynamics, 29*(1), 63–76.

de Groat, W. C., & Yoshimura, N. (2012). Plasticity in reflex pathways to the lower urinary tract following spinal cord injury. *Experimental Neurology, 235*(1), 123–132.

de Groat, W. C., Yoshiyama, M., Ramage, A. G., Yamamoto, T., & Somogyi, G. T. (1999). Modulation of voiding and storage reflexes by activation of alpha1-adrenoceptors. [Review] [20 refs]. *European Urology, 36*(Suppl. 1), 68–73.

Guarneri, L., Angelico, P., Ibba, M., Poggesi, E., Taddei, C., Leonardi, A., & Testa, R. (1996). Pharmacological in vitro studies of the new 1,4-dihydropyridine calcium antagonist lercanidipine. *Arzneimittelforschung, 46*(1), 15–24.

Gu, B., Olejar, K. J., Reiter, J. P., Thor, K. B., & Dolber, P. C. (2004). Inhibition of bladder activity by 5-hydroxytryptamine1 serotonin receptor agonists in cats with chronic spinal cord injury. *Journal of Pharmacology and Experimental Therapeutics, 310*(3), 1266–1272.

Habler, H. J., Janig, W., & Koltzenburg, M. (1990). Activation of unmyelinated afferent fibres by mechanical stimuli and inflammation of the urinary bladder in the cat. *Journal of Physiology, 425*, 545–562.

Harvie, C., Weissbart, S. J., Kadam-Halani, P., Rao, H., & Arya, L. A. (2019). Brain activation during the voiding phase of micturition in healthy adults: A meta-analysis of neuroimaging studies. *Clinical Anatomy, 32*(1), 13–19.

Hashimoto, K., Oyama, T., Sugiyama, T., Park, Y. C., & Kurita, T. (2003). Neuronal excitation in the ventral tegmental area modulates the micturition reflex mediated via the dopamine D1 and D2 receptors in rats. *Journal of Pharmacological Sciences, 92*(2), 143–148.

Heymann, E. W. (2006). Scent marking strategies of New World primates. *American Journal of Primatology, 68*(6), 650–661.

Holstege, G., Griffiths, D., de Wall, H., & Dalm, E. (1986). Anatomical and physiological observations on supraspinal control of bladder and urethral sphincter muscles in the cat. *Journal of Comparative Neurology, 250*(4), 449–461.

Holstege, G., & Mouton, L. J. (2003). Central nervous system control of micturition. *International Review of Neurobiology, 56*, 123–145.

Hou, X. H., Hyun, M., Taranda, J., Huang, K. W., Todd, E., Feng, D., Atwater, E., Croney, D., Zeidel, M. L., Osten, P., & Sabatini, B. L. (2016). Central control circuit for context-dependent micturition. *Cell, 167*(1), 73–86 e12.

Igawa, Y., Mattiasson, A., & Andersson, K. E. (1993). Effects of GABA-receptor stimulation and blockade on micturition in normal rats and rats with bladder outflow obstruction. *The Journal of Urology, 150*(2 Pt 1), 537–542.

Ishiura, Y., Yoshiyama, M., Yokoyama, O., Namiki, M., & de Groat, W. C. (2001). Central muscarinic mechanisms regulating voiding in rats. *Journal of Pharmacology and Experimental Therapeutics, 297*(3), 933–939.

Ishizuka, O., Gu, B. J., Yang, Z. X., Nishizawa, O., & Andersson, K. E. (2002). Functional role of central muscarinic receptors for micturition in normal conscious rats. *The Journal of Urology, 168*(5), 2258–2262.

Ito, T., Sakakibara, R., Nakazawa, K., Uchiyama, T., Yamamoto, T., Liu, Z., Shimizu, E., & Hattori, T. (2006). Effects of electrical stimulation of the raphe area on the micturition reflex in cats. *Neuroscience, 142*(4), 1273–1280.

Ito, H., Sales, A. C., Fry, C. H., Kanai, A. J., Drake, M. J., & Pickering, A. E. (2020). Probabilistic, spinally-gated control of bladder pressure and autonomous micturition by Barrington's nucleus CRH neurons. *Elife, 9*.

Jung, S. Y., Fraser, M. O., Ozawa, H., Yokoyama, O., Yoshiyama, M., De Groat, W. C., & Chancellor, M. B. (1999). Urethral afferent nerve activity affects the micturition reflex; implication for the relationship between stress incontinence and detrusor instability. *The Journal of Urology, 162*(1), 204–212.

Kabat, H., Magoun, H. W., & Ranson, S. W. (1936). Reaction of the bladder to stimulation of points in the forebrain and mid-brain. *Journal of Comparative Neurology, 63*, 211–239.

Kadekawa, K., Yoshimura, N., Majima, T., Wada, N., Shimizu, T., Birder, L. A., Kanai, A. J., de Groat, W. C., Sugaya, K., & Yoshiyama, M. (2016). Characterization of bladder and external urethral activity in mice with or without spinal cord injury–a comparison study with rats. *American Journal of Physiology - Regulatory, Integrative and Comparative Physiology, 310*(8), R752–R758.

Kakizaki, H., Yoshiyama, M., Koyanagi, T., & de Groat, W. C. (2001). Effects of WAY100635, a selective 5-HT1A-receptor antagonist on the micturition-reflex pathway in the rat. *American Journal of Physiology - Regulatory, Integrative and Comparative Physiology, 280*(5), R1407–R1413.

Kanai, A. J. (2011). Afferent mechanism in the urinary tract. *Handbook of Experimental Pharmacology*, (202), 171–205.

Kanie, S., Yokoyama, O., Komatsu, K., Kodama, K., Yotsuyanagi, S., Niikura, S., Nagasaka, Y., Miyamoto, K. I., & Namiki, M. (2000). GABAergic contribution to rat bladder hyperactivity after middle cerebral artery occlusion. *American Journal of Physiology - Regulatory, Integrative and Comparative Physiology, 279*(4), R1230–R1238.

Karnup, S. V., & de Groat, W. C. (2020). Propriospinal neurons of L3-L4 segments involved in control of the rat external urethral sphincter. *Neuroscience, 425*, 12–28.

Keller, J. A., Chen, J., Simpson, S., Wang, E. H., Lilascharoen, V., George, O., Lim, B. K., & Stowers, L. (2018). Voluntary urination control by brainstem neurons that relax the urethral sphincter. *Nature Neuroscience, 21*(9), 1229–1238.

Kihara, K., & de Groat, W. C. (1997). Sympathetic efferent pathways projecting to the bladder neck and proximal urethra in the rat. *Journal of the Autonomic Nervous System, 62*(3), 134–142.

Kimura, Y., Ukai, Y., Kimura, K., Sugaya, K., & Nishizawa, O. (1995). Inhibitory influence from the nucleus reticularis pontis oralis on the micturition reflex induced by electrical stimulation of the pontine micturition center in cats. *Neuroscience Letters, 195*(3), 214–216.

Kitta, T., Chancellor, M. B., de Groat, W. C., Shinohara, N., & Yoshimura, N. (2016). Role of the anterior cingulate cortex in the control of micturition reflex in a rat model of Parkinson's disease. *The Journal of Urology, 195*(5), 1613–1620.

Kitta, T., Kakizaki, H., Furuno, T., Moriya, K., Tanaka, H., Shiga, T., Tamaki, N., Yabe, I., Sasaki, H., & Nonomura, K. (2006). Brain activation during detrusor overactivity in patients with Parkinson's disease: A positron emission tomography study. *The Journal of Urology, 175*(3 Pt 1), 994–998.

Kitta, T., Matsumoto, M., Tanaka, H., Mitsui, T., Yoshioka, M., & Nonomura, K. (2008). GABAergic mechanism mediated via D receptors in the rat periaqueductal gray participates in the micturition reflex: An in vivo microdialysis study. *European Journal of Neuroscience, 27*(12), 3216–3225.

Kitta, T., Mitsui, T., Kanno, Y., Chiba, H., Moriya, K., Yoshioka, M., & Shinohara, N. (2016). Differences in neurotransmitter systems of ventrolateral periaqueductal gray between the micturition reflex and nociceptive regulation: An in vivo microdialysis study. *International Journal of Urology, 23*(7), 593–598.

Kontani, H., Inoue, T., & Sakai, T. (1990). Dopamine receptor subtypes that induce hyperactive urinary bladder response in anesthetized rats. *The Japanese Journal of Pharmacology, 54*, 482–486.

Koshino, K. (1970). Spontaneous potential activities related to the intravesical pressure in the pontine area of the cat. *The Japanese Journal of Physiology, 20*(3), 272–280.

Koyama, Y., Makuya, A., & Kuru, M. (1962). Vesico-motor areas in the cat midbrain. *The Japanese Journal of Physiology, 12*, 63–80.

Koyama, Y., Ozaki, H., & Kuru, M. (1966). Interference between the pontine detrusor nucleus and the pontine urine-storage nucleus. An electromyographical study of the external urethral sphincter. *The Japanese Journal of Physiology, 16*(3), 291–303.

Krhut, J., Holy, P., Tintera, J., Zachoval, R., & Zvara, P. (2014). Brain activity during bladder filling and pelvic floor muscle contractions: A study using functional magnetic resonance imaging and synchronous urodynamics. *International Journal of Urology, 21*(2), 169–174.

Krhut, J., Tintera, J., Holý, P., Zachoval, R., & Zvara, P. (2012). A preliminary report on the use of functional magnetic resonance imaging with simultaneous urodynamics to record brain activity during micturition. *The Journal of Urology, 188*(2), 474–479.

Kruse, M. N., Belton, A. L., & de Groat, W. C. (1993). Changes in bladder and external urethral sphincter function after spinal cord injury in the rat. *American Journal of Physiology, 264*(6 Pt 2), R1157–R1163.

Kruse, M. N., Bennett, B., & De Groat, W. C. (1994). Effect of urinary diversion on the recovery of micturition reflexes after spinal cord injury in the rat. *The Journal of Urology, 151*(4), 1088–1091.

Kruse, M. N., Mallory, B. S., Noto, H., Roppolo, J. R., & de Groat, W. C. (1991). Properties of the descending limb of the spinobulbospinal micturition reflex pathway in the cat. *Brain Research, 556*(1), 6–12.

Kruse, M. N., Mallory, B. S., Noto, H., Roppolo, J. R., & de Groat, W. C. (1992). Modulation of the spinobulbospinal micturition reflex pathway in cats. *American Journal of Physiology, 262*(3 Pt 2), R478–R484.

Kuhtz-Buschbeck, J. P., Gilster, R., van der Horst, C., Hamann, M., Wolff, S., & Jansen, O. (2009). Control of bladder sensations: An fMRI study of brain activity and effective connectivity. *NeuroImage, 47*(1), 18–27.

Kuhtz-Buschbeck, J. P., van der Horst, C., Pott, C., Wolff, S., Nabavi, A., Jansen, O., & Junemann, K. P. (2005). Cortical representation of the urge to void: A functional magnetic resonance imaging study. *The Journal of Urology, 174*(4 Pt 1), 1477–1481.

Kuipers, R., Mouton, L. J., & Holstege, G. (2006). Afferent projections to the pontine micturition center in the cat. *Journal of Comparative Neurology, 494*, 36–53.

Kuru, M. (1965). Nervous control of micturition. *Physiological Reviews, 45*, 425–494.

Langworthy, O. R., & Kolb, L. C. (1933). The encephalic control of tones in the musculature of the urinary bladder. *Brain, 56*, 371–382.

Langworthy, O. R., & Kolb, L. C. (1935). *Demonstration of encephalic control of micturition by electrical stimulation* (Vol 56). Bulletin of the Johns Hopkins Hospital.

Lecci, A., Giuliani, S., Santicioli, P., & Maggi, C. A. (1992). Involvement of 5-hydroxytryptamine1A receptors in the modulation of micturition reflexes in the anesthetized rat. *Journal of Pharmacology and Experimental Therapeutics, 262*, 181–189.

Lee, S. J., Nakamura, Y., & de Groat, W. C. (2003). Effect of (+/-)-epibatidine, a nicotinic agonist, on the central pathways controlling voiding function in the rat. *American Journal of Physiology - Regulatory, Integrative and Comparative Physiology, 285*(1), R84–R90.

Maggi, C. A., & Conte, B. (1990). Effect of urethane anesthesia on the micturition reflex in capsaicin -treated rats. *Journal of the Autonomic Nervous System, 30*, 247–251.

Maggi, C. A., Santicioli, P., & Meli, A. (1986). Postnatal development of micturition reflex in rats. *American Journal of Physiology, 250*(5 Pt 2), R926–R931.

Mallory, B. S., Roppolo, J. R., & de Groat, W. C. (1991). Pharmacological modulation of the pontine micturition center. *Brain Research, 546*, 310–320.

Marson, L. (1997). Identification of central nervous system neurons that innervate the bladder body, bladder base, or external urethral sphincter of female rats: A transneuronal tracing study using pseudorabies virus. *Journal of Comparative Neurology, 389*(4), 584–602.

Martner, J. (1975). Influences on the defecation and micturition reflexes by the cerebellar fastigial nucleus. *Acta Physiologica Scandinavica, 94*(1), 95–104.

Masuda, H., Hayashi, Y., Chancellor, M. B., Kihara, K., de Groat, W. C., de Miguel, F., & Yoshimura, N. (2006). Roles of peripheral and central nicotinic receptors in the micturition reflex in rats. *The Journal of Urology, 176*(1), 374–379.

Matsumoto, S., Levendusky, M. C., Longhurst, P. A., Levin, R. M., & Millington, W. R. (2004). Activation of mu opioid receptors in the ventrolateral periaqueductal gray inhibits reflex micturition in anesthetized rats. *Neuroscience Letters, 363*(2), 116–119.

Matsuta, Y., Yusup, A., Tanase, K., Ishida, H., Akino, H., & Yokoyama, O. (2010). Melatonin increases bladder capacity via GABAergic system and decreases urine volume in rats. *The Journal of Urology, 184*(1), 386–391.

Matsuura, S., Downie, J. W., & Allen, G. V. (2000). Micturition evoked by glutamate microinjection in the ventrolateral periaqueductal gray is mediated through Barrington's nucleus in the rat. *Neuroscience, 101*, 1053–1061.

McGuire, E., Morrissey, S., Zhang, S., & Horwinski, E. (1983). Control of reflex detrusor activity in normal and spinal injured non-human primates. *The Journal of Urology, 129*(1), 197–199.

McKenna, K. E., & Nadelhaft, I. (1986). The organization of the pudendal nerve in the male and female rat. *Journal of Comparative Neurology, 248*(4), 532–549.

McMahon, S. B., & Morrison, J. F. (1982). Spinal neurones with long projections activated from the abdominal viscera of the cat. *Journal of Physiology, 322*, 1–20.

McMahon, S. B., & Spillane, K. (1982). Brain stem influences on the parasympathetic supply to the urinary bladder of the cat. *Brain Research, 234*, 237–249.

Miura, A., Kawatani, M., & De Groat, W. C. (2003). Excitatory synaptic currents in lumbosacral parasympathetic preganglionic neurons evoked by stimulation of the dorsal commissure. *Journal of Neurophysiology, 89*(1), 382–389.

Miyazato, M., Kadekawa, K., Kitta, T., Wada, N., Shimizu, N., de Groat, W. C., Birder, L. A., Kanai, A. J., Saito, S., & Yoshimura, N. (2017). New frontiers of basic science research in neurogenic lower urinary tract dysfunction. *Urologic Clinics of North America, 44*(3), 491–505.

Miyazato, M., Kaiho, Y., Kamo, I., Kitta, T., Chancellor, M. B., Sugaya, K., Arai, Y., de Groat, W. C., & Yoshimura, N. (2009). Role of spinal serotonergic pathways in sneeze-induced urethral continence reflex in rats. *American Journal of Physiology - Renal Physiology, 297*(4), F1024–F1031.

Miyazato, M., Sasatomi, K., Hiragata, S., Sugaya, K., Chancellor, M. B., de Groat, W. C., & Yoshimura, N. (2007). GABA-receptor activation in the lumbosacral spinal cord reduces detrusor overactivity in spinal cord injured rats. *The Journal of Urology, 179*, 1178–1183.

Miyazato, M., Sasatomi, K., Hiragata, S., Sugaya, K., Chancellor, M. B., de Groat, W. C., & Yoshimura, N. (2008a). GABA receptor activation in the lumbosacral spinal cord decreases detrusor overactivity in spinal cord injured rats. *The Journal of Urology, 179*(3), 1178–1183.

Miyazato, M., Sasatomi, K., Hiragata, S., Sugaya, K., Chancellor, M. B., de Groat, W. C., & Yoshimura, N. (2008b). Suppression of detrusor-sphincter dysynergia by GABA-receptor activation in the lumbosacral spinal cord in spinal cord-injured rats. *American Journal of Physiology - Regulatory, Integrative and Comparative Physiology, 295*(1), R336–R342.

Miyazato, M., Sugaya, K., Nishijima, S., Ashitomi, K., Hatano, T., & Ogawa, Y. (2003). Inhibitory effect of intrathecal glycine on the micturition reflex in normal and spinal cord injury rats. *Experimental Neurology, 183*(1), 232–240.

Miyazato, M., Sugaya, K., Nishijima, S., Ashitomi, K., Morozumi, M., & Ogawa, Y. (2005). Dietary glycine inhibits bladder activity in normal rats and rats with spinal cord injury. *The Journal of Urology, 173*(1), 314–317.

Morrison, J. F., Birder, L., & Craggs, M. (2005). Neural control. In P. Abrams, L. Cordozo, S. Khoury, & A. Wein (Eds.), *Incontinence* (pp. 363–422). Plymouth, UK: Health Publication Ltd.

Morrison, J. F., & Spillane, K. (1986). Neuropharmacological Studies on descending inhibitory controls over the micturition reflex. *Journal of the Autonomic Nervous System*, (Suppl. l), 393–397.

Mukhopadhyay, S., & Stowers, L. (2020). Choosing to urinate. Circuits and mechanisms underlying voluntary urination. *Current Opinion in Neurobiology, 60*, 129–135.

Nadelhaft, I., & Booth, A. M. (1984). The location and morphology of preganglionic neurons and the distribution of visceral afferents from the rat pelvic nerve: A horseradish peroxidase study. *Journal of Comparative Neurology, 226*(2), 238–245.

Nadelhaft, I., de Groat, W. C., & Morgan, C. (1980). Location and morphology of parasympathetic preganglionic neurons in the sacral spinal cord of the cat revealed by retrograde axonal transport of horseradish peroxidase. *Journal of Comparative Neurology, 193*(1), 265–281.

Nadelhaft, I., Roppolo, J., Morgan, C., & de Groat, W. C. (1983). Parasympathetic preganglionic neurons and visceral primary afferents in monkey sacral spinal cord revealed following application of horseradish peroxidase to pelvic nerve. *Journal of Comparative Neurology, 216*(1), 36–52.

Nadelhaft, I., & Vera, P. L. (1995). Central nervous system neurons infected by pseudorabies virus injected into the rat urinary bladder following unilateral transection of the pelvic nerve. *Journal of Comparative Neurology, 359*(3), 443–456.

Nadelhaft, I., & Vera, P. L. (1996). Neurons in the rat brain and spinal cord labeled after pseudorabies virus injected into the external urethral sphincter. *Journal of Comparative Neurology, 375*(3), 502–517.

Nadelhaft, I., & Vera, P. L. (2001). Separate urinary bladder and external urethral sphincter neurons in the central nervous system of the rat: Simultaneous labeling with two immunohistochemically distinguishable pseudorabies viruses. *Brain Research, 903*(1–2), 33–44.

Nadelhaft, I., Vera, P. L., Card, J. P., & Miselis, R. R. (1992). Central nervous system neurons labelled following the injection of pseudorabies virus into the rat urinary bladder. *Neuroscience Letters, 143*(1–2), 271–274.

Nakamura, Y., Ishiura, Y., Yokoyama, O., Namiki, M., & De Groat, W. C. (2003). Role of protein kinase C in central muscarinic inhibitory mechanisms regulating voiding in rats. *Neuroscience, 116*(2), 477–484.

Naka, H., Nishijima, S., Kadekawa, K., Sugaya, K., & Saito, S. (2009). Influence of glutamatergic projections to the rostral pontine reticular formation on micturition in rats. *Life Sciences, 85*(21–22), 732–736.

Nardos, R., Gregory, W. T., Krisky, C., Newell, A., Nardos, B., Schlaggar, B., & Fair, D. A. (2014). Examining mechanisms of brain control of bladder function with resting state functional connectivity MRI. *Neurourology and Urodynamics, 33*(5), 493–501.

Nishijima, S., Sugaya, K., Kadekawa, K., Ashitomi, K., & Yamamoto, H. (2012). Effect of chemical stimulation of the medial frontal lobe on the micturition reflex in rats. *The Journal of Urology, 187*(3), 1116–1120.

Nishijima, S., Sugaya, K., Miyazato, M., Shimabukuro, S., Morozumi, M., & Ogawa, Y. (2005). Activation of the rostral pontine reticular formation increases the spinal glycine level and inhibits bladder contraction in rats. *The Journal of Urology, 173*(5), 1812–1816.

Nishizawa, O., Fukuda, T., Matsuzaki, A., Moriya, I., Harada, T., & Tsuchida, S. (1985). Role of the sympathetic nerve in bladder and urethral sphincter function during the micturition cycle in the dog evaluated by pressure flow EMG study. *The Journal of Urology, 134*(6), 1259–1261.

Nishizawa, O., Sugaya, K., Noto, H., Harada, T., & Tsuchida, S. (1987). Pontine urine storage center in the dog. *Tohoku Journal of Experimental Medicine, 153*(1), 77–78.

Nishizawa, O., Sugaya, K., & Shimoda, N. (1995). Pontine and spinal modulation of the micturition reflex. *Scandinavian Journal of Urology & Nephrology - Supplementum, 175*, 15–19.

Noto, H., Roppolo, J. R., Steers, W. D., & de Groat, W. C. (1989). Excitatory and inhibitory influences on bladder activity elicited by electrical stimulation in the pontine micturition center in the rat. *Brain Research, 492*(1–2), 99–115.

Noto, H., Roppolo, J. R., Steers, W. D., & de Groat, W. C. (1991). Electrophysiological analysis of the ascending and descending components of the micturition reflex pathway in the rat. *Brain Research, 549*(1), 95–105.

Nour, S., Svarer, C., Kristensen, J. K., Paulson, O. B., & Law, I. (2000). Cerebral activation during micturition in normal men. *Brain, 123*(Pt 4), 781–789.

Numata, A., Iwata, T., Iuchi, H., Taniguchi, N., Kita, M., Wada, N., Kato, Y., & Kakizaki, H. (2008). Micturition-suppressing region in the periaqueductal gray of the mesencephalon of the cat. *American Journal of Physiology - Regulatory, Integrative and Comparative Physiology, 294*(6), R1996–R2000.

O'Donnell, P. D. (1990). Central actions of bethanechol on the urinary bladder in dogs. *Journal of Urology, 143*(3), 634–637.

Ogawa, T., Sakakibara, R., Kuno, S., Ishizuka, O., Kitta, T., & Yoshimura, N. (2017). Prevalence and treatment of LUTS in patients with Parkinson disease or multiple system atrophy. *Nature Reviews Urology, 14*(2), 79–89.

Patra, S., Valls, L., Heredia, G., Burdette, D., & Elisevich, K. (2019). Human left anterior cingulate stimulation elicits a reproducible micturition response. *Stereotactic and Functional Neurosurgery, 97*(4), 278–281.

Pavlakis, A. J., Siroky, M. B., Goldstein, I., & Krane, R. J. (1983). Neurologic findings in Parkinson's disease. *The Journal of Urology, 129*, 80–83.

Pehrson, R., & Andersson, K. E. (2002). Effects of tiagabine, a gamma-aminobutyric acid re-uptake inhibitor, on normal rat bladder function. *The Journal of Urology, 167*(5), 2241–2246.

Pehrson, R., Lehmann, A., & Andersson, K. E. (2002). Effects of gamma-aminobutyrate B receptor modulation on normal micturition and oxyhemoglobin induced detrusor overactivity in female rats. *The Journal of Urology, 168*(6), 2700–2705.

Pehrson, R., Ojteg, G., Ishizuka, O., & Andersson, K. E. (2002). Effects of NAD-299, a new, highly selective 5-HT1A receptor antagonist, on bladder function in rats. *Naunyn-Schmiedeberg's Archives of Pharmacology, 366*(6), 528–536.

Pikov, V., & McCreery, D. B. (2009). Spinal hyperexcitability and bladder hyperreflexia during reversible frontal cortical inactivation induced by low-frequency electrical stimulation in the cat. *Journal of Neurotrauma, 26*(1), 109–119.

Rocha, I., Burnstock, G., & Spyer, K. M. (2001). Effect on urinary bladder function and arterial blood pressure of the activation of putative purine receptors in brainstem areas. *Autonomic Neuroscience, 88*(1–2), 6–15.

Rouzade-Dominguez, M. L., Miselis, R., & Valentino, R. J. (2003). Central representation of bladder and colon revealed by dual transsynaptic tracing in the rat: Substrates for pelvic visceral coordination. *European Journal of Neuroscience, 18*(12), 3311–3324.

Rouzade-Dominguez, M. L., Pernar, L., Beck, S., & Valentino, R. J. (2003). Convergent responses of Barrington's nucleus neurons to pelvic visceral stimuli in the rat: A juxtacellular labelling study. *European Journal of Neuroscience, 18*(12), 3325–3334.

Ruch, T. C., & Tang, P. C. (1956). Localization of brain stem and diencephalic areas controlling the micturation reflex. *Journal of Comparative Neurology, 106*(1), 213–245.

Ruggiero, D. A., Underwood, M. D., Rice, P. M., Mann, J. J., & Arango, V. (1999). Corticotropic-releasing hormone and serotonin interact in the human brainstem: Behavioral implications. *Neuroscience, 91*(4), 1343–1354.

Sakakibara, R., Nakazawa, K., Shiba, K., Nakajima, Y., Uchiyama, T., Yoshiyama, M., Yamanishi, T., & Hattori, T. (2002). Firing patterns of micturition-related neurons in the pontine storage centre in cats. *Autonomic Neuroscience, 99*(1), 24–30.

Sakakibara, R., Nakazawa, K., Uchiyama, T., Yoshiyama, M., Yamanishi, T., & Hattori, T. (2002). Micturition-related electrophysiological properties in the substantia nigra pars compacta and the ventral tegmental area in cats. *Autonomic Neuroscience, 102*(1–2), 30–38.

Sakakibara, R., Tsunoyama, K., Takahashi, O., Sugiyama, M., Kishi, M., Ogawa, E., Uchiyama, T., Yamamoto, T., Yamanishi, T., Awa, Y., & Yamaguchi, C. (2010). Real-time measurement of oxyhemoglobin concentration changes in the frontal micturition area: An fNIRS study. *Neurourology and Urodynamics, 29*(5), 757–764.

Sasaki, M. (2002). Bladder contractility-related neurons in Barrington's nucleus: Axonal projections to the spinal cord in the cat. *Journal of Comparative Neurology, 449*(4), 355–363.

Sasaki, M. (2004). Feed-forward and feedback regulation of bladder contractility by Barrington's nucleus in cats. *Journal of Physiology, 557*(Pt 1), 287–305.

Sasaki, M. (2005a). Properties of Barrington's neurones in cats: Units that fire inversely with micturition contraction. *Brain Research, 1033*(1), 41–50.

Sasaki, M. (2005b). Role of Barrington's nucleus in micturition. *Journal of Comparative Neurology, 493*(1), 21–26.

Sasaki, M., & Sato, H. (2013). Polysynaptic connections between Barrington's nucleus and sacral preganglionic neurons. *Neurosciences Research, 75*(2), 150–156.

Seki, S., Igawa, Y., Kaidoh, K., Ishizuka, O., Nishizawa, O., & Andersson, K. E. (2001). Role of dopamine D1 and D2 receptors in the micturition reflex in conscious rats. *Neurourology and Urodynamics, 20*(1), 105–113.

Seki, S., Sasaki, K., Igawa, Y., Nishizawa, O., Chancellor, M. B., De Groat, W. C., & Yoshimura, N. (2004). Suppression of detrusor-sphincter dyssynergia by immunoneutralization of nerve growth factor in lumbosacral spinal cord in spinal cord injured rats. *The Journal of Urology, 171*(1), 478–482.

Seseke, S., Baudewig, J., Kallenberg, K., Ringert, R. H., Seseke, F., & Dechent, P. (2006). Voluntary pelvic floor muscle control—an fMRI study. *NeuroImage, 31*(4), 1399–1407.

Shefchyk, S. J. (2001). Sacral spinal interneurones and the control of urinary bladder and urethral striated sphincter muscle function. *Journal of Physiology, 533*(Pt 1), 57–63.

Shimizu, T., Shimizu, S., Higashi, Y., Nakamura, K., Yoshimura, N., & Saito, M. (2016). A stress-related peptide bombesin centrally induces frequent urination through brain bombesin receptor types 1 and 2 in the rat. *Journal of Pharmacology and Experimental Therapeutics, 356*(3), 693–701.

Shimizu, T., Shimizu, S., Wada, N., Takai, S., Shimizu, N., Higashi, Y., Kadekawa, K., Majima, T., Saito, M., & Yoshimura, N. (2017). Brain serotoninergic nervous system is involved in bombesin-induced frequent urination through brain 5-HT(7) receptors in rats. *British Journal of Pharmacology, 174*(18), 3072–3080.

Sillen, U., Rubenson, A., & Hjalmas, K. (1982). Central cholinergic mechanisms in L-DOPA induced hyperactive urinary bladder of the rat. *Urological Research, 10*(5), 239–243.

Skultety, F. M. (1959). Relation to periaqueductal gray matter to stomach and bladder motility. *Neurology, 9*(3), 190–198.

Steers, W. D., & de Groat, W. C. (1989). Effects of m-chlorophenylpiperazine on penile and bladder function in rats. *American Journal of Physiology, 257*, R1441–R1449.

Stone, E., Coote, J. H., Allard, J., & Lovick, T. A. (2011). GABAergic control of micturition within the periaqueductal grey matter of the male rat. *Journal of Physiology, 589*(Pt 8), 2065–2078.

Sugaya, K., & de Groat, W. C. (1994). Micturition reflexes in the in vitro neonatal rat brain stem-spinal cord-bladder preparation. *American Journal of Physiology, 266*(3 Pt 2), R658–R667.

Sugaya, K., Mori, S., Nishizawa, O., Noto, H., & Tsuchida, S. (1987). Chemical stimulation of the pontine micturition center and the nucleus reticularis pontis oralis. *Neurourology and Urodynamics, 6*, 143–144.

Sugaya, K., Nishijima, S., Kadekawa, K., Ashitomi, K., Ueda, T., & Yamamoto, H. (2014). Intravenous or local injections of flavoxate in the rostral pontine reticular formation inhibit urinary frequency induced by activation of medial frontal lobe neurons in rats. *The Journal of Urology, 192*, 1278–1285.

Sugaya, K., Nishijima, S., Miyazato, M., & Ogawa, Y. (2005). Central nervous control of micturition and urine storage. *Journal of Smooth Muscle Research, 41*(3), 117–132.

Sugaya, K., Ogawa, Y., Hatano, T., Koyama, Y., Miyazato, T., & Oda, M. (1998). Evidence for involvement of the subcoeruleus nucleus and nucleus raphe magnus in urine storage and penile erection in decerebrate rats. *The Journal of Urology, 159*(6), 2172–2176.

Sugaya, K., Ogawa, Y., Hatano, T., Nishijima, S., Matsuyama, K., & Mori, S. (2003). Ascending and descending brainstem neuronal activity during cystometry in decerebrate cats. *Neurourology and Urodynamics, 22*(4), 343–350.

Sugaya, K., Roppolo, J. R., Yoshimura, N., Card, J. P., & de Groat, W. C. (1997). The central neural pathways involved in micturition in the neonatal rat as revealed by the injection of pseudorabies virus into the urinary bladder. *Neuroscience Letters, 223*(3), 197–200.

Suzuki, T., Shimizu, T., Kwon, J., Takaoka, E., Yoshikawa, S., Sumino, Y., Kitta, T., Miyazato, M., Miyake, H., & Yoshimura, N. (2018). Role of the serotonergic system in urethral continence reflexes during sneezing in rats. *American Journal of Physiology - Renal Physiology, 315*(1), F79–F85.

Tai, C., Wang, J., Jin, T., Wang, P., Kim, S. G., Roppolo, J. R., & de Groat, W. C. (2009). Brain switch for reflex micturition control detected by FMRI in rats. *Journal of Neurophysiology, 102*(5), 2719–2730.

Takao, T., Tsujimura, A., Miyagawa, Y., Kiuchi, H., Ueda, T., Hirai, T., Komori, K., Takada, S., Nonomura, N., Osaki, Y., Enomoto, K., Hatazawa, J., & Okuyama, A. (2008). Brain responses during the first desire to void: A positron emission tomography study. *International Journal of Urology, 15*(8), 724–728.

Tanaka, Y., Koyama, Y., Kayama, Y., Kawauchi, A., Ukimura, O., & Miki, T. (2003). Firing of micturition center neurons in the rat mesopontine tegmentum during urinary bladder contraction. *Brain Research, 965*(1–2), 146–154.

Tang, P. C. (1955). Levels of brain stem and diencephalon controlling micturition reflex. *Journal of Neurophysiology, 18*(6), 583–595.

Taniguchi, N., Miyata, M., Yachiku, S., Kaneko, S., Yamaguchi, S., & Numata, A. (2002). A study of micturition inducing sites in the periaqueductal gray of the mesencephalon. *The Journal of Urology, 168*(4 Pt 1), 1626–1631.

Testa, R., Guarneri, L., Poggesi, E., Angelico, P., Velasco, C., Ibba, M., Cilia, A., Motta, G., Riva, C., & Leonardi, A. (1999). Effect of several 5-hydroxytryptamine(1A) receptor ligands on the micturition reflex in rats: Comparison with WAY 100635. *Journal of Pharmacology and Experimental Therapeutics, 290*(3), 1258–1269.

Thor, K. B., Blais, D. P., & de Groat, W. C. (1989). Behavioral analysis of the postnatal development of micturition in kittens. *Brain Res Dev Brain Res, 46*(1), 137–144.

Thor, K. B., & de Groat, W. C. (2010). Neural control of the female urethral and anal rhabdosphincters and pelvic floor muscles. *American Journal of Physiology - Regulatory, Integrative and Comparative Physiology, 299*(2), R416–R438.

Thor, K. B., & Donatucci, C. (2004). Central nervous system control of the lower urinary tract: New pharmacological approaches to stress urinary incontinence in women. *The Journal of Urology, 172*(1), 27–33.

Thor, K. B., Hisamitsu, T., & de Groat, W. C. (1990). Unmasking of a neonatal somatovesical reflex in adult cats by the serotonin autoreceptor agonist 5-methoxy-N,N-dimethyltryptamine. *Developmental Brain Research, 54*(1), 35–42.

Thor, K. B., Katofiasc, M. A., Danuser, H., Springer, J., & Schaus, J. M. (2002). The role of 5-HT(1A) receptors in control of lower urinary tract function in cats. *Brain Research, 946*(2), 290–297.

Thor, K. B., Kawatani, M., & de Groat, W. C. (1986). *Plasticity in the reflex pathways to the lower urinary tract of the cat during postnatal development and following spinal cord injury. Development and plasticity of mammalian spinal cord. Fidia Research Series* (Vol 3). Padova, Italy: Fidia Press.

Tish, M. M., & Geerling, J. C. (2020). The brain and the bladder: Forebrain control of urinary (In)Continence. *Frontiers in Physiology, 11*, 658.

Todd, J. K. (1964). Afferent impulses in the pudendal nerves of the cat. *Quarterly Journal of Experimental Physiology & Cognate Medical Sciences, 49*, 258–267.

Valentino, R. J., Kosboth, M., Colflesh, M., & Miselis, R. R. (2000). Transneuronal labeling from the rat distal colon: Anatomic evidence for regulation of distal colon function by a pontine corticotropin-releasing factor system. *Journal of Comparative Neurology, 417*(4), 399–414.

Valentino, R. J., Pavcovich, L. A., & Hirata, H. (1995). Evidence for corticotropin-releasing hormone projections from Barrington's nucleus to the periaqueductal gray and dorsal motor nucleus of the vagus in the rat. *Journal of Comparative Neurology, 363*(3), 402–422.

Verstegen, A. M. J., Klymko, N., Zhu, L., Mathai, J. C., Kobayashi, R., Venner, A., Ross, R. A., VanderHorst, V. G., Arrigoni, E., Geerling, J. C., & Zeidel, M. L. (2019). Non-crh glutamatergic neurons in Barrington's nucleus control micturition via glutamatergic afferents from the midbrain and hypothalamus. *Current Biology, 29*(17), 2775–2789.

Verstegen, A. M. J., Vanderhorst, V., Gray, P. A., Zeidel, M. L., & Geerling, J. C. (2017). Barrington's nucleus: Neuroanatomic landscape of the mouse "pontine micturition center". *Journal of Comparative Neurology, 525*(10), 2287–2309.

Vincent, S. R., & Satoh, K. (1984). Corticotropin-releasing factor (CRF) immunoreactivity in the dorsolateral pontine tegmentum: Further studies on the micturition reflex system. *Brain Research, 308*(2), 387–391.

Vizzard, M. A., Erickson, V. L., Card, J. P., Roppolo, J. R., & de Groat, W. C. (1995). Transneuronal labeling of neurons in the adult rat brainstem and spinal cord after injection of pseudorabies virus into the urethra. *Journal of Comparative Neurology, 355*(4), 629–640.

Wang, Z., Chang, H. H., Gao, Y., Zhang, R., Guo, Y., Holschneider, D. P., & Rodriguez, L. V. (2017). Effects of water avoidance stress on peripheral and central responses during bladder filling in the rat: A multidisciplinary approach to the study of urologic chronic pelvic pain syndrome (MAPP) research network study. *PLoS One, 12*(9), e0182976.

Willette, R. N., Morrison, S., Sapru, H. N., & Reis, D. J. (1988). Stimulation of opiate receptors in the dorsal pontine tegmentum inhibits reflex contraction of the urinary bladder. *Journal of Pharmacology and Experimental Therapeutics, 244*(1), 403–409.

Wood, S. K., McFadden, K., Griffin, T., Wolfe, J. H., Zderic, S., & Valentino, R. J. (2013). A corticotropin-releasing factor receptor antagonist improves urodynamic dysfunction produced by social stress or partial bladder outlet obstruction in male rats. *American Journal of Physiology - Regulatory, Integrative and Comparative Physiology, 304*(11), R940–R950.

Yamamoto, T., Sakakibara, R., Hashimoto, K., Nakazawa, K., Uchiyama, T., Liu, Z., Ito, T., & Hattori, T. (2005). Striatal dopamine level increases in the urinary storage phase in cats: An in vivo microdialysis study. *Neuroscience, 135*(1), 299–303.

Yamamoto, T., Sakakibara, R., Nakazawa, K., Uchiyama, T., Shimizu, E., Hattori, T., & Kuwabara, S. (2010). Neuronal activities of forebrain structures with respect to bladder contraction in cats. *Neuroscience Letters, 473*(1), 42–47.

Yao, J., Zhang, Q., Liao, X., Li, Q., Liang, S., Li, X., Zhang, Y., Li, X., Wang, H., Qin, H., Wang, M., Li, J., Zhang, J., He, W., Zhang, W., Li, T., Xu, F., Gong, H., Jia, H., … Chen, X. (2018). A corticopontine circuit for initiation of urination. *Nature Neuroscience, 21*(11), 1541–1550.

Yokoyama, O., Ootsuka, N., Komatsu, K., Kodama, K., Yotsuyanagi, S., Niikura, S., Nagasaka, Y., Nakada, Y., Kanie, S., & Namiki, M. (2001). Forebrain muscarinic control of micturition reflex in rats. *Neuropharmacology, 41*(5), 629–638.

Yokoyama, O., Yoshiyama, M., Namiki, M., & de Groat, W. C. (1999). Glutamatergic and dopaminergic contributions to rat bladder hyperactivity after cerebral artery occlusion. *American Journal of Physiology, 276*(4 Pt 2), R935–R942.

Yokoyama, O., Yoshiyama, M., Namiki, M., & de Groat, W. C. (2002). Changes in dopaminergic and glutamatergic excitatory mechanisms of micturition reflex after middle cerebral artery occlusion in conscious rats. *Experimental Neurology, 173*(1), 129–135.

Yoshikawa, S., Oguchi, T., Funahashi, Y., de Groat, W. C., & Yoshimura, N. (2012). Glycine transporter type 2 (GlyT2) inhibitor ameliorates bladder overactivity and nociceptive behavior in rats. *European Urology, 62*(4), 704–712.

Yoshimura, N. (1999). Bladder afferent pathway and spinal cord injury: Possible mechanisms inducing hyperreflexia of the urinary bladder. *Progress in Neurobiology, 57*(6), 583–606.

Yoshimura, N., & Chancellor, M. B. (2010). Physiology of micturition. In L. Cordozo, & D. R. Staskin (Eds.), *Textbook of female urology and urogynecology* (3rd ed., Vol 1, pp. 200–211). Informa UK Ltd (Chapter 23).

Yoshimura, N., & de Groat, W. C. (1997). Neural control of the lower urinary tract. *International Journal of Urology, 4*, 111–125.

Yoshimura, N., Erdman, S. L., Snider, M. W., & de Groat, W. C. (1998b). Effects of spinal cord injury on neurofilament immunoreactivity and capsaicin sensitivity in rat dorsal root ganglion neurons innervating the urinary bladder. *Neuroscience, 83*(2), 633–643.

Yoshimura, N., Kaiho, Y., Miyazato, M., Yunoki, T., Tai, C., Chancellor, M. B., & Tyagi, P. (2008). Therapeutic receptor targets for lower urinary tract dysfunction. *Naunyn-Schmiedeberg's Archives of Pharmacology, 377*(4–6), 437–448.

Yoshimura, N., Kuno, S., Chancellor, M. B., de Groat, W. C., & Seki, S. (2003). Dopaminergic mechanisms underlying bladder hyperactivity in rats with a unilateral 6-hydroxydopamine (6-OHDA) lesion of the nigrostriatal pathway. *British Journal of Pharmacology, 139*(8), 1425–1432.

Yoshimura, N., Mizuta, E., Kuno, S., Sasa, M., & Yoshida, O. (1993). The dopamine D_1 receptor agonist SKF 38393 suppresses detrusor hyperreflexia in the monkey with parkinsonism induced by 1-methyl-4-phenyl-1,2,3,6-tetrahydropyridine (MPTP). *Neuropharmacology, 32*, 315–321.

Yoshimura, N., Mizuta, E., Yoshida, O., & Kuno, S. (1998a). Therapeutic effects of dopamine D1/D2 receptor agonists on detrusor hyperreflexia in 1-methyl-4-phenyl-1,2,3,6-tetrahydropyridine-lesioned parkinsonian cynomolgus monkeys. *Journal of Pharmacology and Experimental Therapeutics, 286*(1), 228–233.

Yoshimura, N., Sasa, M., Yoshida, O., & Takaori, S. (1992). Dopamine D-1 receptor-mediated inhibition of micturition reflex by central dopamine from the substantia nigra. *Neurourology and Urodynamics, 11*, 535–545.

Yoshiyama, M., & de Groat, W. C. (2005). Supraspinal and spinal alpha-amino-3-hydroxy-5-methylisoxazole-4-propionic acid and N-methyl-D-aspartate glutamatergic control of the micturition reflex in the urethane-anesthetized rat. *Neuroscience, 132*(4), 1017–1026.

Zago, T., Pea, U., Fumagalli, G. L., Areta, L., Marzorati, G., & Bianchi, F. (2010). Cerebellar pathology and micturitional disorders: Anatomotopographic and functional correlations. *Archivio Italiano di Urologia, Andrologia, 82*(4), 177–180.

Zhang, H., Reitz, A., Kollias, S., Summers, P., Curt, A., & Schurch, B. (2005). An fMRI study of the role of suprapontine brain structures in the voluntary voiding control induced by pelvic floor contraction. *NeuroImage, 24*(1), 174–180.

Chapter 4

The bladder as a readout in neuroscience research

Holly A. Roy[1] and Alexander L. Green[2]
[1]*University of Plymouth and Department of Neurosurgery, Derriford Hospital, Plymouth, United Kingdom;* [2]*ALG Nuffield Department of Surgical Sciences, Oxford & Department of Neurosurgery, John Radcliffe Hospital, Oxford, United Kingdom*

Introduction

The bladder is an example of an organ that operates under integrated somatic and autonomic control. Lower urinary tract physiology, which includes the behavior of the detrusor muscle, urethra, and the associated smooth and striated muscle sphincters, can be manipulated and also measured in the experimental setting to gain a better understanding of the neural systems controlling different aspects of bladder behavior. Core basic neuroscience questions include defining the central and peripheral nervous structures involved in sensing bladder distension, maintaining continence, and facilitating bladder emptying. The unique contribution and role of the autonomic nervous system (ANS) is also of interest, and including clinical neuroscience perspectives broadens the scope. For example, changes in bladder physiology after a pelvic nerve or spinal cord injury (SCI) can be used as an indicator of neural injury, then over time and following treatment an indicator of regeneration, and thus incorporated as an assay into an experimental study of the neuroscience of nervous system repair. Finally, bladder control pathways are integrated with neural centers for cognition and emotion, and there is a growing awareness that bladder physiology may also be an important readout in research into social and affective disorders. Thus, lower urinary tract physiology is of genuine importance in neuroscience research.

The gold standard approach for assessing lower urinary tract physiology is urodynamics. Urodynamics is a form of clinical investigation that enables aspects of lower urinary tract function, particularly those relating to pressure and flow, to be measured and recorded for analysis. The technique requires insertion of a urethral and rectal catheter, each of which is connected to a pressure transducer. This enables measurement of intravesical pressure (p_{ves}) via the urethral catheter and intra-abdominal pressure (p_{abd}) via the rectal catheter. Detrusor pressure (p_{det}) can then be derived by subtracting the intra-abdominal pressure from the intravesical pressure readout ($p_{ves} - p_{abd}$). This allows the urodynamist to distinguish whether changes in intravesical pressure are a result of detrusor contraction or abdominal straining. Various urodynamic approaches are possible, including cystometry, in which detrusor pressure can be evaluated during bladder filling and urinary voiding. For filling cystometry, water or 0.9% saline is usually used as the infusion fluid, and cold fluid should be avoided. Filling rates can vary, but around 50 mL/min is often used, unless the patient has a neurogenic bladder in which case a much slower filling rate of 10 mL/min is preferred. The patient is asked to report when they experience the following sensations:

1. Initial sensation of bladder filling
2. Initial desire to void
3. Normal desire to void
4. Strong desire to void

And to indicate when they reach:

5. Maximal cystometric capacity (the volume at which voiding cannot be deferred).

The patient is also asked to report pain during the bladder filling process. Correlations between the urodynamics trace and patient events can then be explored.

Neuro-Urology Research. https://doi.org/10.1016/B978-0-12-822455-7.00009-X

Pressure-flow studies enable measurements of intravesical pressure and urine flow rate to be obtained during urinary voiding. The relationship of intravesical pressure and urine flow can be used to understand and interpret voiding disorders. For example, if the rate of bladder emptying is reduced or the urine flow rate is reduced, detrusor underactivity may be considered.

It is possible to assess external urethral sphincter (EUS) physiology using invasive EUS electromyography (EMG) recordings. Sphincter EMG can help to understand the electrical behavior of the muscle and may be abnormal in a variety of conditions including Fowler's syndrome, multiple sclerosis, and spina bifida. However, a potential limitation of the technique is the relative shortage of sphincter EMG data in healthy or asymptomatic individuals. These principles can be applied in the context of human or animal urinary physiology.

Additional features are used to enhance the basic urodynamics approach. For example, ambulatory urodynamics is a technique that allows patients to undergo a urodynamics assessment while going about their daily life, albeit with catheters in situ. This enables a more physiologically accurate assessment and may be particularly important in situations where patients have symptoms that are difficult to evoke in the urodynamics lab. Videourodynamics, in which contrast is used as the infusion fluid and serial X-rays taken during the urodynamics process, can help to match morphological features with the urodynamics trace and can be useful for adding weight to the diagnosis of many conditions including vesicoureteral reflux.

Bladder function as a readout in basic neuroscience studies

Electrical stimulation of the brain and measurement of bladder physiology

Electrically stimulating regions of the brain, spinal cord, and peripheral nervous system and recording bladder physiology as an output have been a primary means of exploring the neural control of the lower urinary tract and have led to major insights in this field. Measures of bladder function can also be treated as the manipulated variable while recording changes in neural activity within the brain. In this way, it is possible to study the responses of different brain areas to events relating to bladder physiology. This provides information, for example, about how sensory information from the bladder is "encoded" in the brain or about the nature of the electrical signals that initiate urinary voiding. This information could be useful in the future as input into neural interface systems for the regulation of the bladder or for helping to diagnose disorders that affect bladder function.

There are, broadly speaking, two main contexts in which electrical stimulation can be applied to the brain. Firstly, in the animal model—usually in the "normal" physiological situation—during which activity in various brain areas can be altered and the resultant changes in bladder physiology can be measured. The second context is in humans undergoing *deep brain stimulation* (DBS) where electrodes are implanted to control disease states (such as Parkinson's disease [PD] or chronic pain), but the opportunity can be taken to look at the effects of stimulation on bladder control. DBS was originally developed as a tool to probe brain function in animals but later reemerged as a powerful technique to treat disease. DBS (and indeed inserting electrodes in animal models) is made possible by the use of "stereotaxy." This was first developed over 150 years ago by Dittmar and Ovsjanikov who designed a frame that could be attached to an animal's head (providing a fixed reference point) and brain targets could be inferred using coordinates based on known landmarks. The stereotactic frame was adapted by Spiegel and Wycis in the 1940s and procedures were carried out to precisely insert catheters or electrodes into brain areas such as lesioning of the dorsomedial thalamus (part of the limbic pathway) to replace the popularized prefrontal leucotomy (a form of psychosurgery). Heath is credited with the first human "experiments" using DBS in the 1950s and 1960s, and although DBS for psychiatric conditions abated in the late 20th century, it is now enjoying a renaissance in the context of depression, obsessive-compulsive disorder, and other conditions. To date, DBS has been performed in over 160,000 patients worldwide, mainly for movement disorders such as PD, tremor, and dystonia. The aim of this section is to combine the animal and human reports on electrical stimulation in order to summarize the knowledge that has been gained regarding brain control of bladder physiology and micturition in general. One has to keep in mind the fact that DBS in humans is only used in disease states and that the bladder physiology may not be "normal" although as will be seen, this can sometimes be advantageous in understanding how central control of bladder physiology is disrupted in disease states such as PD.

Pontine micturition centers

Most of the time—at least in humans—the bladder is locked in "storage" mode and only switches to "voiding" mode when it is judged safe and/or socially appropriate to do so. Sensory input from the distended bladder activates a spinal—brainstem—spinal neural circuit, and it is the integration between the feedback sensation from the bladder and feedforward (higher cognitive) mechanisms that determines the threshold for the switch between storage and voiding mode. Barrington first described a region in the dorsolateral pontine tegmentum that is involved in the micturition reflex (termed the *pontine micturition center (PMC))*. His lesion experiments demonstrated that a region ventral and medial to the superior cerebellar peduncle prevented normal emptying of the urinary bladder. Reflex firing of the sacral preganglionic neurons is elicited by distension of the bladder, has a long latency (80—120 ms), and is present in decerebrate or intact animals but not those with spinal transection, also implying that there is a supraspinal pathway. Lalley et al. used electrical stimulation in the brainstem of cats and demonstrated that various areas could be stimulated to produce firing of sacral preganglionic neurons and consequent bladder contractions. One such area was the area of the locus coeruleus (LC) that was in the same general area as described by Barrington as the PMC.

Stimulation of pelvic nerve afferents evoked negative local field potentials (LFPs) in the rostral pontine areas. Stimulation of areas rostral and medial to the PMC/LC regions described depressed bladder contractions and often produced an initial inhibition and late firing of the sacral preganglionic fibers although de Groat argued that this latter phenomenon was in response to a strong stimulus in a wide variety of sites and may represent a nonspecific response of activation of various central pathways. Further lesion experiments by Satoh et al. further pinpointed the PMC to a cell group called the nucleus tegmentalis laterodorsalis (TLD) in the lateral part of the central gray matter, just rostral to the LC. Loewy et al. used horseradish peroxidase (HRP) tracer techniques to demonstrate that the TLD has direct connections to the nucleus of the solitary tract (NTS) and the sacral intermediolateral (IML) cell column. These connections are distinct from those arising from the LC that send projections to the dorsomedial medulla and spinal cord. Loewy et al. concluded from their experiments that the TLD provides a direct excitatory input only to the sacral preganglionic neurons and may influence the lumbar sympathetic preganglionic neurons and pelvic floor motor neurons (important for external sphincter control of continence) via other connections with cell groups in the lateral medullary reticular formation (lMRF) or the NTS. Indeed, electrical stimulation of the lMRF causes bladder contraction, which may be due to TLD fibers projecting to this region. Direct NTS stimulation, however, causes bladder relaxation and it is noteworthy that there are multiple connections between TLD, NTS, and lMRF, and both TLD and NTS as well as spinal cord project rostrally to the lateral hypothalamus. Thus, there exist multilevel interactions between excitatory and inhibitory control.

Further electrical experiments were based on updated anatomical derivations that split the dorsolateral pontine tegmentum into a medial part—the "M-region" and a lateral part—the "L-region." Holstege et al. showed that M-region projects to the sacral intermediomedial and IML cell groups. The latter contains preganglionic parasympathetic neurons that form the motor supply of the detrusor muscle of the bladder (associated with contraction). The L-region projects to Onuf's nucleus (S2-4), which contains motor neurons that innervate the pelvic floor including the urethral (and anal) sphincter (associated with continence). Electrical stimulation in the L-region elicited increases in pelvic floor contraction and urethral pressure but did not influence bladder pressure. Conversely, M-region stimulation caused a decrease in pelvic floor EMG and urethral pressure and a delayed increase (2 s) in intravesical pressure and the start of micturition. Thus, the M-region is now known as the PMC and the L-region as the *pontine continence center* (PCC). It is important to take into account the combination of lesion and electrical studies when determining function of networks within the central nervous system (CNS) and not just electrical stimulation alone. Lesions in general will reduce the activity of a specific region, whereas electrical stimulation may have a variable effect depending on the parameters used (such as frequency, pattern of stimulation, and the size or type of electrodes being used). The lesion experiments of Griffiths et al. confirmed the electrical stimulation experiments on the L- and M-region described above in that bilateral lesions in the M-region led to a 2- to 9-week period of urinary retention associated with reduced detrusor activity and increased bladder activity. Conversely, bilateral L-region lesions led to a 2-month state of urinary incontinence accompanied by a decrease in bladder capacity and overactivity of the detrusor. Brainstem electrical stimulation mapping added to these findings suggested a motor pathway projecting from the M- to the L-region and another descending caudally from the L-region. The investigators conclude that the M-region is the "true" micturition center—facilitating the detrusor voiding contraction and also signaling via its connections to the L-region that the sphincter needs to relax in synergy. Furthermore, the L-region also helps maintain urinary continence by controlling the pelvic floor muscles and sphincters in general (including the anal sphincter important for fecal incontinence). Similar studies performed by Noto et al. used electrical stimulation in anesthetized rats. The investigators found that the optimum sites for evoking bladder contractions were in and close to the TLD and in the PAG just dorsal or dorsolateral to the TLD. Inhibitory responses were found in an area just lateral, consistent with the findings of Holstege et al. Further work by Noto et al. used electrical stimulation in the pelvic nerves and afferent nerves to the bladder with recording in the pons to demonstrate the existence of a spino-bulbo-spinal pathway passing through the dorsal pontine tegmentum that is important for the micturition reflex. The PAG is also part of this system.

The periaqueductal gray area

Fukuda demonstrated that stimulation of both dorsal and ventral PAG in the decerebrate canine inhibited the micturition reflex (and defecation and rhythmic straining) and this inhibition was preserved despite hemisection at C2. This implied that ascending information from the vesical and colorectal elements had been suppressed. Similar findings occurred with dorsal raphe and central tegmental field stimulation. Noto et al. described the PAG as one part of a two-part ascending limb of the micturition reflex, the experiments involving pelvic nerve stimulation and evoked potential recording in the PAG and LDT. They postulated that the PAG may be a primary receiving area projecting to the LDT. Later structural studies showed a paucity of connections between lumbosacral cord to PMC and PCC, but connections to dorsal PAG (dPAG) led the authors to consider that the PAG may be acting as a relay station in the context of wider micturition control. Liu et al. showed that high-frequency stimulation (HFS) of the PAG (100 Hz) inhibited the micturition reflex. This micturition reflex control can occur in the context of defense behaviors or due to stimulation of specific regions in the PAG. Work translating this into humans along with further analogous rodent studies demonstrated that PAG stimulation can be used to delay the switch to micturition. In rats, this manifests as bladder filling until overflow incontinence occurs, that is, the bladder reaches its mechanical storage limit, and in humans, there is an increased functional bladder capacity, that is, the urge to void threshold is increased. The frequencies used in this study were lower than that used by Liu et al. (50 Hz in the rodent and similar frequencies in the human). The authors postulated that the more caudal electrodes were working directly on intrinsic connections important for the micturition reflex, whereas rostral stimulation likely activates connections to the caudal ventrolateral PAG (shown by Liu et al.) originating from forebrain structures (in humans and in animals, these would be involved in the social aspects of micturition control).

There is, in a sense, a hierarchy of areas rostral to the PMC and PCC, and areas in the immediate vicinity include the *nucleus reticularis pontis oralis* (RPO), the pedunculopontine nucleus (PPN), and the PAG. Kimura et al. explored electrical stimulation of the RPO in combination with PMC stimulation. The latter on its own reproduced those of previous investigators (stimulating the micturition reflex), but the addition of RPO stimulation provided an inhibitory effect. This occurred at similar stimulus parameters (20–50 Hz, 20–100 μA, 0.2 ms pulse width).

PPN and rostral pontine areas affecting the micturition reflex

Most of the work on PPN stimulation has been performed as a result of the opportunity afforded by clinical stimulation of the PPN in patients with PD. The PPN is a long nucleus in the rostrocaudal direction, and there is much debate as to what is being stimulated in patients. It consists of a diffuse cell group in the rostral pontine/caudal mesencephalic tegmentum (the latter being equivalent to part of the mesencephalic locomotor region in rodents). In the human, it is not possible to gain postmortem studies in the majority of patients and therefore the exact localization of electrodes is difficult to determine except using imaging. It is therefore possible that at least some of the effects of PPN stimulation in human are due to stimulation of surrounding structures. Candidates include the LC (and PMC) inferiorly, the parabrachial nucleus, and medial lemniscus laterally, among others. PPN stimulation has been used for almost 15 years as a means of symptom control in a specific subset of PD patients with primarily gait instability and "ON" freezing (referring to the patient's movement freezing despite high levels of exogenous dopamine). However, from the earliest clinical reports of PPN stimulation, there have been sporadic reports of stimulation altering urinary function. For example, Aviles-Olmos et al. reported a case of unilateral stimulation causing new-onset lower urinary tract symptoms (LUTS) of detrusor overactivity such as frequency, urgency, and urge incontinence when the most caudal contact was stimulated. The authors postulated that they were stimulating the PMC. Roy et al. conducted a six-patient study to further investigate the effects of PPN stimulation on urinary function and to ascertain the mechanisms involved. Full urodynamics studies were performed in the OFF medication and ON and OFF stimulation conditions to assess the effects of stimulation only. The investigators found that stimulation at 20–40 Hz (pulse width 60–90 μs, 1.5–3.5 v) significantly increased maximal bladder capacity from 131 to 199 mL. DTI analysis in two of the subjects showed that the areas stimulated projected to areas such as the DLT and to higher areas known to be involved in urinary control. The researchers concluded that stimulation of the PPN increased bladder capacity, potentially by altering activity within the PMC or affecting cortical/subcortical bladder networks via ascending white matter pathways. One caveat that should be applied when looking at the effects of human DBS is that the volume of tissue activation (VTA) is typically 1–2 mm but increases with increasing amplitude. Therefore, some of the sites stimulated may include surrounding areas. Further characterization of potential areas for therapeutic stimulation has recently been performed by stimulating the bladder and recording from sites in the PAG (as well as PMC and others) to identify which areas are most active.

Rajneesh et al. recently demonstrated, in rats, that DBS of the pedunculopontine tegmentum (PPTg) can restore voiding inefficiencies that are induced by traumatic brain injury. Thus, it may be that electrical stimulation can exert a greater effect on the micturition pathways in the context of injury to bladder networks.

Chen et al. performed a feasibility study in rats to assess four brainstem areas and their potential suitability for DBS for micturition disorders. These included ventral PAG, ventral PPTg, rostral pontine reticular nucleus (PNO), and ventral LC. Electrical stimulation of PAG, PPTg, and PNO all significantly inhibited the micturition reflex, although the amplitude required with PPTg stimulation was lower and the authors conclude that PPTg stimulation is the optimal target. LC stimulation augmented the micturition reflex, as expected from the aforementioned studies.

Cerebellum

The role of the cerebellum in the micturition reflex is largely unknown although electrical stimulation of the fastigial nucleus depresses the micturition reflex by depressing pelvic motor nerve activity, leaving afferent pelvic nerve activity unaffected.

Thalamus and hypothalamus

While there are a number of fMRI studies and structural studies implicating the thalamus in micturition control, there is a paucity of animal electrical stimulation studies. DBS studies in humans have provided a few insights. For example, Kessler et al. studied seven essential tremor patients undergoing DBS of the ventral intermediate (VIM) nucleus of the thalamus. They found that stimulation at high frequency produced an earlier desire to void and reduced bladder capacity with no changes in detrusor pressure, bladder compliance, or flow rates. In the context of dystonia, Mordasini et al. found that although VIM DBS ameliorated detrusor hyperactivity, it reduced maximal flow rate and increased postvoid residual volume.

Gjone found that there were two areas in the hypothalamus subserving bladder function: a dorsolateral excitatory field and a ventromedial inhibitory field. Excitatory responses were abolished by sectioning the parasympathetic bladder supply and inhibitory responses abolished by sectioning the sympathetic supply. The areas subserving bladder control and cardiovascular control were later shown to be separate.

Subthalamic nucleus

Based on previous observations that the spino-bulbo-spinal micturition reflex is under major inhibitory control by descending nigrostriatal pathways in the basal ganglia, Sakakibara et al. investigated the effects of HFS of the subthalamic nucleus (STN) on the micturition reflex in cats. They found that stimulation between 20 and 100 Hz (50–250 μA) terminated the reflex and inhibited succeeding reflexes. In most responses, the bladder relaxed and the urethral sphincter constricted in a synergistic manner. No facilitatory responses were found. The STN is part of the *indirect* pathway of basal ganglia circuitry and this is dominant in the parkinsonian state, leading to loss of pallidal inhibition of the STN and resultant STN hyperactivity. Sakakibara postulated that HFS of the STN may resume the altered basal ganglia circuitry by inhibiting the indirect pathway (that passes through STN) and facilitating the *direct* pathway although the precise mechanisms are unknown.

Although Sakakibara's study was in animals with an intact (nondegenerative) dopaminergic system, most STN studies have been performed in the context of PD because the STN is the most common target nucleus in DBS and PD is the most common indication. Studies in the Göttingen minipig have been carried out in a PD model using 1-methyl-4-phenyl-1,2,3,6-tetrahydropyridine (MPTP). These studies showed that STN stimulation altered the bladder characteristics in the storage phase although the results were somewhat confusing (stimulation did not increase capacity but chronic stimulation followed by turning the stimulation off resulted in a reduced capacity).

Given that most instances of STN HFS involve humans, most of the work looking at STN and bladder function has occurred in the clinical context. Early studies demonstrated that bilateral HFS increases bladder capacity and reflex volume by two to three times and reduces the amplitude of detrusor hyperreflexic contractions. It is important in the context of PD in which urinary functions are abnormal that such stimulation tends toward a normalization of urinary function. Herzog combined HFS studies with regional cerebral blood flow studies using positron emission tomography (PET) to demonstrate that with STN OFF, there is enhanced activity in the anterior cingulate cortex and lateral frontal cortex and that the modulation of function is likely to be due to facilitated processing of afferent bladder information. There is much information since the study by Herzog et al. that STN DBS can lead to improved quality life due to improvements in symptoms of urgency, frequency, nocturia, and hesitancy. However, Fritsche reported two cases of urinary retention following STN HFS in patients with no previous history of LUTS.

Basal ganglia

The basal ganglia are a group of paired subcortical nuclei important for motor control, but also making contribution to a wide variety of other functions including executive function and emotion. The globus pallidus (composed of the globus pallidus interna [GPi] and globus pallidus externa [GPe]), the caudate, and putamen make up the core basal ganglia nuclei, but these are closely related to other structures including the subthalamic nucleus and PPN, which have already been mentioned. There have been few studies on the role of the entopeduncular nucleus (rodent equivalent of GPi) or the GPi itself in micturition, and most studies in humans and animals relate to functional imaging. However, Pazo investigated electrical stimulation of the rostral striatum in rats and found that stimulation applied to the dorsomedial caudal putamen nucleus elicited vesical contractions and increased excitability of the micturition reflex. Conversely, stimulation of the ventromedial caudal putamen suppressed detrusor contractions and increased the micturition threshold. Similar inhibitory effects were found by electrical stimulation of the globus pallidus. Work by Yamamoto in 2009 found that electrical stimulation (70–400 μA) in posterior ventral caudate nucleus and adjacent putamen reduced inhibition of the spontaneous bladder contraction. Porter et al. investigated the role of the pallidum on bladder control in a feline model and found that stimulation of the pallidum did not alter the resting tone of the bladder but reduced the frequency of spontaneous bladder contractions, suggesting a possible inhibitory influence of the pallidum on the detrusor. Single-unit activity was recorded in the pallidum during detrusor contraction. Of 603 units, 141 units displayed activity that could be seen to correlate with detrusor contraction. The typical response was triphasic with increased firing prior to the contraction, inhibition of firing at the peak of the contraction, and further increased firing during detrusor relaxation. Thus striatal stimulation appears to facilitate micturition, whereas GPi stimulation appears to inhibit it. Regarding human clinical studies, there is some ambiguity in the literature as to the effects of GPi DBS on bladder function although this may be related to disease state. For example, Mock et al. studied 33 patients, of whom 13 had GPi DBS and found no significant differences in urological scores (including QOL) in the GPi group (whereas there was an improvement with STN DBS, i.e., the remaining patients). On the other hand, in a cohort of patients with dystonia, Mordasini found, by using urodynamics, that GPi stimulation (HFS) significantly reduced flow rate, increased postvoid residual volume, and abolished detrusor overactivity. This is in keeping with the aforementioned rodent studies.

Higher (cortical) areas

Gjone and Setekleiv (1963) aimed to explore the effect of cortical stimulation on bladder function and the micturition cycle in the cat. In general, they found that within cortical areas, some neurons produced an excitatory effect on the bladder, some produced an inhibitory effect, and some produced no effect. For example, within the "first sensory-motor area," 28 stimulations produced an excitatory effect, 19 produced an inhibitory effect, and 13 produced no effect. Interestingly when the cingulate cortex was stimulated in this study, it was found that supracallosal stimulation consistently produced bladder excitation in eight cats, whereas when the subcallosal cingulate was exposed in four of the same animals, stimulation led to inhibitory effects on the bladder. Of note, blood pressure and respiratory variations were also noted whenever cortical stimulation evoked a bladder response, and in particular, stimulation of the orbital and anterior cingulate areas led to expiratory arrest followed by reduced respiratory frequency. These findings highlight integrated aspects of regulation of autonomic function at the level of the cortex.

The anterior cingulate cortex has been used as a target for DBS in chronic neuropathic pain. Electrical stimulation of this area in a PD rat model (that has bladder overactivity) significantly increases detrusor intercontraction interval, and there is good evidence that the PAG is involved in this action. In another study, Patra et al. found a specific area in the left ACC in a patient undergoing stereoelectroencephalography that induced micturition when stimulated at 50–150 Hz.

Regarding higher function, specific areas in the M1 (motor cortex) region also have the power to interrupt micturition. These areas correspond to cortical evoked potentials produced with stimulation of the pudendal nerve and are in the sacral, hind leg, and abdominal regions of the homunculus. When these foci were stimulated during the micturition reflex, the reflex discharge of the pelvic vesical (PV) branch was interrupted concomitantly with the urethral branch findings. Pulse train stimulation of the M1 foci reset the cycles that are known to originate in the PMC. Thus, the investigators conclude that the M1 foci inhibit the PMC and concomitantly contract the urethra via the pyramidal tract.

Yamamoto et al. (2009) studied cats under ketamine anesthesia and carried out extracellular single-unit recordings in a variety of "forebrain" locations particularly medially (medial and superior frontal gyrus), but also from some lateral regions within the middle and inferior frontal gyrus. Their aim was to identify neurons whose activity related to contraction and relaxation phases of the bladder. By comparing firing frequency in the contraction versus relaxation phase, they identified 94 neurons that were predominantly active during bladder relaxation and 18 neurons that were active during bladder contraction. The bladder relaxation neurons were more laterally located and the bladder contraction neurons were more medially located.

To the author's knowledge, no amygdaloid stimulations have been carried out in humans while assessing bladder function; however, Koyama stimulated the amygdaloid body and olfactory tubercle in 69 dogs, after sectioning the hypogastric nerve and recording from the PV branch and the pudendal urethral (PU) branch. Stimulation at 2–100 Hz (3 ms, 0.2 mA) induced bladder contraction (increased PV activity and reduced PU activity) at medial parts of the intermediate principle and medial principle nucleus as well as parts of the cortical and pericortical nuclei. Conversely, lateral areas in addition to some cortical and pericortical areas had the opposite effects. Olfactory tubercle stimulation induced a vesicopressor response with PV excitation and PU inhibition. This latter response was abolished by section of the lateral preoptic area, lateral hypothalamus, and lateral mesencephalic tegmentum. It was also abolished by partial section of the lateral part of the ipsilateral pontine reticular formation.

Recording of neural activity within the brain during physiological changes in bladder activity

Lalley et al. used electrical stimulation in the brainstem of cats and demonstrated that various areas could be stimulated to produce firing of sacral preganglionic neurons and consequent bladder contractions. One such area was the area of the LC that was in the same general area as described by Barrington. Electrical stimulation of the bladder has been shown to activate LC in addition to the PMC (Meriaux et al., 2017), and neuronal activity within the LC shows temporal correlation with micturition initiation (Manohar et al., 2017). A recent rodent study found, by recording LFPs in the medial prefrontal cortex, that STN stimulation alters alpha power, also associated with reductions in levodopa, dopamine, and serotonin. Recording of LFPs within the STN during urinary voiding in humans was carried out in patients who had undergone STN DBS for PD. In this group of patients, beta oscillations in the STN did not change significantly during urinary voiding. However, the group did not comment on changes in other frequency bands. In the same study, Roy et al. recorded LFP activity from the GPi in patients with PD during urinary voiding and found that beta oscillations were inhibited during voiding compared with the resting state. As beta suppression is typically associated with voluntary movement, the study concluded that the GPi could be contributing to the motor aspects of the voiding process.

More recently, the bladder has been used as the manipulated variable in studies designed to understand the PAG contribution to bladder control. Meriaux et al., 2017 described an approach in rats using implanted electrodes within the bladder wall, delivering stimulation via parameters able to preferentially stimulate sensory afferents but without evoking tissue damage or other adverse effect. The group showed, using cFos quantification, that neuronal activation occurred in the caudal ventrolateral PAG as well as the LC, PMC, spinal cord, superficial dorsal horn, sacral parasympathetic nucleus, and central canal, following electrical stimulation of the bladder. A similar approach was used by Zare et al. (2018). In this study, the bladder was stimulated using electrodes implanted at the bladder neck and dome, selecting parameters that would evoke the micturition reflex but not pain. The study not only identified a significant reduction in firing rate across the units recorded but also found that there was not a direct excitatory or inhibitory effect of bladder stimulation in 24 out of the 26 neurons tested, leaving questions open about the specific role of neuronal populations in the PAG in response to this sort of stimulation. PAG single unit recordings in a decerebrate cat preparation during isovolumetric bladder contraction identified different neuronal populations including tonic storage neurons, which were tonically active during storage and voiding but preferentially active during storage, tonic voiding neurons (tonically active across storage and voiding but preferentially active during voiding), phasic storage neurons, which were active only during storage, and phasic voiding neurons, active only during voiding. Tonic storage neurons were the most common (36/84), followed by tonic micturition (24/84), phasic storage (13/84), and phasic micturition (11/84). These recording studies illustrate the complex organizational nature of the PAG with regard to bladder sensation and control, with much still left to be understood.

In a single patient, Roy et al. recorded LFPs within the sensory thalamus while filling the urinary bladder. They found that sensory thalamic oscillatory activity in the high gamma frequency range was inversely correlated with bladder volume, implying that the sensory thalamus may have a role in encoding bladder volume. They also showed that stimulating the PAG disrupted this correlated activity, implying that PAG stimulation could affect higher processing of bladder afferent information. Further work is needed to better delineate the oscillatory responses to bladder filling within the presumed bladder sensory network.

Optogenetics and pharmacological manipulation of brain circuits with measurement of bladder function

Progress has been made in recent years on the role of the brainstem in bladder control by combining novel approaches including optogenetics and calcium imaging with more traditional tract-tracing, electrical stimulation, and recording approaches. Hou et al. (2016) studied a population of corticotropin-releasing hormone (Crh)−positive neurons within the PMC of the mouse using optogenetic approaches. Activity of this neuronal population correlated with the occurrence of bladder contractions. Moreover, these neurons were identified to have spinal cord projecting glutamatergic fibers, which supported a potentially excitatory role. Labeling studies showed that the Crh-positive PMC population exhibited connectivity with higher brain regions, centers including the anterior cingulate cortex, hypothalamus, amygdala, and the PAG, demonstrating the potential for modulation by these centers. Keller et al. (2018) isolated a neuronal population within Barrington's nucleus, which expressed estrogen receptor 1 (ESR1), and used optogenetics to demonstrate their role in relaxation of the EUS. Verstegen et al. (2019) used optogenetics to confirm excitatory glutamatergic inputs to Barrington's nucleus from the lateral hypothalamic area (LHA) and the PAG and demonstrated that activation of Bar neurons via PAG stimulation resulted in "incontinent voids," that is, small volume urine release wherever the mouse was at the time, whereas stimulation of the LHA afferents led to voiding behavior in which the mouse would move to the corner of the cage and void in a behaviorally and physiologically appropriate manner. These researchers also compared the effects of optogenetic stimulation of the Bar^{Vglut2} population with the smaller $Bar^{Crh/Vglut2}$ population and demonstrated incontinent voids (Bar^{Vglut2}) compared with bladder contraction typically without voiding in the case of $Bar^{Crh/Vglut2}$ activation. They also used fiber photometry to indicate that the Bar^{Vglut2} activity preceded the rise in bladder pressure associated with urinary voiding and continued for some time, whereas $Bar^{Crh/Vglut2}$ activity corresponded to the increase in bladder pressure, indicating a role for Bar^{Vglut2} in initiating voiding and $Bar^{Crh/Vglut2}$ for sustaining or supporting the contraction.

Ito et al. (2020) found that within the Crh-expressing neuronal population in Barrington's nucleus (Bar^{Crh}) and also within a neighboring population ($Bar^{Crh\text{-}like}$), neuronal firing rates corresponded to bladder pressure, with a sigmoid relationship between the two, in which the increase in firing rate preceded the increase in bladder pressure. However, the group also found that the intrinsic excitability of $Barr^{CRH}$ neurons was not related to the micturition cycle.

The use of optogenetic techniques to probe the contribution of separate cell groups to the process of urinary voiding and bladder storage has begun to provide exquisite clarity about the control of micturition in an animal model, particularly with regard to cell populations within the brainstem. We anticipate much progress in the future as these approaches are expanded to address other cellular groups which are hypothesized to have a role in the micturition network.

Bladder function as a readout in clinical neuroscience including developmental, regenerative, and degenerative neuroscience

Two major areas of neuroscience research are the study of regenerative neuroscience, namely the capacity of the nervous system to repair and regenerate along damaged pathways and the study of degenerative neuroscience, understanding the mechanisms of neurodegenerative disease. Developmental neuroscience, closely related to the science of regeneration and degeneration, also has some questions that may be addressed by studying the behavior of the bladder.

Indeed, bladder function has been used as a readout in clinical studies investigating changes in brain and spinal networks in pathological conditions, including those associated with neurodevelopment, regeneration, and degeneration. This is a vast field covering numerous conditions; here we select only a few illustrative cases to discuss.

Developmental neuroscience

The developmental neuroscience of the lower urinary tract refers to the processes relevant to the normal development of the neural systems controlling the lower urinary tract. This includes development in utero but also in postnatal life. Various developmental factors can impact on the healthy development and maturation of the lower urinary tract, with molecular factors most relevant in prenatal life, but social and environmental factors playing an increasingly important role in postnatal life. Bladder function can be an important readout in this context, but may also become a focus for treatment if a developmental problem has led to problems with urinary function.

Spina bifida and tethered cord

Spina bifida is a neurodevelopmental condition in which neural tube closure does not occur completely during neural development leaving the spinal cord and meninges protruding through a defect within the vertebral column and/or the skin. The neural material remains within a sac containing cerebrospinal fluid (CSF) and exposed to amniotic fluid. Many associated clinical features can occur alongside spina bifida including hydrocephalus, Arnold-Chiari II malformation, limb weakness, and bowel and bladder dysfunction including incontinence. In terms of bladder and pelvic organ function, peripheral innervation of the bladder and rectum is reduced in myelomeningocele (MMC) cases, and the composition of the EUS in some cases shows increased fibrosis between muscle bundles; however, the factors driving variability between patients are not well understood. Prenatal repair of MMC demonstrated significant clinical benefits compared with postnatal repair supporting a two-hit hypothesis, whereby the failure of neural tube formation contributes to symptoms, but an additional component of neurological damage to spinal cord structures occurs as a result of prolonged in utero exposure of neural elements to amniotic fluid components including ammonia and urea or mechanical trauma of the herniated cord structures resulting in what is essentially a progressive SCI. Benefits of prenatal repair include reduced need to carry out CSF diversion via VP shunt, improved lower limb function, and reduced rate of Chiari malformation. However, urological function appears to show less immediate benefits from early versus late MMC repair and work is still needed to better understand the mechanisms of bladder dysfunction and restoration of function in this condition. For example, Koh et al. described complete EUS denervation in children with MMC who had undergone prenatal repair, compared with only 39% of the postnatal cohort lacking sphincter activity. On the other hand, urologic follow-up data of patients with pre- and postnatal MMC repairs did show in some studies that certain outcome measures were improved following prenatal repair, including a higher rate of volitional bladder control and urinary continence in patients undergoing prenatal repair compared with a historic postnatal repair group, and in long-term retrospective follow-up of the "Management of Myelomeningocele Study" (MOMS), Brock et al. found reduced rates of clean intermittent catheterization as an index of urinary function, in the prenatal compared with the postnatal repair group. However, a clear biological understanding of the basis of neural injury in MMC and the potential for neural repair and how to encourage this after repair is lacking. Larger studies focusing more closely on pelvic organ function in this patient group is needed, combining state-of-the-art imaging with physiologic lower urinary tract assessment, and perhaps utilizing animal models to investigate the time course of development of sphincter and detrusor control in utero and the organization of sensory, autonomic, and motor pathways for these functions.

Postnatal emergence of continence in animals and humans

Many mammalian species are not born with mature pathways for regulation of bladder control. As discussed in a later section, rodents such as rats are born unable to void urine independently and the voiding reflex must be triggered by anogenital licking carried out by the mother for the first 3 weeks of the rat pup's life. Human infants are born with an intact micturition reflex, but continence is generally not achieved until many months after birth. This is a highly culturally dependent process, and toilet training may begin any time from weeks after birth to after the age of 24 months, with a trend toward toilet training at older ages in some Western countries (Joinson et al., 2009). Relatively little is known about the neuroscience underlying achievement of continence although research has been done to investigate the association between childhood incontinence and developmental, psychological, and parental factors. Joinson et al. (2009) found in their prospective study of a birth cohort of 8334 children that children whose toilet training started after 24 months were more likely to fall into a developmental trajectory group associated with a higher prevalence of daytime wetting. They also found that children whose toilet training was begun before 6 months also had a greater likelihood of having problems with daytime wetting. Although the group did not identify specific neural or cognitive factors responsible for these associations, they suggested that both late and early toilet training could expose the child to stressors that could interfere with the necessary developmental processes important for bladder control. For example, the early toilet training group tended to have younger mothers with lower levels of educational attainment and a higher chance of being lone parents. The authors suggested that these factors might predispose the child to exposure to stressful life events. In the older age group, they suggested that later toilet training increases the time span over which the child might be exposed to stressful life events prior to training, which could then impair their ability to achieve continence. Stressful life events, particularly divorce or parental separation, have also been found to be linked to childhood enuresis. One Swiss study found no association between prematurity and mild-to-moderate neurological impairment with the age at which continence was achieved; however, another study of 170 controls and 156 enuretic children found some impairments in perceptual and motor development in the enuretics compared with the control participants. A larger population-based cohort study, which included data from over 10,000 children, found an association between daytime wetting and child developmental milestones (communication, fine motor, gross motor, social, and total developmental score), as well as maternal depression/anxiety scores (Joinson et al., 2008).

Exploration of cognitive abilities that correlate with achievement of continence, and neuroimaging-based assessments to look at maturation of brain areas happening alongside the development of continence, would shed more light on the neural changes linked with the development of bladder control in childhood. As an example of a neuro-urological developmental area which has received recent attention, there is a growing literature around brain imaging in primary nocturnal enuresis (PNE). Abnormalities have been found in gray matter and white matter in the brains of patients with PNE compared with controls. Sun et al. applied diffusion-weighted MRI techniques to describe significant differences in neurite dispersion and orientation in various white matter tracts of children with PNE, including the internal capsule and cingulum. Gray matter changes include increased gray matter volumes in the supplementary motor area and medial prefrontal cortex. Resting state fMRI changes have also been seen in PNE patients, affecting left medial superior frontal gyrus and left superior occipital gyrus, with also decreased functional connectivity between the left thalamus and left medial superior frontal gyrus. A major question is whether these changes are related to the pathophysiology of the enuresis itself or more closely related to the effect of sleep disruption or the psychological consequences of the condition.

Degenerative neuroscience

Many neurodegenerative conditions are associated with LUTS. Studying the relationship between urinary symptoms and the neuropathological changes seen in these conditions can help understand more about the brain and spinal cord pathways for bladder control. It is also an important route to developing treatment options for the urinary symptoms in these conditions, which are often neglected in the face of more prominent cognitive and motor symptoms.

Parkinson's disease

PD is a neurodegenerative disease resulting from loss of dopaminergic neurons within the substantia nigra, and aggregation of alpha-synuclein deposits as Lewy bodies in neurons. Although PD is predominantly associated with motor symptoms, autonomic dysfunction also occurs including urinary storage symptoms such as frequency, urgency, and nocturia (Chaudhuri et al., 2006; Jain, 2011; Sakakibara et al., 2001). Indeed, some form of bladder symptom occurs in 38%–71% of patients with PD (Sakakibara et al., 2018b, 2018b). Urodynamic assessment demonstrates a range of symptoms including reduced bladder capacity, detrusor overactivity (Sakakibara et al., 2001), uninhibited sphincter relaxation, and reduced detrusor contractility during voiding (Terayama et al., 2012). Reduction in detrusor overactivity and increase in bladder capacity occurs following DBS of the STN. PET imaging of the brain combined with urodynamics in the STN DBS OFF versus ON state suggests that STN improves sensory processing of bladder afferent information with enhanced activity in the anterior cingulate cortex and lateral frontal cortex.

Normal pressure hydrocephalus

Bladder function can be affected in hydrocephalus, a multietiologic disease spectrum associated with enlargement of the fluid spaces within the brain. In particular, urinary symptoms can occur in normal pressure hydrocephalus (NPH), a condition more common in the elderly in which there is enlargement of the ventricular spaces within the brain but normal intracranial pressure on pressure measurement either via lumbar puncture and manometry or by intracranial pressure monitoring. Treatment of NPH is with diversion of CSF, normally by insertion of a ventriculoperitoneal shunt (VP shunt). Hypertension and cardiovascular risk factors are commonly associated with NPH, and poor venous sinus compliance is also noted; however, the pathophysiology is as yet unclear.

Krzastek et al. carried out a study of 55 patients with newly diagnosed NPH and invited those with questionnaire scores indicating moderate-to-severe symptoms or bother for urodynamic assessment. All had evidence of detrusor overactivity on urodynamics and 78.9% had detrusor overactivity–related incontinence. It would have been interesting to see urodynamic results across the whole spectrum of patients recruited to this study, not just those reporting a higher level of symptoms on questionnaires. This study did not look at changes in urodynamic parameters following VP shunt surgery, the standard treatment for NPH.

Aruga et al. (2018) retrospectively analyzed notes for 48 patients who had undergone shunt surgery for management of NPH. These patients participated in urodynamics assessments pre and postoperatively. They found that maximum cystometric capacity and bladder compliance increased significantly following surgery. They also found that of 37 patients who had detrusor overactivity preoperatively, 7 had no detrusor overactivity detected postoperatively. In the 30 patients who had remaining detrusor overactivity, the bladder volume at which the first detrusor contraction occurred increased significantly after surgery. On the other hand, voiding parameters were not significantly affected by VP shunt insertion.

Sakakibara et al. used SPECT to investigate changes in brain perfusion before and after VP shunt insertion in NPH patients who reported (i) no change in bladder symptoms, (ii) improvement in bladder symptoms, and (iii) worsening of bladder symptoms after shunt insertion. They found that patients whose bladder function was reported as improved had significantly increased blood flow in the midcingulate cortex, parietal, and left frontal regions compared with preoperatively. Analysis was done based on patient self-reporting of symptoms rather than urodynamic evaluation pre- and postshunt surgery, which introduces the possibility of confounding factors such as improved mobility leading to faster access to the toilet and therefore fewer episodes of incontinence. Further work to link bladder physiology with neuroimaging findings and intracranial pressure and brain perfusion in NPH and other intracranial disorders of CSF pressure and flow are much needed to help better understand the links between CSF disorders and urinary symptoms, in particular the links between bladder control and cognition.

Multiple system atrophy

Multiple system atrophy (MSA) is another neurodegenerative condition that has prominent LUTS, and indeed, bladder symptoms tend to occur early in the disease process and to become severe over the course of the disease. Erectile dysfunction is another pelvic organ symptom that features in MSA. Underactive bladder and voiding difficulties are prominent features of MSA and degenerative changes occur both in the brain and also in Onuf's nucleus, which innervates the striated muscle of the EUS.

Regenerative neuroscience

In the field of regenerative neuroscience, physiological measurements from the lower urinary tract have been used to provide information about regeneration and the underlying physiology and molecular biology of the regenerative processes within key neural structures important for lower urinary tract function, including particularly the peripheral nerve and spinal pathways involved in bladder control.

The EUS and the external anal sphincter are innervated by the pudendal nerve, which originates in Onuf's nucleus in the spinal cord, and is formed by fibers from S2-S4 in humans (in the rat, the motoneurons of the pudendal nerve are located at L6-S1). The sphincter is normally tonically contracted, and during urinary voiding, the sphincter relaxes to facilitate voiding before contracting again to maintain continence. Injury to the pudendal nerve or its nerve roots can result in urinary or fecal incontinence due to loss of control over the sphincter. Direct nerve damage can occur in various conditions including obstetric trauma (occurring in around 32% of all vaginal deliveries). Sphincter dysfunction can also occur secondary to spinal nerve root injury, such as in cauda equina syndrome. Other conditions such as diabetic peripheral neuropathy and MSA can also affect the pudendal nerve.

Following pudendal nerve injury, regeneration can occur to reinnervate the sphincter muscle, and the process is guided by trophic factors released from the denervated muscle. A measure of the resistance of the urethra to leakage from the bladder, known as leak point pressure (LPP), and EUS sphincter EMG have been used alongside pudendal nerve recordings as readouts in animal studies to investigate pudendal nerve regeneration. It has been shown that activity of the EUS muscle following pudendal nerve injury is at its lowest 4 days postpudendal nerve crush. By 3 weeks postinjury, there is some recovery of both pudendal nerve activity and EUS EMG. Following transection of the pudendal nerve, there is much less recovery than in the nerve crush model. Balog et al. (2020) looked at the effect of TrkB on the inhibition of brain-derived neurotrophic factor (BDNF) in rats following pudendal nerve injury and found that administration of TrkB was associated with reduced expression of the regeneration associated gene beta 2 tubulin in the dorsal lateral motoneurons of the pudendal nerve complex and was also associated with reduced firing rate along the motor branch of the pudendal nerve; however, functional recovery in terms of LPP and EUS EMG firing rate was no different between the groups.

Cauda equina/conus medullaris compression

Bladder function is also an important readout in studies of cauda equina/conus medullaris injury in humans and animal models. Cauda equina and conus medullaris compression are rare and can result from a prolapsed intervertebral disc, fracture, or tumor in the lower spinal cord; however, the results can be catastrophic, with permanent sensory changes (including saddle anesthesia and chronic pain), bowel, bladder, and sexual dysfunction. Animal models of cauda equina injury include the ventral root avulsion injury model, which has been used to explore the effect of these injuries on bladder function. Bilateral L5-S2 ventral root avulsion produced a flaccid bladder with no contractile response to filling and also complete absence of EUS contraction on EMG. Bilateral root reimplantation was found to result in bladder and EUS reinnervation and reacquisition of bladder reflexes, albeit with a degree of bladder sphincter dyssynergia and reduced voiding time. Unilateral L5-S2 injuries resulted in an underactive bladder with reduced intravesical pressure and reduced EUS activation. These studies have helped to explore the regenerative possibilities around bladder function. In the human, objective quantification of nerve injury in cauda equina syndrome and deriving a correlation between this and the severity of bladder outcomes have not been carried out at the time of writing, to the authors' knowledge. However, it is currently believed, based on (mostly retrospective) clinical series of patients with cauda equina syndrome, that earlier decompression of the cauda equina (within 48 h of the onset of urinary symptoms) leads to better outcomes in terms of LUTS.

Bladder outcome is also an important measure in SCI. Bladder physiology has been used as a readout to evaluate efficacy of novel repair strategies following SCI including stem cell–based approaches and early sacral neuromodulation, both of which appear to show promise for improving bladder function after SCI. Clinical trials to investigate some of these areas further are planned.

The autonomic nervous system and bladder control

The organization of autonomic pathways innervating the bladder

The ANS is defined as those neural pathways whose efferent nerves have a ganglionic synapse outside the CNS (Brading, 1999). Strictly speaking, the afferent neurons that run from the visceral organs are not part of the ANS, although their axons may travel with the autonomic nerves. Functionally, the ANS can be divided into sympathetic, parasympathetic, and enteric components and it tends to play a role in contributing to homeostatic maintenance. Although some autonomic reflexes operate entirely outside of conscious awareness (to our knowledge), such as the baroreceptor reflex, micturition requires the coordinated activity of autonomic and somatic nerves and involves sensory information from the bladder entering conscious awareness. These functions are controlled by centers in the spinal cord and the brainstem and modulated by descending inputs from cortical and subcortical brain regions.

Preganglionic nerves of the ANS travel from the brainstem or spinal cord to ganglia. Sympathetic ganglia tend to be located some distance from the target organ, whereas parasympathetic ganglia tend to be located close to the target organ. Preganglionic autonomic nerves are usually small myelinated fibers (approximately 3 μm in diameter), whereas post-ganglionic nerves are generally unmyelinated. In the case of the lower urinary tract, autonomic innervation consists of the parasympathetic preganglionic axons that travel in the S2-S4 nerve roots and synapse in the pelvic ganglia to give rise to postganglionic fibers. The cell bodies of the preganglionic neurons are in the sacral parasympathetic nucleus of the spinal cord at L5-S1 level. The postganglionic nerves (which can be found in the pelvic plexus and the bladder wall) innervate the smooth muscle of the bladder detrusor muscle and release acetylcholine, which acts on cholinergic muscarinic receptors resulting in bladder contraction. Sacral parasympathetic efferents also release nitric oxide (NO) at the urethral smooth muscle, which leads to relaxation of the muscle during bladder contraction to facilitate urine flow (Fowler et al., 2010). Recent developmental studies have raised questions about whether the sacral parasympathetic outflow should indeed be described as parasympathetic due to dissimilarities between markers expressed by sacral neurons compared with the cranial parasympathetic nerves; however, from a functional point of view, the sacral outflow remains generally regarded as parasympathetic; further research may lead to more granular definition of population types. Sympathetic innervation of the bladder originates at the IML cell column of the spinal cord at T11-L2 level. Sympathetic efferents innervate the base of the bladder and the smooth muscle of the urethra. Release of noradrenaline at alpha receptors produces contraction of the urethral and bladder base and helps to maintain continence. Sensory fibers from the bladder travel predominantly with the pelvic nerves, although a smaller proportion travel with sympathetic fibers and enter the thoracolumbar cord—these thoracolumbar afferents are not involved in the micturition reflex per se and are activated mainly by significant bladder distension. Having reached the spinal cord, ascending information about bladder sensation travels in the lateral spino-thalamic tracts and is thought to be processed in the PAG and higher cortical areas.

The central autonomic network and the bladder

Although the mechanics of the peripheral innervation of the bladder by autonomic nerves is well established, the interaction between autonomic effectors and the central regions for bladder control is less well understood. In particular, is there a reciprocal relationship between bladder function and global ANS activity, or is bladder function and its regulation mostly independent of the general workings of the ANS? The network of brain regions that act as an interface with the ANS is known as the central autonomic network (CAN). The ventromedial prefrontal cortex, anterior cingulate cortex, and insula represent key cortical regions in the CAN, and within the diencephalon, the hypothalamus is of importance (Thome et al., 2017). The brainstem contributes a number of regions to the CAN, including the periaqueductal gray area, LC, parabrachial nucleus of the pons, nucleus ambiguus, dorsal motor nucleus, and the NST (Roy & Green, 2019). Sympathetic outflow pathways exist in the ventral, caudal, and rostroventral medulla. The nucleus ambiguous and dorsal motor nucleus represent important sites for parasympathetic outflow. Although it is recognized that a number of regions involved in autonomic control, such as the anterior cingulate cortex, insular cortex, hypothalamus, and PAG, are already well-established hubs within the bladder control network, it is not clear how much their autonomic functions overlap with their bladder control functions. For example, the PAG and ACC have been shown to play a role in central command or preparation for exercise, a process whereby cardio-respiratory changes occur in anticipation of exercise (Green et al., 2012. Gillies et al., 2019). It may be that their role in bladder control is analogous to this, and it may also be found that modulation of bladder state occurs via these brain regions during their central command functions, though it is thought that bladder excitability/physiology has been monitored during central command experiments.

Heart rate variability (HRV), the beat-to-beat variation in time interval between heart beats, is a measure of vagal activity and an index of autonomic function and gives an indication of the balance of parasympathetic compared to sympathetic activity. Studies have shown that HRV varies according to stage in the bladder filling cycle (Mehnert et al., 2009; Oladosu et al., 2019). Mehnert et al. (2009) found that parasympathetic and sympathetic activity was evenly balanced in the early stages of bladder filling but that as "strong desire to void" was reached, there was an increase in relative activity of the sympathetic nervous system, with increased low frequency activity. Increased heart rate was also detected at strong desire to void. These findings imply that the bladder interacts with systemic autonomic dynamics and can influence changes in autonomic activity. Bladder filling could influence HRV and sympathovagal balance by modulating activity within the central autonomic network or through local interactions at the paravertebral sympathetic ganglia or at a postganglionic level, resulting in excitation of multiple components of the lumbar and thoracic sympathetic outflow (Weaver et al., 2012). Autonomic dysreflexia in patients with high spinal cord injuries is an example of how viscerosympathetic reflexes initiated by bladder filling can drive systemic autonomic responses when uncoupled from top-down inhibition (Wallin et al., 2013). Further investigation is needed to understand the mechanisms underlying this effect and its functional relevance, including the neural regions that prevent excessive autonomic responses to bladder filling in healthy individuals. Moreover, HRV is known to be altered in various disorders including posttraumatic stress disorder (Thome et al., 2017), depression (Stein et al., 2000), and bladder pain syndrome (Williams et al., 2015); further research is needed to understand whether longstanding changes in HRV affect the threshold for bladder excitability or elicit clinical symptoms.

The bladder as a readout in affective and social neuroscience: understanding the cognition of voiding

Many studies use bladder function as an output to achieve broader neuroscience goals. Some have been touched on in previous sections, such as those describing the approach of combining bladder physiology with advanced neuroscience approaches such as optogenetics and calcium imaging to identify specific cell populations with roles in lower urinary tract control. One of the most exciting areas of bladder-neuroscience research involves the study of the cognitive aspects of bladder control and bladder physiology. Very little is understood as yet of the neurocognitive mechanics underlying the processing of bladder sensory information within the brain or of the regulatory networks required to maintain continence. Furthermore, there is strong overlap between brain areas for affective processing and bladder regulation, such as the PAG, anterior cingulate cortex, and insula. It is possible that emotional/affective disorders that influence these networks may have a subsequent impact on bladder function. Some early work has begun to explore this connection. This final section describes the links between bladder function and cognitive, social, and affective structures and functions.

Urinary physiology is profoundly connected with various aspects of social and affective neuroscience. Urination, and defecation, in certain neonatal mammals such as rat pups, is entirely dependent on a social factor, namely maternal anogenital licking. Without this stimulation from the mother in the first 3 weeks of life, rat pups will be unable to void or defecate and will therefore not survive. In many species of mammals, urination also represents an important form of social communication, through the presence of scent markers known as pheromones. Urinary pheromones, such as darcin, can drive complex behavioral responses and act as an unconditioned stimulus for associative learning. Darcin is present in the urine of male mice, and the behavioral responses evoked in female mice who are exposed to the pheromone are mediated by a brain network, which includes sensory receptors in the vomeronasal organ, the accessory olfactory bulb, and the medial amygdala. Darcin exposure evokes, as part of the behavioral response, (i) attraction to physical locations associated with the presence of darcin, (ii) responsive urinary scent marking (which is distinctly different to the normal pattern of micturition displayed by female mice: darcin-responsive urination occurring closer to the darcin-enriched area, smaller in volume than normal voids), and (iii) ultrasonic vocalizations (associated temporally with the urinary scent marking). Photoactivation of darcin-responsive neurons in the medial amygdala elicited urine scent marking and vocalizations.

Urinary voiding can also be used to express social hierarchy. In their landmark paper in 1973, Desjardins et al. described marked modulation of urinary voiding patterns in male house mice depending on how they were housed in regard to socially dominant mice. Isolated male mice would typically create 30–80 pools of urine around the cage during a 12-h overnight period, usually around the outer parts of the cage. If housed in a single cage separated from a protagonist mouse by a wire barrier (seen but not brought into physical contact), both mice in the pair would void up to 1000 small pools throughout their cage. The males were then introduced and social dominance was established. After separation back to their own part of the enclosure, separated again by the wire barrier, the dominant mouse continued to mark his cage in the previous pattern (many small pools of urine through the area), while the subordinate mouse urinated only in 2–4 pools at the far corner of the cage. This behavior pattern was maintained over a 5-day period.

Further work has been done specifically to explore the relationship between social stress and bladder function in animal models. For example, Mann et al. (2015) showed that voiding dysfunction, increased bladder wall thickness, and increased bladder to body weight ratio developed in mice who experienced social defeat but not restraint stress, although corticosterone levels were elevated in all stressed mice. In another study of social stress, Mingin et al. (2014) found that mice exposed to social stress displayed increased voiding frequency and reduced bladder capacity compared with control subjects, linking this with increased inflammatory mediators including histamine and nerve growth factor. Also, Chang et al. reported abnormal voiding patterns and bladder hypertrophy in mice exposed to social stressors.

One possible mechanism by which stress could impact on psychological, emotional, and affective functioning as well as bladder function is via stress hormones such as corticotrophin-releasing factor (CRF), which is increased in Barrington's nucleus in response to stress and has a known inhibitory effect on urinary voiding. CRF is synthesized in the neuroendocrine cells of the paraventricular nucleus of the hypothalamus and released via the median eminence into the hypophyseal capillaries to cause ACTH release at the anterior pituitary. CRF administration increases micturition threshold in awake rats, but not sleeping rats. Chronic stress is known to increase CRF mRNA in Barrington's nucleus neurons, and thus it could be hypothesized that chronic stress could lead to urinary retention or difficulty initiating voiding via this pathway. However, to counter this hypothesis, eliminating expression of the Crh gene from Barrington's nucleus has no effect on micturition in mice (Versteegen et al., 2019).

The PAG has been implicated in the development of posttraumatic stress disorder, a condition associated with extreme exposure to stress. dPAG stimulation leads to fight or flight responses and sympathetic nervous system activation. Ventrolateral PAG activity on the other hand has been linked with the dissociative subtype (DS) of PTSD and vagal activity. Functional MRI research has shown that in patients with PTSD + DS, there is increased functional connectivity not only between the ventral PAG and forebrain regions involved in threat vigilance common to all forms of PTSD, such as the dorsal anterior cingulate cortex and the fusiform gyrus, but also with those specifically linked with depersonalization such as the temporoparietal junction and the Rolandic operculum. This subtype of PTSD manifests with emotional blunting and suppression of arousal responses. It is more closely linked with a passive response, including freezing behavior, sympathetic inactivation, and micturition. A single exposure to an adverse fear stimulus induces long-term changes in the PAG. For example, in rats, maternal separation in early life can lead to lower stimulation thresholds in the dPAG for triggering panic behaviors in adulthood. Given the intrinsic role of the PAG in initiating micturition, and its involvement in PTSD, it is possible that the long-term connectivity changes linked to social stress, maternal separation, and trauma may influence micturition along with other PAG functions. To our knowledge, this has not been systematically investigated although it has been shown in a clinical study that patients with interstitial cystitis/bladder pain syndrome were more likely to have posttraumatic stress disorder symptoms as patients with other chronic pain conditions although both groups had similar exposure to traumatic events.

Bladder disorders have been linked with exposure to psychological stress in humans and childhood sexual trauma, but no mechanism described. In a retrospective database review of female patients in a urology clinic, Nault et al. found that women with a history of bullying or abuse had higher scores on the urological distress inventory (UDI-6) and the pelvic floor distress inventory (PFDI-20) as well as higher overall urogenital pain scores (Nault et al., 2016). It is not clear whether pain was controlled for as a confounding factor in evaluating the PFDI-20 and UDI-6 scores. Children with voiding disorders are more likely to have behavioral and psychological disorders than children with normal voiding patterns. In a longitudinal study of female veterans, Bradley et al. (2017) found that new symptoms of overactive bladder occurred more frequently in patients with a baseline history of anxiety, posttraumatic stress disorder, and lifetime sexual assault, with anxiety and sexual assault predicting 1-year incidence of overactive bladder. Another study by Chiu et al. (2017) found that adult patients with interstitial cystitis/bladder pain syndrome had higher rates of physical abuse in childhood or adulthood and higher rates of childhood trauma by close others compared with healthy controls presenting with acute cystitis. Furthermore, in a study of children who had been exposed to a natural disaster (flood), a research group who had been carrying out a neurodevelopmental study in a rural setting in Bangladesh identified an increase in sphincter dysfunction from 16.8% at baseline to 40.4% following the flood, along with a significant increase in aggressive behavior.

Also linked with social communication and stress is another key neurobiological factor that seems to influence the physiology of the bladder—pain and opioid use—and we note here that social pain involves similar brain networks to physical pain (Eisenberger & Lieberman, 2004). In a urodynamics study in human subject, Malinovsky et al. showed that opioids altered bladder sensation and postvoid residual in young adult males, and in the case of fentanyl, morphine, and buprenorphine, led to an inability to void urine. Herman et al. showed that intrathecal morphine increased bladder capacity and reduced uninhibited detrusor contractions in patients with spinal cord injuries, and Pandita et al., (2003) demonstrated reduced micturition in freely moving female rats following administration of tramadol. The link between chronic pain and bladder function is not as clearly demonstrated as the effects of opiates on bladder function. However, anecdotally it is often stated that pain can cause problems with bladder function and however there are a number of conditions, including dysmenorrhea and low back pain/sciatica in which patients experience chronic pain and may also present with urinary symptoms, including reduced bladder capacity, urinary incontinence, and urinary retention. The complex relationship between opioids, pain, and bladder function requires further investigation.

Conclusion

Bladder physiology is used as a readout in a wide variety of studies ranging from basic electrophysiological studies aimed at understanding the neural basis of bladder control to studies that use bladder function as an index of damage in a disease process or regeneration. There is also growing interest in the role of the bladder in social and emotional disorders, and accordingly, there is an emerging literature in both animal and human studies looking at the links between bladder physiology and some of these conditions. Capitalizing on new noninvasive imaging techniques and further expansion of optogenetic approaches combined with effective clinical and physiological assays of bladder function will help to bring further progress in the future.

References

Aruga, S., et al. (2018). Effect of cerebrospinal fluid shunt surgery on lower urinary tract dysfunction in idiopathic normal pressure hydrocephalus. *Neurourology and Urodynamics, 37*(3), 1053–1059.

Balog, B. M., et al. (2020). The pudendal nerve motor branch regenerates via a brain derived neurotrophic factor mediated mechanism. *Experimental Neurology, 334*, 113438.

Brading, A. (1999). *The autonomic nervous system and its effectors*. Wiley-Blackwell.

Bradley, C. S., et al. (2017). Longitudinal associations between mental health conditions and overactive bladder in women veterans. *American Journal of Obstetrics and Gynecology, 217*(4), 430 e431–430 e438.

Chaudhuri, K. R., et al. (2006). Non-motor symptoms of Parkinson's disease: Diagnosis and management. *The Lancet Neurology, 5*(3), 235–245.

Chiu, C. D., et al. (2017). Childhood trauma perpetrated by close others, psychiatric dysfunction, and urological symptoms in patients with interstitial cystitis/bladder pain syndrome. *Journal of Psychosomatic Research, 93*, 90–95.

Eisenberger, N. I., & Lieberman, M. D. (2004). Why rejection hurts: A common neural alarm system for physical and social pain. *Trends in Cognitive Sciences, 8*(7), 294–300.

Fowler, C. J., Panicker, J. N., & Emmanuel, A. (2010). *Pelvic organ dysfunction in neurological disease*. Cambridge University Press.

Gillies, M. J., et al. (2019). Direct neurophysiological evidence for a role of the human anterior cingulate cortex in central command. *Autonomic Neuroscience, 216*, 51–58.

Gjone, R., & Setekleiv, J. (1963). Excitatory and inhibitory bladder responses to stimulation of the cerebral cortex in the cat. *Acta Physiologica Scandinavica, 59*, 337–348.

Green, A. L., et al. (2012). Switching off micturition using deep brain stimulation at midbrain sites. *Annals of Neurology, 72*(1), 144–147.

Hou, X. H., et al. (2016). Central control circuit for context-dependent micturition. *Cell, 167*(1), 73–86 e12.

Ito, H., et al. (2020). Probabilistic, spinally-gated control of bladder pressure and autonomous micturition by Barrington's nucleus CRH neurons. *Elife, 9*, e56605.

Jain, S. (2011). Multi-organ autonomic dysfunction in Parkinson disease. *Parkinsonism & Related Disorders, 17*(2), 77–83.

Joinson, C., et al. (2008). Early childhood risk factors associated with daytime wetting and soiling in school-age children. *Journal of Pediatric Psychology, 33*(7), 739–750.

Joinson, C., et al. (2009). A prospective study of age at initiation of toilet training and subsequent daytime bladder control in school-age children. *Journal of Developmental and Behavioral Pediatrics, 30*(5), 385–393.

Keller, J. A., et al. (2018). Voluntary urination control by brainstem neurons that relax the urethral sphincter. *Nature Neuroscience, 21*(9), 1229–1238.

Mann, E. A., et al. (2015). Chronic social defeat, but not restraint stress, alters bladder function in mice. *Physiology & Behavior, 150*, 83–92.

Manohar, A., et al. (2017). Brainstem network dynamics underlying the encoding of bladder information. *Elife, 6*, e29917.

Mehnert, U., et al. (2009). Heart rate variability: An objective measure of autonomic activity and bladder sensations during urodynamics. *Neurourology and Urodynamics, 28*(4), 313–319.

Meriaux, C., Hohnen, R., Schipper, S., Zare, A., Jahanshahi, A., Birder, L. A., Temel, Y., & van Koeveringe, G. A., et al. (2017). Neuronal activation in the periaqueductal gray matter upon electrical stimulation of the bladder. *Frontiers in Cellular Neuroscience, 12*, 133. https://doi.org/10.3389/fncel.2018.00133

Mingin, G. C., et al. (2014). Social stress induces changes in urinary bladder function, bladder NGF content, and generalized bladder inflammation in mice. *American Journal of Physiology - Regulatory, Integrative and Comparative Physiology, 307*(7), R893–900.

Nault, T., et al. (2016). Does a history of bullying and abuse predict lower urinary tract symptoms, chronic pain, and sexual dysfunction? *International Urology and Nephrology, 48*(11), 1783–1788.

Oladosu, F. A., et al. (2019). Persistent autonomic dysfunction and bladder sensitivity in primary dysmenorrhea. *Scientific Reports, 9*(1), 2194.

Pandita, R. K., et al. (2003). Actions of tramadol on micturition in awake, freely moving rats. *British Journal of Pharmacology, 139*(4), 741–748.

Roy, H. A., & Green, A. L. (2019). The central autonomic network and regulation of bladder function. *Frontiers in Neuroscience, 13*, 535.

Sakakibara, R., et al. (2001). Questionnaire-based assessment of pelvic organ dysfunction in Parkinson's disease. *Autonomic Neuroscience, 92*(1–2), 76–85.

Sakakibara, R., et al. (2018a). Parkinson's disease and prostate enlargement: Both contribute to overactive bladder in the elderly. *International Journal of Urology, 25*(11), 982–983.

Sakakibara, R., et al. (2018b). Urological dysfunction in synucleinopathies: Epidemiology, pathophysiology and management. *Clinical Autonomic Research, 28*(1), 83–101.

Stein, P. K., Carney, R. M., & Freedland, K. E., et al. (2000). Severe depression is associated with markedly reduced heart rate variability in patients with stable coronary heart disease. *Journal of Psychosomatic Research, 48*(4-5), 493–500. https://doi.org/10.1016/s0022-3999(99)00085-9

Terayama, K., et al. (2012). Weak detrusor contractility correlates with motor disorders in Parkinson's disease. *Movement Disorders, 27*(14), 1775–1780.

Thome, J., et al. (2017). Desynchronization of autonomic response and central autonomic network connectivity in posttraumatic stress disorder. *Human Brain Mapping, 38*(1), 27–40.

Verstegen, A. M. J., et al. (2019). Non-Crh glutamatergic neurons in Barrington's nucleus control micturition via glutamatergic afferents from the midbrain and hypothalamus. *Current Biology, 29*(17), 2775–2789. e2777.

Wallin, B. G., et al. (2013). Viscerosympathetic reflexes in human spinal cord injury: Relationships between detrusor pressure, blood pressure and skin blood flow during bladder distension. *Experimental Physiology, 98*(6), 1081–1091.

Weaver, F. M., Follett, K., Stern, M., Luo, P., Harris, C. L., Hur, K., Marks, W. J., Jr., Rothlind, J., Sagher, O., Moy, C., Pahwa, R., Burchiel, K., Hogarth, P., Lai, E. C., Duda, J. E., Holloway, K., Samii, A., Horn, S., Bronstein, J. M., … CSP 468 Study Group. (2012). Randomized trial of deep brain stimulation for Parkinson disease: Thirty-six-month outcomes. *Neurology, 79*(1), 55–65.

Williams, D. P., et al. (2015). Effects of chronic pelvic pain on heart rate variability in women. *The Journal of Urology, 194*(5), 1289–1294.

Yamamoto, T., et al. (2009). Effects of electrical stimulation of the striatum on bladder activity in cats. *Neurourology and Urodynamics, 28*(6), 549–554.

Zare, A., et al. (2018a). Glutamatergic cells in the periaqueductal gray matter mediate sensory inputs after bladder stimulation in freely moving rats. *International Journal of Urology, 25*(6), 621–626.

Zare, A., et al. (2018b). Electrophysiological responses of the ventrolateral periaqueductal gray matter neurons towards peripheral bladder stimulation. *Brain Research Bulletin, 142*, 116–121.

Further reading

Abrams, P. (2005). *Urodynamics.* Springer.

Adzick, N. S., et al. (2011). A randomized trial of prenatal versus postnatal repair of myelomeningocele. *New England Journal of Medicine, 364*(11), 993–1004.

Ahn, U. M., et al. (2000). Cauda equina syndrome secondary to lumbar disc herniation: A meta-analysis of surgical outcomes. *Spine (Phila Pa 1976), 25*(12), 1515–1522.

Albanese, A., et al. (1988). Bladder hyperreflexia induced in marmosets by 1-methyl-4-phenyl-1,2,3,6-tetrahydropyridine. *Neuroscience Letters, 87*(1–2), 46–50.

Alonso, P., et al. (2015). Deep brain stimulation for obsessive-compulsive disorder: A meta-analysis of treatment outcome and predictors of response. *PLoS One, 10*(7), e0133591.

Aviles-Olmos, I., Foltynie, T., Panicker, J., Cowie, D., Limousin, P., Hariz, M., Fowler, C. J., & Zrinzo, L. (2011). Urinary incontinence following deep brain stimulation of the pedunculopontine nucleus. *Acta Neurochirurgica, 153*(12), 2357–2360.

Barrington, F. (1925). The effect of lesions of the hind- and midbrain on micturition in the cat. *Quarterly Journal of Experimental Physiology, 15*, 81–102.

Blok, B. F., et al. (1995). Ultrastructural evidence for a paucity of projections from the lumbosacral cord to the pontine micturition center or M-region in the cat: A new concept for the organization of the micturition reflex with the periaqueductal gray as central relay. *Journal of Comparative Neurology, 359*(2), 300–309.

Blomstedt, P., et al. (2007). Dittmar and the history of stereotaxy; or rats, rabbits, and references. *Neurosurgery, 60*(1), 198–201. discussion 201-192.

Boccard, S. G. J., et al. (2017). Long-term results of deep brain stimulation of the anterior cingulate cortex for neuropathic pain. *World Neurosurgery, 106*, 625–637.

Bradley, W. E., & Teague, C. T. (1969). Cerebellar influence on the micturition reflex. *Experimental Neurology, 23*(3), 399–411.

Brandao, M. L., & Lovick, T. A. (2019). Role of the dorsal periaqueductal gray in posttraumatic stress disorder: Mediation by dopamine and neurokinin. *Translational Psychiatry, 9*(1), 232.

Brock, J. W., 3rd, et al. (2019). Effect of prenatal repair of myelomeningocele on urological outcomes at school age. *The Journal of Urology, 202*(4), 812–818.

Brouette-Lahlou, I., et al. (1992). Role of pups' ultrasonic calls in a particular maternal behavior in Wistar rat: Pups' anogenital licking. *Behavioural Brain Research, 50*(1–2), 147–154.

Brunoni, A. R., et al. (2013). Heart rate variability is a trait marker of major depressive disorder: Evidence from the sertraline vs. electric current therapy to treat depression clinical study. *The International Journal of Neuropsychopharmacology, 16*(9), 1937–1949.

Carr, M. C. (2015). Urological results after fetal myelomeningocele repair in pre-MOMS trial patients at the Children's Hospital of Philadelphia. *Fetal Diagnosis and Therapy, 37*(3), 211–218.

Chang, A., et al. (2009). Social stress in mice induces voiding dysfunction and bladder wall remodeling. *American Journal of Physiology - Renal Physiology, 297*(4), F1101–1108.

Chang, H. H., et al. (2018a). EMG characteristics of the external anal sphincter guarding reflex and effects of a unilateral ventral root avulsion injury in rhesus macaques (*Macaca mulatta*). *Journal of Neurophysiology, 120*(6), 2710–2718.

Chang, H. H., et al. (2018b). Mapping and neuromodulation of lower urinary tract function using spinal cord stimulation in female rats. *Experimental Neurology, 305*, 26–32.

Chang, H. Y., & Havton, L. A. (2008). Re-established micturition reflexes show differential activation patterns after lumbosacral ventral root avulsion injury and repair in rats. *Experimental Neurology, 212*(2), 291–297.

Chang, H. H., & Havton, L. A. (2013). Serotonergic 5-HT(1A) receptor agonist (8-OH-DPAT) ameliorates impaired micturition reflexes in a chronic ventral root avulsion model of incomplete cauda equina/conus medullaris injury. *Experimental Neurology, 239*, 210–217.

Chang, H. H., & Havton, L. A. (2016). A ventral root avulsion injury model for neurogenic underactive bladder studies. *Experimental Neurology, 285*(Pt B), 190–196.

Chen, S. C., et al. (2017). Feasibility of deep brain stimulation for controlling the lower urinary tract functions: An animal study. *Clinical Neurophysiology, 128*(12), 2438–2449.

Chennu, S., et al. (2014). Spectral signatures of reorganised brain networks in disorders of consciousness. *PLoS Computational Biology, 10*(10), e1003887.

Clayton, D. B., et al. (2020). Fetal repair of myelomeningocele: Current status and urologic implications. *Journal of Pediatric Urology, 16*(1), 3–9.

Cote, K. A., et al. (1999). Scalp topography of the auditory evoked K-complex in stage 2 and slow wave sleep. *Journal of Sleep Research, 8*(4), 263–272.

Cruse, D., Thibaut, A., Demertzi, A., Nantes, J. C., Bruno, M. A., Gosseries, O., Vanhaudenhuyse, A., & Bekinschtein, T. A. (2013). Actigraphy assessments of circadian sleep-wake cycles in the vegetative and minimally conscious states. *BMC Medicine, 11*, 18.

Dalmose, A. L., et al. (2004). Effects of high frequency deep brain stimulation on urine storage and voiding function in conscious minipigs. *Neurourology and Urodynamics, 23*(3), 265–272.

De Groat, W. C. (1975). Nervous control of the urinary bladder of the cat. *Brain Research, 87*(2–3), 201–211.

DeLong, M. R. (1990). Primate models of movement disorders of basal ganglia origin. *Trends in Neurosciences, 13*(7), 281–285.

Demir, E., et al. (2020). The pheromone darcin drives a circuit for innate and reinforced behaviours. *Nature, 578*(7793), 137–141.

Desjardins, C., et al. (1973). Social rank in house mice: Differentiation revealed by ultraviolet visualization of urinary marking patterns. *Science, 182*(4115), 939–941.

Drake, M. J., et al. (2018). Fundamentals of urodynamic practice, based on International Continence Society good urodynamic practices recommendations. *Neurourology and Urodynamics, 37*(S6), S50–S60.

Duong, T. H., et al. (2013). Vietnamese mothers' experiences with potty training procedure for children from birth to 2 years of age. *Journal of Pediatric Urology, 9*(6 Pt A), 808–814.

Durkin, M. S., et al. (1993). The effects of a natural disaster on child behavior: Evidence for posttraumatic stress. *American Journal of Public Health, 83*(11), 1549–1553.

van Erp, W. S., et al. (2014). The vegetative state/unresponsive wakefulness syndrome: A systematic review of prevalence studies. *European Journal of Neurology, 21*(11), 1361–1368.

Espinosa-Medina, I., et al. (2016). The sacral autonomic outflow is sympathetic. *Science, 354*(6314), 893–897.

Finazzi-Agro, E., et al. (2003). Effects of subthalamic nucleus stimulation on urodynamic findings in patients with Parkinson's disease. *The Journal of Urology, 169*(4), 1388–1391.

Fisher, R., Salanova, V., Witt, T., Worth, R., Henry, T., Gross, R., Oommen, K., Osorio, I., Nazzaro, J., Labar, D., Kaplitt, M., Sperling, M., Sandok, E., Neal, J., Handforth, A., Stern, J., DeSalles, A., Chung, S., Shetter, A., … SANTE Study Group. (2010). Electrical stimulation of the anterior nucleus of thalamus for treatment of refractory epilepsy. *Epilepsia, 51*, 899–908.

Fowler, C. J., Griffith, D., & de Groat, W. C. (2008). The neural control of micturition. *Nature Reviews Neuroscience, 9*, 453–466.

Fritsche, H. M., et al. (2009). Acute urinary retention in two patients after subthalamic nucleus deep brain stimulation (STN-DBS) for the treatment of advanced Parkinson's disease. *Movement Disorders, 24*(10), 1553–1554.

Fukuda, H., & Koga, T. (1991). Midbrain stimulation inhibits the micturition, defecation and rhythmic straining reflexes elicited by activation of sacral vesical and rectal afferents in the dog. *Experimental Brain Research, 83*(2), 303–316.

Fukuda, H., & Koga, T. (1992). Stimulation of three areas of the primary motor cortex interrupts micturition in dogs. *Journal of the Autonomic Nervous System, 38*(3), 177–190.

Gammie, A., & Drake, M. J. (2018). The fundamentals of uroflowmetry practice, based on International Continence Society good urodynamic practices recommendations. *Neurourology and Urodynamics, 37*(S6), S44–S49.

Garcia-Rill, E., Tackett, A., Byrum, S. D., Lan, R. S., Mackintosh, S. G., Hyde, J. R., Bisagno, V., & Urbano, F. J. (2019). Local and relayed effects of deep brain stimulation of the pedunculopontine nucleus. *Brain Sciences, 9*(3), 64.

Gebretsadik, T., et al. (2020). Rates of hospitalization for urinary tract infections among medicaid-insured individuals by spina bifida status, Tennessee 2005–2013. *Disability and Health Journal, 13*(4), 100920.

Giacino, J. T., Whyte, J., Bagiella, E., Kalmar, K., Childs, N., Khademi, A., Eifert, B., Long, D., Katz, D. I., Cho, S., Yablon, S. A., Luther, M., Hammond, F. M., Nordenbo, A., Novak, P., Mercer, W., Maurer-Karattup, P., & Sherer, M. (2012). Placebo-controlled trial of amantadine for severe traumatic brain injury. *NEJM, 366*(9), 819–826.

Gill, B. C., et al. (2013). Stress incontinence in the era of regenerative medicine: Reviewing the importance of the pudendal nerve. *The Journal of Urology, 190*(1), 22–28.

Gjone, R. (1966). Excitatory and inhibitory bladder responses to stimulation of 'limbic', diencephalic and mesencephalic structures in the cat. *Acta Physiologica Scandinavica, 66*(1), 91–102.

Green, A. L., et al. (2007). Identifying cardiorespiratory neurocircuitry involved in central command during exercise in humans. *Journal of Physiology, 578*(Pt 2), 605–612.

Griffiths, D., Holstege, G., Dalm, E., & de Wall, H. (1990). Control and coordination of bladder and urethral function in the brainstem of the cat. *Neurourology and Urodynamics, 9*, 63–82.

Groenendijk, I. M., et al. (2020). Whole brain 7T-fMRI during pelvic floor muscle contraction in male subjects. *Neurourology and Urodynamics, 39*(1), 382–392.

Haddad, A. R., L, V., & Green, A. L. (2019). Deep brain stimulation for recovery of consciousness in minimally conscious patients after traumatic brain injury: A systematic review. *Neuromodulation, 22*(4), 373–379.

Halim, A., et al. (2011). Effect of deep brain stimulation on autonomic dysfunction in patients with Parkinson's disease. *Journal of Clinical Neuroscience, 18*(6), 804–806.

Hamid, R., et al. (2018). Epidemiology and pathophysiology of neurogenic bladder after spinal cord injury. *World Journal of Urology, 36*(10), 1517–1527.

Harricharan, S., et al. (2016). fMRI functional connectivity of the periaqueductal gray in PTSD and its dissociative subtype. *Brain and Behavior, 6*(12), e00579.

Heath, R. G., & Norman, E. C. (1946). Electroshock therapy by stimulation of discrete cortical sites with small electrodes. *Proceedings of The Society for Experimental Biology and Medicine, 63*(3), 496–502.

Heffez, D. S., et al. (1990). The paralysis associated with myelomeningocele: Clinical and experimental data implicating a preventable spinal cord injury. *Neurosurgery, 26*(6), 987–992.

Hellman, K. M., et al. (2018). Identification of experimental bladder sensitivity among dysmenorrhea sufferers. *American Journal of Obstetrics and Gynecology, 219*(1), 84 e81–84 e88.

Herman, R. M., et al. (1988). The effect of a low dose of intrathecal morphine on impaired micturition reflexes in human subjects with spinal cord lesions. *Anesthesiology, 69*(3), 313–318.

Herzog, J., et al. (2006). Subthalamic stimulation modulates cortical control of urinary bladder in Parkinson's disease. *Brain, 129*(Pt 12), 3366–3375.

Hilz, M. J., et al. (2019). Autonomic dysfunction in the neurological intensive care unit. *Clinical Autonomic Research, 29*(3), 301–311.

Hoang, T. X., et al. (2003). Autonomic and motor neuron death is progressive and parallel in a lumbosacral ventral root avulsion model of cauda equina injury. *Journal of Comparative Neurology, 467*(4), 477–486.

Hoeritzauer, I., et al. (2021). Scan-negative cauda equina syndrome: A prospective cohort study. *Neurology, 96*(3), e433–e447.

Holstege, G. (2010). The emotional motor system and micturition control. *Neurourology and Urodynamics, 29*(1), 42–48.

Holstege, G., et al. (1986). Anatomical and physiological observations on supraspinal control of bladder and urethral sphincter muscles in the cat. *Journal of Comparative Neurology, 250*(4), 449–461.

Hu, M., et al. (2019). Electrical stimulation enhances neuronal cell activity mediated by Schwann cell derived exosomes. *Scientific Reports, 9*(1), 4206.

Hyam, J. A., et al. (2019). Cardiovascular autonomic responses in patients with Parkinson disease to pedunculopontine deep brain stimulation. *Clinical Autonomic Research, 29*(6), 615–624.

Janik, K., et al. (2020). Spinal cord injury in myelomeningocele: Prospects for therapy. *Frontiers in Cellular Neuroscience, 14*, 201.

Jarvelin, M. R. (1989). Developmental history and neurological findings in enuretic children. *Developmental Medicine and Child Neurology, 31*(6), 728–736.

Jarvelin, M. R., et al. (1990). Life changes and protective capacities in enuretic and non-enuretic children. *Journal of Child Psychology and Psychiatry, 31*(5), 763–774.

Jiang, H. H., et al. (2009). Dual simulated childbirth injuries result in slowed recovery of pudendal nerve and urethral function. *Neurourology and Urodynamics, 28*(3), 229–235.

Jiang, H. H., et al. (2018). Electrical stimulation of the pudendal nerve promotes neuroregeneration and functional recovery from stress urinary incontinence in a rat model. *American Journal of Physiology - Renal Physiology, 315*(6), F1555–F1564.

JL, B. (2006). Chronic disorders of consciousness. *Lancet, 367*, 1181–1192.

Jonnalagadda, K., et al. (2019). Cardiac autonomic neuropathy in type 2 diabetes mellitus. *Journal of the Association of Physicians of India, 67*(3), 100.

Kavia, R. B., et al. (2005). Functional imaging and the central control of the bladder. *Journal of Comparative Neurology, 493*(1), 27–32.

Keller, E. E., et al. (2020). Early sacral neuromodulation ameliorates urinary bladder function and structure in complete spinal cord injury minipigs. *Neurourology and Urodynamics, 39*(2), 586–593.

Kiddoo, D. A., et al. (2006). Impact of state of arousal and stress neuropeptides on urodynamic function in freely moving rats. *American Journal of Physiology - Regulatory, Integrative and Comparative Physiology, 290*(6), R1697–1706.

Kim, J. H., et al. (2015). Bladder recovery by stem cell based cell therapy in the bladder dysfunction induced by spinal cord injury: Systematic review and meta-analysis. *PLoS One, 10*(3), e0113491.

Kimura, Y., et al. (1995). Inhibitory influence from the nucleus reticularis pontis oralis on the micturition reflex induced by electrical stimulation of the pontine micturition center in cats. *Neuroscience Letters, 195*(3), 214–216.

Kisely, S., et al. (2018). A systematic review and meta-analysis of deep brain stimulation for depression. *Depression and Anxiety, 35*(5), 468–480.

Kitta, T., et al. (2016). Role of the anterior cingulate cortex in the control of micturition reflex in a rat model of Parkinson's disease. *The Journal of Urology, 195*(5), 1613–1620.

Koh, C. J., et al. (2006). Bladder and external urethral sphincter function after prenatal closure of myelomeningocele. *The Journal of Urology, 176*(5), 2232–2236.

Kohles, S. S., et al. (2004). Time-dependent surgical outcomes following cauda equina syndrome diagnosis: Comments on a meta-analysis. *Spine (Phila Pa 1976), 29*(11), 1281–1287.

Korse, N. S., et al. (2017). The long term outcome of micturition, defecation and sexual function after spinal surgery for cauda equina syndrome. *PLoS One, 12*(4), e0175987.

Koyama, K. (1991). Effects of olfactory tubercle stimulation on efferent activities in the vesical branch of the pelvic nerve and in the urethral branch of the pudendal nerve in dogs. *Nippon Hinyokika Gakkai Zasshi - Japanese Journal of Urology, 82*(6), 922–931.

Kringelbach, M. L., et al. (2010). Sing the mind electric - principles of deep brain stimulation. *European Journal of Neuroscience, 32*(7), 1070–1079.

Lai, H. H., et al. (2016). Impact of childhood and recent traumatic events on the clinical presentation of overactive bladder. *Neurourology and Urodynamics, 35*(8), 1017–1023.

Lalley, P. M., Groat, W., & McLain, P. L. (1972). Activation of the sacral parasympathetic pathway to the urinary bladder by brain stem stimulation. *Federation Proceedings, 31*, 386.

Largo, R. H., et al. (1999). Development of bladder and bowel control: Significance of prematurity, perinatal risk factors, psychomotor development and gender. *European Journal of Pediatrics, 158*(2), 115–122.

Lewin, R. J., et al. (1967). Extrapyramidal inhibition of the urinary bladder. *Brain Research, 4*(4), 301–307.

Lewin, R. J., & Porter, R. W. (1965). Inhibition of spontaneous bladder activity by stimulation of the globus pallidus. *Neurology, 15*(11), 1049–1052.

Liu, Z., et al. (2004). Micturition-related neuronal firing in the periaqueductal gray area in cats. *Neuroscience, 126*(4), 1075–1082.

Loewy, A. D., et al. (1979). Descending projections from the pontine micturition center. *Brain Research, 172*(3), 533–538.

Lozano, A. M., et al. (2019). Deep brain stimulation: Current challenges and future directions. *Nature Reviews Neurology, 15*(3), 148–160.

Maggi, C. A., et al. (1986). Postnatal development of micturition reflex in rats. *American Journal of Physiology, 250*(5 Pt 2), R926–931.

Malinovsky, J. M., et al. (1998). The urodynamic effects of intravenous opioids and ketoprofen in humans. *Anesthesia & Analgesia, 87*(2), 456–461.

McKernan, L. C., et al. (2019). Posttraumatic stress disorder in interstitial cystitis/bladder pain syndrome: Relationship to patient phenotype and clinical practice implications. *Neurourology and Urodynamics, 38*(1), 353–362.

Miyakoda, K., & Watanabe, K. (1993). A study of brain MRI findings and clinical response of bladder emptying failure in brain bladder. *Nippon Hinyokika Gakkai Zasshi - Japanese Journal of Urology, 84*(2), 280–290.

Miyazato, M., et al. (2013). The other bladder syndrome: Underactive bladder. *Reviews in Urology, 15*(1), 11–22.

Mock, S., et al. (2016). The impact of pallidal and subthalamic deep brain stimulation on urologic function in Parkinson's disease. *Neuromodulation, 19*(7), 717–723.

Mordasini, L., et al. (2014). Bladder function in patients with dystonia undergoing deep brain stimulation. *Parkinsonism & Related Disorders, 20*(9), 1015–1017.

Noto, H., et al. (1989). Excitatory and inhibitory influences on bladder activity elicited by electrical stimulation in the pontine micturition center in the rat. *Brain Research, 492*(1–2), 99–115.

Noto, H., et al. (1991). Electrophysiological analysis of the ascending and descending components of the micturition reflex pathway in the rat. *Brain Research, 549*(1), 95–105.

Nour, S., et al. (2000). Cerebral activation during micturition in normal men. *Brain, 123*(Pt 4), 781–789.

Numata, A., et al. (2008). Micturition-suppressing region in the periaqueductal gray of the mesencephalon of the cat. *American Journal of Physiology - Regulatory, Integrative and Comparative Physiology, 294*(6), R1996–2000.

O'Neal, C. M., et al. (2017). Dr. Robert G. Heath: A controversial figure in the history of deep brain stimulation. *Neurosurgical Focus, 43*(3), E12.

Olds, J., & Milner, P. (1954). Positive reinforcement produced by electrical stimulation of septal area and other regions of rat brain. *Journal of Comparative & Physiological Psychology, 47*(6), 419–427.

Otake, K., et al. (1994). Afferents to the midline thalamus issue collaterals to the nucleus tractus solitarii: An anatomical basis for thalamic and visceral reflex integration. *Journal of Neuroscience, 14*(9), 5694–5707.

Patra, S., et al. (2019). Human left anterior cingulate stimulation elicits a reproducible micturition response. *Stereotactic and Functional Neurosurgery, 97*(4), 278–281.

Pazo, J. H. (1976). Caudate-putamen and globus pallidus influences on a visceral reflex. *Acta Physiologica Latinoamericana, 26*(4), 260–266.

Pereira, E. A., et al. (2008). Deep brain stimulation of the pedunculopontine nucleus in Parkinson's disease. Preliminary experience at Oxford. *British Journal of Neurosurgery, 22*(Suppl. 1), S41–44.

Pietraszko, W., et al. (2013). Efficacy of deep brain stimulation of the subthalamic nucleus on autonomic dysfunction in patients with Parkinson's disease. *Folia Medica Cracoviensia, 53*(2), 15–22.

Pisa, F. E., et al. (2014). The prevalence of vegetative and minimally conscious states: A systematic review and methodological appraisal. *The Journal of Head Trauma Rehabilitation, 29*(4), E23–30.

Porter, R. W. (1967a). A pallidal response to detrusor contraction. *Brain Research, 4*(4), 381–383.

Porter, R. W. (1967b). Visceral manifestations of extrapyramidal dysfunction. *Proceedings. Clinical Spinal Cord Injury Conference, 16*, 120–123.

Potenzoni, D., et al. (1982). Urodynamic assessment in multiple sclerosis and Parkinson's disease. *Acta Bio-Medica de l'Ateneo Parmense, 53*(3), 121–128.

Praveen Rajneesh, C., Liou, J.-C., Hsieh, T.-H., Chin, H.-Y., & Peng, C.-W. (2020). Efficacy of deep brain stimulation on the improvement of the bladder functions in traumatic brain injured rats. *Brain Sciences, 10*(11), 850.

Redshaw, J. D., et al. (2018). Protocol for a randomized clinical trial investigating early sacral nerve stimulation as an adjunct to standard neurogenic bladder management following acute spinal cord injury. *BMC Urology, 18*(1), 72.

Rocha, I., et al. (2004). Effect of stimulation of anterior hypothalamic area on urinary bladder function of the anesthetized rat. *Clinical Autonomic Research, 14*(4), 264–269.

Roy, H. A., et al. (2018a). Beta oscillations and urinary voiding in Parkinson disease. *Neurology, 90*(17), e1530–e1534.

Roy, H. A., et al. (2018b). Effects of pedunculopontine nucleus stimulation on human bladder function. *Neurourology and Urodynamics, 37*(2), 726–734.

Roy, H. A., et al. (2019). Sensory thalamus and periaqueductal grey area local field potential signals during bladder filling. *Journal of Clinical Neuroscience, 68*, 342–343.

Sakakibara, R., et al. (2002). Micturition-related electrophysiological properties in the substantia nigra pars compacta and the ventral tegmental area in cats. *Autonomic Neuroscience-Basic & Clinical, 102*(1–2), 30–38.

Sakakibara, R., et al. (2003). Effects of subthalamic nucleus stimulation on the micturation reflex in cats. *Neuroscience, 120*(3), 871–875.

Sakakibara, R., et al. (2008). Mechanism of bladder dysfunction in idiopathic normal pressure hydrocephalus. *Neurourology and Urodynamics, 27*(6), 507–510.

Sakamoto, K., et al. (2000). Neuroregeneration and voiding behavior patterns after pudendal nerve crush in female rats. *Neurourology and Urodynamics, 19*(3), 311–321.

Sato, F., et al. (1996). Three-dimensional analysis of cerebellar terminals and their postsynaptic components in the ventral lateral nucleus of the cat thalamus. *Journal of Comparative Neurology, 371*(4), 537–551.

Satoh, K., S, N., Tohyama, M., & Maeda, T. (1978). Localization of the micturition reflex center at dorsolateral pontine tegmentum of the rat. *Neuroscience Letters, 8*, 27–33.

Seif, C., et al. (2004). Effect of subthalamic deep brain stimulation on the function of the urinary bladder. *Annals of Neurology, 55*(1), 118–120.

Smith, A. L., et al. (2011). The effects of acute and chronic psychological stress on bladder function in a rodent model. *Urology, 78*(4), 967 e961–967.

Spiegel, E. A., et al. (1947). Stereotaxic apparatus for operations on the human brain. *Science, 106*(2754), 349–350.

Steadman, C. J., & Grill, W. M. (2020). Spinal cord stimulation for the restoration of bladder function after spinal cord injury. *Healthcare Technology Letters, 7*(3), 87–92.

Sun, H., et al. (2020). Abnormal neurite orientation dispersion and density imaging of white matter in children with primary nocturnal enuresis. *NeuroImage: Clinical, 28*, 102389.

Suzuki, K., et al. (2018). The prevalence, course and clinical correlates of migraine in Parkinson's disease: A multicentre case-controlled study. *Cephalalgia, 38*(9), 1535–1544.

Swinn, M. J., & Fowler, C. J. (2001). Isolated urinary retention in young women, or Fowler's syndrome. *Clinical Autonomic Research, 11*(5), 309–311.

Sysoev, Y., et al. (2020). Site-specific neuromodulation of detrusor and external urethral sphincter by epidural spinal cord stimulation. *Frontiers in Systems Neuroscience, 14*, 47.

Tai, C., et al. (2009). Brain switch for reflex micturition control detected by FMRI in rats. *Journal of Neurophysiology, 102*(5), 2719–2730.

Taniguchi, N., et al. (2002). A study of micturition inducing sites in the periaqueductal gray of the mesencephalon. *The Journal of Urology, 168*(4 Pt 1), 1626–1631.

Tawadros, C., et al. (2015). External urethral sphincter electromyography in asymptomatic women and the influence of the menstrual cycle. *BJU International, 116*(3), 423–431.

Thevathasan, W., et al. (2012). A spatiotemporal analysis of gait freezing and the impact of pedunculopontine nucleus stimulation. *Brain, 135*(Pt 5), 1446–1454.

Tiwari, E., et al. (2019). Nerve transfer for restoration of lower motor neuron-lesioned bladder and urethra function: Establishment of a canine model and interim pilot study results. *Journal of Neurosurgery: Spine, 32*(2), 258–268.

Tokunaga, S., & Kuru, M. (1959). Vesico-constrictor center and vesico-relaxer center in the medulla. *The Japanese Journal of Physiology, 9*, 365–374.

Toth, R., Z, M., Ottaway, J., Gillbe, I., Martin, S., Benjaber, M., Lamb, G., Noone, T., Taylor, B., Deli, A., Kremen, V., Worrell, G., Constandinou, T. G., De Wachter, I. G. S., Knowles, C., Sharott, A., Velentin, A., Green, A. L., & Denison, T. (2020). DyNeuMo Mk-2: An investigational circadian-locked neuromodulator with responsive stimulation for applied chronobiology. *IEEE: Transactions on Systems, Man, and Cybernetics: Systems, 9283187*, 3433–3440.

Uchiyama, T., et al. (2003). Short-term effect of a single levodopa dose on micturition disturbance in Parkinson's disease patients with the wearing-off phenomenon. *Movement Disorders, 18*(5), 573–578.

Valentino, R. J., et al. (2011). The bladder-brain connection: Putative role of corticotropin-releasing factor. *Nature Reviews Urology, 8*(1), 19–28.

Vargas, L. C., et al. (2000). Micturition and defensive behaviors are controlled by distinct neural networks within the dorsal periaqueductal gray and deep gray layer of the superior colliculus of the rat. *Neuroscience Letters, 280*(1), 45–48.

Wang, M., et al. (2018). Morphometric magnetic resonance imaging study in children with primary monosymptomatic nocturnal enuresis. *Frontiers in Pediatrics, 6*, 103.

Weaver, L. C. (1985). Organization of sympathetic responses to distension of urinary bladder. *American Journal of Physiology, 248*(2 Pt 2), R236–240.

Webb, S., et al. (2014). Managing perineal trauma after childbirth. *BMJ, 349*, g6829.

Wood, S. K., et al. (2009). Social stress-induced bladder dysfunction: Potential role of corticotropin-releasing factor. *American Journal of Physiology - Regulatory, Integrative and Comparative Physiology, 296*(5), R1671–1678.

Woodfield, J., et al. (2021). Suspected cauda equina syndrome: No reduction in investigation, referral and treatment during the COVID-19 pandemic. *Annals of the Royal College of Surgeons of England, 103*(6), 432–437.

Wyndaele, M., & Rosier, P. (2018). Basics of videourodynamics for adult patients with lower urinary tract dysfunction. *Neurourology and Urodynamics, 37*(S6), S61–S66.

Yamamoto, T., et al. (2010). Neuronal activities of forebrain structures with respect to bladder contraction in cats. *Neuroscience Letters, 473*(1), 42–47.

Yamamoto, T., S, R., Uchiyama, T., & Kuwabara, S. (2020). Subthalamic stimulation inhibits bladder contraction by modulating the local field potential and catecholamine level of the medial prefrontal cortex. *Frontiers in Neuroscience, 14*, 917.

Yoshimura, N., et al. (1992). Inhibitory effects of Hachimijiogan on micturition reflex via the locus coeruleus. *Nippon Yakurigaku Zasshi - Folia Pharmacologica Japonica, 99*(3), 161–166.

Zare, A., et al. (2019). The role of the periaqueductal gray matter in lower urinary tract function. *Molecular Neurobiology, 56*(2), 920–934.

Zhu, W., et al. (2019). Study on neuropathological mechanisms of primary monosymptomatic nocturnal enuresis in children using cerebral resting-state functional magnetic resonance imaging. *Scientific Reports, 9*(1), 19141.

Zrinzo, L., & Zrinzo, L. V. (2008). Surgical anatomy of the pedunculopontine and peripeduncular nuclei. *British Journal of Neurosurgery, 22*(Suppl. 1), S19–24.

Section II

Fundamental and translational neuro-urology research

Chapter 5

How treatment of lower urinary tract symptoms can benefit from basic research

Jason P. Van Batavia

Division of Pediatric Urology, Children's Hospital of Philadelphia, University of Pennsylvania Perelman School of Medicine, Philadelphia, PA, United States

Introduction

Lower urinary tract symptoms (LUTS) encompass a wide range of clinical symptoms that affect the storage and emptying of urine by the bladder and lower urinary tract (LUT). LUTS include urinary urgency (i.e., the sudden and unexpected need to void), frequency (i.e., increased number of voids per day), incontinence (i.e., leakage), hesitancy, and straining (Austin et al., 2014). All of these LUTS can occur as isolated symptoms or in combination to cause LUT dysfunction, which are divided into specific conditions or disorders. The combination of LUTS can suggest an underlying etiology, and for simplicity, LUT dysfunction is often broken down into disorders of urine storage or urine emptying. Identification of the true underlying condition is essential to guide treatment, many of which are based on our understanding of the normal physiology and pathophysiology of the lower urinary tract.

The drive to better understand the basic science behind normal and abnormal LUT function and brain–bladder connections behind volitional voiding stems from the high prevalence of LUTS and LUT dysfunction in both children and adults. In fact, LUTS affect 17%–22% of school-aged children across all ethnicities, >20% of adults ages 20–60, and >40% of adults over the age of 60 (Boyle et al., 2003; Coyne et al., 2011; Sureshkumar et al., 2009; Vaz et al., 2012). LUTS are estimated to account for ~40% of all pediatric urology outpatient referrals and lead to decreased quality of life, social humiliation, and isolation in both children and adults (Landgraf et al., 2004). Children affected with LUTS are at increased risk of developing interstitial cystitis (IC) and chronic pelvic pain as adults (Fitzgerald et al., 2006). Despite the high prevalence, morbidity, and healthcare cost of LUTS, our ability to effectively treat patients with specific LUTS is still limited by our knowledge of how the nervous system interacts with the bladder to control volitional voiding. In fact, the American Urological Association (AUA) in their 2015 publication entitled "National Urology Research Agenda (NURA)" prioritized basic science and translational study of neuro-urology to further our understanding of both normal micturition and the pathophysiology of abnormal micturition characterized by urinary incontinence and overactive bladder (OAB) (AUA-UCF, 2018). The authors note that "Neurological, neuromuscular, inflammatory and neoplastic conditions are among the causes of urinary incontinence, and investigators in all of these fields should be involved in understanding its pathogenesis. **Neurophysiological investigation of both central and peripheral nervous mechanisms** will be central to our understanding and the deployment of synaptically acting pharmaceuticals."

In this chapter, we will focus on bladder afferent signaling and how this activity relates to our understanding of LUTS. Beginning with a general overview of afferent bladder pathways, we then review the unique role of the urothelium (i.e., the special epithelial barrier layer of the bladder) as a mechanosensory organ. Next, the role of neuronal afferents in two specific clinical conditions—overactive (i.e., urgency) and underactive bladder—is explored along with how knowledge of the neurotransmitters and receptors involved has influenced clinical treatment of these disorders. Finally, a particularly debilitating clinical condition called painful bladder syndrome (PBS) and interstitial cystitis (PBS/IC) is described along with the importance of basic science research for future potential treatment breakthroughs.

Neuro-Urology Research. https://doi.org/10.1016/B978-0-12-822455-7.00006-4

Afferent bladder pathways

Control of bladder function is a highly coordinated process involving nearly all aspects of the nervous system including both the peripheral and central nervous systems, autonomic and somatic nerves, and parasympathetic and sympathetic pathways. Specifically, the LUT, including the bladder and urethra, is innervated by three distinct sets of peripheral nerves from the parasympathetic, sympathetic, and somatic nervous systems (Chai & Birder, 2016). Both efferent and afferent nerves travel through shared mixed peripheral nerves between the LUT and spinal cord. Parasympathetic nerves (S2—S4) travel through the pelvic nerve, sympathetic nerves (T11-L2) travel through the hypogastric nerve, and somatic nerves (S2—S4) travel through the pudendal nerve. While efferent nerves coordinate the precise events to lead to bladder emptying and voiding, sensory information via afferent nerves are critical during storage of urine and to alerting the CNS that it is time to void. LUT afferent nerves monitor bladder volume during the storage phase as well as bladder contraction amplitude during voiding (Kanai & Andersson, 2010). In fact, the vast majority of time is spent during bladder filling, and storage and dysfunction of these sensory signals can result in various clinical LUTS such as urgency and pain.

Location of afferent neurons in lower urinary tract

Afferent sensory nerves are located throughout the bladder wall and nerve endings are found in all layers of the bladder wall including the serosa, mucosa (i.e., detrusor muscle), and lamina propria. Afferent nerve axons even extend all the way to the urothelium, the highly specialized epithelial layer that lines the bladder, ureters, and urethra. Interestingly, the concentration of afferent nerves in the lamina propria is highest in the bladder neck and proximal urethra region and becomes less dense as one progresses cranially such that the lamina propria of the dome of the bladder has few if any afferent axons (Chai & Birder, 2016). In the urethra, afferent nerves have been identified between muscle fibers, surrounding blood vessels, and throughout all layers of the urothelium and suburothelium. Interestingly, from a clinical standpoint, the location of these afferent nerves correlates to where patients tend to have the most pain or bother from foreign bodies (i.e., catheters) or stones in their bladders. For instance, a suprapubic tube is a catheter that is inserted into the anterior aspect of the bladder percutaneously through the lower abdominal wall; the tube has a curl at the end, and if this curl sits on the trigone of the bladder, patients often report pain and discomfort; repositioning of the tube with the curl pointed toward the dome of the bladder often leads to symptom improvement and less bother. This is just one example of how an understanding of the basic science of bladder innervation has led to clinical improvement for patients.

Afferent nerves in the deeper layers of the bladder, especially in the detrusor muscle, are more uniformly distributed throughout the bladder and not asymmetrically concentrated like the more superficial nerves. Afferent somatic sensory nerves project from the external urethral sphincter.

Pelvic and pudendal nerve afferents

Axons from pelvic nerve afferents that innervate the LUT originate from the sacral dorsal root ganglia (DRG) from roots S2 to S4. Tracer labeling studies in cats and rats have shown that these preganglionic neurons travel rostral in Lissauer's tract after entering the sacral spinal cord and give off collaterals laterally via lamina I (De Groat, 1986; Jancso & Maggi, 1987; Morgan et al., 1981; Steers et al., 1991). These lateral projections terminate in the sacral parasympathetic nucleus (SPN), as well as giving off additional collaterals medially to the dorsal commissure (see Fig. 5.1). Pudendal nerve afferents traverse from the external urethral sphincter and also have their cell bodies in the DRG at S2—4, the same DRGs that house the cell bodies of pelvic nerve afferents. Animal studies show that these first-order neurons terminate in lateral laminae I and V—VII and in lamina X (De Groat, 1986; de Groat and Yoshimura, 2009). Stimulation of the bladder and urethra in rats leads to increased *fos* protein in both the dorsal commissure, superficial dorsal horn and the SPN, and indicated synaptic activity (Birder and de Groat, 1992, 1993; Vizzard, 2000). While some of these interneurons project to the brain and brainstem, others synapse locally in the spinal cord and are involved in spinal reflexes (Birder et al., 1999).

Hypogastric nerve afferents

Sympathetic first-order afferent neurons from the LUT pass through the DRG of the thoracolumbar segments T11—L2 and have similar termination sites in laminae I, V—VII, and X. Interestingly, while the vast majority of afferent projections are to the ipsilateral side of the spinal cord, approximately 10%—20% project to the contralateral side of the cord (de Groat and Yoshimura, 2009; Morgan et al., 1986). In both the cat and rat, the hypogastric nerves contain mostly smaller diameter, myelinated axons, compared to the pelvic nerves which are larger in diameter and consist of mixed myelinated and

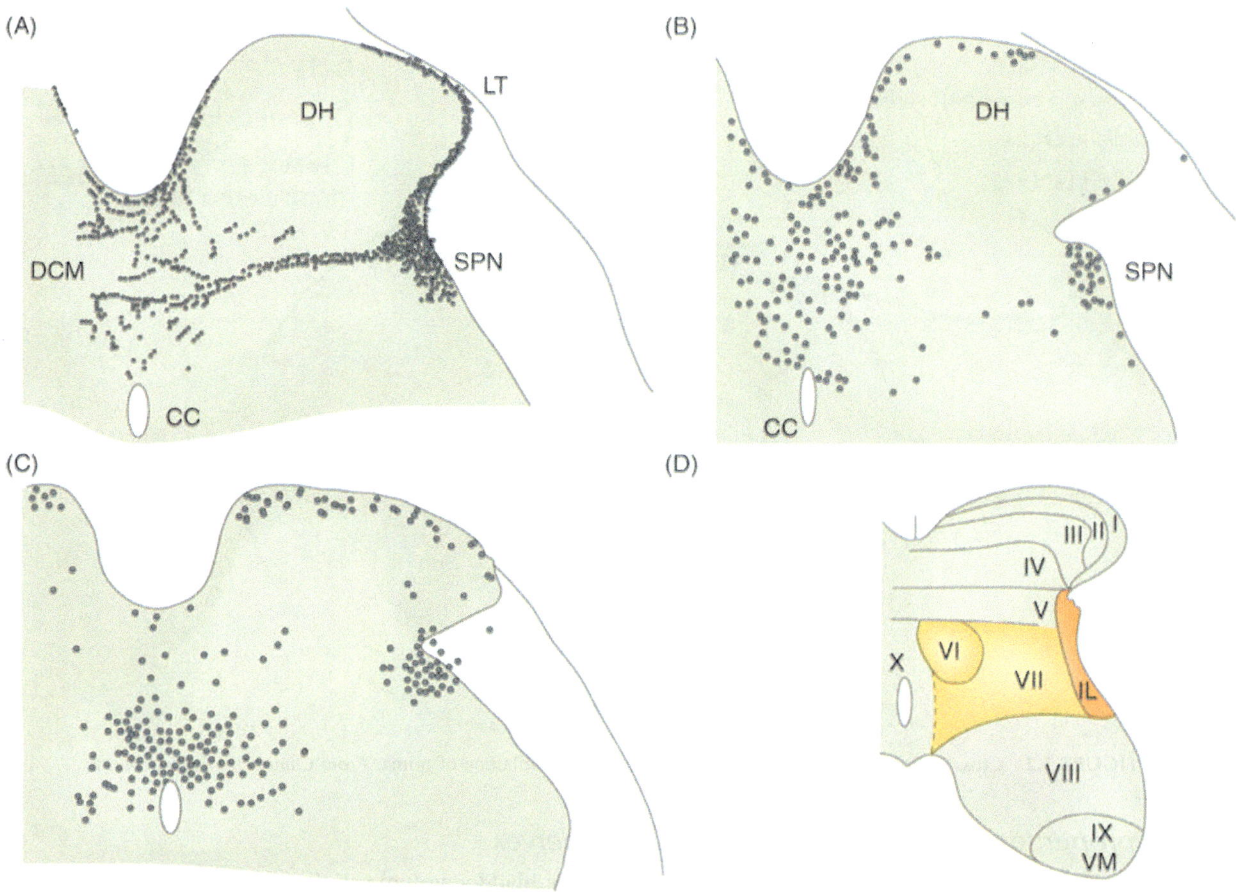

FIGURE 5.1 - Primary afferent and spinal interneuronal pathways involved in micturition. (A) Primary afferent pathways to the L6 spinal cord of the rat project to regions of the dorsal commissure (DCM), the superficial dorsal horn (DH), and the sacral parasympathetic nucleus (SPN) that contain parasympathetic preganglionic neurons. The afferent nerves consist of myelinated (Aδ) axons, which respond to bladder distension and contraction, and unmyelinated (C) axons, which respond to noxious stimuli. (B) Spinal interneurons that express c-Fos following the activation of bladder afferents by a noxious stimulus (acetic acid) to the bladder are located in similar regions of the L6 spinal segment. (C) Spinal interneurons involved in bladder reflexes (labeled by transneuronal transport of pseudorabies virus injected into the urinary bladder) are localized to the regions of the spinal cord that contain primary afferents and c-Fos. Some of these interneurons provide excitatory and inhibitory inputs to the parasympathetic preganglionic neurons located in the SPN. (D) The laminar organization of the cat sacral spinal cord, showing the location of parasympathetic preganglionic neurons in the intermediolateral region of laminae V and VII (shaded area). *CC*, central canal; *IL*, intermediolateral nucleus; *LT*, Lissauer's tract; *VM*, ventromedial nucleus (Onuf's nucleus). *From De Groat, W. C., Griffiths, D., & Yoshimura, N. (2015). Neural control of the lower urinary tract.* Comprehensive Physiology, *5, 327–396.*

unmyelinated axons. Of note, the total number of axons in all three of these peripheral nerves is quite larger than the number of afferent neurons in the DRG and efferent neurons from the spinal cord, suggesting that these afferent axons undergo substantial branching as they traverse peripherally (Langford & Coggeshall, 1981).

Spinal interneurons

Retrograde neuronal labeling with pseudorabies virus (PRV) injection into the bladder and/or urethra has identified numerous spinal interneuron pathways. PRV-labeled spinal interneurons from brain efferent neurons are spatially located in the same areas of the spinal cord that received afferent input from the LUT (see Figs. 5.1 and 5.2). These spinal interneurons likely play a role in both bladder reflexes (see later in this chapter) and providing both excitatory and inhibitory feedback to the supraspinal centers that control micturition (i.e., periaqueductal gray, Pontine micturition center [PMC], and hypothalamus) (Fowler et al., 2008).

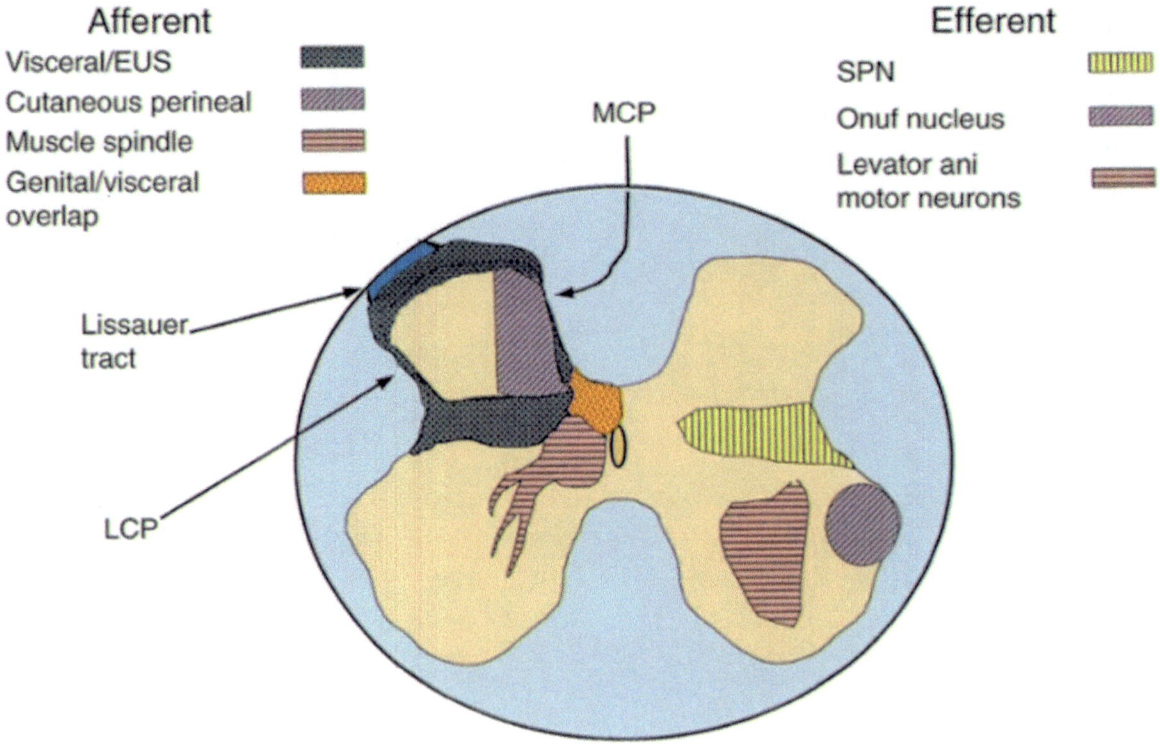

FIGURE 5.2 Cross section of sacral spinal cord; neuroanatomic distribution of prima. *From Campbell-Walsh Urology.*

Role and properties of lower urinary tract afferent nerves

The main roles of LUT afferent neurons appear to be to monitor bladder volume during storage and the amplitude of bladder contractions during voiding, and therefore provide both sensory input to initiate micturition and feedback to maintain the bladder contraction. In these roles, afferent nerves act as tension mechanoreceptors and studies have shown that these afferents are silent when the bladder is empty but show a gradual increase in activity and firing frequency as the bladder fills under physiologic conditions (Kanai & Andersson, 2010). In fact, during filling, there is sequential recruitment of groups of afferent neurons each with different thresholds in response to the increasing intravesical pressures. Importantly, in animal models, these thresholds occur on the bladder–pressure volume curve between 25% and 75% of where the curve becomes steep and compliance shifts, and these findings correspond nicely to intravesical pressures at which humans first note a sensation of bladder filling. Of note, ~20–25% of bladder afferents do not appear to respond to bladder distention (volume or pressure) and have been referred to as "mechano-silent" or silent afferents (Kanai & Andersson, 2010).

While some bladder afferents neurons are mechanoreceptors, others act more as chemoreceptors and are responsive to stimuli from various sources including chemicals in urine and those in the bladder wall from urothelial cells, detrusor cells, inflammatory cells, and other nerves. Chemical stimuli include potassium, capsaicin, low pH, and high osmolality. Cell injury and other pathological states can lead to increased release of these chemical or noxious stimuli, which in turn can alter afferent neuron excitability and even change some of these neurons from "mechano-silent" (i.e., nonresponsive to bladder pressure and bladder wall tension) into "mechano-sensitive" (de Groat et al., 2015). The implications this has on LUT dysfunction and disease states are discussed in later sections of this chapter.

Urethral afferent nerves from the pelvic, hypogastric, and pudendal nerves appear to have different characteristics. Hypogastric urethral afferents have little or no response to bladder stimulation and only some show low firing rates to high bladder pressures. In animal models, both pelvic and pudendal urethral afferent neurons respond to the fluid traveling through the urethra, although pudendal fibers appear to be more sensitive (Talaat, 1937). Talaat also found that distension of the urethra to high intra-urethral pressures (>50 cm H20) evoked discharge in afferent fibers in both the pelvic and hypogastric nerves. Interestingly, in rats, stimulation of the flow-sensitive urethral afferents via saline infusion has been shown to enhance volume-dependent reflex bladder contractions (Jung et al., 1999).

Two types of afferent neurons

There are two types of afferent fibers found in the LUT: myelinated Aδ-fibers and unmyelinated C-fibers (Table 5.1).

(i). *Aδ-fibers*

Myelinated Aδ-fibers make up approximately one-third of bladder afferent fibers and are the primary mechanosensitive fibers in the LUT with nerve endings throughout the detrusor smooth muscle. These fibers respond to both stretching of the bladder wall during filling/storage of urine and bladder contraction during bladder emptying. Studies in both rats and cats have shown that myelinated Aδ-fibers mediate the normal micturition reflex and have conduction velocities between 2.5 and 15 m/s. Aδ-fibers have been shown to have firing rates around 15−30 Hz. While these fibers are silent when the bladder is empty, with progressive bladder filling, their activity increases as more neurons are recruited at different pressure/volume thresholds. In mice, studies have shown at least two threshold fiber groups, a lower threshold comprises ~75% of all detrusor muscle afferents and a higher threshold comprises stretch-sensitive population. The mechanoreceptor properties of Aδ-fibers in guinea pigs have identified two populations of stretch-sensitive afferents: one that terminates in the detrusor muscle only and the other that terminates in the muscle and urothelium (Zagorodnyuk et al., 2007). Interestingly, detrusor muscle−only mechanoreceptors are activated only by stretch but not mucosal stroking, while muscle-urothelium mechanoreceptors are activated by stretch, mucosal stroking, and hypertonic solution, but not capsaicin (thus identifying these fibers as Aδ-fibers). For muscle-only mechanoreceptors, removal of the urothelium does not affect stretch-induced activation (Zagorodnyuk et al., 2007).

(ii). *C-fibers*

Unmyelinated C-fibers, unlike Aδ-fibers, are found more widespread in the bladder wall with many concentrated in the lamina propria and urothelium. Ultrastructural studies of the human bladder have identified unmyelinated C-fibers only in the urothelium and suburothelial layer, while myelinated Aδ-fibers have been seen in the detrusor smooth muscle layers (Wiseman et al., 2002). C-fibers have slower conduction velocity (<2 m/s) when compared to Aδ-fibers and are more numerous than myelinated fibers in bladder afferent pathways (Hulsebosch & Coggeshall, 1982). While studies in cats suggest that bladder C-fibers are mechano-insensitive or "silent," studies in rats have found that approximately half of the unmyelinated C-fibers are mechano-sensitive. In these rat studies, C-fiber activity in response to slow bladder distention was much lower than that seen in myelinated Aδ-fibers and these C-fiber afferents were found to be volume sensitive but not responsive to bladder contractions or pressure itself (Dmitrieva and McMahon, 1996).

The population of "silent" mechano-insensitive C-fibers have been shown to respond to nociceptive and cold stimuli or chemical/noxious stimuli. These C-fibers, especially those with nerve endings in the urothelium, can thus be activated by urine directly or by the various neurotransmitters and chemicals released by cells in the urothelium or detrusor. Interestingly, after exposure to these noxious stimuli in both cats and rats, these "silent" C-fiber afferents can become mechanoreceptive and sensitive to bladder distention. This pathophysiologic finding in animal studies may have implications for management and treatment of LUT dysfunction in humans. For instance, studies in cats and rats have shown that capsaicin exposure (the C-fiber afferent neurotoxin) does not block normal micturition reflexes, thus suggesting that C-fibers are not essential for normal voiding. Conversely, however, capsaicin exposure

TABLE 5.1 Bladder afferent properties.

Fiber type	Location	Normal function	Inflammation effect
Aδ (finely myelinated axons)	Smooth muscle	Sense bladder fullness (wall tension)	Increase discharge at lower pressure threshold
C-fiber (unmyelinated axons)	Mucosa	Respond to stretch (bladder volume sensors)	Increase discharge at lower threshold
C-fiber (unmyelinated axons)	Mucosa muscle	Nociception to overdistention	Sensitive to irritants
		Silent afferent	Become mechanosensitive and unmask new afferent pathway during inflammation

From Campbell's chapter.

can reduce bladder overactivity induced by noxious stimuli in animal models, which suggests that C-fiber afferents are involved in the pathophysiology of some types of LUT dysfunction (Maggi et al., 1993).

Clinical correlation—The finding in animal models that capsaicin exposure, particularly via instillation in the bladder (i.e., intravesical), results in decreased detrusor overactivity (DO), or detrusor hyperreflexia (older terminology for neurogenic DO), led to clinical trials in humans using intravesical capsaicin or resiniferatoxin (another chemical that is toxic to C-fibers). Several studies used these treatments for nonneurogenic DO (i.e., OAB), neurogenic DO, and/or IC/chronic bladder pain. Giannantoni et al. performed a prospective, randomized trial of intravesical capsaicin versus resiniferatoxin in 24 spinal cord injury patients with refractory nonneurogenic DO, and while there was no benefit from intravesical capsaicin, intravesical resiniferatoxin led to statistically significant increases in maximum bladder capacity and decreases in incontinent episodes (Giannantoni et al., 2002). Of concern, however, was that most patients who received capsaicin developed side effects including autonomic dysreflexia, limb spasms, suprapubic discomfort, and hematuria. Interestingly, none of the patients who received resiniferatoxin reported side effects. The authors concluded that chemical defunctionalization of C-fiber bladder afferents may have a role in treating debilitating neurogenic bladder conditions. Later studies found that standardization of the solute used for instillation between capsaicin and resiniferatoxin (previously capsaicin was diluted in 30% alcohol, compared to 10% for resiniferatoxin) led to equivalent improvement in neurogenic DO with similar and few side effects from either treatment (de Seze et al., 2004). Despite these promising early results, neither intravesical capsaicin nor resiniferatoxin has been approved by the FDA for treatment of any neurogenic or nonneurogenic LUT dysfunction. Furthermore, a recent systematic review and meta-analysis from the International Continence Society on the use of intravesical vanilloids for treating neurogenic LUT dysfunction in patients with multiple sclerosis found that while preliminary data suggested the treatments might be effective, the safety profile was overall unfavorable and overall quality of evidence was low (Phe et al., 2018).

From a rudimentary standpoint, the basic roles of the Aδ-fibers and C-fibers can be thought of as functioning in the normal physiological functioning of the LUT and the abnormal pathophysiology of LUT dysfunction, respectively. In the normal micturition reflex, the Aδ-fibers respond to bladder filling as mechanoreceptors. Once bladder filling is at capacity, in-series recruitment of Aδ-fibers leads to a strong sense of urgency by relaying this information to the brain and brainstem via spinal tracts. These Aδ-fibers are predominant in this pathway and C-fibers are for the most part silent. In pathologic conditions, however, whether from neurologic diseases (i.e., spinal cord injury), aging, or inflammatory diseases, there is recruitment of C-fibers which become predominant over Aδ-fibers. This leads to an aberrant functional afferent pathway that can cause urinary urgency, incontinence, and possibly bladder pain (Chai & Birder, 2016).

Chemical properties of afferent nerves

(i). *Neuropeptides*: Studies in various species have identified numerous neuropeptides in afferent neurons including substance P (SP), calcitonin gene–related peptide (CGRP), pituitary adenylate cyclase-activating polypeptide (PACAP), leucine enkephalin, corticotropin-releasing hormone (CRH), and vasoactive intestinal polypeptide (VIP). Neuropeptides in afferent pathways have been identified in various locations including afferent neurons in the lumbosacral DRG, afferent neuron axons and terminals in the lumbosacral spinal cord, and afferent fibers in the bladder and urethra. In rats, greater than 70% of bladder DRG neurons contain multiple neuropeptides, especially SP, PACAP, and CGRP (de Groat and Yoshimura, 2009). VIP has been also seen in bladder DRG neurons in cats. Interestingly, while neuropeptide containing afferent nerve axons are found throughout each bladder layer, they are more concentrated in the lamina propria that sits just underneath the urothelium, the location of many C-fibers. In the spinal cord of various species, peptide-containing afferents have been found in Lissauer's tract, lamina I, and in the parasympathetic nucleus (de Groat and Yoshimura, 2009; Vizzard, 2001).

Many of the neuropeptides described above have been shown to be released in the bladder by noxious stimulation. For instance, acute treatment of rat bladders with C-fiber afferent neurotoxins (i.e., vanilloids such as capsaicin or resiniferatoxin) leads to release of SP, CGRP, and PACAP into the bladder wall, which triggers inflammatory responses, including plasma extravasation, vasodilation, and alternations in bladder smooth muscle activity (Chai & Birder, 2016; Maggi, 1993). Chronic treatment with vanilloids leads to reduction in peptidergic afferent staining in the bladder wall of both animals and humans, suggesting that the majority of peptidergic bladder afferent nerves are C-fibers (de Groat and Yoshimura, 2009).

(ii). *Specific modulators of afferent sensitivity:*

(i). Putative excitatory amino acid transmitters—Afferent nerve fibers have been shown in immunohistochemistry studies to contain both glutamate and aspartic acid. While there has been much focus on the role of glutamate as an important neurotransmitter in the regulation of descending pathways from the PMC and in spinal reflex pathways that control both bladder and urethral sphincter function, glutamate has also been shown to have a role as an excitatory neurotransmitter in the afferent limb of the micturition reflex. Glutamate is present in the primary afferent neurons in the spinal cord and in interneurons. Overall, glutamatergic neurons tend to be excitatory compared with the inhibitory effects of glycinergic neurons. Spinal interneurons in the micturition reflex pathway that are normally activated by bladder afferents have been shown to be suppressed after the administration of both NMDA and non-NMDA glutamatergic receptor antagonists (de Groat et al., 2015). For a potential clinical correlation, aging in rats leads to a decrease in the density of glutamatergic synaptic inputs in pelvic neurons, which may play a role in the decreased LUT, specifically detrusor, function associated with aging.

(ii). Nitric oxide—In contrast to the excitatory role of the amino acid transmitters, specifically glutamate, nitric oxide has been shown to play an inhibitory role in bladder afferent activity. Aizawa and associates found that intravesical instillation of a nitric oxide synthase (NOS) inhibitor in rats resulted in significantly increased afferent activity in both Aδ- and C-fibers, while intravenous injection of L-arginine (a substrate for NOS) significantly decreased the activities of both afferent fibers (Aizawa et al., 2011). Interestingly, the investigators also noted that pretreatment with L-arginine prevented the increased afferent firing of both Aδ- and C-fibers that is normally seen in a well characterized rat model of PBS/IC (Aizawa et al., 2011). From a clinical standpoint, the effects of NO are mediated through intracellular increase in the secondary messenger cGMP. cGMP is inactivated by phosphodiesterases (PDE), and the PDE5 isoenzyme is known to have high concentrations in the bladder and prostate. Inhibition of PDE5 leads to increased activity of NO and thus PDE5 inhibitors, which are commonly used for erectile dysfunction, may also have a role in the treatment of bladder hypersensitivity disorders such as DO and PBS/IC.

(iii). *Neuronal receptors*: Bladder afferent fibers express numerous receptors that have been the focus of study over the previous decades and remain an active area of research today. Both afferent neuronal bodies and axons, especially C-fibers, have been shown to express a wide range of receptors including transient receptor potential vanilloid (TRPV, capsaicin receptor, various types have been described including TRPV1, TRPV2, TRPV4, and TRPM8 known to be a cold receptor), transient receptor potential ankyrin 1 receptor (TRPA1), tyrosine kinase receptor A (TrkA, which responds to nerve growth factor [NGF]) and B (TrkB, which responds to brain-derived neurotrophic factor [BDNF]), α and β estrogen receptors, glial cell line—derived neurotrophic factor (GNDF) receptors, muscarinic receptors, endothelin receptors, and purinergic receptors (P2X2, P2X3, and P2Y, which are activated by ATP) (de Groat et al., 2015). Both pelvic nerve afferent fibers from the lumbosacral DRG and hypogastric nerve afferents from the thoracolumbar DRG have been shown to respond to ATP, protons, and/or capsaicin, showing that these various neuronal receptors are present not only in axons in the bladder wall itself but also in the cell bodies and within the spinal cord.

In rats, C-fibers in the LUT can be characterized based on the expression of isolectin B4 binding sites (IB4). Nerves that are IB4-negative (i.e., do not bind lectin) are peptidergic and make up approximately 70%—80% of all bladder C-fiber afferents. IB4-positive fibers are nonpeptidergic and only account for ~20% of bladder C-fiber afferents in the bladder and proximal urethra. Interestingly, these IB4-positive C-fibers make up ~50% of the C-fiber afferents in the distal urethra, similar to percentages of IB4-positive C-fibers found in somatic afferents throughout the skin. In 2004, Nishiguchi and colleagues investigated the role of IB4-binding afferent pathways in the micturition reflex in rats by intrathecally injecting an IB4-saporin conjugate that targets and destroys IB4-binding neurons (Nishiguchi et al., 2004). The investigators found that 3 weeks after IB4-saporin delivery, cystometry revealed no effect on normal voiding function, but the rats did have a reduced bladder overactivity in response to intravesical capsaicin or ATP, suggesting that targeting of IB4-positive, nonpeptidergic afferent pathways may be an effect treatment for overactivity and/or pain responses in the bladder.

Over the past decade, the role of cannabinoid (CB) receptors in sensory fibers has sparked interest. Cannabis-based extracts have been shown clinically to decrease symptoms of urgency, urge incontinence, and DO in patients with multiple sclerosis (Chai & Birder, 2016). Histologic studies of human bladder specimens have detected CB1 and CB2 receptors in the urothelium and detrusor muscle. Gratzke et al. found that CB2 receptors were the predominant subtype in sensory nerve fiber in both the urothelium and suburothelium and that CB2 agonists inhibited nerve-induced bladder contractions, suggesting that CB receptors have a modulatory role in sensory afferent signaling in the bladder (Gratzke et al., 2009). These receptors may represent a novel therapeutic target for LUT disorders.

One important takeaway from the large amount of data from bladder receptor studies is that given the vast array of overlapping receptors present in bladder afferents, the sensory pathways in the bladder are complex and involve a delicate interplay between various chemical and mechanical signaling mechanisms. In vivo and in vitro studies have shown that intravesical administration or direct application of receptor agonists for TRPV1, TRPA1, TRPM8, TrkA, P2X, endothelin, nicotinic, and/or muscarinic receptors leads to enhanced bladder voiding reflex activity, increased afferent fiber activity, and/or release of neurotransmitters. Conversely, other neuromodulators such as nitric oxide, nicotinic, and muscarinic agonists have an inhibitory effect on afferent fiber activity and the voiding reflex (de Groat and Yoshimura, 2009). From a clinical standpoint, an understanding of how these receptors and pathways interact to produce bladder sensation and excitation versus suppression of afferent activity is critical to pave the way toward novel therapies for both bladder overactivity and underactivity.

(iv). *Ion channels*: Afferent neurons in the LUT express a variety of ion channels that can work in synergy or dyssynergy to regulate fiber activity. These channels may respond to substances in the urine or in the bladder wall itself or to mechanical stimuli such as stretch or pressure. Studies have determined that the most important channels present in afferent fibers are sodium channels, potassium channels, calcium channels, purinergic channels, and acid-sensing ion channels (ASICs). As mentioned previously, TRPV receptors are also present in these afferent neurons and evoke an inward current when bond to their substrate, namely capsaicin. Sodium currents measured via voltage clamp recordings in bladder afferents can be divided into two types: tetrodotoxin (a Na+ channel blocker) resistant and tetrodotoxin sensitive. Tetrodotoxin-resistant currents are more prevalent in smaller bladder neurons, particular C-fibers, and have higher spike thresholds. Studies have revealed that two Na^+ channel subunits ($Na_v1.8$ and $Na_v1.9$) are primarily accountable for the tetrodotoxin-resistant Na^+ currents in DRG neurons (de Groat and Yoshimura, 2009). $Na_v1.8$ tends to be expressed in C-fibers that are peptidergic, IB4 negative, while $Na_v1.9$ channels are more common in C-fibers that are nonpeptidergic, IB4 positive. Overall, $Na_v1.8$ channels are more prevalent in DRG neurons and play a role in bladder nociceptive pathways.

Potassium currents in bladder sensory neurons follow a similar distribution to Na^+ currents, with two types of currents predominant: tetrodotoxin-resistant slow-inactivating current in small neurons and a tetrodotoxin-sensitive fast-inactivating current in larger neurons. The tetrodotoxin-resistant slow-inactivating currents are found in most small C-fibers and thus contribute along with the high-threshold Na^+ currents to the high thresholds for neuronal activation. Interestingly, SP, which can be released in the bladder, causes a decrease in potassium current and increased tonic firing of afferent nerve fibers, thus increasing the excitability at C-fiber terminals. Targeting SP may represent another potential therapeutic option for bladder hypersensitivity or pain.

Voltage-sensitive Ca^{2+} channels are also found in bladder afferent neurons. High-voltage activated Ca^{2+} channels are responsible for neurotransmitter release from the nerve terminal and their activity itself is regulated by neurotransmitters. In particular, nitric oxide inhibits N-type high voltage Ca^{2+} channels, which are found in both C- and Aδ-fibers. Thus, nitric oxide from urothelial cells likely has an inhibitory effect on neurotransmitter release and excitability at bladder afferent terminals (de Groat and Yoshimura, 2009). Low-voltage activated T-type Ca^{2+} channels play a role in controlling cell excitability and are expressed in the somatic afferent fibers that innervate the urethra and pelvic floor muscles, but are lacking in afferent neurons from the bladder.

Given that the large majority of DRG nerves are activated in response to exposure to pH 5 solutions, studies have investigated whether this response is secondary to TRPV1 channels or ASICs. The findings that amiloride, a nonselective ASIC blocker, can nearly completely block transient currents in these neurons confirmed the importance of ASIC channels in transient activating currents in bladder afferent neurons in the DRG (Dang et al., 2005).

Clinical correlation: Recently, Marshall and associates published their findings on the mechanosensitive ion channel PIEZO2 in the LUT (Marshall et al., 2020). The authors found that PIEZO2 is expressed in both the urothelium and innervating sensory neurons where it acts as a sensor for low-threshold bladder stretch and urethral micturition reflexes. In mice PIEZO2 knockouts, the micturition interval was noted to be irregular with, on average, longer time between bladder contractions that resulted in voiding and had higher peak pressures during voiding, leading the authors to conclude that PIEZO2-deficient mice were less sensitive to bladder filling and the detrusor muscle had to work harder to accomplish voiding. In long-term studies of PIEZO2-knockout mice, the authors identified larger bladders compared to wildtype mice suggesting bladder wall remodeling/hypertrophy secondary to chronic voiding dysfunction. Interestingly, Marshall et al. also studied humans with PIEZO2 deficiency via questionnaires and found that all 12 human patients (aged 5–42 years old) had decreased voiding frequency and deficient bladder-filling sensation (Marshall et al., 2020). The authors conclude that while PIEZO2 is a critical mediator of LUT function, it is not the only critical mechanotransduction ion channel as the PIEZO2 knockout mice were still able to void and the knockout in the bladder sensory nerves and urothelium was not lethal. Nevertheless, PIEZO2 is an enticing new target for future therapeutics aimed at either increasing bladder sensation and voiding (PIEZO2 agonists) or decreasing bladder sensation and overactivity (PIEZO2 antagonist).

Afferent neurons in bladder reflexes

(i). Sympathetic storage reflexes

Sympathetic input to the LUT is not thought to be essential for micturition or voiding, but does play a role in urinary storage. Any method to decrease sympathetic innervation or activity to the LUT leads to a reduction in urethral outflow resistance, decreased bladder capacity, and increased frequency and amplitude of bladder contractions (de Groat and Yoshimura, 2009). The sympathetic storage reflex is triggered by Aδ-fiber activity in the pelvic nerve that leads to increased sympathetic efferent nerve activity and thus facilitates bladder storage and larger storage volumes. This reflex arc is controlled via a sacrolumbar intersegmental spinal pathway and is inhibited when bladder pressure increases above the threshold for micturition (Fig. 5.3 from de Groat and Yoshimura, 2009). The observation that this inhibitory response is blocked by transection of the spinal cord at the thoracic level is evidence that it originates from a supraspinal site.

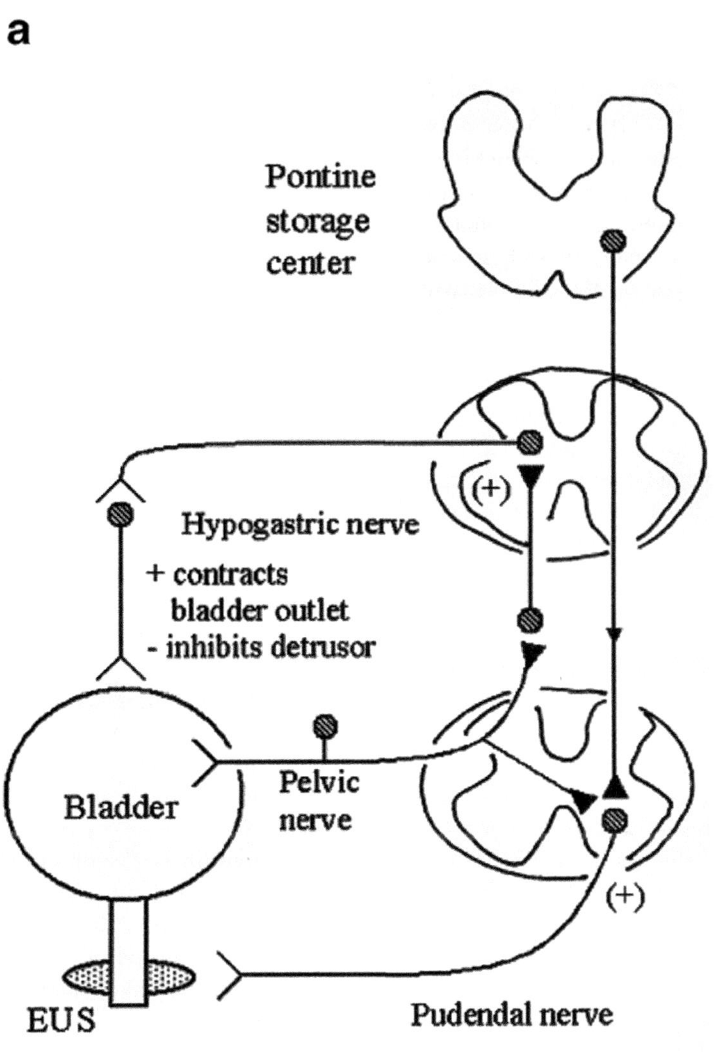

FIGURE 5.3 Neural circuits controlling continence. (A) Urine storage reflexes. During the storage of urine, distension of the bladder produces low-level afferent firing in the pelvic nerve, which in turn stimulates (1) the sympathetic outflow to the bladder outlet (base and urethra) and (2) pudendal outflow to the external urethral sphincter. These responses occur by spinal reflex pathways and represent guarding reflexes, which promote continence. Sympathetic firing also inhibits detrusor muscle and modulates transmission in bladder ganglia. A region in the rostral pons (the pontine storage center) increases external urethral sphincter activity. *From De Groat, W.C., & Yoshimura, N. (2009). Afferent nerve regulation of bladder function in health and disease.* Handbook of Experimental Pharmacology, *91-138.*

(ii). Urethral sphincter storage reflexes

During bladder filling, the striated muscles of the external urethral sphincter (innervated via the pudendal nerve) demonstrate an increase in tonic firing. Often called the "guarding reflex," this activity leads to clamping down of the urethral outlet to prevent urinary leakage. This reflex is mediated by a pathway in the lumbosacral spinal cord that is activated by low-level afferent firing from the bladder via the pelvic nerve. Studies in cats and rodents have pinpointed Aδ-fibers as the afferent nerves that initiate the guarding reflex. Cold stimulation also leads to an increase in tonic external urethral sphincter activity that is dependent on bladder C-fiber afferents.

Evidence from animal studies (cats and monkeys) also supports a sphincter-to-bladder reflex wherein afferent input from the striated external urethral sphincter via the pudendal nerve during bladder contractions leads to suppression of bladder activity and increased bladder capacity (de Groat and Yoshimura, 2009).

(iii). Voiding reflexes

Reflex activation of the bladder for micturition is mediated by a spinobulbospinal reflex pathway that likely passes through the PMC (also known as Barrington's nucleus). This reflex arc is activated by afferent Aδ-fibers that travel from the bladder via the pelvic nerve to the sacral spinal cord that activates sacral parasympathetic efferent nerves to the bladder and urethra. This reflex pathway also induces inhibition of the spinal motoneurons that go to the external urethral sphincter (i.e., guarding reflex) and inhibition of the sympathetic storage reflex. Micturition is thus accomplished by relaxation of the urethral smooth muscle mediated by activation of sacral parasympathetic efferent fibers and release of nitric oxide, an inhibitory transmitter, in the urethra.

Role of urothelium

The urothelium is a specialized epithelium that lines the urinary tract from the renal pelvis all the way to the urethra. Classically, the urothelium has been thought to be composed of at least three layers of distinct cell subtypes (from basement membrane to lumen): basal cells, intermediate cells, and superficial (i.e., umbrella) cells. While initially believed to function primarily as a watertight, impermeable barrier to prevent reabsorption or passage of urine contents into the bladder wall, the urothelium is now understood to play an important role in bladder sensation and acts as a mechanosensory organ itself. Urothelial cells have been shown to behave in many ways like sensory neurons and can detect physiologic stimuli, activate signal transduction pathways, and release chemical mediators, which act directly on afferent nerve fibers in the both the urothelium and suburothelium. Approximately 10% of afferents innervating the bladder are found in the urothelium (Kanai & Andersson, 2010).

As a mechanosensor, urothelial cells can respond to distension, low pH, high potassium concentrations, increased urine osmolality, and chemical irritation. The mechanism by which urothelium responds to these physiologic stimuli is likely through a variety of ways and urothelial cells have been shown to contain Na^+ channels, several subtypes of TRP channels, including TRPA1, TRPV1, 2, and 4, and receptors for bradykinin neurotrophin, purines, norepinephrine, acetylcholine, protease activated receptors, and prostaglandin E2. Activation of these receptors/channels by mechanical or chemical stimuli results in release of sensory nerve–specific modulators mentioned earlier in this chapter, including NO, ATP, acetylcholine, prostaglandins, and SP (de Groat et al., 2015). Each of these signaling molecules or neurotransmitters in turn has either an excitatory or inhibitory effect on sensory afferent nerves located in the urothelium or suburothelium. In addition to directly acting on afferent nerves, these chemicals can also act on myofibroblasts in the suburothelium that lie adjacent to the afferent nerves. Myofibroblasts are often referred to as "interstitial cells" and are connected intricately together by gap junctions. Chemicals and neurotransmitters from myofibroblast cells act on the adjacent afferent nerves and thus both myofibroblast and urothelial cells likely interact in regulation of bladder sensory pathways.

NO has been shown to be released from urothelial cells during trauma and inflammation. Animal studies suggest that the effect of urothelial NO is on sensory nerves as the molecule has minimal direct effect on the detrusor muscle itself. ATP is also released from urothelial cells, most often as a result of stretch, and activates P2X2 and P2X3 receptors on sub-urothelial afferent nerves. Both of these P2X receptor subtypes are found on unmyelinated afferent fibers. Pandita et al. found that intravesical ATP could induce DO in a rate model, and Cockayne et al. noted that P2X3 knockout mice have urinary bladder hyporeflexia (i.e., an underactive/hypoactive bladder) (Cockayne et al., 2000; Pandita et al., 2000). These and other data suggest that purinergic signaling via ATP released from urothelial cells during stretch and bladder distension act on P2X2 and P2X3 receptors on suburothelial bladder afferents to initiate the voiding reflex. In the normal bladder, the interplay and balance between the inhibitory effects of NO and the excitatory effects of ATP is thought to regulate micturition thresholds, bladder capacity, and urinary frequency. Disruption of this balance may play a role in LUT dysfunction and disorders. For instance, elevated ATP levels have been identified in PBS/IC patients (Kumar et al., 2007). Additionally, using a TRPV1 knockout mouse model, Charrua et al. found decreased NO and ATP release from urothelial cells, which was associated with decreased DO in response to bladder irritation, again suggesting that the balance between NO and ATP is critical for normal and abnormal bladder function (Charrua et al., 2007).

Muscarinic receptors have also been shown to contribute to bladder function and sensation. All five subtypes (M1, M2, M3, M4, and M5) of muscarinic receptors, as well as nicotinic receptors, have been identified in the urothelium, some in high densities. In fact, Hawthorn et al. found a greater muscarinic receptor density in the urothelium compared to the detrusor smooth muscle in pigs (Hawthorn et al., 2000). Bladder stretch/distension and aging both lead to release of acetylcholine from the urothelium, which activates urothelial muscarinic receptors resulting in the release of nerve modulators such as ATP. These substances then have effects on bladder sensory nerves and detrusor smooth muscle. Hawthorn et al. also showed that muscarinic agonists lead to the release of a substance called urothelium-derived inhibitory factor. This factor decreases the strength of bladder contractions via effect on detrusor smooth muscle, and the unknown substance was determined to not be NO, a catecholamine, a prostaglandin, adenosine, or GABA (Hawthorn et al., 2000). Future studies to identify this unknown substance could lead to a novel target for LUT dysfunction.

PIEZO2 has recently been identified in superficial/umbrella cells in the urothelium by Marshall et al. using RNAscope; the authors suggest that PIEZO2 is the basis for the mechanosensory role of these cells and help set bladder stretch sensitivity and initiation of appropriate micturition reflexes and bladder relaxation during filling (Marshall et al., 2020). This proposed role of PIEZO2 in superficial cells in mediating bladder relaxation during filling is intriguing and warrants further investigation.

Neuronal afferents role in LUTS

1) *Idiopathic DO or OAB (urgency):*

OAB is often used synonymously with idiopathic or nonneurogenic detrusor overactivity (IDO), although the recognized definition of OAB is urinary urgency, with or without urinary incontinence, usually accompanied by frequency and nocturia (Austin et al., 2014). While many patients with OAB have IDO, IDO itself is a urodynamic finding of involuntary detrusor contractions during filling in the absence of a neurogenic cause, and use of the term OAB does not require documentation of IDO. Many patients with IDO have urgency incontinence, and this condition is a large healthcare burden and leads to decreased quality of life in children and adults. Some estimates of the economic burden of OAB and IDO in the United States reach as high as $65 billion per year and these estimates are rising annually.

While the underlying pathophysiologic cause of IDO is likely multifactorial and may vary depending on the specific patient, two commonly proposed theories evolve around either a "myogenic" or a neurogenic basis for IDO. From a myogenic standpoint, studies of patients with IDO have noted an increase in excitability and ease of contraction spreading in the detrusor muscle. The neurogenic origin of IDO can be further divided between the central and peripheral nervous systems. In terms of peripheral nervous system, recent evidence has led to a focus on aberrant bladder afferent activity as an important underlying cause of IDO. As mentioned earlier, DO may have a pathophysiologic basis in the emergence of C-fiber pathways that are normally silent. Animal studies have shown that in response to noxious stimuli, bladder neck outlet obstruction, or spinal cord injury, unmyelinated C-fiber afferents, rather than Aδ-fibers, become mechanosensitive and initiate autonomic micturition. Translational research supports this hypothesis as human studies have also found increased density of TRPV1 and P2X3 in suburothelial nerves and in the urothelial basal layer in patients with neurogenic DO from multiple sclerosis or spinal cord injury (de Groat and Yoshimura, 2009). Similar to these findings in patients with neurogenic DO, Smet et al. noted an increased density of suburothelial afferent nerves that are immunoreactive to SP and CGRP in women with IDO, while Apostolidis et al. found that bladder biopsies from patients with IDO had increased expression of TRPV1 and P2X3 that returned to normal after intradetrusor injection of botulinum A neurotoxin (Apostolidis et al., 2005; Fowler et al., 2008; Smet et al., 1997). Furthermore, others have shown that intravesical administration of TRPV1 agonists, such as resiniferatoxin, can delay or suppress involuntary detrusor activity in IDO patients.

As mentioned earlier, CB receptors are present in afferent nerve fibers in both the urothelium and suburothelium (Chai & Birder, 2016). In particular, expression of CB1 receptors was found to be increased in the urothelium of patients with BPS/IC and IDO, while other studies have shown the CB2 receptor agonism leads to inhibition of nerve-induced bladder contractions (Gratzke et al., 2009). Therefore, CB receptors in the bladder may have a modulatory role in bladder sensation, overactivity, and sense of urgency, and further understanding of how these receptors function and interact may lead to novel treatment strategies for IDO.

Numerous animal models for DO have been described and many involve transgenic animals with knockouts of genes important in mechanosensing and afferent pathways. Some examples of transgenic models of DO include serotonin re-uptake transporter knockout, urothelially restricted NGF knock in, urothelially restricted β1-integrin knockout, muscarinic receptor knockouts, purinergic receptor (P2X2, P2X3) knockouts, and TRPV1 knockout (Chai & Birder, 2016). These models have the potential to identify novel treatment targets for IDO and to also help further define the mechanism of action of current therapies for IDO.

Clinical correlation: For instance, one of the most used treatments for IDO is antimuscarinic therapy. While there are multiple antimuscarinic medications clinically available, some which have been in use for decades, the long-standing belief about the mechanism of action of these drugs is that they inhibit muscarinic receptors on the detrusor muscle directly and thereby prevent involuntary bladder contractions. Muscarinic receptors have also been shown to be present prejunctionally on the parasympathetic efferent nerves and these receptors likely play a role in either inhibiting (M2 or M4) or facilitating (M1) release of acetylcholine. Recent studies and accumulating evidence, however, suggests that muscarinic receptors in the urothelium and suburothelial layers play an important role in the sensory function of the bladder. Activation of muscarinic receptors on urothelial cells leads to the release of ATP, which in turn activated afferent nerves and alters the excitability of bladder sensing (Hegde, 2006). Therefore, antimuscarinic therapies may, in part, achieve their benefit in patients with IDO via decreasing bladder sensation and urgency.

A similar multifaceted mechanism of action has been proposed for β3-adrenoceptor agonists, a more recently FDA-approved therapy for IDO. Like muscarinic receptors, β3-adrenoceptors have been found in the detrusor myocytes, urothelial cells, suburothelial interstitial cells, and afferent nerve fibers (Igawa et al., 2019). While much initial focus was on the role of these receptors in the detrusor muscle itself in promoting relaxation, the importance of β3-adrenoceptors in urothelial cells, particularly in regard to NO release, and in Aδ- and C-fibers, has been a recent focus of research.

2) *Underactive bladder in diabetes mellitus and aging:*

In humans, diabetes mellitus has several well-known urologic complications including impaired bladder sensation, increased bladder capacity, reduced detrusor contractility, and inefficient emptying with large postvoid residuals. Studies have shown that some of these abnormalities in urine storage and voiding can be explained by alterations in the bladder afferent pathways. For instance, Nadelhaft and Vera found reduced afferent nerve conduction velocity in a rat diabetic model in which bladders were noted to be thicker and heavier (Nadelhaft & Vera, 1992). Using the same streptozotocin-induced diabetic rat model, Sasaki et al. noted decreased NGF levels in the bladder and lumbosacral DRG of diabetic rats; furthermore, the investigators used NGF gene therapy directed to the bladder wall and were able to reverse these changes and improve voiding dynamics (reduced bladder capacity and reduced postvoid residual volumes), findings that suggest a role of reduced axonal transport of NGF in afferent bladder nerves (i.e., impaired bladder sensation) in the diabetic bladder phenotype (Sasaki et al., 2004).

Similar findings of decreased bladder afferent activity have been identified in older bladders and may be a consequence of aging. Aging in humans and animals, in general, is associated with increased prevalence of LUTS, and in rat aging studies, impaired bladder function is characterized by increased voided volumes, high micturition pressure threshold, and reduced sensitivity of pelvic nerve afferents in response to increased bladder volume. Similarly, aging mice have been shown to have normal bladder contractility but diminished bladder afferent firing, again suggesting that aging bladder dysfunction is, at least in part, caused by reduced bladder sensation. Recently, the importance of underactive or hypoactive bladder, particularly as a result of aging, as a cause of LUT dysfunction has come into the foreground and even the NIH/NIDDK has recognized the need for research in this area via specific funding opportunity announcements for "basic, clinical, or translational research on underactive bladder (UAB) and detrusor underactivity (DU) and its consequences in aging and older persons."

Neuronal afferents' role in painful bladder syndrome and interstitial cystitis

As briefly mentioned previously, PBS/IC is a debilitating condition whose diagnosis is often driven by symptoms of chronic bladder pain and urinary frequency and urgency. PBS/IC is characterized by unusual hypersensitivity to bladder distension, bladder wall edema, vasodilation, proliferation of nerve fibers, and infiltration of inflammatory cells, particularly mast cells (Chai & Birder, 2016). While there are some known causes of bladder cystitis including urinary tract infection, radiation, and bladder irritants, in PBS/IC, the underlying cause remains elusive, and the diagnosis is often one of the exclusion. Basic science research using both human bladder specimens and animal models has identified activation of bladder afferent nerves, by either mechanical or chemical stimuli, as an important factor in inducing painful bladder sensations and in releasing chemicals in the bladder wall that induce or enhance inflammation. In particular, there appears to be sensitization of the normally unexcitable (i.e., silent) bladder wall and urothelial C-fiber afferents and thus an emergence of abnormal C-fiber firing in PBS/IC bladders.

Numerous animal models of cystitis have been used to study bladder and nerve changes including chemically induced cystitis from cyclophosphamide, turpentine oil, acid solutions, acrolein, and mustard oil. Chemical irritation leads to release of many pro-inflammatory factors including tachykinins, serotonin, histamine, adenosis, prostaglandin E2, and NGF. Bladder changes also include increased expression of NOS, PACAP, and protease-activated receptors, as well as an increased density of peptidergic afferent nerve fibers in both the urothelium and detrusor muscle layers. These findings all support the activation of C-fibers normally silent or unresponsive to mechanical stimulation and the growth of new afferent fibers during chronic cystitis (de Groat and Yoshimura, 2009).

Clinical correlation: In aggregate, animal and human studies of cystitis and PBS/IC have identified the emergence of C-fiber afferent activity as an important contributor to bladder pain. Given this mechanism, suppression of these normally silent C-fibers would, in theory at least, be an ideal option for treatment of bladder pain. Early studies of intravesical capsaicin or resiniferatoxin for patients with PBS/IC showed promising benefits in terms of treating painful bladder symptoms (Lazzeri et al., 2000); however, these results were not replicated in a prospective, randomized clinical trial of intravesical resiniferatoxin for PBS/IC patients (Payne et al., 2005).

In addition to the role of C-fiber afferents, there is also some evidence that Aδ-fibers may undergo neuroplasticity in PBS/IC bladders. Aδ-fibers in a feline model of PBS/IC were found to be more sensitive to bladder pressure changes than afferent nerves in normal cats (Chai & Birder, 2016). Thus Aδ-fibers afferent, in addition to C-fibers, may play a role via abnormal mechanosensing in PBS/IC. More research on Aδ-fibers and these afferents as a possible target for therapy for patients with PBS/IC is warranted.

Qiao and Vizzard utilizing the cyclophosphamide-induced cystitis model in rats found that affected bladders had increased expression of neurotrophic growth factors, particularly NGF, BDNF, and CTNF (Qiao & Vizzard, 2002). In addition, the investigators noted increased phosphorylation of tyrosine kinase receptors TrkA and TrkB, which are the targets of NGF and BDNF, respectively, and increased phosphorylation of cyclic AMP response element binding protein (CREB) in bladder afferent nerves. Interestingly, resiniferatoxin, a C-fiber neurotoxin, reduced the amount of phosphorylated CREB in DRG neurons in this cyclophosphamide-induced cystitis model, suggesting that chronic cystitis can reveal an altered CREB phosphorylation that is mediated by capsaicin-sensitive (i.e., C-fiber) bladder afferents and plays a role as a transcription factor in LUT plasticity (Qiao & Vizzard, 2004). In rat models, exogenous NGF can induce bladder nociceptive response when applied into the bladder lumen or bladder wall. Application of NGF-sequestering molecules, such as TrkA-IgG or REN1820, can reduce thermal hyperalgesia elicited by bladder inflammation from intravesically applied turpentine oil or cyclophosphamide-induced cystitis (Chai & Birder, 2016).

Clinical correlation: Clinical studies on patients with PBS/IC have found elevated levels of neurotrophic factors as well, including NGF, NT-3, and GDNF in urine, and increased expression of NGF in bladder biopsies. These findings, coupled with findings from animal studies, suggest that NGF expression is directly involved in the emergency of bladder-associated pain sensation in cystitis, and therefore, NGF and its pathway could serve as a potential target for treatment of pain in PBS/IC. Further translational studies are warranted.

Animal models of PBS/IC have also indicated a role of purinergic mechanisms for bladder pain/dysfunction in chronic cystitis. In a feline model of PBS/IC, enhanced ATP release from the urothelium was noted, as well as an increase in cell size and firing rates in capsaicin-sensitive C-fibers in DRG neurons. In rat models, exogenous purinergic agonists mimic the effects of cyclophosphamide-induced cystitis and P2X receptor antagonists suppress the effects of purinergic agonists and cystitis (de Groat and Yoshimura, 2009). These findings are consistent with a role of endogenous released purinergic agonists in enhancing the excitability of bladder afferent nerves in inflamed bladders via activation of the P2X receptors. Studies of human bladder specimens in patients with PBS/IC have supported these animal findings and stretch of cultured PBS/IC urothelial cells resulted in increased ATP release and increased P2X3 expression (Chai & Birder, 2016).

Chronic bladder inflammation and cystitis has also been shown to result in changes in the function of chemosensitive receptors in sensory afferent nerves. TRPV1 receptor activity is altered in a feline model of PBS/IC and this altered activity is reversed by inhibition of PKC, suggesting that TRPV1 activity changes in PBS/IC is caused by endogenous PKC activity. Similarly, using a rat model of cyclophosphamide-induced cystitis, Dinis et al. found that bladder overactivity was due, in part, to increased TRPV1 receptor activity, which led to increased anandamide expression in the bladder (Dinis et al., 2004). Mouse studies also support a role of TRPV1 receptors as a contributor to bladder pain in cystitis. Mouse models of PBS/IC with either cyclophosphamide or intravesical acrolein show bladder hyperactivity, but this response is blocked in TRPV1 knockout mice. Furthermore, Charrua et al. studied the effects of GRC-6211, an orally active TRPV1 antagonist, on the function of normal bladders and inflamed bladders using cystitis rat and mice models (Charrua et al., 2009). The investigators found that GRC-6211 at high doses suppressed normal bladder activity via a TRPV1-dependent mechanism and that GRC-6211 counteracted the bladder hyperactivity and noxious input induced by cystitis. These findings suggest a role of TRPV1 antagonism as a treatment for bladder pain and/or cystitis.

Conclusion

Basic science research into afferent neuronal pathways, including the receptors and neurotransmitters involved in bladder sensation, has the potential to lead to novel targets for therapy of common LUTS. LUTS are an extremely common clinical problem that affect all age groups and can lead to chronic conditions such as PBS/IC. The huge healthcare burden from LUTS and need for novel therapeutics and alternative therapies is evident in the drive for future research by the AUA-UCF in their "National Urology Research Agenda" and by the NIH/NIDDK via specialized grant funding opportunities focused on LUTS. The majority of bladder afferent pathways involve Aδ-fibers and C-fibers, and many of these nerve endings are present in the urothelium and suburothelial layer of the bladder, consistent with the notion of the urothelium as a sensory organ. These afferent nerves play an important role in both overactive and underactive bladder. Under normal conditions, Aδ-fibers act as mechanosensors and lead to normal voiding reflexes, but in pathologic states, the emergence of C-fiber pathways plays a role in abnormal bladder hyperactivity and increased bladder sensation and pain. Understanding the specific mechanisms by which C-fibers and Aδ-fibers are altered in LUT dysfunction, including identification and characterization of the receptors and neurotransmitters involved, may lead to future potential treatment breakthroughs.

References

Aizawa, N., Igawa, Y., Nishizawa, O., & Wyndaele, J. J. (2011). Effects of nitric oxide on the primary bladder afferent activities of the rat with and without intravesical acrolein treatment. *European Urology, 59*, 264–271.

Apostolidis, A., Popat, R., Yiangou, Y., Cockayne, D., Ford, A. P., Davis, J. B., Dasgupta, P., Fowler, C. J., & Anand, P. (2005). Decreased sensory receptors P2X3 and TRPV1 in suburothelial nerve fibers following intradetrusor injections of botulinum toxin for human detrusor overactivity. *The Journal of Urology, 174*, 977–982. discussion 982-3.

AUA-UCF. (2018). *American urological association*. Linthicum, MD: AUA - National Urology Research Agenda (NURA). Available: http://www.auanet.org/research/resources-and-initiatives/national-urology-research-agenda-(nura. (Accessed 16 April 2018).

Austin, P. F., Bauer, S. B., Bower, W., Chase, J., Franco, I., Hoebeke, P., Rittig, S., Vande Walle, J., Von Gontard, A., Wright, A., Yang, S. S., & Neveus, T. (2014). The standardization of terminology of lower urinary tract function in children and adolescents: Update report from the Standardization Committee of the International Children's Continence Society. *The Journal of Urology, 191*, 1863–1865.e13.

Birder, L. A., & De Groat, W. C. (1992). Increased c-fos expression in spinal neurons after irritation of the lower urinary tract in the rat. *Journal of Neuroscience, 12*, 4878–4889.

Birder, L. A., & De Groat, W. C. (1993). Induction of c-fos expression in spinal neurons by nociceptive and nonnociceptive stimulation of LUT. *American Journal of Physiology, 265*, R326–R333.

Birder, L. A., Roppolo, J. R., Erickson, V. L., & De Groat, W. C. (1999). Increased c-fos expression in spinal lumbosacral projection neurons and preganglionic neurons after irritation of the lower urinary tract in the rat. *Brain Research, 834*, 55–65.

Boyle, P., Robertson, C., Mazzetta, C., Keech, M., Hobbs, F. D., Fourcade, R., Kiemeney, L., & Lee, C. (2003). The prevalence of lower urinary tract symptoms in men and women in four centres. The UrEpik study. *BJU International, 92*, 409–414.

Chai, T., & Birder, L. (2016). Chapter 69: Physiology and pharmacology of the bladder and urethra. In A. Wein, L. Kavoussi, A. Partin, & C. Peters (Eds.), *Campbell-walsh urology* (11 ed.). Philadelphia, PA: Elsevier.

Charrua, A., Cruz, C. D., Cruz, F., & Avelino, A. (2007). Transient receptor potential vanilloid subfamily 1 is essential for the generation of noxious bladder input and bladder overactivity in cystitis. *The Journal of Urology, 177*, 1537–1541.

Charrua, A., Cruz, C. D., Narayanan, S., Gharat, L., Gullapalli, S., Cruz, F., & Avelino, A. (2009). GRC-6211, a new oral specific TRPV1 antagonist, decreases bladder overactivity and noxious bladder input in cystitis animal models. *The Journal of Urology, 181*, 379–386.

Cockayne, D. A., Hamilton, S. G., Zhu, Q. M., Dunn, P. M., Zhong, Y., Novakovic, S., Malmberg, A. B., Cain, G., Berson, A., Kassotakis, L., Hedley, L., Lachnit, W. G., Burnstock, G., Mcmahon, S. B., & Ford, A. P. (2000). Urinary bladder hyporeflexia and reduced pain-related behaviour in P2X3-deficient mice. *Nature, 407*, 1011–1015.

Coyne, K. S., Sexton, C. C., Vats, V., Thompson, C., Kopp, Z. S., & Milsom, I. (2011). National community prevalence of overactive bladder in the United States stratified by sex and age. *Urology, 77*, 1081–1087.

Dang, K., Bielefeldt, K., & Gebhart, G. F. (2005). Differential responses of bladder lumbosacral and thoracolumbar dorsal root ganglion neurons to purinergic agonists, protons, and capsaicin. *Journal of Neuroscience, 25*, 3973–3984.

De Groat, W. C. (1986). Spinal cord projections and neuropeptides in visceral afferent neurons. *Progress in Brain Research, 67*, 165–187.

De Groat, W. C., Griffiths, D., & Yoshimura, N. (2015). Neural control of the lower urinary tract. *Comprehensive Physiology, 5*, 327–396.

De Groat, W. C., & Yoshimura, N. (2009). Afferent nerve regulation of bladder function in health and disease. *Handbook of Experimental Pharmacology*, 91–138.

De Seze, M., Wiart, L., De Seze, M. P., Soyeur, L., Dosque, J. P., Blajezewski, S., Moore, N., Brochet, B., Mazaux, J. M., Barat, M., & Joseph, P. A. (2004). Intravesical capsaicin versus resiniferatoxin for the treatment of detrusor hyperreflexia in spinal cord injured patients: A double-blind, randomized, controlled study. *The Journal of Urology, 171*, 251–255.

Dinis, P., Charrua, A., Avelino, A., Yaqoob, M., Bevan, S., Nagy, I., & Cruz, F. (2004). Anandamide-evoked activation of vanilloid receptor 1 contributes to the development of bladder hyperreflexia and nociceptive transmission to spinal dorsal horn neurons in cystitis. *Journal of Neuroscience, 24*, 11253–11263.

Dmitrieva, N., & Mcmahon, S. B. (1996). Sensitisation of visceral afferents by nerve growth factor in the adult rat. *Pain, 66*, 87–97.

Fitzgerald, M. P., Thom, D. H., Wassel-Fyr, C., Subak, L., Brubaker, L., Van Den Eeden, S. K., & Brown, J. S. (2006). Childhood urinary symptoms predict adult overactive bladder symptoms. *The Journal of Urology, 175*, 989–993.

Fowler, C. J., Griffiths, D., & DE Groat, W. C. (2008). The neural control of micturition. *Nature Reviews Neuroscience, 9*, 453–466.

Giannantoni, A., Di Stasi, S. M., Stephen, R. L., Navarra, P., Scivoletto, G., Mearini, E., & Porena, M. (2002). Intravesical capsaicin versus resiniferatoxin in patients with detrusor hyperreflexia: A prospective randomized study. *The Journal of Urology, 167*, 1710–1714.

Gratzke, C., Streng, T., Park, A., Christ, G., Stief, C. G., Hedlund, P., & Andersson, K. E. (2009). Distribution and function of cannabinoid receptors 1 and 2 in the rat, monkey and human bladder. *The Journal of Urology, 181*, 1939–1948.

Hawthorn, M. H., Chapple, C. R., Cock, M., & Chess-Williams, R. (2000). Urothelium-derived inhibitory factor(s) influences on detrusor muscle contractility in vitro. *British Journal of Pharmacology, 129*, 416–419.

Hegde, S. S. (2006). Muscarinic receptors in the bladder: From basic research to therapeutics. *British Journal of Pharmacology, 147*(Suppl. 2), S80–S87.

Hulsebosch, C. E., & Coggeshall, R. E. (1982). An analysis of the axon populations in the nerves to the pelvic viscera in the rat. *The Journal of Comparative Neurology, 211*, 1–10.

Igawa, Y., Aizawa, N., & Michel, M. C. (2019). beta3 -Adrenoceptors in the normal and diseased urinary bladder-what are the open questions? *British Journal of Pharmacology, 176*, 2525–2538.

Jancso, G., & Maggi, C. A. (1987). Distribution of capsaicin-sensitive urinary bladder afferents in the rat spinal cord. *Brain Research, 418*, 371–376.

Jung, S. Y., Fraser, M. O., Ozawa, H., Yokoyama, O., Yoshiyama, M., De Groat, W. C., & Chancellor, M. B. (1999). Urethral afferent nerve activity affects the micturition reflex; implication for the relationship between stress incontinence and detrusor instability. *The Journal of Urology, 162*, 204–212.

Kanai, A., & Andersson, K. E. (2010). Bladder afferent signaling: Recent findings. *The Journal of Urology, 183*, 1288–1295.

Kumar, V., Chapple, C. R., Surprenant, A. M., & Chess-Williams, R. (2007). Enhanced adenosine triphosphate release from the urothelium of patients with painful bladder syndrome: A possible pathophysiological explanation. *The Journal of Urology, 178*, 1533–1536.

Landgraf, J. M., Abidari, J., Cilento, B. G.,, JR., Cooper, C. S., Schulman, S. L., & Ortenberg, J. (2004). Coping, commitment, and attitude: Quantifying the everyday burden of enuresis on children and their families. *Pediatrics, 113*, 334–344.

Langford, L. A., & Coggeshall, R. E. (1981). Branching of sensory axons in the peripheral nerve of the rat. *The Journal of Comparative Neurology, 203*, 745–750.

Lazzeri, M., Beneforti, P., Spinelli, M., Zanollo, A., Barbagli, G., & Turini, D. (2000). Intravesical resiniferatoxin for the treatment of hypersensitive disorder: A randomized placebo controlled study. *The Journal of Urology, 164*, 676–679.

Maggi, C. A. (Ed.). (1993). *The dual, sensory and efferent function of the capsaicin-sensitive primary sensory nerves in the bladder and urethra*. London, UK: Harwood Acadmic Publishers.

Maggi, C. A., Lecci, A., Santicioli, P., Del Bianco, E., & Giuliani, S. (1993). Cyclophosphamide-induced cystitis in rats: Involvement of capsaicin-sensitive primary afferents. *Agents Actions*, 38 Spec No C28-C30.

Marshall, K. L., Saade, D., Ghitani, N., Coombs, A. M., Szczot, M., Keller, J., Ogata, T., Daou, I., Stowers, L. T., Bönnemann, C. G., Chesler, A. T., & Patapoutian, A. (2020). PIEZO2 in sensory neurons and urothelial cells coordinates urination. *Nature, 588*, 290–295.

Morgan, C., Nadelhaft, I., & De Groat, W. C. (1981). The distribution of visceral primary afferents from the pelvic nerve to Lissauer's tract and the spinal gray matter and its relationship to the sacral parasympathetic nucleus. *The Journal of Comparative Neurology, 201*, 415–440.

Morgan, C., Nadelhaft, I., & Degroat, W. C. (1986). The distribution within the spinal cord of visceral primary afferent axons carried by the lumbar colonic nerve of the cat. *Brain Research, 398*, 11–17.

Nadelhaft, I., & Vera, P. L. (1992). Reduced urinary bladder afferent conduction velocities in streptozocin diabetic rats. *Neuroscience Letters, 135*, 276–278.

Nishiguchi, J., Sasaki, K., Seki, S., Chancellor, M. B., Erickson, K. A., De Groat, W. C., Kumon, H., & Yoshimura, N. (2004). Effects of isolectin B4-conjugated saporin, a targeting cytotoxin, on bladder overactivity induced by bladder irritation. *European Journal of Neuroscience, 20*, 474–482.

Pandita, R. K., Mizusawa, H., & Andersson, K. E. (2000). Intravesical oxyhemoglobin initiates bladder overactivity in conscious, normal rats. *The Journal of Urology, 164*, 545–550.

Payne, C. K., Mosbaugh, P. G., Forrest, J. B., Evans, R. J., Whitmore, K. E., Antoci, J. P., Perez-Marrero, R., Jacoby, K., Diokno, A. C., O'reilly, K. J., Griebling, T. L., Vasavada, S. P., Yu, A. S., Frumkin, L. R., & Group, I. R. S. (2005). Intravesical resiniferatoxin for the treatment of interstitial cystitis: A randomized, double-blind, placebo controlled trial. *The Journal of Urology, 173*, 1590–1594.

Phe, V., Schneider, M. P., Peyronnet, B., Abo Youssef, N., Mordasini, L., Chartier-Kastler, E., Bachmann, L. M., & Kessler, T. M. (2018). Intravesical vanilloids for treating neurogenic lower urinary tract dysfunction in patients with multiple sclerosis: A systematic review and meta-analysis. A report from the Neuro-Urology Promotion Committee of the International Continence Society (ICS). *Neurourology and Urodynamics, 37*, 67–82.

Qiao, L. Y., & Vizzard, M. A. (2002). Cystitis-induced upregulation of tyrosine kinase (Trka, TrkB) receptor expression and phosphorylation in rat micturition pathways. *The Journal of Comparative Neurology, 454*, 200–211.

Qiao, L. Y., & Vizzard, M. A. (2004). Up-regulation of phosphorylated CREB but not c-Jun in bladder afferent neurons in dorsal root ganglia after cystitis. *The Journal of Comparative Neurology, 469*, 262–274.

Sasaki, K., Chancellor, M. B., Goins, W. F., Phelan, M. W., Glorioso, J. C., De Groat, W. C., & Yoshimura, N. (2004). Gene therapy using replication-defective herpes simplex virus vectors expressing nerve growth factor in a rat model of diabetic cystopathy. *Diabetes, 53*, 2723–2730.

Smet, P. J., Moore, K. H., & Jonavicius, J. (1997). Distribution and colocalization of calcitonin gene-related peptide, tachykinins, and vasoactive intestinal peptide in normal and idiopathic unstable human urinary bladder. *Laboratory Investigation, 77*, 37–49.

Steers, W. D., Ciambotti, J., Etzel, B., Erdman, S., & De Groat, W. C. (1991). Alterations in afferent pathways from the urinary bladder of the rat in response to partial urethral obstruction. *The Journal of Comparative Neurology, 310*, 401–410.

Sureshkumar, P., Jones, M., Cumming, R., & Craig, J. (2009). A population based study of 2,856 school-age children with urinary incontinence. *The Journal of Urology, 181*, 808–815. discussion 815-6.

Talaat, M. (1937). Afferent impulses in the nerves supplying the urinary bladder. *The Journal of Physiology, 89*, 1–13.

Vaz, G. T., Vasconcelos, M. M., Oliveira, E. A., Ferreira, A. L., Magalhaes, P. G., Silva, F. M., & Lima, E. M. (2012). Prevalence of lower urinary tract symptoms in school-age children. *Pediatric Nephrology, 27*, 597–603.

Vizzard, M. A. (2000). Alterations in spinal cord Fos protein expression induced by bladder stimulation following cystitis. *American Journal of Physiology - Regulatory, Integrative and Comparative Physiology, 278*, R1027–R1039.

Vizzard, M. A. (2001). Alterations in neuropeptide expression in lumbosacral bladder pathways following chronic cystitis. *Journal of Chemical Neuroanatomy, 21*, 125–138.

Wiseman, O. J., Brady, C. M., Hussain, I. F., Dasgupta, P., Watt, H., Fowler, C. J., & Landon, D. N. (2002). The ultrastructure of bladder lamina propria nerves in healthy subjects and patients with detrusor hyperreflexia. *The Journal of Urology, 168*, 2040–2045.

Zagorodnyuk, V. P., Gibbins, I. L., Costa, M., Brookes, S. J., & Gregory, S. J. (2007). Properties of the major classes of mechanoreceptors in the Guinea pig bladder. *The Journal of Physiology, 585*, 147–163.

Chapter 6

"Translational effects of neuro-urology research on clinical practice"; Patient population–specific lower urinary tract symptoms

Ingrid Hoeritzauer
Department of Clinical Neurosciences, Royal Infirmary Edinburgh, NHS Lothian and Centre for Clinical Brain Sciences, University of Edinburgh, Edinburgh, United Kingdom

Bladder problems such as urinary incontinence and difficulty passing urine affect 75%–93% of patients with common neurological diseases such as dementia, multiple sclerosis (MS), and Parkinson's disease and negatively affect social functioning and physical and mental health (Kessler, 2020). Despite the high prevalence of lower urinary tract symptoms, the majority of current investigation and management of bladder problems focuses on pathomechanical outflow dysfunction such as pelvic floor weakness or benign prostatic hypertrophy. In patients with neurological disease, bladder problems are usually due to coordination dysfunction between the brain, spinal cord, and lower urinary tract, uroneurological dysfunction. In this chapter, we aim to familiarize the reader with neurological patient populations commonly affected by lower urinary tract symptoms.

We will explore the currently understood models of urological control in health, look at nomenclature and localization-based uroneurological dysfunction, and then look at how lower urinary tract dysfunction occurs in several neurological populations. We will describe the current treatments and finish with ongoing basic research into new treatment modalities. There are varying levels of animal studies in each condition and I approach this chapter as a clinician rather than an animal-based researcher. I am hopeful that the insights from human and current animal models will provide the basis of a helpful understanding of the problems caused by lower urinary tract dysfunction in neurological disease and explore our current human and animal understanding of uroneurological dysfunction.

Brain–bladder axis in health

It is unsurprising that lower urinary tract symptoms occur in so many neurological disorders. In health, urinary continence relies on a complex brain–bladder axis encompassing the urethral sphincters, detrusor, sacral and thoracic cord, Pontine micturition center (PMC), and periaqueductal gray as well as higher brain network. Parasympathetic, sympathetic, and somatic pathways are involved (see Fig. 6.1).

The lower urinary tract spends 99% of the time in a low-pressure storage phase with passive bladder filling to between 400 and 600 mL. During the storage phase, the detrusor is sympathetically inhibited and therefore remains in a low-pressure relaxed state as the involuntary internal urethral sphincter is progressively contracted. Both rely on sympathetic innervation from the inferior mesenteric ganglia at T11-L1/2. The voluntary external urethral sphincter is contracted during storage by somatic innervation from the nucleus of Onuf in the S2–4 pudendal nerve region.

Centrally, the PMC is inhibited by activity from cortical and subcortical centers such as the periaqueductal gray, insula, anterior cingulate gyrus, and prefrontal cortex. The periaqueductal gray is thought to play a significant role linking the higher centers and lower urinary tract and has projections to the amygdala, thalamus, and hypothalamus as well as

Neuro-Urology Research. https://doi.org/10.1016/B978-0-12-822455-7.00008-8

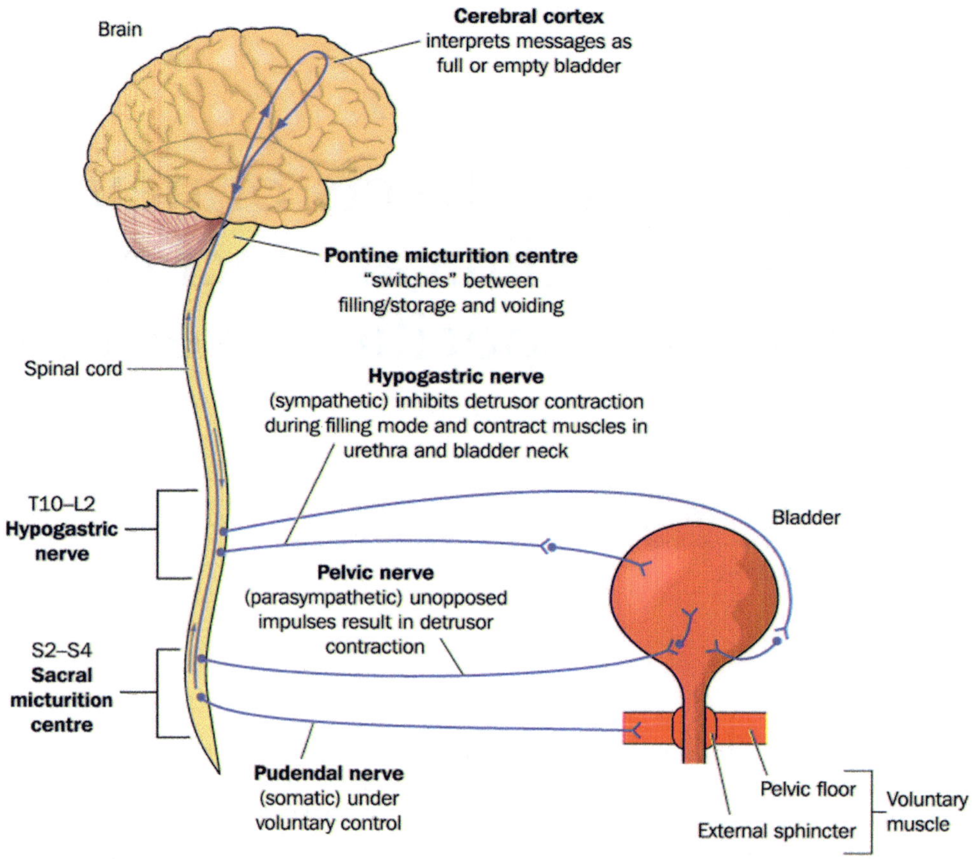

FIGURE 6.1 An overview of the brain–bladder axis. *From (2017) https://www.acquiredbraininjury-education.scot.nhs.uk/impact-of-abi/bladder-bowel-dysfunction/bladder-dysfunction/.*

receiving information from the bladder. In health, voiding only occurs when it is safe and socially appropriate. Therefore, as the bladder fills, increasing signals are sent from the detrusor and urethral sphincters to the sacral afferents and on to the PMC. These are monitored largely unconsciously via the periaqueductal gray and parahippocampal regions of the temporal cortex.

Every 3–4 h when it is deemed socially appropriate and safe to void, the PMC is released from its tonic inhibition. Signals are sent down the spinal cord to the sacral cord. Parasympathetic nuclei from S2–4 cause detrusor contraction and pelvic floor and internal and external urethral sphincter relaxation. If it is socially inappropriate or unsafe to void, the periaqueductal gray will increase the micturition threshold above 400 mL and reduce the awareness of needing to void.

We aim to create links between understanding the areas affected by common neurological diseases and the lower urinary tract symptoms that result from these.

Nomenclature

The International Continence Society categorizes lower urinary tract symptoms into storage symptoms, voiding symptoms, and postmicturition symptoms (Table 6.1).

Storage symptoms are

- urinary urgency or urgency incontinence, which is a sudden desire to pass urine,
- frequency, which is voiding >8 times per day, and
- nocturia, the need to wake up overnight to pass urine.

These are sometimes collectively termed as overactive bladder syndrome. On urodynamics, various findings can be associated with these symptoms including detrusor overactivity and increased or decreased bladder sensation. These symptoms can be seen in many neurological disease, and overactive bladder syndrome has been called "a symptom in

TABLE 6.1 Overview of lower urinary tract symptoms in neurological disease.

Type of symptoms	Symptoms		Urodynamics	CNS sites causing symptoms	Associated neurological diseases
Storage symptoms	Urinary urgency or urgency incontinence	Also known collectively as overactive bladder syndrome	*Various:* Detrusor overactivity	Suprapontine	Alzheimer's disease
			Increased bladder sensation	Spinal cord	
	Frequency		Decreased bladder sensation		Multiple sclerosis
	Nocturia				
Voiding symptoms	Poor stream		*Various:* Detrusor sphincter dyssynergia	Spinal cord	Traumatic spinal cord lesion
	Hesitancy		Detrusor underactivity		Spina bifida
	Intermittency			Conus lesion	Inflammatory conditions. e.g., infection: herpes simplex virus; inflammatory: neuromyelitis optica spectrum
	Straining			Cauda equina lesion	Compressive, e.g., disc tumor
	Dysuria				
	Feeling of needing to revoid		Unclear	Unclear periaqueductal gray	Fowler's syndrome/idiopathic urinary retention

search of a disease" (Van Batavia et al., 2017). In patients with overactive bladder symptoms and neurological disorders, further testing such as urodynamics is often required to identify the mechanism of symptom production.

Voiding symptoms related to difficulty passing urine are:

- poor stream,
- hesitancy,
- intermittency,
- straining,
- terminal dribble,
- dysuria, and
- a feeling of needing to revoid.

Urodynamic findings may include detrusor sphincter dyssynergia, which is contraction of the urethra and the detrusor at the same time, or detrusor underactivity.

Postmicturition symptoms are:

- a feeling of incomplete emptying and postmicturition dribble.

On urodynamics, a postvoid residual is often seen.

Localization-related symptoms in patients with neurological disease

In patients with common neurological disorders and lower urinary tract dysfunction, the symptoms may be similar, but the mechanism can often be understood based on where the neurological insult has occurred, be that suprapontine, spinal, or sacral (Panicker et al., 2015).

In suprapontine lesions, there is damage to the higher-order networks and impairment of the signals to the PMC. The PMC neurons normally inhibit micturition. When micturition is felt to be safe and socially appropriate, higher-order signals delay the PMC neurons. This allows micturition to occur. When damage to the higher-order networks occurs, this inhibition can be impacted and incontinence can occur. There is reduced bladder capacity and detrusor overactivity. Patients with suprapontine lesions will describe storage symptoms in keeping with overactive bladder syndrome: frequency, urgency, nocturia, and incontinence. On urodynamics, an overactive detrusor will be seen. Detrusor and urethral sphincter contraction/relaxation patterns remain in sync, so full bladder evacuation will occur. There will be no or minimal postvoid residual, although without the usual ability to fully inhibit urination until it is safe and socially appropriate urgency incontinence may occur.

In spinal cord injury (SCI), there is also a lack of bladder inhibition as links from higher-order brain networks to the thoracic and sacral cord are damaged. After an acute vascular or traumatic SCI, there is an acute phase of spinal shock lasting around 6 weeks (Goldmark et al., 2014) During this, the detrusor is hypocontractile or acontractile and patients will have urinary retention. Following this, previously silent c fibers become active and cause involuntary, spontaneous detrusor contractions. When slowly progressive or chronic spinal cord damage occurs, patients will experience overactive bladder symptoms including urinary incontinence. Patients will also have limited voluntary control due to lack of somatic input, so there can be incomplete emptying with >100 mL left in the bladder after voiding. This can lead to recurrent urinary tract infections (UTIs) and voiding symptoms such as straining and needing to revoid. On urodynamics detrusor overactivity, >100 mL postvoid residual and a small volume overactive bladder can be seen. Additionally, in some etiologies of SCI such as spina bifida and traumatic SCI, the coordination of detrusor contraction at the same time as internal and external urethral sphincter relaxation is lost. This results in *simultaneous* contraction of the detrusor and external urethral sphincter. This is called detrusor sphincter dyssynergia and can be seen on urodynamics. Simultaneous contraction of the external urethral sphincter and detrusor causes a high-pressure state in the bladder. This can result in upper urinary tract damage due to urinary reflux. Damage to the kidneys causing renal impairment is a major cause for concern in patients with SCIs and spina bifida. Patients with detrusor sphincter dyssynergia experience voiding symptoms including being able to void small amounts, intermittency, straining, and a feeling of needing to revoid. On urodynamics, detrusor sphincter dyssynergia can be seen. On the contrary, patients with inflammatory conditions affecting the spinal cord such as MS are very unlikely to have high detrusor pressure or develop upper urinary tract damage (Bacsu et al., 2012). The reasons for this are not clear.

In sacral lesions, such as conus medullaris inflammatory disorders or discogenic cauda equina syndrome, patients develop a large, flaccid bladder with urinary retention. This is due to loss of sacral parasympathetic innervation, which usually causes relaxation of the detrusor and internal urethral sphincter and sacral somatic innervation, which results in relaxation of the external urethral sphincter, which is under voluntary control. Tonic activity of the sympathetic fibers from the upper spine and PMC continue to provide inhibition of the detrusor and internal urethral sphincter. Patients experience inability to pass urine, and urodynamics will demonstrate a hypocontractile bladder.

Neurological populations with frequent lower urinary tract dysfunction

In this section, we will discuss the neurological conditions that commonly have lower urinary tract symptoms as a comorbidity: dementia, stroke, Parkinson's disease, MS, SCI, cauda equina syndrome, Fowler's syndrome, and functional neurological disorders.

The section will briefly describe each disease and then focus on the uroneurological symptoms and mechanisms, which occur in each disease. It is important clinically to consider that some of these conditions, such as dementia and stroke, primarily affect patients over 65 years old. While we will not focus on the pathomechanical comorbidities here, in older patients, age-related lower urinary tract conditions such as pelvic organ prolapse, stress urinary incontinence, and benign prostatic hypertrophy or age-related pathophysiological contributors such as chronic pelvic ischemia and sex hormone deficiency should always be considered and are important to bear in mind when investigating lower urinary tract symptoms. In the community-dwelling elderly population, factors that increase the risk of urinary incontinence are poor physical function, obesity, depression, and poor general health. These should be sought in general history and examination. In neurological conditions, especially those with a significant overlap with chronic pain, medications should be considered, which may alter lower urinary tract function, particularly analgesia such as opiates.

Dementia

Dementia is a progressive neurodegenerative disease which affects around 6% of all patients over the age of 65 years old (Urological Infections EAU Guidelines On, 2018). Dementia causes progressive deterioration in cognitive function affecting memory, praxis, language, and executive function (Kester & Scheltens, 2009). As dementia occurs most commonly in older adults, there is overlap with urinary incontinence due to aging related urological conditions. However, urinary incontinence is more likely to occur in patients older than 65 with dementia than in age matched controls (Averbeck et al., 2017). Rates of urinary incontinence vary greatly from 11% to 93% of patients depending on the study, the type and severity of dementia. Alzheimer's disease is the most common type of dementia and is responsible for roughly half of all dementias in patients over 65 years old. Vascular dementia is the second most common type, accounting for around 16% of dementias. A further 30% is made up of different forms of dementia such as frontotemporal dementia, dementia with Lewy bodies, and normal pressure hydrocephalus. In patients younger than 65 years old, frontotemporal dementia, alcohol-related dementia, and dementia secondary to systematic disease such as MS follow Alzheimer's disease in prevalence.

Alzheimer's affects 1 in 14 people over 65 years old. The disease typically causes early progressive memory loss, executive dysfunction, and behavioral change including agitation. Senile plaques, neurofibrillary tangles, and hippocampal and cortical neurodegeneration are the pathological processes causing this clinical deterioration. In Alzheimer's disease, incontinence develops later in the disease and correlates with disease severity (Lee et al., 2014). In severe Alzheimer's disease, the prevalence of urinary incontinence is over 50%. Urinary incontinence leads to reduced quality of life, increased caregiver burden, and institutionalization.

In most kinds of dementia including Alzheimer's disease, the most common cause of incontinence in those with lower urinary tract symptoms is detrusor overactivity. The proposed mechanism of urinary incontinence in dementia is cerebral dysfunction, leading to a loss of the higher brain centers' inhibitory influence. Detrusor overactivity is thought to be the result of damage to inhibitory brain—bladder networks in the anterior and medial frontal lobe, the anterior and middle cingulate cortex, insular cortex, and supplementary motor area.

A study of 215 patients with Alzheimer's disease followed up for 1 year demonstrated frontal lobe function was associated with urinary incontinence even after adjustment for potential confounders (Sugimoto et al., 2017). The authors suggested that in early Alzheimer's disease, urinary incontinence should be thought of as a clinical manifestation of abnormal frontal lobe function rather than due to physical or behavioral problems. As the disease progresses, frontal predominant networks fail and there is a loss of bladder inhibition with resulting urinary incontinence. Some studies suggest that acetylcholinesterase inhibitors, which are used in mild to moderate dementia, worsen urinary incontinence and that the dose should be reduced if patients develop new or worsening urinary incontinence rather than presuming urinary incontinence is a sign of worsening disease activity (Averbeck et al., 2017).

While lower urinary tract dysfunction occurs in many patients with Alzheimer's disease but is not a key feature, there are some dementias in which urinary incontinence is always part of the diagnostic phenotype, such as normal pressure hydrocephalus.

Normal pressure hydrocephalus is a brain disease diagnosed based clinically on the triad of incontinence, gait dysfunction with magnetic-type gait, and deteriorating memory (Williams & Malm, 2016). Patients have dilated cerebral ventricles on imaging but normal intracerebral pressure. Incontinence is present in 93% of patients and predates gait disorder and dementia (Williams & Malm, 2016). Imaging studies have shown that severe bladder dysfunction is predicted by right frontal hypoperfusion (Sakakibara et al., 2016). Improvement in bladder function after shunting is linked to bilateral increased midcingulate perfusion. Urodynamics show detrusor overactivity (Nagata et al., 2017). The only effective treatment for patients is a CSF shunt, although the effectiveness of this deteriorates over time (Takeuchi & Yajima, 2019).

Stroke

Stroke affects 450 people per 100,000/year and is responsible for 10% of all cardiovascular mortality. Stroke can be ischemic, hemorrhagic, or due to hypoperfusion. Neurological dysfunction is caused by focal brain damage due to ischemia and hemorrhage. Ischemic stroke is due to thrombosis, commonly from atherosclerotic disease, or embolism, commonly from a cardiac clot, causing obstructed blood flow within the vessel. Strokes are commonly described based on the vessel territory, which has been occluded: the middle cerebral artery, anterior cerebral artery, or posterior cerebral artery. In patients who have had a stroke, recognizing and treating urinary incontinence is important for rehabilitation and quality of life (Mehdi et al., 2013). Depression is twice as common in stroke survivors who are incontinent than those who are continent. Continuing incontinence is associated with poor outcome and institutionalization in stroke survivors (Panicker, 2017).

The majority of patients (57%–83%) will have some lower urinary tract symptoms in the first month poststroke (Urological Infections EAU Guidelines On, 2018). A Cochrane review found urinary incontinence in roughly half (53%) of stroke survivors in the first month after stroke (Thomas et al., 2019). There is some improvement over time. At 3 months, 44% of patients were still incontinent and at 1 year, 38% remained incontinent. The more severe the stroke, the greater the likelihood of urinary incontinence.

Mehdi et al. (2013) helpfully split up the causes of poststroke incontinence into mechanistic problems caused by direct damage to neuromicturition pathways causing detrusor hyperreflexia; impaired awareness of bladder signals; functional incontinence due to direct cognitive damage or inability to communicate; and transient causes of incontinence due to UTIs, medications, or delirium. In keeping with our understanding of the bladder–brain axis, strokes affecting the anteromedial frontal lobe, its descending pathways, and the basal ganglia are most likely to cause lower urinary tract dysfunction due to mechanistic problems. This is in keeping with loss of the tonic inhibition of the PMC and thereby control over the detrusor, resulting in symptoms of overactive bladder syndrome due to detrusor hyperreflexia (Sakakibara, 2015). Impaired awareness of bladder signals was associated less with frontal damage, but rather with total or partial anterior circulation strokes and new parietal or subcortical damage (Pettersen et al., 2007). This is postulated to be due to failure to correctly identify and validate the bladder signals due to breakdown of the higher bladder–brain networks secondary to parietal damage. In keeping with studies of the mechanism of incontinence in normal pressure hydrocephalus, when the ischemic lesion was in an area in the right frontal region, bladder overactivity and urgent urinary incontinence were the predominant symptoms. This is due to interrupted pathways from the brainstem to the cortex and vice versa.

In the rat model of middle cerebral artery occlusion, there is evidence of detrusor overactivity due to reduced bladder capacity during awake cystometry. Similarly to humans, when the ischemic lesion was in a small area in right frontal region, bladder overactivity and urgent urinary incontinence were the predominant symptoms. Ischemic lesions may interrupt pathways from the brainstem to the cortex or vice versa. At the neurotransmitter level, NMDA glutamatergic excitatory mechanisms appear to play an important role in overactivity due to stroke. In rat models, pretreatment with an NMDA receptor agonist can prevent bladder overactivity.

In the first 3 days after stroke in a mouse model within the brain, there may be impaired balance between GABA, which is inhibitory, and glutamate, which is excitotoxic (Ward, 2017). These mechanisms are involved in the micturition reflex and are thought to contribute to bladder overactivity and urethral dysfunction. Ascending and descending pathways through raphe nuclei and locus coeruleus are major sources of spinal serotonergic and noradrenergic pathways in the brainstem and may be disrupted by a stroke lesion resulting in impaired urethral continence.

Urodynamics most commonly show detrusor overactivity, often with impaired detrusor contractility (Pizzi et al., 2014).

Parkinson's disease

Parkinson's disease is one of the most common neurodegenerative conditions, affecting 100–180 per 100,000 people (Vurture et al., 2019). It is caused by abnormal deposition of the protein alpha-synuclein in the central and peripheral nervous system, leading to degeneration of the nigrostriatal dopaminergic neurons and insufficient formation and action of the neurotransmitter dopamine. This type of condition is referred to as an alpha-synucleinopathy. Diagnosis is based on the patient's history and clinical signs. Patients with Parkinson's disease have a triad of motor symptoms: bradykinesia (slowness of movement), asymmetrical tremor, and rigidity. It is now also recognized that there are many nonmotor manifestations of Parkinson's disease including depression, loss of smell, REM sleep behavior disorder, orthostatic hypotension, constipation, sexual dysfunction, and lower urinary tract symptoms. Gastrointestinal symptoms and orthostatic hypotension are thought to be due to peripheral dopaminergic denervation.

Lower urinary tract symptoms, particularly overactive bladder symptoms, affect between 27% and 85% of patients with Parkinson's disease (McDonald et al., 2017). Urinary incontinence occurs later in the disease, usually after 5–7 years (Winge, 2015). In the PRIAMO study, 1072 patients with Parkinson's disease were investigated for nonmotor symptoms of Parkinson's disease and 57% were found to have lower urinary tract symptoms (Barone et al., 2009). The wide range of prevalence across studies may be due to the lack of validated questionnaires for lower urinary tract symptoms in Parkinson's disease, or comorbid age-related lower urinary tract symptoms. Medication regimens for patients with Parkinson's disease often become complex and include L-DOPA, dopamine receptor agonists, and other medications, which may affect lower urinary tract function. The effects of dopaminergic medications on lower urinary tract symptoms are unclear, but at least some studies suggest improvement in bladder capacity when medicated for their Parkinson's disease. There are conflicting accounts of whether lower urinary tract symptoms are associated with disease severity or disease duration or whether lower urinary tract symptoms become more bothersome due to deteriorating motor and cognitive symptoms. Lower urinary tract symptoms in Parkinson's disease are associated with increased falls risk, poorer quality of life, and increased admission to care.

Urodynamic investigations are not commonly carried out in patients with Parkinson's disease. Urodynamic studies available have shown the majority of patients with lower urinary tract symptoms have detrusor overactivity (36%–81%). Detrusor overactivity, bladder capacity, and bladder volume at first desire to void correlate with disease severity in most studies. There is evidence of detrusor weakness in 66% of women and 40% of men, which correlates with motor disorders (Sakakibara et al., 2001). These findings suggest lower urinary tract symptoms occur due to central and peripheral effects of the disease.

Centrally, there is thought to be less inhibitory input from the periaqueductal gray to the PMC due to disease-specific altered signaling in the nigrostriatal dopaminergic system. Animal models suggest that activation of D1 receptors tonically inhibit the micturition reflex and that detrusor overactivity is, in part, due to failure of D1 receptors to be stimulated. PET brain imaging has found a lack of activation of the pons or anterior cingulate cortex in patients with Parkinson's disease and detrusor overactivity during bladder filling compared to health healthy controls. Anterior cingulate cortex neurons have an inhibitory effect on detrusor control and play a role in the executive function required to create micturition. Peripherally, there is thought to be alpha-synuclein–related dysfunction of the autonomic nervous system including interrupting the pelvic visceral afferents input into the lower urinary tract.

Postvoid residual would not be expected in Parkinson's disease and should make a clinician look carefully for other signs of autonomic dysfunction and consider multiple system atrophy (MSA).

In rat models, findings of improved bladder volume at first desire to void and reduced prevalence of overactive bladder syndrome with subthalamic nucleus and globus pallidus interna deep brain stimulation suggest stimulation of the subthalamic nucleus improves the processing of afferent bladder information.

Multiple system atrophy

MSA is an alpha-synucleinopathy, which usually begins in middle age, affecting both sexes equally and is a progressive neurodegenerative disorder with a mean survival of 6–9 years. Patients have autonomic dysfunction and/or parkinsonian, cerebellar, or pyramidal features. Lower urinary tract symptoms are the initial presenting complaint in nearly 20% of patients and affect up to 95% of patients (Winge, 2015). Early symptoms are predominantly storage symptoms such as urgency, nocturia, and urinary incontinence. Voiding difficulties and postvoid residual occur gradually over time, often increasing to >100 mL from the second year onward. Incomplete emptying can lead to recurrent UTIs. Postvoid residual should be checked every year.

Urodynamics demonstrate changes over time from detrusor overactivity to low compliance and eventually to acontractile detrusor. An open bladder neck can be seen along with weakness of the striated urethral sphincter. This is due to damage to the lateral columns in the thoracic cord, which convey the descending sympathetic innervation and supply the sympathetically innervated hypogastric nerve. Denervation of the external urethral sphincter is a relatively specific finding, particularly early in the disease course of MSA. It is due to the loss of anterior horn cells in Onuf's nucleus of the sacral spinal cord (Winge, 2015). This suggests a progressive central to peripheral dysfunction mediated by widespread neurodegeneration and transsynaptic degeneration (Ito et al., 2006).

Treatment can be complicated by antimuscarinic medication required to manage other symptoms, causing increased postvoid residuals and storage symptoms.

Multiple sclerosis

MS is an autoimmune condition caused by inflammatory demyelination and axonal loss throughout the brain and spinal cord. MS has an incidence of 400–700 per 100,000 people worldwide (Al Dandan et al., 2020). It is three times as common in women and typically onset is in patients between 20 and 40 years old. It is more prevalent in white females in northern countries. The most common type of MS is relapsing-remitting MS affecting 85%–90% of patients with MS. Patients will have episodes of clinical symptoms caused by active demyelination, then complete or partial recovery. After 10 years of relapsing-remitting MS, patients may develop secondary progressive MS. This involves gradual accumulation of neurological deficits without clear relapses. Primary progressive MS also exists, in which patients have a gradual progressive course of worsening without relapses or the relapse/remission pattern seen in relapsing-remitting MS. Diagnosis is based upon the McDonald 2017 criteria findings of demyelination separated in time and space, using clinical, radiology, $\pm$ cerebrospinal fluid findings to make a diagnosis. At present, all disease-modifying drugs except ocrelizumab, a CD-20 monoclonal antibody which is licensed for early, inflammatory primary progressive MS, are only efficacious in relapsing-remitting MS.

Six years after diagnosis, approximately 40% of patients with MS have lower urinary tract symptoms. By 10 years, most patients have lower urinary tract symptoms (Nazari et al., 2020). With increasing age, the incidence of moderate and severe lower urinary tract symptoms increases (Nazari et al., 2020). Duration of disease, more aggressive clinical course, and severity of disability are associated with severity of lower urinary tract symptoms. A recent meta-analysis of 12 studies involving 2507 patients found a pooled prevalence of 68% of patients with MS experiencing lower urinary tract symptoms and 64% had objective evidence of abnormal function on urodynamics (Al Dandan et al., 2020). Lower urinary tract symptoms have a major impact on quality of life for patients with MS. Lower urinary tract symptoms are associated with increased falls risk and are a barrier to engaging in physical activity (Al Dandan et al., 2020). UTIs may result in discomfort, hospital admissions, and transient worsening of MS symptoms (pseudorelapse), and there is some evidence that UTIs may precipitate an acute inflammatory relapse (Panicker & Fowler, 2015).

Patients usually describe both storage and voiding symptoms. Urinary frequency is the most common symptom affecting up to 73%, followed by urgency 64% and a feeling of incomplete bladder emptying 61%. Nocturia 59% and incontinence 43% are commonly seen. On urodynamic studies, detrusor overactivity (43% pooled prevalence) and detrusor sphincter dyssynergia (5% pooled prevalence) are the most common findings (Al Dandan et al., 2020). Unfortunately, patients with MS may not always be aware if they develop incomplete bladder emptying, so a postvoid residual should be checked in all patients with lower urinary tract symptoms (Panicker & Fowler, 2015). Despite how common lower urinary tract symptoms are, patients with MS are no more likely to develop renal failure than the general UK population.

Lower UTIs in MS are thought to be largely due to demyelinating lesions that affect the spinal cord. Studies have demonstrated severity of lower urinary tract symptoms correlating with extent of spinal cord involvement. Lesions of the spinal cord result in both detrusor overactivity and detrusor sphincter dyssynergia. Inflammatory lesions are thought to interrupt the connection between the PMC and the parasympathetic and sympathetic nerves from the sacral and thoracic cord, which receive and transmit afferent bladder signals (see Fig. 6.2).

Rodent models of MS are most commonly used to study the autoimmunity and inflammatory precipitated neurodegeneration seen in MS. Animals are either actively immunized with experimental autoimmune encephalitis or can be passively given it by transfer to encephalitogenic T cells (da Silva et al., 2020). In active immunization, encephalitogenic antigens such as proteolipid protein, myelin basic protein, and myelin-associated glycoprotein along with complete Freund's adjunct are used to elicit the autoimmune encephalomyelitis. In addition to spinal cord demyelination causing lower urinary tract symptoms, there is evidence of bladder urothelial dysfunction in animal autoimmune encephalomyelitis models of MS. The cholinergic component of bladder contraction and pannexin 1, a member of the gap junction protein expressed in bladder urothelium, may cause sensitized afferent nerve terminals and a positive feedback loop for inflammatory responses in bladder dysfunction (Miyazato et al., 2017).

Spinal cord injury

SCI has a worldwide incidence of 250,000–500,000 new cases per year and predominantly affects males (78%) (Teplitsky et al., 2019). It occurs in a bimodal distribution with a peak in young adults, predominantly related to driving accidents,

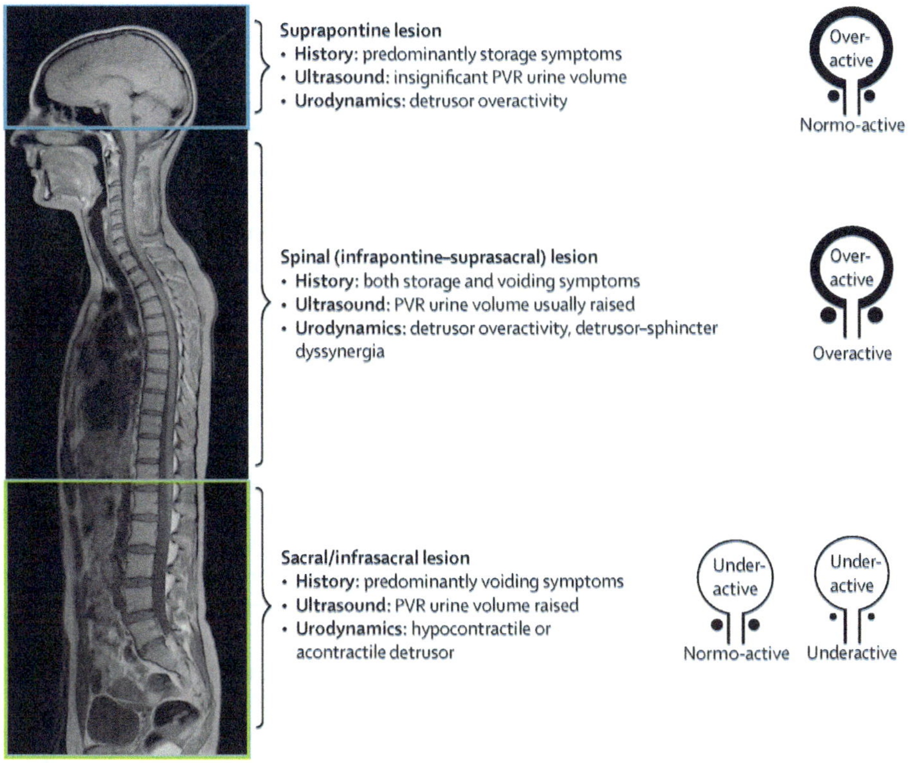

FIGURE 6.2 Correlating lower urinary tract symptoms with anatomical dysfunction. *From Panicker, J. N., Fowler, C. J., & Kessler, T. M. (2015). Lower urinary tract dysfunction in the neurological patient: Clinical assessment and management. The Lancet Neurology, 14(7), 720–732. https://doi.org/10.1016/S1474-4422(15)00070-8.*

alcohol use, and participation in high risk sports and a second peak in patients over 60 years old with a more equal male/female ratio.

Approximately 80% of patients with SCIs have lower urinary tract symptoms at 1 year and few make a full recovery (<1%) (Goldmark et al., 2014). Prior to the 1970s, up to 75% of patients with SCIs died from urosepsis or renal failure (Schurch et al., 2015). An increasing recognition of lower urinary tract dysfunction and pyelonephritis as a source of morbidity and mortality led to research and a change in management, focusing on preventing renal failure and aggressively treating septicemia. Currently urosepsis or renal failure account for 2% of deaths in patients with SCI. However, risk of urosepsis continues to affect patients with SCIs at many times the rate of the general population. Additionally, patients with SCI have higher mortality ratios from bladder cancer, with a standardized mortality ratio of 6.7 compared to the general population (Nahm et al., 2015). This may be due to the presence of long-term indwelling catheters. Patients with SCI and incontinence have worse quality of life, and incontinence is one of the major issues affecting sexuality of women with SCIs (Elmelund et al., 2018).

After an SCI, there is an acute phase of spinal shock lasting around 6 weeks (Goldmark et al., 2014). During this time, patients lose awareness of their bladder fullness and experience urinary retention and overflow incontinence as the detrusor is hypocontractile or acontractile and the internal and external urethral sphincters are closed. If untreated, patients can develop UTIs and urosepsis. The mechanism for this is not known.

Following spinal shock, bladder sensations may partially return in incomplete spinal cord lesions. However, voluntary inhibition of voiding is lost. Involuntary detrusor contraction occurs, maximal voiding pressure increases, leading to bladder hypertrophy and reduced voiding efficiency. Urge and urge incontinence occur due to the neurogenic detrusor overactivity. Patients can also develop detrusor sphincter dyssynergia, which is urethral sphincter contraction during bladder contraction. Untreated, patients have incomplete bladder emptying and develop high intravesicular voiding pressures to overcome the urethral sphincter contraction. This leads to back pressure, recurrent UTIs, stone formation, and hydronephrosis with eventual renal failure if unchecked (Goldmark et al., 2014; Schurch et al., 2015).

Detrusor sphincter dyssynergia is seen in urodynamics in 96% of patients with SCIs (Goldmark et al., 2014). However, although knowledge of the location of spinal damage is helpful in understanding the potential mechanism of lower urinary tract symptoms, urodynamics are required to provide precise classification of the lower urinary tract dysfunction because the picture can be complicated by multiple injury levels and how complete an SCI is. Furthermore, whether detrusor sphincter dyssynergia is continuous or intermittent can assist with classification. Patients with continuous detrusor sphincter dyssynergia, which is present during the voiding phase and hence obstructs the urine outflow, are at higher risk of upper urinary tract damage.

Feline and rodent animal models have demonstrated that normally silent, capsaicin- and cold-sensitive afferent c fibers become mechanosensitive and induce automatic micturition, detrusor overactivity, and detrusor sphincter dyssynergia after SCI. Intravesical capsaicin treatment in humans and systemic capsaicin treatment in rodents reduces or inhibits detrusor overactivity after chronic SCI (Lazzeri et al., 1999). Additionally, rodent models demonstrate a combination of peripheral and central neurotrophic hormones such as nerve growth factor (NGF), which are produced at increased rates after SCI. Suppression of NGF reduced neurogenic detrusor overactivity and detrusor sphincter dyssynergia in rats with SCIs (Hamid et al., 2018).

Spina bifida

Spina bifida refers to the incomplete closure of the caudal end of the neural tube (the structure from which the spinal cord and brain form) during embryonic growth (Mitchell et al., 2004). Lumbar and lumbosacral forms are the most common location (60%). Myelomeningoceles or "open" spina bifida are the most severe form as there is no dermal covering over the spinal cord. Worldwide prevalence of open and closed forms of spina bifida varies from 0.31 to 2.92 per 1000. Lower urinary tract dysfunction occurs in 96% of patients. Urological goals of treatment are focused on preserving renal function and, as the patient ages, maintaining continence. Urological issues cause death in almost one third of patients with open spina bifida and over 50% of patients are incontinent. Patients with spina bifida who are continent are significantly more likely to be employed (Wiener et al., 2018).

Lower urinary tract dysfunction is caused by incomplete or damaged innervation by the somatic, parasympathetic, and sympathetic innervation of the bladder afferents depending on the site of incomplete spinal cord closure (Snow-Lisy et al., 2015). In patients with myelomeningoceles, neurological and urological deficits are thought to be due both to defective neurulation with myelodysplasia and then subsequently also exposure to amniotic fluid, hydrodynamic pressure, or direct trauma.

Since the randomized controlled study comparing prenatal versus postnatal repair of myelomeningoceles (the MOMS study: Management of Myelomeningocele), evidence emerged that patients treated prenatally had improved lower extremity function, improved mental and motor function, less need for CSF diversion, and decreased Chiari malformation. Unfortunately, the outcomes for lower urinary tract dysfunction and upper tract damage do not appear to be different between those treated pre- and postnatally with closure.

Urodynamics demonstrate detrusor overactivity in about 45% of patients, which is often accompanied by detrusor sphincter dyssynergia (Weiss et al., 2019) or underactive or noncontractile detrusor, which may be accompanied by incompetent (continuously open) bladder neck. At least a quarter appear to have reduced detrusor compliance. Patients with open and closed spina bifida can have similar levels of lower urinary tract dysfunction and lower urinary tract dysfunction can change with age, so recurrent urodynamics are required. Urodynamics should be done 2 and 6 months after birth, then yearly if function is normal. Urodynamics are undertaken to assess for upper tract damage and evidence of new detrusor overactivity, which may point to development of tethered cord syndrome. Tethered cord syndrome occurs when there is fixation of the cord to the spinal column and can be a late complication of spina bifida. Strain and stretching of the cord occur resulting in bladder dysfunction, pain, and lower limb motor and sensory deficits in more than 50% of patients in a recent systematic review (O'Connor, 2020).

Cauda equina syndrome

Cauda equina syndrome is caused by compression of the cauda equina nerve roots, which lie at the bottom of the spinal cord. The most common cause of compression is an acute prolapse of the L4/5 disc. Cauda equina syndrome predominantly affects middle-aged patients. The cauda equina nerve roots provide motor and sensory innervations to the pelvic organs and lower limbs. Patients present with acute pain in the back or legs with one or more of new bladder, bowel or sexual dysfunction often with altered perianal sensation, lower limb weakness, or sensory changes (Fraser et al., 2009). The incidence of CES is 0.3–0.6 per 100,000 per year in asymptomatic populations. While uncommon, it is one of the true spinal emergencies and can have major clinical and medicolegal ramifications if missed or operative care occurs late. Clinical symptoms and signs of cauda equina syndrome do not correlate well with whether radiological cauda equina compression exists or not so an MRI is required in any cases when CES is suspected. Recent work has suggested acute multichannel urodynamic studies along with postvoid residual, perianal sensation, and bulbocavernosus reflex may help improve diagnostic accuracy pre-MRI imaging (Kalidindi et al., 2020). Patients are treated in an emergency fashion with transfers to achieve urgent MRI scans (Demetriades, 2009) at great cost due to a small meta-analysis of 322 patients (Ahn et al., 2000). The current best evidence is that patients require emergency surgery within 48 h of symptom onset to prevent chronic pain, bladder, bowel, and sexual dysfunction. In a follow-up study of patients with cauda equina syndrome, more than 3 years postsurgery, the prevalence of bladder dysfunction was 76%, bowel dysfunction in 13%, and sexual dysfunction in 39% (Hazelwood et al., 2019).

Cauda equina compression damages the S1–S4 sacral micturition center. This leads to loss of parasympathetic innervation of the bladder and internal urethral sphincter and loss of somatic innervation to the external urethral sphincter. Due to loss of afferent pathways, patients loose awareness of bladder fullness. However, pain sensation may be preserved as the hypogastric nerve which leaves the cord at the thoracic region will remain intact. Patients may experience abdominal fullness or pain when bladder distension occurs. Patients will experience urinary retention ± overflow incontinence due to detrusor acontractility as well as stress incontinence due to incompetent internal and external urethral closure. On urodynamics, the typical picture is of detrusor underactivity with reduced awareness of bladder filling resulting in urinary retention. Some patients will, however, develop detrusor overactivity with poor compliance. The bladder neck may remain open due to lack of parasympathetic input.

Patients with "scan-negative" cauda equina syndrome are those patients who attend A&E with the clinical symptoms of cauda equina syndrome: acute bladder, bowel, and sexual dysfunction but have negative or nonexplanatory MRI imaging. They make up 70%–80% of patients who are scanned with cauda equine syndrome symptoms. Recent research suggests that the mechanism of acute bladder dysfunction is one of pain, medications, and panic, affecting both bottom-up and top-down bladder–brain axis processing (Hoeritzauer et al., 2021).

Animal work on cauda equina syndrome can be challenging as it does not usually affect quadrupeds. An award winning study of cauda equina compression on dogs was not felt to be applicable to humans as there is often a less favorable outcome than with this Beagle animal model of cauda equina syndrome (Delamarter et al., 1991): 1991 Volvo Award in experimental studies: Cauda equina syndrome: neurologic recovery following immediate, early, or late decompression (Delamarter et al., 1991). Data have been extrapolated from SCI data in porcine and canine models. Better animal models would be helpful as debate still rages about whether time of compression or mean arterial intraspinal pressure due to disc prolapse is associated with poor outcomes. A meta-analysis of animal studies suggested worse outcomes were associated with poor predecompression function and cauda equina compression exceeding the mean arterial blood pressure for ≥1 h (Pronin et al., 2019).

Idiopathic urinary retention: Fowler's syndrome and functional urological disorders

There are several syndromes that cause idiopathic urinary retention: Fowler's syndrome, paruresis, scan-negative cauda equina syndrome, and functional urological disorders. These syndromes are of increasing interest as understanding of the impact of the brain on urological function grows.

Fowler's syndrome occurs in women in their 20s and 30s (Swinn & Fowler, 2001). These patients present with painless urinary retention of over 1 L due to primary failure of external urethral sphincter relaxation. They often describe a "gripping" sensation on intermittent self-catheterization, and urethral sphincter EMG has two unusual findings: complex repetitive discharges and decelerating bursts (Swinn et al., 2002). Recent work has demonstrated these EMG findings in women without lower urinary tract symptoms and similar levels of gynae pathology in patients without Fowler's syndrome (Karmarkar et al., 2015; Tawadros et al., 2015). Patients with Fowler's syndrome have up to a 50% comorbidity with chronic pain and an elevated rate of functional neurological disorders, with almost one quarter of patients being affected (Hoeritzauer et al., 2016).

Paruresis, or shy bladder syndrome, is a condition with a prevalence of up to 16% and causes an inability to void when others are present (usually in a public toilet) (Kuoch et al., 2017). Patients have intermittent urinary retention when away from home. Untreated, it can have major effects on patient's social, romantic, and occupational lives. It is underrecognized but is a treatable condition with cognitive behavioral therapy providing significant symptom improvement even in 1 year follow-up (Soifer et al., 2010).

In neurology, functional neurological disorders (previously called psychogenic disorders) are positively diagnosed disorders, which have specific neurological signs but without a pathophysiological correlate such as functional weakness or dissociative seizures. Motor components are caused by corruption of the preconscious phases of motor planning. There have only been four studies systematically investigating urinary dysfunction and functional neurological disorders. A study of lower urinary tract symptoms in patients with functional movement disorders found that 20% of patients self-reported lower urinary tract symptoms. The majority of patients described overactive bladder symptoms (63.6%). Patients with fixed dystonia had the most severe bladder symptoms and were more likely to be on opiates and have low stream symptoms (Batla et al., 2016). In a study of patients with psychogenic urinary dysfunction, most urodynamics were normal, although some patients demonstrated underactive/acontractile detrusor or had increased bladder sensation.

Mechanistic hypotheses as to how functional urological disorders occur come predominantly from understanding of older literature describing "psychogenic" urinary retention, Fowler's syndrome, and paruresis. In Fowler's syndrome, fMRI imaging has demonstrated the probable mechanism of painless urinary retention (Kavia et al., 2010). fMRI imaging suggested lower urinary tract inhibition is caused by a permanently contracted external urethral sphincter sending inhibitory messages to the sacral cord and on to the periaqueductal gray. Further human and animal studies are required.

Treatment options for patients with neurological disorders

The overall aims of treatment are:

- to achieve urinary continence and improve quality of life,
- to prevent UTIs, and
- to preserve upper urinary tract function.

Management should address both storage symptoms (urinary frequency, nocturia, urgency incontinence, and stress incontinence) and voiding dysfunction (urinary retention). The severity of symptoms and the risk of developing upper urinary tract dysfunction will often determine the most appropriate form of treatment. Patients with spina bifida or SCI are much more likely to develop upper urinary tract damage than patients with slowly progressive disorders such as Parkinson's disease or MS. In patients in whom renal function is at risk of compromise, the aim of treatment is to convert the high-pressure detrusor into a low-pressure reservoir despite the possibility of this potentially necessitating intermittent self-catheterization due to postvoid residual urine. For many patients, a single medical treatment is insufficient to produce continence and complimentary therapies such as medication plus intermittent self-catheterization may be required.

Treatment for storage dysfunction

Behavioral treatment

Although this book is predominantly about translational animal to human models and neuro-urology research, it is important to briefly touch on physical/behavioral treatments, which may be initial forms of treatment for many patients. These can be appropriate even for patients who have cognitive or motor difficulties depending on the level of caregiver or health professional support.

Habit retraining requires preemptive bladder emptying through learning the natural voiding pattern of a patient and developing an individual schedule. Timed voiding involves toilet visits at fixed time intervals. Pelvic floor exercises for 1 month have been found to be beneficial in reducing urinary frequency and number of times incontinent in a study of patients with urinary incontinence due to MS and SCI (Elmelund et al., 2018). These work by increasing pelvic muscle strength and increasing the inhibitory effect of pelvic floor contraction on the detrusor.

Antimuscarinic drugs

Antimuscarinics are recommended by International Consultation on Incontinence guidelines, National Institute of Clinical Excellence (NICE), and the European Association of Urologists (EAU) for neurogenic detrusor overactivity (Jaggi & Fatoye, 2017). Their method of action is antagonism of the muscarinic receptors causing detrusor relaxation and lowering intravesical pressures. There are several different medications, which all appear similarly efficacious. However, concern is raised in all major guidelines about the possibility of worsening cognitive dysfunction, constipation, and UTIs caused by increased postvoid residual. Medications should be reviewed and postvoid residual checked before commencing antimuscarinics. If recurrent UTIs occur, then postvoid residual should be rechecked. Some antimuscarinics, which are unable to cross the blood—brain barrier and which have a high affinity for bladder muscarinic receptors, such as trospium chloride or tolterodine, should theoretically have a better side effect profile in patients with neurological lower urinary tract symptoms at risk of cognitive dysfunction. However, there is limited evidence to support this as yet. There is a high rate of discontinuation of antimuscarinics due to side effects such as dry mouth, blurred vision, and constipation (Hsu et al., 2019).

Mirabegron

Mirabegron is a highly selective B3 receptor agonist that relaxes the detrusor muscle. Due to the novel nature of mirabegron and the short study follow-up of between 4 and 12 weeks, it is not yet FDA approved for neurogenic detrusor overactivity. In the European Association of Urology 2019 guidelines, the lack of long-term outcomes or significant reduction in detrusor pressure has led them to stipulate that the role of B3 receptor agonists is still unclear. The best current evidence for mirabegron in patients with uroneurological dysfunction comes from a recent systematic review (El Helou et al., 2020). Seven studies and 302 patients were included. Most studies were of patients with MS or SCI. All studies included patients who had previously tried antimuscarinics. Two studies included either additional antimuscarinics or desmopressin. Mirabegron was safe (3%–5% adverse event rate) and well tolerated in patients who had not benefited from antimuscarinics. Side effects which did occur affected the cardiovascular system and included raised blood pressure, palpitations, and rarely atrial fibrillation. There was an improvement in both clinical and urodynamic findings. On urodynamics testing, bladder compliance and capacity improved, suggesting an improvement in the detrusor filling phase.

Desmopressin

Desmopressin promotes water reabsorption at the collecting and distal tubules of the kidney, thereby temporarily diminishing urine production. It can be helpful in managing nocturia. However, it should only be used once per day due to risk of hyponatraemia and should be carefully considered in patients who are older than 65 or with peripheral edema. Adverse events are common.

Alpha-blockers

Alpha-adrenoceptor antagonists work by blocking alpha-1A receptors in the bladder neck, urethra, and prostate. They may also work on alpha-1D adrenoreceptors in the detrusor and spinal cord causing improved bladder relaxation and reducing detrusor overactivity or at the prejunctional alpha-1 adrenoceptors in the bladder reducing detrusor contractions by decreasing acetylcholine release (Nickel, 2003). While the NICE guidelines do not recommend alpha-blockers, both the International Consultation on Incontinence and EAU guidelines recommend them for treatment of bladder outlet obstruction.

Intravesical drug treatment

Intravesical botulinum toxin

Intravesical injection of onabotulinumtoxin A is effective, well tolerated, and safe for the treatment of neurogenic detrusor overactivity causing incontinence. Effects last up to 9 months. It is a minimally invasive treatment requiring cystoscopy and the VC-30 small areas of intravesicular injection. The procedure can be done as a day case under local anesthesia for most patients. Good evidence exists for its use in SCIs and MS (Jaggi et al., 2018; Panicker & Fowler, 2015). Guidelines differ with the ICI recommending it for any neurological disorder, whereas NICE and the EAU recommend it only for SCI and MS. Up to 42% of patients develop urinary retention requiring de novo self-catheterization following the procedure. This is an important side effect which needs to be thoroughly discussed with patients prior to treatment. Assessing for the dexterity to carry out intermittent self-catheterization is part of the treatment decision-making process.

Intravesical antimuscarinics

A small randomized controlled study of 35 patients demonstrated improved tolerability of intravesical 0.1% oxybutynin. These target the muscarinic receptors on the urothelium where they cause the release of a factor that inhibits detrusor contraction. Intravesical antimuscarinics have been recommended for patients who cannot tolerate oral antimuscarinics in the EAU guidelines.

Tibial neuromodulation

Tibial nerve stimulation has had positive results in several trials for lower urinary tract symptoms in various neurological conditions especially MS, Parkinson's disease, and SCI. Percutaneous tibial nerve stimulation involves placing a needle 4–5 cm above the medial malleolus and electrically stimulating the posterior tibial nerve. A systematic review including three trials of percutaneous tibial nerve stimulation for 30 min per week for 12 weeks found 47%–100% of patients had improvement in urinary symptoms. There was increased bladder capacity or improvement in symptoms of urgency in two studies involving patients with MS and Parkinson's disease (Gaziev et al., 2013). In a randomized control study of 100 patients with SCI, there was similar efficacy with improvement in all bladder diary parameters for patients who had solifenacin (an antimuscarinic used in approximately a quarter of patients with SCI) (Jaggi & Fatoye, 2017) and percutaneous tibial nerve stimulation but without any adverse effects, which occurred in 5% of the patients in the solifenacin group and led to two patients withdrawing from treatment (Chen et al., 2015). Transcutaneous tibial nerve stimulation is more patient friendly than the typical tibial nerve stimulation treatment, which requires patients to attend hospital and have a fine gauge steel needle inserted.

Original evidence for tibial nerve stimulation came from animal studies showing detrusor inhibition in felines. From rat and feline studies, the method of action of this is thought to be by electrical recruitment of unmyelinated C-fibers and small nociceptive fibers which produce increased levels of endogenous CNS opioid peptides (Paquette & Yoo, 2019).

Tibial nerve stimulation is well tolerated and has few adverse effects. It does not cause increased postvoid residual; however, the effect is short lived and studies while encouraging are not large enough for this to be a mainstay of treatment. An ongoing multicenter randomized controlled trial will provide further information on short- and long-term outcomes in patients with SCI (Liechti et al., 2020).

Sacral neuromodulation

Sacral neuromodulation has been approved since 1997 and has been used for idiopathic urinary retention, neurogenic detrusor overactivity from a variety of conditions, interstitial cystitis, and overactive bladder syndrome, which is refractory to medical treatments. It is usually a two-stage procedure and involves stimulation of the sacral nerve, primarily S3 via tined electrode, which is placed through the S3 foramen. In the first-stage test, stimulation is done for 3–5 days after a small percutaneous lead is placed under local anesthetic. If the test improves bladder symptoms by at least 50%, then a permanent lead is placed under local anesthetic with a subcutaneous battery. A 2010 meta-analysis of sacral neuromodulation was overall positive but included only 6 studies, of which half had groups of 10 or less patients. Sparse literature means it is not currently included in the EAU guidelines (Barboglio Romo & Gupta, 2017; Hu et al., 2019). However, there is increasing interest in the use of sacral neuromodulation early after SCI to prevent neurogenic detrusor overactivity and small-volume high-pressure detrusor (Redshaw et al., 2018). Ways around the side effects from sacral neuromodulation such as pain and lack of organ specificity are being sought in tests of animal models with closed-loop optogenetic neuromodulation of bladder sensory afferent nerves (Mickle et al., 2019).

Treatment for voiding dysfunction

All urological symptoms can impact significantly on quality of life, and incontinence is associated with worse quality of life even in patients with similar levels of disease. Voiding dysfunction has the additional concern for neurologists and urologists that patients may develop recurrent UTIs or even back pressure and upper urinary tract damage. Patients with SCI and spina bifida are at much higher risk of upper tract damage due to detrusor sphincter dyssynergia, higher detrusor pressures, and lower urinary tract symptoms. Urodynamic findings in these patient groups should be monitored closely over the patient's lifetime.

Intermittent self-catheterization

Once postvoid residual is over 100 mL, there is a general consensus that intermittent catheterization should occur. Intermittent self-catheterization should occur once to three times per day for patients who incompletely void and 4–6 times per 24 h for patients who have complete retention. Ideally, bladder volume at catheterization should not exceed 400 or 500 mL. When undertaking aseptic intermittent self-catheterization, hydrophilic 12-16Fr catheters are often used. Hydrophilic catheters seem to be associated with a lower rate of UTIs (Phé et al., 2016; Taweel & Seyam, 2015).

There is a low incidence of symptomatic UTIs for patients who regularly use intermittent self-catheterization, but various factors such as weakness, dexterity, rigidity, spasticity, and impaired visual acuity or cognition all impact on the success of the intermittent self-catheterization. On studies of patients with SCI and spinal bifida, around 30% use intermittent self-catheterization over a long term (Hoen et al., 2017).

Suprapubic catheters and indwelling transurethral catheters should be avoided when possible due to enhanced UTI, poorer quality of life, and increased risk of bladder cancer. If long-term catheters are required, silicone catheters are preferred for neuro-urology patients due to high incidence of latex allergy and the reduced susceptibility to encrustation.

Surgical options

Bladder augmentation

Bladder augmentation involves replacing or expanding the bladder, usually with detubularized intestine. The aim is to reduce detrusor pressure by improving compliance or reducing detrusor overactivity. On follow-up, renal function stabilizes, and there is a low rate of side effects and an improved quality of life, which exceeds outcomes with botulinum toxin treatment. However, on long-term follow-up, 10% can have stone formation, bladder perforation can occur, and bowel dysfunction can be found in 15% of cases (Hoen et al., 2017). Lifelong follow-up is required for any patients who have undergone bladder augmentation.

Urinary diversion

Urinary diversion is undertaken when there is threat to the upper urinary tract and when other treatments have been unsuccessful. Continent and incontinent diversions are possible. For patients who can catheterize, continent diversions are first choice. The augmentation is often formation of an ileal conduit. However, there are long-term complications which include perforation (up to 40%), tubal stenosis, bladder stones, stomal bleeding, metabolic acidosis, and recurrent UTIs (Phé et al., 2017).

Bladder neck and urethral procedures

An artificial urinary sphincter or urethral sling should only be undertaken when there is no evidence of upper tract damage and the risk of upper tract damage is felt to be low. Urologists may consider bladder augmentation (see above) in patients with uroneurological disease at the same time if there is felt to be any likelihood of raised detrusor pressure.

Artificial urinary sphincter

Urethra strictures can occur in men with uroneurological lower urinary tract dysfunction due to repeated trauma from intermittent self-catheterization, prostatic hypertrophy, or both. Artificial urinary sphincters can be used to manage urethral strictures and to prevent urinary incontinence in men with some neurological conditions. The main problem with artificial urinary sphincters is a high revision rate, which can reach up to 60% in patients with uroneurological disorders (Groen et al., 2016).

Urethral sling

A urethral sling involves a sling of tissue, usually the abdominal muscle covering the rectus fascia, placed behind the midurethra. It should be considered in women with neurogenic stress incontinence as it is the most successful operative procedure (Fusco et al., 2017). The most common side effects include pain, bladder overactivity with urinary urgency, and voiding symptoms.

Urinary tract infections

UTIs can worsen many types of neurological conditions such as causing a deterioration in confusion in patients with Alzheimer's disease or triggering a pseudorelapse in MS. Recurrent UTIs occur in approximately 20% of patients with neurogenic lower urinary tract dysfunction (Pannek & Wöllner, 2017). UTIs may present in different ways in patients with neurological disease including incontinence, fatigue, increased spasticity, or leaking around the catheter.

As in other conditions, asymptomatic bacteriuria should not be treated. UTIs should be treated according to local microbiology guidelines for 5 to 7 days. Recurrent UTIs suggest an underlying problem such as incomplete voiding, bladder stones, or raised intravesicular pressure, and urological investigation should be undertaken (Groen et al., 2016; Pannek & Wöllner, 2017; Phé et al., 2016).

Hydrophilic catheters are associated with a lower rate of UTIs. Previously used noninvasive treatments such as cranberry juice and probiotics and long-term prophylactic antibiotics are no longer recommended due to lack of efficacy and increased bacterial resistance after long-term antibiotic usage (Pannek & Wöllner, 2017; Phé et al., 2016). There are some positive results with weekly cycling of antibiotic prophylaxis, with UTI incidence decreasing from 9 to 2 per patient year. No new resistant bacterial strains were found, but concern remains about bacterial resistance. Given the lack of large study evidence, this is not currently recommended (Urological Infections EAU Guidelines On, 2018).

New investigation and treatment possibilities

Lower urinary tract classifications and urinary biomarkers

Some researchers such as Powell have argued in favor of more detailed neurogenic lower urinary tract dysfunction criteria including urinary biomarkers to indicate the presence and severity of lower urinary tract dysfunction (Powell, 2016).

Urinary biomarkers are a noninvasive test of the urinary proteome made up of urinary tract desquamated cells and plasmatic proteins. Recent studies of patients with various neurological conditions have suggested some differences in neurotrophins (NGF and brain-derived neurotrophic factor), which appeared higher in MS, and mediators of inflammation (TGF beta-1) are significantly higher in patients with spina bifida than patients with SCI (Richard et al., 2020). These may be a useful and noninvasive way to further understand the pathogenesis of uroneurological symptoms in different neurological diseases.

Deep brain stimulation

Deep brain stimulation involves the insertion of an electrode into a targeted brain area. In Parkinson's disease, this is the subthalamic nuclei or the globus pallidus interna. It is hypothesized to work by overriding the pathological network activity and regulating axonal output. In animals and humans who had deep brain stimulation of the subthalamic nuclei, there was increased bladder capacity and an inhibitory effect on micturition (Sakakibara et al., 2018). There is interest and some evidence for deep brain stimulation in the treatment of pain and mood disorders, although this is not a widespread practice. In animals and six humans with periaqueductal gray deep brain stimulation for chronic neuropathic pain, there has been evidence of reversible inhibition in urinary voiding and increased urinary volume before leakage (Green et al., 2012). A study of deep brain stimulation on 11 rats with traumatic brain injury from a 2 meter weight height drop found a 28% decrease in voiding efficiency (Praveen Rajneesh et al., 2019). In the traumatic brain-injured rats, residual bladder volume increased and detrusor pressure during voiding decreased when compared with 12 healthy controls. There was reduced bladder volume perception as measured by the micturition threshold, which was the volume of saline which caused the first voiding detrusor contraction. Additionally, there was no higher-level inhibition, so when a detrusor contraction occurred, the urethral sphincters relaxed. These rats therefore had evidence of both urinary retention and urinary incontinence. A deep brain stimulation electrode was inserted into the rostral pontine reticular nucleus. This resulted in improvement in bladder volume perception as measured by the volume of saline, which caused the first voiding detrusor contraction, reduced residual volume, and increased detrusor pressure. Further work is required on the best locations, safety profile, and ways this can be used to actively improve patient care and uroneurological symptoms.

Cannabinoids

Cannabis is the most popular recreational drug in the world. It is a generic term for a heterogenous group of metabolites of phytocannabinoids. About 2.5% of the world's population, 147 million people, consume cannabis in some form per year (WHO | Cannabis. WHO, 2010). The most commonly studied compound is tetrahydrocannabinol (THC). There is moderate quality evidence for cannabinoids in spasticity for MS, chronic neuropathic and cancer pain, and sleep disorders (Campbell, 2001). However, the mechanism of action of cannabinoids and their long risk/benefit ratio requires further study.

The effect of cannabis on the lower urinary tract is due to the endocannabinoid system, which animal models have demonstrated has an effect on both the central and peripheral brain—bladder axis (Kazui, 2016). In healthy rodents, cannabinoids seems to play a role in normal micturition. Experimentally, intrathecal or intravenous cannabinoids affecting cannabinoid receptors 1 and 2 resulted in increased bladder volumes, increased micturition threshold, and decreased micturition frequency in rodent models of inflammatory disease (Hedlund & Gratzke, 2016).

A systematic review of cannabinoids for MS-related lower urinary tract symptoms found three studies involving 426 patients treated with a mixture of cannabidiol and THC (Abo Youssef et al., 2017). There were improvements in incontinence rates, nocturia, and number of voids per day. Limited severe adverse events were seen. The small number of studies means caution must be urged but findings were positive. Replicate studies of patients who smoke cannabis recreationally have found some benefit to their lower urinary tract symptoms (Consroe et al., 1997).

Prophylaxis for UTI bacteriophages

Bacteriophages are viruses that replicate inside a bacterial host causing cell lysis (Wojciuk et al., n.d.). During cell lysis, the phages will be released and will dock on specific bacteria and introduce their DNA. This commences the cycle of replication and then lysis. If the specific bacterial cells are not present, the additional phages will be excreted safely from the body. Large-scale trials are required, but there is significant positivity around this treatment especially given the likelihood of further worldwide antibiotic resistance.

Early tibial nerve stimulation after SCI

Stopping neurogenic lower urinary tract dysfunction is a central goal of treatment. Early neuromodulation has produced some promising results leading to a recent randomized controlled trial (TASCI) (Kessler, 2020; Liechti et al., 2020). This is investigating whether early transcutaneous tibial nerve stimulation in the early phase of SCI could prevent high-pressure, detrusor sphincter dyssynergia and detrusor overactivity. In a trial of 114 patients, all commenced on tibial nerve stimulation within 40 days of their SCI. The trials aim to prevent unhelpful up-activation of the c fibers, which would prevent neurogenic detrusor overactivity and detrusor sphincter dyssynergia.

Conclusion

Urological dysfunction is a common part of many neurological conditions. Better control of bladder function is often listed as one of the top 10 research priorities for patients (Van Middendorp et al., 2016). Understanding of the brain—bladder axis and control of lower urinary tract function has increased significantly in the last 2 decades. Despite this increased understanding, treatment remains limited and many patients have reduced quality of life due to lower urinary tract symptoms. The chapter summarizes neurological conditions that commonly have comorbid lower urinary tract dysfunction, the understanding of mechanism and treatment, areas where additional (animal) and clinical research studies are needed, and some areas for future development.

References

Abo Youssef, N., Schneider, M. P., Mordasini, L., Ineichen, B. V., Bachmann, L. M., Chartier-Kastler, E., Panicker, J. N., & Kessler, T. M. (2017). Cannabinoids for treating neurogenic lower urinary tract dysfunction in patients with multiple sclerosis: A systematic review and meta-analysis. *BJU International, 119*(4), 515—521. https://doi.org/10.1111/bju.13759

Ahn, U. M., Ahn, N. U., Buchowski, J. M., Garrett, E. S., Sieber, A. N., & Kostuik, J. P. (2000). Cauda equina syndrome secondary to lumbar disc herniation. *Spine, 25*(12), 1515—1522. https://doi.org/10.1097/00007632-200006150-00010

Al Dandan, H. B., Coote, S., & McClurg, D. (2020). Prevalence of lower urinary tract symptoms in people with multiple sclerosis: A systematic review and meta-analysis. *International Journal of MS Care, 22*(2), 91—99. https://doi.org/10.7224/1537-2073.2019-030

Averbeck, M. A., Altaweel, W., Manu-Marin, A., & Madersbacher, H. (2017). Management of LUTS in patients with dementia and associated disorders. *Neurourology and Urodynamics, 36*(2), 245—252. https://doi.org/10.1002/nau.22928

Bacsu, C. D., Chan, L., & Tse, V. (2012). Diagnosing detrusor sphincter dyssynergia in the neurological patient. *BJU International, 109*(3), 31—34. https://doi.org/10.1111/j.1464-410X.2012.11042.x

Barboglio Romo, P. G., & Gupta, P. (2017). Peripheral and sacral neuromodulation in the treatment of neurogenic lower urinary tract dysfunction. *Urologic Clinics of North America, 44*(3), 453—461. https://doi.org/10.1016/j.ucl.2017.04.011

Barone, P., Antonini, A., Colosimo, C., Marconi, R., Morgante, L., Avarello, T. P., Bottacchi, E., Cannas, A., Ceravolo, G., Ceravolo, R., Cicarelli, G., Gaglio, R. M., Giglia, R. M., Iemolo, F., Manfredi, M., Meco, G., Nicoletti, A., Pederzoli, M., Petrone, A., et al. (2009). The PRIAMO study: A multicenter assessment of nonmotor symptoms and their impact on quality of life in Parkinson's disease. *Movement Disorders, 24*(11), 1641—1649. https://doi.org/10.1002/mds.22643

Batla, A., Pareés, I., Edwards, M. J., Stamelou, M., Bhatia, K. P., & Panicker, J. N. (2016). Lower urinary tract dysfunction in patients with functional movement disorders. *Journal of the Neurological Sciences, 361*, 192—194. https://doi.org/10.1016/j.jns.2015.12.048

Campbell, F. A. (2001). Are cannabinoids an effective and safe treatment option in the management of pain? A qualitative systematic review. *BMJ*, 13. https://doi.org/10.1136/bmj.323.7303.13

Chen, G., Liao, L., & Li, Y. (2015). The possible role of percutaneous tibial nerve stimulation using adhesive skin surface electrodes in patients with neurogenic detrusor overactivity secondary to spinal cord injury. *International Urology and Nephrology, 47*(3), 451—455. https://doi.org/10.1007/s11255-015-0911-6

Consroe, P., Musty, R., Rein, J., Tillery, W., & Pertwee, R. (1997). The perceived effects of smoked cannabis on patients with multiple sclerosis. *European Neurology, 38*(1), 44—48. https://doi.org/10.1159/000112901

Delamarter, R. B., Sherman, J. E., & Carr, J. B. (1991). 1991 Volvo award in experimental studies: Cauda equina syndrome: Neurologic recovery following immediate, early, or late decompression. *Spine, 16*(9), 1022—1029. https://doi.org/10.1097/00007632-199109000-00002

Demetriades, A. K. (2009). Out of hours MRI scanning for cauda equina syndrome (CES): What is the positive pick-up rate and what are the final diagnoses in those with a negative scan? *British Journal of Neurosurgery, 23*.

El Helou, E., Labaki, C., Chebel, R., El Helou, J., Abi Tayeh, G., Jalkh, G., & Nemr, E. (2020). The use of mirabegron in neurogenic bladder: A systematic review. *World Journal of Urology, 38*(10), 2435—2442. https://doi.org/10.1007/s00345-019-03040-x

Elmelund, M., Biering-Sørensen, F., Due, U., & Klarskov, N. (2018). The effect of pelvic floor muscle training and intravaginal electrical stimulation on urinary incontinence in women with incomplete spinal cord injury: An investigator-blinded parallel randomized clinical trial. *International Urogynecology Journal, 29*(11), 1597—1606. https://doi.org/10.1007/s00192-018-3630-6

Fraser, S., Roberts, L., & Murphy, E. (2009). Cauda equina syndrome: A literature review of its definition and clinical presentation. *Archives of Physical Medicine and Rehabilitation, 90*(11), 1964—1968. https://doi.org/10.1016/j.apmr.2009.03.021

Fusco, F., Abdel-Fattah, M., Chapple, C. R., Creta, M., La Falce, S., Waltregny, D., & Novara, G. (2017). Updated systematic review and meta-analysis of the comparative data on colposuspensions, pubovaginal slings, and midurethral tapes in the surgical treatment of female stress urinary incontinence. *European Urology, 72*(4), 567—591. https://doi.org/10.1016/j.eururo.2017.04.026

Gaziev, G., Topazio, L., Iacovelli, V., Asimakopoulos, A., Di Santo, A., De Nunzio, C., & Finazzi-Agrò, E. (2013). Percutaneous tibial nerve stimulation (PTNS) efficacy in the treatment of lower urinary tract dysfunctions: A systematic review. *BMC Urology, 13*. https://doi.org/10.1186/1471-2490-13-61

Goldmark, E., Niver, B., & Ginsberg, D. A. (2014). Neurogenic bladder: From diagnosis to management. *Current Urology Reports, 15*(10). https://doi.org/10.1007/s11934-014-0448-8

Green, A. L., Stone, E., Sitsapesan, H., Turney, B. W., Coote, J. H., Aziz, T. Z., Hyam, J. A., & Lovick, T. A. (2012). Switching off micturition using deep brain stimulation at midbrain sites. *Annals of Neurology, 72*(1), 144–147. https://doi.org/10.1002/ana.23571

Groen, J., Pannek, J., Castro Diaz, D., Del Popolo, G., Gross, T., Hamid, R., Karsenty, G., Kessler, T. M., Schneider, M., Hoen, L., & Blok, B. (2016). Summary of European association of urology (EAU) guidelines on neuro-urology. *European Urology, 69*(2), 324–333. https://doi.org/10.1016/j.eururo.2015.07.071

Hamid, R., Averbeck, M. A., Chiang, H., Garcia, A., Al Mousa, R. T., Oh, S. J., Patel, A., Plata, M., & Del Popolo, G. (2018). Epidemiology and pathophysiology of neurogenic bladder after spinal cord injury. *World Journal of Urology, 36*(10), 1517–1527. https://doi.org/10.1007/s00345-018-2301-z

Hazelwood, J. E., Hoeritzauer, I., Pronin, S., & Demetriades, A. K. (2019). An assessment of patient-reported long-term outcomes following surgery for cauda equina syndrome. *Acta Neurochirurgica, 161*(9), 1887–1894. https://doi.org/10.1007/s00701-019-03973-7

Hedlund, P., & Gratzke, C. (2016). The endocannabinoid system - a target for the treatment of LUTS? *Nature Reviews Urology, 13*(8), 463–470. https://doi.org/10.1038/nrurol.2016.110

Hoen, L., Ecclestone, H., Blok, B. F. M., Karsenty, G., Phé, V., Bossier, R., Groen, J., Castro-Diaz, D., Padilla Fernández, B., Del Popolo, G., Musco, S., Pannek, J., Kessler, T. M., Gross, T., Schneider, M. P., & Hamid, R. (2017). Long-term effectiveness and complication rates of bladder augmentation in patients with neurogenic bladder dysfunction: A systematic review. *Neurourology and Urodynamics, 36*(7), 1685–1702. https://doi.org/10.1002/nau.23205

Hoeritzauer, I., Carson, A., Statham, P., Panicker, J. N., Granitsiotis, V., Eugenicos, M., Summers, D., Demetriades, A. K., & Stone, J. (2021). Scan-negative cauda equina syndrome. *Neurology, 96*(3), e433–e447. https://doi.org/10.1212/wnl.0000000000011154. PMID: 33177221.

Hoeritzauer, I., Stone, J., Fowler, C., Elneil-Coker, S., Carson, A., & Panicker, J. (2016). Fowler's syndrome of urinary retention: A retrospective study of co-morbidity. *Neurourology and Urodynamics, 35*(5), 601–603. https://doi.org/10.1002/nau.22758

Hsu, F. C., Weeks, C. E., Selph, S. S., Blazina, I., Holmes, R. S., & McDonagh, M. S. (2019). Updating the evidence on drugs to treat overactive bladder: A systematic review. *International Urogynecology Journal, 30*(10), 1603–1617. https://doi.org/10.1007/s00192-019-04022-8

Hu, M., Lai, S., Zhang, Y., Liu, M., & Wang, J. (2019). Sacral neuromodulation for lower urinary tract dysfunction in spinal cord injury: A systematic review and meta-analysis. *Urologia Internationalis, 103*(3), 337–343. https://doi.org/10.1159/000501529

Ito, T., Sakakibara, R., Yasuda, K., Yamamoto, T., Uchiyama, T., Liu, Z., Yamanishi, T., Awa, Y., Yamamoto, K., & Hattori, T. (2006). Incomplete emptying and urinary retention in multiple-system atrophy: When does it occur and how do we manage it? *Movement Disorders, 21*(6), 816–823. https://doi.org/10.1002/mds.20815

Jaggi, A., Drake, M., Siddiqui, E., & Fatoye, F. (2018). A comparison of the treatment recommendations for neurogenic lower urinary tract dysfunction in the national institute for health and care excellence, European Association of Urology and international consultations on incontinence guidelines. *Neurourology and Urodynamics, 37*(7), 2273–2280. https://doi.org/10.1002/nau.23581

Jaggi, A., & Fatoye, F. (2017). Real world treatment patterns in the neurogenic bladder population: A systematic literature review. *Translational Andrology and Urology, 6*(6), 1175–1183. https://doi.org/10.21037/tau.2017.09.17

Kalidindi, K. K. V., Chhabra, H. S., Suman, D., Mannem, A., & Bhat, M. R. (2020). Cauda equina syndrome: False-positive diagnosis of neurogenic bladder can be reduced by multichannel urodynamic study. *European Spine Journal, 29*(6), 1236–1247. https://doi.org/10.1007/s00586-019-06277-7

Karmarkar, R., Abtahi, B., Saber-Khalaf, M., Gonzales, G., & Elneil, S. (2015). Gynaecological pathology in women with Fowler's syndrome. *European Journal of Obstetrics and Gynecology and Reproductive Biology, 194*, 54–57. https://doi.org/10.1016/j.ejogrb.2015.08.009

Kavia, R., DasGupta, R., Critchley, H., Fowler, C., & Griffiths, D. (2010). A functional magnetic resonance imaging study of the effect of sacral neuromodulation on brain responses in women with Fowler's syndrome. *BJU International, 105*(3), 366–372. https://doi.org/10.1111/j.1464-410X.2009.08819.x

Kazui, H. (2016). Current state of diagnosis and treatment of idiopathic normal pressure hydrocephalus. *Brain and nerve = Shinkei kenkyū no shinpo, 68*(4), 429–440. https://doi.org/10.11477/mf.1416200416

Kessler, T. M. (2020). Neuro-urology, Quo Vadis? *European Urology Focus, 6*(5), 801–803. https://doi.org/10.1016/j.euf.2020.04.006

Kester, M. I., & Scheltens, P. (2009). Dementia: The bare essentials. *Practical Neurology, 9*(4), 241–251. https://doi.org/10.1136/jnnp.2009.182477

Kuoch, K. L. J., Meyer, D., Austin, D. W., & Knowles, S. R. (2017). A systematic review of paruresis: Clinical implications and future directions. *Journal of Psychosomatic Research, 98*, 122–129. https://doi.org/10.1016/j.jpsychores.2017.05.015

Lazzeri, M., Spinelli, M., Beneforti, P., Zanollo, A., & Turini, D. (1999). Urodynamic assessment during intravesical infusion of capsaicin for the treatment of refractory detrusor hyperreflexia. *Spinal Cord, 37*(6), 440–443. https://doi.org/10.1038/sj.sc.3100854

Lee, S. H., Cho, S. T., Na, H. R., Ko, S. B., & Park, M. H. (2014). Urinary incontinence in patients with Alzheimer's disease: Relationship between symptom status and urodynamic diagnoses. *International Journal of Urology, 21*(7), 683–687. https://doi.org/10.1111/iju.12420

Liechti, M. D., van der Lely, S., Stalder, S. A., Anderson, C. E., Birkhäuser, V., Bachmann, L. M., Brinkhof, M. W. G., Curt, A., Jordan, X., Leitner, L., Mehnert, U., Möhr, S., Pannek, J., Schubert, M., & Kessler, T. M. (2020). Update from TASCI, a Nationwide, randomized, sham-controlled, double-blind clinical trial on transcutaneous tibial nerve stimulation in patients with acute spinal cord injury to prevent neurogenic detrusor overactivity. *European Urology Focus, 6*(5), 877–879. https://doi.org/10.1016/j.euf.2019.09.019

McDonald, C., Winge, K., & Burn, D. J. (2017). Lower urinary tract symptoms in Parkinson's disease: Prevalence, aetiology and management. *Parkinsonism and Related Disorders, 35*, 8–16. https://doi.org/10.1016/j.parkreldis.2016.10.024

Mehdi, Z., Birns, J., & Bhalla, A. (2013). Post-stroke urinary incontinence. *International Journal of Clinical Practice, 67*(11), 1128–1137. https://doi.org/10.1111/ijcp.12183

Mickle, A. D., Won, S. M., Noh, K. N., Yoon, J., Meacham, K. W., Xue, Y., McIlvried, L. A., Copits, B. A., Samineni, V. K., Crawford, K. E., Kim, D. H., Srivastava, P., Kim, B. H., Min, S., Shiuan, Y., Yun, Y., Payne, M. A., Zhang, J., Jang, H., et al. (2019). A wireless closed-loop system for optogenetic peripheral neuromodulation. *Nature, 565*(7739), 361–365. https://doi.org/10.1038/s41586-018-0823-6

Mitchell, L. E., Scott Adzick, N., Melchionne, J., Pasquariello, P. S., Sutton, L. N., & Whitehead, A. S. (2004). Spina bifida. *Lancet, 364*(9448), 1885–1895. https://doi.org/10.1016/S0140-6736(04)17445-X. Elsevier B.V.

Miyazato, M., Kadekawa, K., Kitta, T., Wada, N., Shimizu, N., de Groat, W. C., Birder, L. A., Kanai, A. J., Saito, S., & Yoshimura, N. (2017). New frontiers of basic science research in neurogenic lower urinary tract dysfunction. *Urologic Clinics of North America, 44*(3), 491–505. https://doi.org/10.1016/j.ucl.2017.04.014

Nagata, Y., Bundo, M., Sugiura, S., Kamita, M., Ono, M., Hattori, K., Yoshida, S., Goto, Y. I., Urakami, K., & Niida, S. (2017). PTPRQ as a potential biomarker for idiopathic normal pressure hydrocephalus. *Molecular Medicine Reports, 16*(3), 3034–3040. https://doi.org/10.3892/mmr.2017.7015

Nahm, L. S., Chen, Y., Devivo, M. J., & Lloyd, L. K. (2015). Bladder cancer mortality after spinal cord injury over 4 decades. *The Journal of Urology, 193*(6), 1923–1928. https://doi.org/10.1016/j.juro.2015.01.070

Nazari, F., Shaygannejad, V., Mohammadi Sichani, M., Mansourian, M., & Hajhashemi, V. (2020). The prevalence of lower urinary tract symptoms based on individual and clinical parameters in patients with multiple sclerosis. *BMC Neurology, 20*(1). https://doi.org/10.1186/s12883-019-1582-1

Nickel, J. C. (2003). The use of α1-adrenoceptor antagonists in lower urinary tract symptoms: Beyond benign prostatic hyperplasia. *Urology, 62*(3), 34–41. https://doi.org/10.1016/S0090-4295(03)00472-2

O'Connor, K. P. (2020). Surgical treatment of tethered cord syndrome in adults: A systematic review and meta-analysis. *World Neurosurgery, 137*.

Panicker, J. N. (2017). Urogenital symptoms in neurologic patients. *Continuum Lifelong Learning in Neurology, 23*(2), 533–552. https://doi.org/10.1212/CON.0000000000000448

Panicker, J. N., & Fowler, C. J. (2015). Lower urinary tract dysfunction in patients with multiple sclerosis. In *Handbook of clinical neurology* (Vol. 130, pp. 371–381). Elsevier B.V. http://www.sciencedirect.com/science/journal/00729752.

Panicker, J. N., Fowler, C. J., & Kessler, T. M. (2015). Lower urinary tract dysfunction in the neurological patient: Clinical assessment and management. *The Lancet Neurology, 14*(7), 720–732. https://doi.org/10.1016/S1474-4422(15)00070-8

Pannek, J., & Wöllner, J. (2017). Management of urinary tract infections in patients with neurogenic bladder: Challenges and solutions. *Research and Reports in Urology, 9*, 121–127. https://doi.org/10.2147/RRU.S113610

Paquette, J. P., & Yoo, X. P. B. (2019). Recruitment of unmyelinated c-fibers mediates the bladder-inhibitory effects of tibial nerve stimulation in a continuous-fill anesthetized rat model. *American Journal of Physiology Renal Physiology, 317*(1), F163–F171. https://doi.org/10.1152/ajprenal.00502.2018

Pettersen, R., Saxby, B. K., & Wyller, T. B. (2007). Poststroke urinary incontinence: One-year outcome and relationships with measures of attentiveness. *Journal of the American Geriatrics Society, 55*(10), 1571–1577. https://doi.org/10.1111/j.1532-5415.2007.01396.x

Phé, V., Boissier, R., Blok, B. F. M., Del Popolo, G., Musco, S., Castro-Diaz, D., Padilla Fernández, B., Groen, J., Hamid, R., Hoen, L., Ecclestone, H., Kessler, T. M., Gross, T., Schneider, M. P., Pannek, J., & Karsenty, G. (2017). Continent catheterizable tubes/stomas in adult neuro-urological patients: A systematic review. *Neurourology and Urodynamics, 36*(7), 1711–1722. https://doi.org/10.1002/nau.23213

Phé, V., Pakzad, M., Curtis, C., Porter, B., Haslam, C., Chataway, J., & Panicker, J. N. (2016). Urinary tract infections in multiple sclerosis. *Multiple Sclerosis, 22*(7), 855–861. https://doi.org/10.1177/1352458516633903

Pizzi, A., Falsini, C., Martini, M., Rossetti, M. A., Verdesca, S., & Tosto, A. (2014). Urinary incontinence after ischemic stroke: Clinical and urodynamic studies. *Neurourology and Urodynamics, 33*(4), 420–425. https://doi.org/10.1002/nau.22420

Powell, C. R. (2016). Not all neurogenic bladders are the same: A proposal for a new neurogenic bladder classification system. *Translational Andrology and Urology, 5*(1), 12–21. https://doi.org/10.3978/j.issn.2223-4683.2016.01.02

Praveen Rajneesh, C., Lai, C. H., Chen, S. C., Hsieh, T. H., Chin, H. Y., & Peng, C. W. (2019). Improved voiding function by deep brain stimulation in traumatic brain-injured animals with bladder dysfunctions. *International Urology and Nephrology, 51*(1), 41–52. https://doi.org/10.1007/s11255-018-2028-1

Pronin, S., Koh, C. H., Bulovaite, E., Macleod, M. R., & Statham, P. F. (2019). Compressive pressure versus time in cauda equina syndrome: A systematic review and meta-analysis of experimental studies. *Spine, 44*(17), 1238–1247. https://doi.org/10.1097/BRS.0000000000003045

Redshaw, J. D., Lenherr, S. M., Elliott, S. P., Stoffel, J. T., Rosenbluth, J. P., Presson, A. P., & Myers, J. B. (2018). Protocol for a randomized clinical trial investigating early sacral nerve stimulation as an adjunct to standard neurogenic bladder management following acute spinal cord injury. *BMC Urology, 18*(1). https://doi.org/10.1186/s12894-018-0383-y

Richard, C., Bendavid, C., Hascoet, J., Alimi, Q., Khene, Z.e., Kerdraon, J., Voiry, C., Brochard, C., Bouguen, G., Siproudhis, L., Manunta, A., Gamé, X., & Peyronnet, B. (2020). Urinary biomarkers profiles in patients with neurogenic detrusor overactivity according to their neurological condition. *World Journal of Urology, 38*(9), 2261–2268. https://doi.org/10.1007/s00345-019-03016-x

Sakakibara, Ryuji (2015). *Lower urinary tract dysfunction in patients with brain lesions* (pp. 269–287). Elsevier BV. https://doi.org/10.1016/b978-0-444-63247-0.00015-8

Sakakibara, R., Hattori, T., Uchiyama, T., & Yamanishi, T. (2001). Videourodynamic and sphincter motor unit potential analyses in Parkinson's disease and multiple system atrophy. *Journal of Neurology Neurosurgery and Psychiatry, 71*(5), 600–606. https://doi.org/10.1136/jnnp.71.5.600

Sakakibara, R., Tateno, F., Yamamoto, T., Uchiyama, T., & Yamanishi, T. (2018). Urological dysfunction in synucleinopathies: Epidemiology, pathophysiology and management. *Clinical Autonomic Research, 28*(1), 83–101. https://doi.org/10.1007/s10286-017-0480-0

Sakakibara, R., Uchida, Y., Ishii, K., Hashimoto, M., Ishikawa, M., Kazui, H., Yamamoto, T., Uchiyama, T., Tateno, F., Kishi, M., Tsuyusaki, Y., Aiba, Y., Tateno, H., Nagao, T., Terada, H., & Inaoka, T. (2016). Bladder recovery relates with increased mid-cingulate perfusion after shunt surgery in idiopathic normal-pressure hydrocephalus: A single-photon emission tomography study. *International Urology and Nephrology, 48*(2), 169–174. https://doi.org/10.1007/s11255-015-1162-2

Schurch, B., Tawadros, C., & Carda, S. (2015). Dysfunction of lower urinary tract in patients with spinal cord injury. In *Handbook of clinical neurology* (Vol. 130, pp. 247–267). Elsevier B.V. http://www.sciencedirect.com/science/journal/00729752.

da Silva, A. P. B., Silva, R. B. M., Goi, L. D. S., Molina, R. D., Machado, D. C., & Sato, D. K. (2020). Experimental models of neuroimmunological disorders: A review. *Frontiers in Neurology*. https://doi.org/10.3389/fneur.2020.00389

Snow-Lisy, D. C., Yerkes, E. B., & Cheng, E. Y. (2015). Update on urological management of spina bifida from prenatal diagnosis to adulthood. *The Journal of Urology, 194*(2), 288–296. https://doi.org/10.1016/j.juro.2015.03.107

Soifer, S., Himle, J., & Walsh, K. (2010). Paruresis (shy bladder syndrome): A cognitive-behavioral treatment approach. *Social Work in Health Care, 49*(5), 494–507. https://doi.org/10.1080/00981381003684898

Sugimoto, T., Yoshida, M., Ono, R., Murata, S., Saji, N., Niida, S., Toba, K., & Sakurai, T. (2017). Frontal lobe function correlates with one-year incidence of urinary incontinence in elderly with Alzheimer disease. *Journal of Alzheimer's Disease, 56*(2), 567–574. https://doi.org/10.3233/JAD-160923

Swinn, M. J., & Fowler, C. J. (2001). Isolated urinary retention in young women, or Fowler's syndrome. *Clinical Autonomic Research, 11*(5), 309–311. https://doi.org/10.1007/BF02332976

Swinn, M. J., Wiseman, O. J., Lowe, E., & Fowler, C. J. (2002). The cause and natural history of isolated urinary retention in young women. *The Journal of Urology, 167*(1), 151–156. https://doi.org/10.1016/S0022-5347(05)65402-X

Takeuchi, T., & Yajima, K. (2019). Long-term 4 years follow-up study of 482 patients who underwent shunting for idiopathic normal pressure hydrocephalus -course of symptoms and shunt efficacy rates compared by age group-. *Neurologia Medico-Chirurgica, 59*(7), 281–286. https://doi.org/10.2176/nmc.oa.2018-0318

Tawadros, C., Burnett, K., Derbyshire, L. F., Tawadros, T., Clarke, N. W., & Betts, C. D. (2015). External urethral sphincter electromyography in asymptomatic women and the influence of the menstrual cycle. *BJU International, 116*(3), 423–431. https://doi.org/10.1111/bju.13042

Taweel, W. A., & Seyam, R. (2015). Neurogenic bladder in spinal cord injury patients. *Research and Reports in Urology, 7*, 85–99. https://doi.org/10.2147/RRU.S29644

Teplitsky, S., Murphy, A., & Shenot, P. J. (2019). Knowledge gaps in urologic care of female spinal cord injury patients. *Current Urology Reports, 20*(5). https://doi.org/10.1007/s11934-019-0884-6

Thomas, L. H., Coupe, J., Cross, L. D., Tan, A. L., & Watkins, C. L. (2019). Interventions for treating urinary incontinence after stroke in adults. *Cochrane Database of Systematic Reviews, 2019*(2). https://doi.org/10.1002/14651858.CD004462.pub4

Urological infections EAU guidelines on.(2018).

Van Batavia, J. P., Combs, A. J., Fast, A. M., & Glassberg, K. I. (2017). Overactive bladder (OAB): A symptom in search of a disease – its relationship to specific lower urinary tract symptoms and conditions. *Journal of Pediatric Urology*, 277.e1–277.e4. https://doi.org/10.1016/j.jpurol.2017.02.010

Van Middendorp, J. J., Allison, H. C., Ahuja, S., Bracher, D., Dyson, C., Fairbank, J., Gall, A., Glover, A., Gray, L., El Masri, W., Uttridge, A., & Cowan, K. (2016). Top ten research priorities for spinal cord injury: The methodology and results of a British priority setting partnership. *Spinal Cord, 54*(5), 341–346. https://doi.org/10.1038/sc.2015.199

Vurture, G., Peyronnet, B., Palma, J. A., Sussman, R. D., Malacarne, D. R., Feigin, A., Palmerola, R., Rosenblum, N., Frucht, S., Kaufmann, H., Nitti, V. W., & Brucker, B. M. (2019). Urodynamic mechanisms underlying overactive bladder symptoms in patients with Parkinson disease. *International Neurourology Journal, 23*(3), 211–218. https://doi.org/10.5213/inj.1938086.043

Ward, N. S. (2017). Restoring brain function after stroke—bridging the gap between animals and humans. *Nature Reviews Neurology, 13*(4), 244–255. https://doi.org/10.1038/nrneurol.2017.34

Weiss, D. A., Lee, A. S., Flanders, T. M., Long, C. J., van Batavia, J. P., Zderic, S. A., & Heuer, G. G. (2019). Neurological dysfunction of the bladder from myelomeningocele. *Neurosurgical Focus, 47*(4), 1–8. https://doi.org/10.3171/2019.7.FOCUS19468

WHO | Cannabis. (2010). WHO.

Wiener, J. S., Suson, K. D., Castillo, J., Routh, J. C., Tanaka, S. T., Liu, T., Ward, E. A., Thibadeau, J. K., & Joseph, D. B. (2018). Bladder management and continence outcomes in adults with spina bifida: Results from the national spina bifida patient registry, 2009 to 2015. *The Journal of Urology, 200*(1), 187–194. https://doi.org/10.1016/j.juro.2018.02.3101

Williams, M. A., & Malm, J. (2016). Diagnosis and treatment of idiopathic normal pressure hydrocephalus. *Continuum Lifelong Learning in Neurology, 22*(2), 579–599. https://doi.org/10.1212/CON.0000000000000305

Winge, K. (2015). Lower urinary tract dysfunction in patients with parkinsonism and other neurodegenerative disorders. In *Handbook of clinical neurology* (Vol. 130, pp. 335–356). Elsevier B.V. https://doi.org/10.1016/b978-0-444-63247-0.00019-5

Wojciuk, B., Salabura, A., & Ciechanowski, K. (n.d.). Urobiome: In sickness and in health (pp. 1–13).

Chapter 7

Effect of androgens and estrogens on bladder/lower urinary tract function

Anne E. Turco[1] and Chad M. Vezina[2]
[1]University of Wisconsin-Madison, Molecular and Environmental Toxicology, Madison, WI, United States; [2]University of Wisconsin-Madison, Department of Comparative Biosciences, George M. O'Brien Center for Benign Urology Research, Madison, WI, United States

Aging is inextricably linked to benign urinary voiding dysfunction, including stress incontinence, overactive bladder, and prostate-related disease. Aging-related changes in sex hormone concentrations are clear contributors to urinary disease processes. In men, testosterone plasma concentrations decrease while estradiol concentrations increase with age. In women, estradiol concentrations decrease while testosterone concentrations increase with age (Bélanger et al., 1994). Here, we outline known impacts of testosterone and estradiol on female and male bladder anatomy, physiology, and benign bladder-related disease. We also summarize emerging evidence that environmental chemical exposures modulate sex hormones to influence bladder structure and function.

Sex differences in bladder histology

Sex differences in male and female lower urinary tracts are typically classified as anatomical or physiological, but differences also exist at the cellular level. Understanding these differences is critical because sex and gender can impact diagnosis and treatment of urinary dysfunction ("Board on Health Sciences Policy, and Committee on Understanding the Biology of Sex and Gender Differences," 2001). Many bladder diseases are rooted in cellular irregularities. We begin by describing bladder histology and influence of sex and sex hormones on bladder cell behavior and function.

Sex differences in anatomy and physiology

The bladder is divided into multiple anatomical regions that differ with respect to innervation, muscle thickness, and muscle expression of neurotransmitter receptors. The trigone is a triangular region near the junction of the bladder and urethra that contributes to urine storage. Left and right lateral walls straddle the trigone, the dome constitutes the roof (dorsal aspect) of the bladder, and the posterior wall (bladder apex) is joined by the umbilical ligament to the umbilicus (Shermadou et al., 2020). A cross section of the bladder reveals a multilayered bladder wall comprised of a specialized epithelium that forms a barrier between urine and underlying tissue, lamina propria, which is enriched in blood vessels and sensory axons, and smooth muscle (also known as detrusor) (Abelson, et al., 2018).

Sex differences in histology

The mature detrusor contains three layers. Longitudinal and circular cells make up the layers and collagen fibers are distributed throughout (Karl-Erik & Anders, 2004). Sex differences exist in adult detrusor. The muscle is thicker in men and the pressure required for bladder emptying is also greater in adult men than in adult women (Altaf et al., 2013). Even though sex hormones are present at the onset of gonadal steroidogenesis in the fetus (Preslock, 1980a, 1980b), and in sufficient concentrations to masculinize or feminize internal and external genitalia, bladder muscle sexual dimorphism does not become grossly apparent until later in life. Questions remain about when bladder sex differences emerge. One possibility is hormones gradually impact the bladder and when directly exposed to high concentration of sex hormones,

Neuro-Urology Research. https://doi.org/10.1016/B978-0-12-822455-7.00011-8

divergence occurs. Another possibility is bladder sexual dimorphism may be driven by secondary processes. The factors that support sex differences in adult bladder muscle function have not been fully identified, and further studies are needed to identify if direct or indirect exposure is required.

Muscle and collagen

Exogenous estrogens and androgens influence additional aspects of bladder histology beyond the detrusor. Here, we discuss the impact of estrogens and androgens, individually and in combination, on bladder muscle and collagen. Table 7.1 summarizes the impact of sub- and supraphysiological estrogens and testosterone described in this section. Estrogens are clearly important for the maintenance of bladder muscle and wall homeostasis. Ovariectomy (OVX) in mice and rats degenerates the female rat detrusor: gaps appear between muscle fascicles and bladder weight decreases (Chen et al., 2008; Eika et al., 1990) (Table 7.1). OVX also predisposes to bladder to inflammation, characterized by infiltrating leukocytes, and collagen accumulation. OVX furthermore causes detrusor atrophy and decreases bladder wall thickness (Chen et al., 2008; Eika et al., 1990; Yiloren et al., 2011) (Table 7.1). Supplemental estrogens increase rat bladder collagen fiber density and bladder cell proliferation and reduce bladder cell apoptosis (Liang et al., 2013; Madeiro et al., 2002) (Table 7.1). The bladder response to estrogens is nonmonotonic, and a single 1 mg/kg dose of estradiol benzoate (a relatively high dose for rats) reduces detrusor thickness in OVX rats, suggesting very high concentrations of supplemental estrogens may be deleterious to bladder function in aging women (Chen et al., 2008). These studies are important because they support the idea that estrogens are required for bladder homeostasis and supplementation can protect from physiological decline due to an aging-related loss in circulating estrogens in postmenopausal women (Liang et al., 2013; Tanidir et al., 2011). Importantly, the relationship between estrogen dose and bladder response appears to be nonmonotonic, suggesting the need to achieve an optimal estrogen concentration for homeostatic bladder maintenance.

Table 7.1 describes the impact of supra- or subphysiological estrogen and testosterone on bladder cell types.

An aging-related decline in circulating testosterone concentrations in men is linked to a decline in bladder muscle function, a reduction in bladder muscle abundance, and an increase in bladder collagen content (Fraga et al., 2007) (Table 7.1). The concentration of circulating testosterone in men declines by 1% a year after 40 years of age (Bravo et al., 2017). There is experimental evidence that testosterone supplementation may be beneficial to bladder health in aging men. Castration of young mice, to simulate aging-related hypogonadism, causes significant bladder atrophy, decreases smooth muscle thickness, and increases bladder collagen content (Table 7.1). Testosterone treatment of castrated young adult rats increases bladder capacity, increases bladder smooth muscle content, and reduces bladder collagen abundance (Tek et al., 2010) (Table 7.1). One study examined bladder histology and mechanical properties in 18- to 20-month-old control rats and in rats treated with testosterone undecanoate, an androgen and anabolic steroid medication (50 mg/kg) (Barros et al., 2013). The outcome of this study supports the idea that testosterone replacement may afford therapeutic benefit in men with voiding dysfunction by increasing bladder smooth muscle (Barros et al., 2013; Tek et al., 2010). However, another study reported a lower testosterone dose (10 mg/kg) contributed to bladder fibrosis after bladder outlet obstruction in

TABLE 7.1 Impact of estradiol and testosterone modulation on bladder cells.

Bladder cell type	Sex hormone modulation	Characteristic	References
Detrusor smooth muscle	Supraphysiological testosterone	1. Thicker in men and increases bladder capacity. 2. Increases smooth muscle thickness. 3. Reduces or increases collagen depending on concentration.	Mangera et al. (2013) and Tek et al. (2010)
Detrusor smooth muscle	Subphysiological estrogens	1. Decrease detrusor thickness. 2. Gaps in muscle cells. 3. Bladder weight decreases. 4. Increased collagen accumulation.	Chen et al. (2008) and Eika et al. (1990)
Detrusor smooth muscle	Subphysiological testosterone	Bladder atrophy, smooth muscle thickness. Increases collagen.	Tek et al. (2010)
Detrusor function	Supraphysiological estrogens	Increase bladder capacity and compliance.	Celayir (2003)
Urothelium	Supraphysiological testosterone	Male Upk3 null mice have urinary retention, increased bladder pressure.	

castrated mice, indicating the bladder responds to testosterone at the cellular level (Flum et al., 2017) (Table 7.1). The relationship between testosterone concentration and male bladder response is likely to be nonmonotonic, such that high doses may be therapeutic, whereas low doses are deleterious.

Changes in basal plasma concentrations of testosterone and estradiol cause the bladder to remodel. Male rabbits were castrated or sham operated and administered daily injections of testosterone, polyestradiol, or progesterone (Celayir, 2003). Polyestradiol increases bladder capacity and compliance, while testosterone and progesterone reduce both parameters in male rabbits (Celayir, 2003). Another study assessed the impact of testosterone on voiding in female and male mice. Females typically void in a stream-like pattern, while males exhibit more droplet-like voids. Mice were castrated and treated with exogenous testosterone or estradiol to determine the influence on voiding function. Castration feminized male urinary physiology, while exogenous testosterone masculinized female urinary voiding physiology (Ruetten et al., 2019). The urethral histology of these mice was analyzed in a follow-up study. Exogenous testosterone increased urethral epithelial cell count in female mice but did not masculinize female urethral histology (Ruetten et al., 2020). None of the treatments feminized male urethral histology or increased urethral epithelial area. Another study found that the combination of androgen and estrogen increased bladder blood vessel density, urothelial thickness, and number of muscle fibers (Madeiro et al., 2002). Estrogen and testosterone and the concentration ratio of both hormones in circulation influence bladder muscle health.

Urothelium: The urothelium is a specialized epithelium of the bladder and consists of at least three cell types: superficial cells (also called umbrella cells), intermediate cells, and basal cells. Bladder superficial epithelial cells synthesize proteins called uroplakins, which assemble to form hexagonal structures contributing to the bladder permeability barrier (Walz et al., 1995). A role for urothelial cells and their uroplakin products in conferring sexually dimorphic bladder function was revealed by depleting uroplakin III (*Upk3*) in mice. Male *Upk3* null mice are prone to urinary retention and increased bladder pressure and are susceptible to bladder decompensation (Table 7.1). Yet, bladder function in female *Upk3* null mice is remarkably normal (Table 7.1). It is unclear how sex differences in *Upk3* null mice derive and whether they are driven by differential responses to a loss of barrier function or by a cell-specific signaling mechanism.

Male and female bladders differ in responsiveness to changes in urinary pH. Acid-sensing ion channels (ASIC) and H^+ gated ion channels maintain pH within the urothelium (Kobayashi et al., 2009). ASIC1 is more abundant in male mouse urothelium, while ASIC2 is more abundant in female mouse urothelium. Although the functional significance of sex differences in bladder ASIC expression is not fully known, it has been hypothesized that ASICs control sensitivity to pathogens. The acidity of urine and the presence of some small molecules may facilitate the growth of bacteria in the urinary tract to cause urinary tract infections. Recent studies using female dogs found the more acidic the urine, the greater the *Escherichia coli* growth, and the more alkaline the urine, the lower the *E. coli* bacterial growth (Thornton et al., 2018). Clinically significant urinary tract infections are more common in women than men and sex hormones are at least partially responsible. Supplemental estrogen mediates a urothelial defense against uropathogenic *E. coli* (Lüthje et al., 2013; Wang et al., 2013). Conversely, depletion of gonadal estrogens by OVX sensitizes the bladder to endotoxin and increases urinary frequency (Acevedo-Alvarez et al., 2018).

Hormone receptors: Male canine estrogen receptor (ER) immunostaining intensity within the lower urinary tract is most abundant in the prostatic urethra (Schulze & Barrack, 1987). The ER staining pattern in males is comparable to females (Schulze & Barrack, 1987). However, males have a higher concentration of ER staining in the urethra than the prostate (Schulze & Barrack, 1987). The ER staining pattern in male and female dogs is consistent with the location of histological changes caused by estrogen administration (Schulze & Barrack, 1987). Androgen receptor (AR) expression is primarily located in urothelial cells of human tissue (Yasui et al., 2019). AR expression has been widely characterized in bladder cancer, but few studies have examined sex-specific expression in benign or normal tissue. Further studies are necessary to fill this gap and build the foundation for understanding sex differences in AR receptor density.

Single-cell RNA sequencing of bladder has helped identify transcriptionally unique cell types in human and animal models. Using this technique, it will be possible to define bladder cell populations that define normal and abnormal. It will also be possible to test whether testosterone and estradiol change bladder resident and inflammatory cell populations, as these changes may underly hormone-mediated changes in bladder function.

Impact of testosterone and estradiol on smooth muscle physiology

Smooth muscle

There is evidence for a role of testosterone and estradiol in smooth muscle physiology. Muscle cells in bladder are long spindle-shaped cells with a central nucleus, similar to those in other muscular organs. During bladder filling, smooth muscle cells relax, elongate, and rearrange to accommodate volumetric expansion (Dixon & Gosling, 1990). The micturition reflex generates a force and synchronously shortens muscle fibers quickly over a large area. The bladder contains both single and multiunit muscle units. Single units have sheets or bundles and cell membranes that have many gap junctions. Gap junctions are formed by connexin subunits and are low resistance pathways through which ions flow between cells to spread electrical signal rapidly through the tissue (Brink, 1998). Single-unit muscle cells can sometimes exhibit spontaneous pace maker activation. Multiunit smooth muscle cells are comprised of muscle fibers or fiber bundles that operate independently. They are innervated by autonomic fibers and controlled through direct nerve stimulation. Multiunit smooth muscle cells exhibit no spontaneous activity (Andersson & Arner, 2004).

During force generation and shortening of muscle, myosin cross-bridges with actin and hydrolyzes MgATP to produce MgADP and inorganic phosphate (Pi). Smooth muscle contraction is initiated by an increase in intracellular Ca^{2+} concentration. Ca^{2+} enters the cytoplasm through the cell membrane via Ca^{2+} channels or can be released from the sarcoplasmic reticulum (Rüegg, 2012). Calcium release is triggered by inositol trisphosphate via IP3 receptors and ryanodine receptors (Brading, 2002). Interestingly, stretch in rabbit urinary bladder smooth muscle cells can also activate Ca^{2+} release (Ji et al., 2002a, 2002b). In addition to Ca^{2+} storage and release, the sarcoplasmic reticulum has another important function which involves modulation of K^+ channel activity to induce relaxation. Calcium activates contractile proteins via a phosphorylation pathway where Ca^{2+} binds to calmodulin and the Ca^{2+}/calmodulin complex activates the myosin light chain kinase to phosphorylate the myosin regulator light chains (Arner & Pfitzer, 1999). Relaxation is caused by dephosphorylation of the regulatory light chain via the myosin light chain phosphatase (Andersson & Arner, 2004).

Testosterone and estradiol inhibit voltage-operated Ca^{2+} channels essential for detrusor function (Ratz et al., 1999). Estradiol decreases spontaneous phasic contractions from single unit smooth muscle cells and reduces nerve-evoked contractions in human detrusor strips. These activities are linked to a rise in depolarization-induced whole-cell K^+ currents (Hristov et al., 2017), and because they are rapid and abolished by paxilline, they suggest a nongenomic mechanism on conductance voltage and Ca^{2+}-activated K^+ (BK) channels. However, estradiol and testosterone can act alone to influence bladder physiology as described in the sections below. Fig. 7.1 summarizes the impacts of testosterone and estradiol on smooth muscle contraction and relaxation.

Estradiol influences the bladder's response to cholinergic stimulation. OVX significantly diminishes atropine-resistant stimulation in rats (Eika et al., 1990). OVX also increased the contractile response to carbachol, a response that is reversed by estradiol (Tanidir et al., 2011). OVX increases the bladder relaxation response to isoproterenol in female rats, but it is unknown whether testosterone or estradiol is driving these changes (Tanidir et al., 2011). OVX + testosterone increases the maximal contractile response of isolated bladder preparations to carbachol and electrical field stimulation. OVX + estrogen decreases urothelial damage, inflammatory cell infiltration, and collagen accumulation caused by OVX (Tanidir et al., 2011). The smooth muscle contraction inhibitory effects of sex steroids may prove therapeutic in treating overactive bladder from elevated muscarinic or purinergic receptor activity (Ratz et al., 1999). These phenotypes suggest

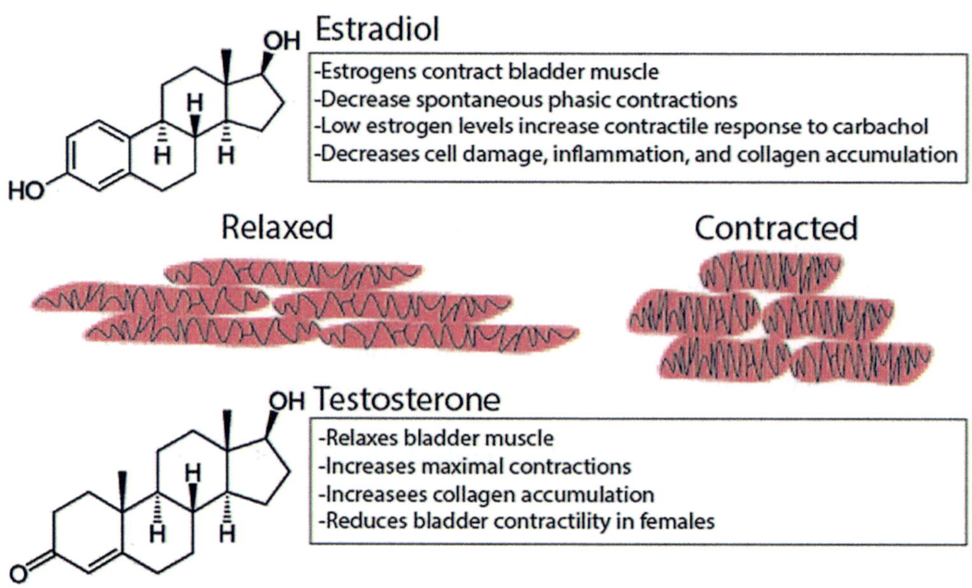

FIGURE 7.1 The impact of estradiol and testosterone on bladder muscle contraction and relaxation. The figure describes how prostate smooth muscle contraction is impacted by supra- or subphysiological testosterone or estradiol levels.

menopause may cause significant structural and functional bladder problems that can be treated by combined estrogen and testosterone supplementation (Eika et al., 1990).

Testosterone concentrations are also critical to bladder muscle health, but testosterone may impact bladder function more than it influences structure (Bravo et al., 2017; Coit et al., 2008). A study delivered letrozole, a potent aromatase inhibitor, to adult female rabbits (Lin et al., 2007). Aromatase metabolizes testosterone into estradiol, and aromatase inhibition increases the plasma concentration of testosterone. A week of letrozole treatment is sufficient to reduce female bladder smooth muscle contractile responses to ATP, carbachol, and electrical stimuli; increase bladder collagen density; and reduce bladder smooth muscle and bladder compliance (Lin et al., 2007). Bladder contractile responses returned to baseline 2–3 s week after letrozole treatment. Overall, this study shows that testosterone may be just as important to the contractility of female bladder as estrogens and its actions are reversible. Testosterone appears to influence male and female bladder differently. In females, it reduces bladder contractility and response to stimuli and increases bladder collagen content. In males, testosterone relaxes bladder muscle, specifically the trigone, and reduces collagen content. The sexually dimorphic influence of testosterone on male and female bladders may help explain gendered bladder disease progression and indicate an interaction with other endogenous hormones or a concentration-dependent impact on bladder function or histology.

Bladder outlet obstruction in aging males leads to bladder remodeling characterized by a loss of detrusor smooth muscle, an increase in bladder collagen content, and a decrease in voiding power (Hughes et al., 2017; Kranse & van Mastrigt, 2003; Li et al., 2017; Stanton et al., 2006). There is some evidence that testosterone replacement in aging males could protect against a decline in bladder function or even restore bladder function. Supraphysiological testosterone given in adulthood or during the prepubertal period reduces bladder collagen content and increases bladder capacity compared to castrated male rats (Shortliffe et al., 2014; Tek et al., 2010). Supplementary testosterone increases bladder capacity in aging men (Fernandes et al., 2012). The mechanism of testosterone activity is linked in part to its action on bladder tone, specifically in the bladder neck region where it serves to relax smooth muscle and promote bladder emptying. A study using isolated preparations of pig bladder neck showed that testosterone relaxes the bladder neck via mechanisms independent of the urothelium, AR, aromatase, 5α-reductase, NO synthase, COX and K (+) channels, and guanylyl cyclase but dependent on inhibition of extracellular Ca^{2+} L-type voltage-operated channels (Fernandes et al., 2012).

Nerves

The bladder is innervated by many nerve subtypes, and the composition and distribution of nerve fibers in the bladder differ between females and males. Innervation of the fetal urinary bladder is established by 13 weeks postconception (Gosling et al., 1999). In humans, the inferior hypogastric plexus contains sympathetic fibers originating from the hypogastric nerve and the sacral splanchnic nerve branching from the T11-L2 ganglia (Baader & Herrmann, 2003). The inferior hypogastric plexus also contains sympathetic fibers that derive from pelvic splanchnic nerves of the ventral rami of S2–4 (Baader & Herrmann, 2003). Nerve branches to the bladder in women originate from the inferior hypogastric plexus and distribute to the lateral surface of the bladder while independent branches follow the ureters from the hypogastric nerve to terminate in the trigone region (Yamaguchi et al., 2011). In males, the inferior hypogastric plexus is located on the dorsolateral side of the prostate and bladder, making the location of the hypogastric plexus lower in male than in female pelvis (Yamaguchi et al., 2011). Further, the independent branches running along the ureter are smaller and less present in males than females (Yamaguchi et al., 2011).

Pelvic ganglia provide motor nerve supply to reproductive organs and are sexually dimorphic. There are more pelvic neurons in males than females (Greenwood et al., 1985). In addition, males possess a higher proportion of noradrenergic neurons in the pelvic ganglia than females (Dail et al., 1975). Males also have a larger number of preganglionic neurons that exist in the lumbar spinal cord. Lastly, lumbar dorsal root ganglia are larger in males than females as a result of greater neuronal death in females (Mills & Sengelaub, 1993). Rat dorsal root ganglia neurons express ERs, and estrogens have been shown to up- and downregulate one sensory axon subtype, calcitonin gene-related peptide (CGRP) (Yang et al., 1998). Bladder-projecting afferent neurons in lumbosacral dorsal root ganglia of female rats express ER (Bennett et al., 2003). ER (ERα and ERβ colocalization) is expressed in many sensory neurons in the upper lumbar or sacral ganglia (Bennett et al., 2003). ER-positive bladder-projecting neurons that innervate the lower urinary tract are likely targets for circulating estrogens (Bennett et al., 2003).

There are sex differences in the type and pattern of axons innervating the bladder. Although the bladder receives sparse noradrenergic innervation, the terminal ureter and its junction with the bladder (vesicoureteral junction) is richly innervated by noradrenergic axons. A study of male and female canine bladder revealed that noradrenaline is more abundant in the proximal urethra and bladder dome than other bladder regions (Dohkita et al., 1987). Noradrenaline is more abundant in proximal urethra of males than females, but in the bladder base and distal urethra, noradrenaline is more abundant in females than in males (Dohkita et al., 1987). These data suggest that bladder sympathetic innervation is greatest in the proximal urethra and there is a sex difference in innervation (Dohkita et al., 1987). The male rabbit bladder is more sensitive to high-frequency electrical stimulation than female bladder (Lee et al., 1994). Even though there are no sex differences in rabbit bladder sensitivity to phenylephrine (alpha-1 adrenoreceptor agonist) or bethanechol (cholinergic receptor agonist) stimulation (Lee et al., 1994), phentolamine (alpha-adrenoreceptor antagonist) stimulates female rabbit bladder neck more than male bladder neck (Lee et al., 1994). Female rabbit bladders are more sensitive to isoproterenol-mediated relaxation than male rabbit bladders (Felsen, 1998; Morita et al., 1998). Overall, the differential responses to neuromodulatory drugs suggest sex differences in alpha adrenoceptors and beta-adrenoceptors but not alpha-1 adrenoceptors or cholinergic receptors.

Micturition reflexes can be influenced by circulating testosterone and estradiol concentrations. Noradrenergic pelvic neurons are dramatically impacted by castration. Androgens do not appear to affect bladder cholinergic axon activity, even though androgens do affect cholinergic axon activity in reproductive tissues such as penis and glandular tissues (Keast & Saunders, 1998). A more recent study contradicts this and found testosterone reduces male rat bladder responsiveness to carbachol (cholinergic agonist) in a dose-dependent manner (100–300 μM) and has a greater inhibitory response in older males than younger males, indicating receptor densities may change with age (Hall et al., 2002). This recent study also reported sex differences in testosterone's impact on cholinergic stimuli. Testosterone more potently (1000X) inhibits bladder responsiveness to carbachol (cholinergic agonist) in females than in males (Hall et al., 2002). Taken together, these studies reveal clear sex differences in cholinergic receptors (muscarinic and nicotinic) in rat bladder that may be driven by testosterone.

The human bladder exhibits sex differences in abundance of alpha-1 adrenergic receptors (ADRA1) (Sigala et al., 2004). Alpha adrenergic receptors are the receptors for noradrenaline released by noradrenergic axons. The abundance of ADRA1 in the bladder neck and detrusor is similar between males and females, but in the trigone, women have significantly more ADRA1 than men. The expression of ADRA subtypes also differs between men and women. Men express more ADRA1A1 and ADRA1D than women (Sigala et al., 2004). Castration changes ADRA1, ADRA2, ADRAB, and muscarinic cholinergic receptor densities in the bladder and proximal urethra (Takyu, 1993). Castration decreases ADRA1 abundance in the proximal urethra and muscarinic cholinergic receptor abundance in the bladder dome (Takyu, 1993). The response of bladder muscle strips to autonomic drugs parallels the densities of the receptors they target, indicating the changes in receptor densities are functionally relevant (Takyu, 1993).

Pelvic neurons are also impacted by changes in circulating testosterone and estradiol concentrations. The autonomic nervous system contributes to male reproductive organ smooth muscle contraction, exocrine secretion, and blood flow. Castration of prepubescent male rats affects the structure, chemistry, and function of pelvic autonomic neurons. Noradrenergic pelvic neurons supplying the prostate and urinary bladder are reduced in size compared to controls and cholinergic pelvic neurons are unchanged. Castration of male rats also reduces the size of pre- and paravertebral sympathetic neurons that supply the pelvic visceral (Keast & Saunders, 1998). Overall, circulating androgens are essential for maturation and maintenance of autonomic neurons and pelvic ganglia supplying the prostate and bladder and testosterone may have direct effects on neuronal gene expression (Keast & Saunders, 1998).

Testosterone and estradiol in benign bladder diseases

Lower urinary tract symptoms

Urinary disorders are classified as problems with storage, voiding, or postmicturition diagnosed according to criteria established by the International Continence Society. Lower urinary tract symptoms (LUTS) are a manifestation of urinary dysfunction and occur in men and women (Lepor, 2005). LUTS are characterized by increased urinary frequency, dribbling, and incomplete bladder emptying, among others, but causes are thought to be multifactorial (Wang et al., 2018). Some hypothesized causes may be increased prostate size, smooth muscle dysfunction, or increased prostate fibrosis. Testosterone and estradiol have been long thought to contribute to disease progression. Here, the potential causative roles of testosterone and estradiol in LUTS is examined. Table 7.2 summarizes the impacts of testosterone and estradiol together and independently on LUTS susceptibility.

Table 7.2 describes the impact of supra- or subphysiological testosterone or estradiol levels on bladder peripheral innervation.

Aging and testosterone and estradiol

Men and women experience changes in plasma testosterone and estradiol concentrations as they age. In men, the plasma concentration of testosterone decreases and estradiol concentration increases. In women, the concentration of estradiol

TABLE 7.2 Impact of testosterone and estradiol modulation on bladder innervation.

Sex hormone modulation	Impact on LUTS	References
Subphysiological testosterone	1. Decreased with age and correlated with LUTS severity. 2. Decreased volume threshold. 3. Decreased mean flow rate. 4. Does not impact LUTS severity.	Cheng and de Groat (2016)
Supraphysiological testosterone	1. Decreased LUTS and postvoid residual urine. 2. Decreases smooth muscle excitability via BK channels. 3. Increases bladder capacity and compliance. 4. Reduces IPSS scores long term.	Haider et al. (2010, 2018), Hristov et al. (2016) and Karazindiyanoğlu and Cayan (2008)
Subphysiological estrogen	Improved IPSS after pharmacological treatment.	Egan et al. (2015)
Estrogen supplementation	1. Ineffective for human LUTS. 2. Clomiphene decreased LUTS phenotype in mice.	Hayashi et al. (2007), Nicholson et al. (2015) and Roehrborn (2008)
Testosterone + estradiol supplementation	Causes progressive bladder outlet obstruction, fibrosis, and decompensation in mice, rats, and dogs.	Aumüller et al. (1982), Mahapokai et al. (2000), Nicholson et al. (2012) and Walsh and Wilson (1976)

decreases and that of testosterone increases (Bélanger et al., 1994; Kaufman & Vermeulen, 2005). LUTS occur in 60% of men and 57.7% of women over the age of 60 (Wang et al., 2018). LUTS in men and women appear to be driven by testosterone and estradiol; therefore, many researchers are examining how changes in ratio of these sex hormones impact disease.

Body mass index (BMI) influences the ratio of testosterone to estradiol and predisposes men and women to LUTS. Circulating testosterone concentrations in aging men with benign prostatic hyperplasia (BPH) are inversely associated with BMI (Kaplan et al., 2013). In addition, estradiol concentrations correlate with BMI in men (Fejes et al., 2006). This observation is not surprising since aromatase activity in adipose tissue converts testosterone to estradiol. Interestingly, when compressed pellets of testosterone and estradiol are delivered to mice to mimic the hormone environment of aging men, these mice develop severe urinary dysfunction characterized by increased prostate size, bladder volume, urinary frequency, and risk of hydronephrosis (Nicholson et al., 2012).

An aging-related decline in plasma testosterone concentrations can drive LUTS in elderly men. Exposure of human bladders to a castration-relevant concentration of testosterone negatively influences active and passive properties of human bladders by decreasing volume threshold for micturition and mean flow rate, both of which contribute to LUTS (Cheng & de Groat, 2016) (Table 7.2). Subphysiological testosterone concentrations in men also correlate with risk of metabolic syndrome and erectile dysfunction (Yassin et al., 2008). One study reported that administration of parenteral testosterone undecanoate decreased patient-reported LUTS and postvoid residual urine volumes (Haider et al., 2010). These studies suggest that an aging-related decrease in testosterone concentration has the capacity to influence LUTS severity in aging men, but because testosterone can be aromatized to estradiol, it is unclear which hormone is responsible.

Estrogens have a role in BPH and LUTS in aging men and women, but no current therapies directly target estrogen action for these diseases. Estrogens act through two receptor subtypes ERα (also known as Esr1) and Erβ (also known as Esr2). GPCR 30 (G protein–coupled ER, GPER) can also bind estrogens and mediate its response through nongenomic actions (Comeglio et al., 2014). Few studies are reported on GPCR 30 in bladder, but one study found that while GPCR 30 is important to mediate estrogen's action in prostate, ERα mainly acts in bladder (Comeglio et al., 2014). Estrogens are hypothesized to be important in LUTS because lower baseline serum estradiol levels are associated with improved International Prostate Symptom Score (IPSS) following pharmaceutical treatment (Egan et al., 2015) (Table 7.2). Some clinical trials have investigated the impact of selective ER modulators (SERMs) in treating LUTS, but results have been mixed. For example, one study administered tamoxifen to patients with BPH and found tamoxifen was not effective enough to increase max flow or decrease symptoms in men (Hanus & Matousková, 1993) (Table 7.2). Further, an ERβ agonist LY500307 was not effective in treating LUTS in patients with enlarged prostates and BPH (Roehrborn et al., 2015) (Table 7.2). The benefits of estradiol treatment for LUTS is not fully understood and may still present a benefit clinically if varied estrogen doses or routes of treatment are studied.

Both testosterone and estradiol seem to contribute to LUTS, indicating it is likely the interaction of both that contributes to severity. The administration of sustained release implants of testosterone and estradiol to mice, to mimic the hormonal environment of aging men, causes progressive bladder outlet obstruction, characterized by prostate hypertrophy, detrusor compensation and decompensation, and bladder fibrosis (Nicholson et al., 2012) (Table 7.2). Rats and dogs generally respond to exogenous testosterone and estradiol the same as mice (Aumüller et al., 1982; Constantinou & Omata, 1996; Mahapokai et al., 2000; Nicholson et al., 2012; Walsh & Wilson, 1976; Winter et al., 1995; Zhou et al., 2009) (Table 7.2). The progressive voiding dysfunction mediated by exogenous testosterone and estradiol requires ERα but not ERβ (Nicholson et al., 2015). This finding led to an examination of SERMs to prevent bladder complications in T + E2-treated mice (Nicholson et al., 2015) (Table 7.2). Clomiphene is a SERM approved for treatment of ovulatory dysfunction in women and has also been used to increase serum T and E2 concentrations in men suffering from LUTS with successful alleviation of some symptoms (Shabsigh et al., 2005).

Women experiencing menopause experience a reduction in plasma estradiol concentrations and are at a greater risk of experiencing LUTS (Varella et al., 2016). Supplemental estrogens are effective for the treatment of urge urinary incontinence, overactive bladder, and stress urinary incontinence and reduce the risk of urinary tract infections in postmenopausal women (Nappi & Davis, 2012).

LUTS, testosterone and estradiol, and innervation

An aging-related change in circulating concentrations of testosterone and estradiol in part causes BPH and detrusor overactivity in mice. Alpha-1 adrenoceptor antagonists relax prostatic and bladder neck smooth muscle to suppress irritative symptoms in humans. The alpha-1 adrenoceptor antagonist silodosin reduces T + E2-mediated detrusor overactivity, indicating that excessive response to sympathetic nerve stimulation mediated via alpha-1A adrenoceptor in prostate gives rise to detrusor overactivity (Tatemichi et al., 2006). Autonomic nervous system overactivity has also been associated with subphysiological testosterone concentrations and aging (Yassin et al., 2008). This overactivity may contribute to the severity of LUTS, but further studies are needed.

Muscle sensitivity

A role for smooth muscle in the onset or progression of LUTS is supported by the fact that smooth muscle relaxing agents provide benefit to some men with LUTS. Interestingly, the concentration of circulating sex hormones influences detrusor sensitivity to contraction. Estradiol reduces spasmogenity, antagonizes bladder muscle contractions, and reduces the amplitude and frequency of spontaneous Ca^{2+} flashes of isolated preparations of male rat bladder (Valeri et al., 2009). Testosterone can decrease bladder smooth muscle excitability via activation of Ca^{2+}-activated (BK) channels (Hristov et al., 2016). Working together, estradiol and testosterone concentration modulation can influence bladder smooth muscle contractility to contribute to bladder muscle contraction.

Phosphodiesterase type 5 (PDE-5) inhibitors are prescribed for erectile dysfunction and LUTS (Brasure et al., 2016). Their exact mechanism is unclear, but they inhibit PDE-5, increase cyclic guanosine monophosphate, and mediate smooth muscle relaxation. PDE-5 inhibitors function as vasodilators by impeding break down of cyclic guanosine monophosphate, a second messenger involved in smooth muscle relaxation. PDE-5 is highly expressed in the bladder (Filippi et al., 2007). PDE-5 activity is more pronounced in males versus female rabbit bladders, testosterone increases PDE-5 abundance in male but not female rat bladder, and the testosterone to estradiol ratio regulates the bladder responsiveness to PDE-5 inhibitors (Vignozzi et al., 2012).

Testosterone as a therapy for LUTS

The role of testosterone as a therapy for LUTS is controversial. Some studies report testosterone as a therapeutic. Testosterone replacement therapy has been used clinically to treat male LUTS. Although many studies failed to observe significant improvement in symptoms among healthy aging men, men with metabolic syndrome may be specifically responsive (Kathrins et al., 2016; Saad et al., 2007; Yassin et al., 2014). One study found that 1 year of transdermal testosterone improved sexual function and bladder function characterized by increased bladder capacity and compliance and reduced voiding pressure (Karazindiyanoğlu & Cayan, 2008). Long-term testosterone therapy reduces IPSS scores and postvoid bladder residual in hypogonadal men (Haider et al., 2018). Further, aromatase inhibitors are an effective treatment of BPH-related voiding symptoms in some men and reduce prostatic and bladder volumes and increase in the testosterone/estradiol ratio (Schweikert & Tunn, 1987). In addition, testosterone has been considered a therapy in females with overactive bladder, which is often associated with LUTS (Hristov et al., 2016).

Some studies report that testosterone is harmful to bladder structure and function. As mentioned above, castration causes significant bladder atrophy and detrusor remodeling. Treatment with a 100 mg/kg testosterone significantly increases urothelial thickness and smooth muscle to collagen ratio (Abdel-Hamid & Ali, 2015). Increased collagen content stiffens the bladder and reduces its compliance. One study found that testosterone did not improve urodynamics or reduce bladder obstruction in patients with irritable or obstructive bladder symptoms due to BPH (Flamm et al., 1979). Patients treated with the luteinizing hormone—releasing hormone (LHRH) analogue, buserelin, for 6 months exhibited castration-like testosterone levels but had no significant effect on bladder outflow obstruction (Keane et al., 1988).

Stress incontinence

Stress incontinence is defined as "involuntary leakage on effort or exertion or on sneezing or coughing" by the International Continence Society. Incontinence results from a variety of causes including bladder dysfunction, urethral dysfunction, pelvic floor dysfunction, sphincter dysfunction, or a combination (Nitti, 2001). The prevalence of urinary incontinence is lower in men than in women (Nitti, 2001). The prevalence of stress and mixed (stress and urge) incontinence is higher than urge incontinence alone (Nitti, 2001). The following section will describe the impacts of testosterone and estradiol on bladder cells in men and women with stress incontinence.

Aging and T and E2

Severe incontinence is not prevalent in young women but increases with age and the incidence typically peaks around menopause (Nitti, 2001). Stress urinary incontinence affects approximately 15%—30% of women (Legendre et al., 2012; Mäkinen et al., 1995). The impact of declining estrogen concentrations on bladder function was examined in young (2 month) and mature (10 month) rats. Estrogen deprivation reduces young rat bladder responsiveness to cholinergic stimuli (Diep & Constantinou, 1999). Mature OVX rats exhibited the most significant decrease in response to cholinergic stimulation (Diep & Constantinou, 1999). Overall, these findings suggest that supplemental estradiol partially restores bladder function in young adult OVX rats and fully restores bladder function in aged OVX rats (Diep & Constantinou, 1999).

Estrogens and innervation

Estrogens have pronounced trophic effects on neurons throughout the peripheral and central nervous system. Most studies support a therapeutic role for supplemental estrogens after neurological injury. ERs can directly modulate neurotransmitters and act like antioxidants (Garcia-Segura et al., 2001). The pudendal nerve innervates the pelvic floor and the external urethral sphincter. This nerve is often injured during childbirth. Pudendal nerve injury reduces external urethral sphincter activity, causing atrophy and leading to stress urinary incontinence. The impact of estradiol treatment on pudendal nerve damage was examined in a rat model of stress incontinence. OVX rats received a bilateral pudendal nerve crush and a subcutaneous slow release estradiol capsule. Tissues were examined 2 weeks later. Estradiol supplementation after pudendal nerve crush causes unmyelinated axons to be spared and regenerated (Kane et al., 2004). This indicates a therapeutic role of estrogen supplementation in women with stress urinary incontinence. Clearly, estrogen plays an important role in modifying nerves in stress incontinence and future studies should identify whether nerves are remodeled as estrogen concentrations change.

Estradiol as a therapy

There is a wealth of experimental evidence supporting the benefits of supplemental estrogens on stress urinary incontinence. Estrogen treatment is effective in treating spay incontinence in dogs and is used clinically to treat stress incontinence in women (Hirai & Tsuda, 2009; Kim et al., 2006; Veronesi et al., 2009). Though there are some reported differences in clinical outcome based on the type of estrogen and route of administration, clinical studies historically supports a beneficial effect of supplemental estrogen with respect to patient-reported symptoms, urine leakage, and urodynamic parameters including maximum urethral pressure, urethral closure pressure, and volume to sense bladder fullness (Cody et al., 2012; Sacco et al., 1990). Recent studies indicate a disputed role of estrogens in treating urinary incontinence and do not recommend its prescription (Hendrix, 2005; Townsend et al., 2009). Route of administration may contribute to disputes in estrogens efficacy.

Overactive bladder

Overactive bladder can be defined as "urgency with or without urgency incontinence usually with increased daytime frequency and nocturia". Overactive bladder is more common in both women and men (16% in men and 16.9% in women)

(Stewart et al., 2003). With advancing age, overactive bladder becomes much more prevalent in women than men (Stewart et al., 2003). Despite the sex differences in prevalence, there are no clinically relevant differences in circulating hormone concentrations relative to improvement in overactive bladder, indicating sex hormones may not play as important of a role in this bladder disease (Kallner et al., 2016).

TRPs in overactive bladder

Transient receptor potential (TRP) subfamily is comprised of cation channels involved in various cellular functions. They are encoded by 27 genes and grouped into six subfamilies based on amino acid sequence and homology (Pan et al., 2011). TRP receptors are thermosensors with most responding to heat, but some responding to cold. TRP receptors are expressed in the lower urinary tract in neuronal fibers and urothelial and muscle layers of bladder and urethra including TRPV1, TRPV2, TRPV4, TRM4, TRPM8, and TRPA1 (Andersson, 2019). TRP receptors are expressed by bladder urothelial cells and by demyelinated C-fibers that project onto the urothelium (Xu et al., 2008). The urinary bladder contains a high density of TRPV1 receptors that regulate pain perception and bladder contraction. TRPV4 receptors are mechanoreceptors that regulate bladder excitability and contractability. TRPV4 is expressed in rat and mouse urothelial and detrusor muscle cells (Girard et al., 2013). Genetic depletion of TRPV4 in mice significantly increases urinary spotting, reduces intervoid intervals, and increases nonvoiding contractions (Gevaert et al., 2007). These findings suggest TRPV4 plays a unique role in maintaining normal urination and may be involved in regulating the mechanical pressure of bladder filling (Gevaert et al., 2007).

Our understanding of hormonal regulation of TRP receptor expression in the bladder is incomplete, but possible lines of new research are guided by knowledge of how gonadal hormones influence TRP abundance. Estrogens, progesterone, and androgens change TRP channel expression in the brain, lung, and mammary gland (Artero-Morales et al., 2018). For example, progesterone decreases the expression of TRPV4 in human airway and mammary gland epithelia cells (Jung et al., 2009). In endometrium, estrogens increase TRPA1 and TRPV1 abundance in rat sensory neurons (Pohóczky et al., 2016). The hormonal control of TRP abundance is physiologically relevant, as 17β-estradiol interferes with TRPV1 activation by capsaicin in sensory neurons (Xu et al., 2008) and increases TRPV1 abundance in endometrium (Pohóczky et al., 2016). Future research opportunities include assessment of how TRP receptor density and distribution in the bladder changes with age, in response to changes in circulating testosterone and estradiol concentrations, and with disease processes such as overactive bladder.

Environmental factors that mediate changes in testosterone and estradiol concentrations

Plasma concentrations of testosterone and estradiol can be modified by physiological factors such as age and pregnancy, but also by environmental factors such as diet or chemical exposure. This section describes how the environment modifies testosterone and estrogen concentrations or actions in bladder disease. Table 7.3 summarizes specific environmental chemicals, how they modulate testosterone and estradiol, and their impact on bladder.

Table 7.3 describes the impact of various environmental estrogens on bladder structure and function.

TABLE 7.3 Environmental estrogens can impact bladder structure and function.

Chemical	Modulate testosterone or estradiol?	Impact on bladder	References
Folic acid	No	Protects against LUTS in mice	Keil et al. (2015)
Soy phytoestrogens	Estrogen	Decrease rat detrusor contractions	Valeri et al. (2012)
Panax ginseng	No	Protects against LUTS in mice	Kim et al. (2018)
2,3,7,8-tetrachlorodibenzo-p-dioxin (TCDD)	Estrogen	Worsens testosterone- and estradiol-mediated LUTS after developmental exposure in mice	Ricke et al. (2016) and Turco et al. (2020)
BPA	Estrogen	Increases spontaneous voiding and urinary retention in mice	Nicholson et al. (2018)
UV filters	Estrogen	Decrease prostate and testis weight, impact on bladder unknown	Nakamura et al. (2018)

Diet

Diet can influence bladder function in many ways, some of which are related to fluid volume regulation, others to neurogenic mechanisms, and yet others to direct actions. One study showed that a folic acid–enriched diet protects against progression of urinary voiding dysfunction in mice exposed to testosterone and estradiol (Keil et al., 2015) (Table 7.3). A potential mechanism involves a folic acid diet-dependent increase in dorsal root ganglia *Trpv1* abundance and increased bladder sensitivity to filling (Keil et al., 2015). Though the study described above was conducted in male mice, its results may be relevant in the context of pregnancy-induced urinary incontinence, as folic acid supplements are recommended to pregnant women to support fetal development and may also support maternal bladder health.

Soy products such as tofu, edamame, miso, and soy milk are increasingly consumed by individuals worldwide for their health effects. Soy has estrogenic properties and can influence bladder function. The soy phytoestrogens genistein and daidzein protect against rat bladder hyperactivity from ischemia/reperfusion injury (Valeri et al., 2012) (Table 7.3). Genistein and daidzein also decrease rat detrusor contractions from electrical field stimulation, indicating potential benefit for overactive or hyperactive bladder or nerve damage (Valeri et al., 2012). One study compared the ability of endogenous, pharmacological, and dietary estrogens to antagonize purinergic- or muscarinic-mediated bladder contractions. 17β-estradiol and progesterone inhibit purinergic contractions, estradiol and ethinyl estradiol inhibit muscarinic contractions, but dietary estrogens daidzein and genistein are minimally effective, reducing muscarinic contractions by a meager 7% (Ratz et al., 1999). This study shows that dietary estrogens have weak estrogenic activity but may be minimally effective in treating overactive bladder, especially overactive bladder from elevated muscarinic or purinergic activity. Importantly, soy isoflavones do not significantly alter circulating testosterone or estradiol concentrations, so they are not likely to predispose to LUTS or other bladder diseases influenced by testosterone and estradiol (Valeri et al., 2012).

The Korean plant panax ginseng (200 mg/kg) protects against the urinary symptoms of BPH in rats through a neurogenic mechanism (Kim et al., 2018) (Table 7.3). Panax ginseng downregulates nerve growth factor in the rat pontine nucleus, the key regulatory region for micturition (Kim et al., 2018). Dietary supplements can clearly influence sex hormone concentrations or bladder function so further research on other estrogenic foods is needed to determine its role in bladder disease. Studying patient's diets may provide important insights into why variability in disease severity presents in bladder.

Environmental chemical exposure

Some environmental chemicals can modulate testosterone and estradiol concentrations to predispose to bladder disease. Here, we describe some examples of chemicals known to impact bladder and hormone concentrations. 2,3,7,8-tetrachlorodibenzo-*p*-dioxin (TCDD) is a ubiquitous environmental contaminant that persists in the environment. The primary mechanism of exposure for humans is through diet by eating contaminated dairy and meats (Nauman & Schaum, 1987). TCDD is released into today's environment primarily through waste incineration (Nauman & Schaum, 1987). Interestingly, TCDD has been found to have estrogenic properties. One study found maternal exposure to low concentrations of TCDD (10 ng/kg) increased the abundance of ERα and ERβ in male rat offspring (Kakeyama et al., 2003). Environmental contaminant exposure during fetal, neonatal, or pubertal development is a risk modifier of many diseases of advancing age, including urinary voiding dysfunction. For example, perinatal exposure to TCDD worsens T + E2-mediated urinary voiding dysfunction in adulthood (Ricke et al., 2016; Turco et al., 2020) (Table 7.3). The mechanism through which TCDD acts to worsen LUTS in mice is unknown. TCDD is an endocrine disruptor, so perhaps its modulation of estrogen concentrations can predispose to LUT development.

Bisphenol A (BPA), a plasticizer in polycarbonate plastics, has also been implicated in susceptibility to male voiding dysfunction. The primary mechanism for human exposure is through the diet, specifically leaching of BPA from resins in food and beverage container linings and dental implants. BPA was shown to have estrogenic activity in 1936 (Dodds & Lawson, 1936). Testosterone + BPA treatment of adult male mice significantly changes spontaneous voiding behaviors by inducing an intermittent voiding pattern and causing urinary retention and bladder dysfunction, reminiscent of the T + E2 phenotype (Nicholson et al., 2018) (Table 7.3). Exposure to BPA during the fetal and neonatal periods sensitizes to T + E2-mediated voiding dysfunction in adult mice (Taylor et al., 2020).

The impact of environmental chemicals on urinary function is only beginning to be unraveled. There are a host of natural chemicals and pesticides that are known to influence reproductive hormone concentrations and thereby have the potential to influence urinary voiding function. Lead (100 or 1000 ppm in drinking water for 7 or 13 weeks) reduces the mass of rat seminal vesicles, epididymis, and reproductive capacity (McMurry et al., 1995). Cadmium (0.5 mg/kg) impairs the regeneration of rat Leydig cells (Wu et al., 2017). DDT exposure in adulthood increases free and bioavailable testosterone and lowers follicle-stimulating hormone (FSH) concentrations in men (Bornman et al., 2018). Deltamethrin, a pyrethroid insecticide, decreases sperm quality and impairs libido and decreases inhibin B abundance and reproductive performance (Ben Slima et al., 2017).

Industries have recently developed an interest in ultraviolet (UV) filters present in sunscreens. Several reports of negative health effects have inspired further studies investigating which chemicals are toxic to humans. UV filter exposure during development can have a significant impact on prostate, bladder, and testosterone and estradiol concentrations. Exposure to UV filters in adulthood can have additionally concerning impacts. The urinary concentrations of BPA, parabens, and UV filters were observed after 2 weeks of exposure (Janjua et al., 2004). The organic UV filter, octocrylene, level was found to be significantly more concentrated in the urine of women with polycystic ovarian syndrome in women with a BMI >24 kg/m^2 (Gu et al., 2019). BPA has been previously shown to cause urinary dysfunction in mice, as mentioned above, so it is possible UV filter exposure may influence bladder similarly, but more studies are needed.

The UV filter 4-MBC (0.7, 7, 24, 47 mg/kg/day delays male rat puberty and decreases adult prostate weight and increases testis weight (Durrer et al., 2005). 2-hydroxy-4-methoxybenzophenone is also a UV filter used in sunscreens, cosmetics, and plastics. It has weak estrogenic activity and activates candidate biomarker genes for testicular and prostatic toxicity in rats (Nakamura et al., 2018). Benzophenone-3, 4-MBC, octocrylene, benzyl salicylate, and homosalate are all UV filters that may act similarly and cause similar phenotypes (Krause et al., 2012). They are present in 85% of Swiss breast milk samples, which make them a risk to developing children (Krause et al., 2012). There are increasing numbers of studies that have reported the endocrine-disrupting effects of UV filters, so future research into these chemicals are necessary to determine safety (Krause et al., 2012).

Summary and conclusions

Bladder histology, innervation, function, and physiology are clearly influenced by endogenous and exogenous testosterone and estradiol. Most bladder diseases such as LUTS, stress urinary incontinence, and overactive bladder occur in aging men and women. Aging promotes changes in exogenous testosterone and estradiol concentrations that can be correlated with bladder diseases. Sex differences in bladder structure, function, and disease prevalence further emphasize the importance of testosterone and estradiol in the urinary bladder. Environmental factors, including diet and dietary and airborne chemicals, can also influence testosterone and estradiol levels and bladder and prostate physiology and histology.

References

Abdel-Hamid, A. A., & Ali, E. M. (2015). Effect of testosterone therapy on the urinary bladder in experimental hypogonadism of rats. *The Journal of Molecular Histology, 46*, 263–272.

Abelson, B., Sun, D., Que, L., Nebel, R. A., Baker, D., Popiel, P., Amundsen, C. L., Chai, T., Close, C., DiSanto, M., Fraser, M. O., Kielb, S. J., Kuchel, G., Mueller, E. R., Palmer, M. H., Parker-Autry, C., Wolfe, A. J., & Damaser, M. S. (2018). Sex differences in lower urinary tract biology and physiology. *Biology of Sex Differences, 9*(1), 45. https://doi.org/10.1186/s13293-018-0204-8

Acevedo-Alvarez, M., Yeh, J., Alvarez-Lugo, L., Lu, M., Sukumar, N., Hill, W. G., & Chai, T. C. (2018). Mouse urothelial genes associated with voiding behavior changes after ovariectomy and bladder lipopolysaccharide exposure. *Neurourology and Urodynamics, 37*(8), 2398–2405. https://doi.org/10.1002/nau.23592

Altaf, M., Osman, N. I., & Chapple, R. C. (2013). Anatomy of the lower urinary tract. *Surgery*, 319–325. https://doi.org/10.1016/j.mpsur.2013.04.013

Andersson, K.-E. (2019). TRP channels as lower urinary tract sensory targets. *Medical Sciences, 7*(5). https://doi.org/10.3390/medsci7050067

Andersson, K.-E., & Arner, A. (2004). Urinary bladder contraction and relaxation: Physiology and pathophysiology. *Physiological Reviews, 84*(3), 935–986. https://doi.org/10.1152/physrev.00038.2003

Arner, A., & Pfitzer, G. (1999). Regulation of cross-bridge cycling by Ca^{2+} in smooth muscle. *Reviews of Physiology, Biochemistry & Pharmacology, 134*, 63–146. https://doi.org/10.1007/3-540-64753-8_3

Artero-Morales, M., González-Rodríguez, S., & Ferrer-Montiel, A. (2018). TRP channels as potential targets for sex-related differences in migraine pain. *Frontiers in Molecular Biosciences, 5*. https://doi.org/10.3389/fmolb.2018.00073

Aumüller, G., Funke, P. J., Hahn, A., Hoffbauer, G., Tunn, U., & Neumann, F. (1982). Phenotypic modulation of the canine prostate after long-term treatment with androgens and estrogens. *The Prostate, 3*(4), 361–373. https://doi.org/10.1002/pros.2990030406

Baader, B., & Herrmann, M. (2003). Topography of the pelvic autonomic nervous system and its potential impact on surgical intervention in the pelvis. *Clinical Anatomy, 16*(2), 119–130. https://doi.org/10.1002/ca.10105

Barros, C. A. V. D., Lorenzetti, F., Ortiz, V., & Dambros, M. (2013). Testosterone supplementation's effects on age-related bladder remodeling—experimental study in rats. *The Aging Male, 16*(3), 102–107. https://doi.org/10.3109/13685538.2013.807426

Bélanger, A., Candas, B., Dupont, A., Cusan, L., Diamond, P., Gomez, J. L., & Labrie, F. (1994). Changes in serum concentrations of conjugated and unconjugated steroids in 40- to 80-year-old men. *The Journal of Clinical Endocrinology and Metabolism, 79*(4), 1086–1090. https://doi.org/10.1210/jcem.79.4.7962278

Ben Slima, A., Chtourou, Y., Barkallah, M., Fetoui, H., Boudawara, T., & Gdoura, R. (2017). Endocrine disrupting potential and reproductive dysfunction in male mice exposed to deltamethrin. *Human & Experimental Toxicology, 36*(3), 218–226. https://doi.org/10.1177/0960327116646617

Bennett, H. L., Gustafsson, J.-Å., & Keast, J. R. (2003). Estrogen receptor expression in lumbosacral dorsal root ganglion cells innervating the female rat urinary bladder. *Autonomic Neuroscience, 105*(2), 90–100. https://doi.org/10.1016/S1566-0702(03)00044-4

Bornman, M., Delport, R., Farías, P., Aneck-Hahn, N., Patrick, S., Millar, R. P., & de Jager, C. (2018). Alterations in male reproductive hormones in relation to environmental DDT exposure. *Environment International, 113*, 281–289. https://doi.org/10.1016/j.envint.2017.12.039

Brading, A. F. (2002). The sarcoplasmic reticulum in disease and smooth muscle dysfunction: Therapeutic potential. *Novartis Foundation symposium, 246*, 244–254. https://doi.org/10.1002/0470853050.ch18 (discussion 254–257, 272–276).

Brasure, M., MacDonald, R., Dahm, P., Olson, C. M., Nelson, V. A., Fink, H. A., Risk, M., Rwabasonga, B., & Wilt, T. J. (2016). Table 1, medications used to treat LUTS attributed to BPH. *Text. Agency for Healthcare Research and Quality (US)*. https://www.ncbi.nlm.nih.gov/books/NBK368432/table/introduction.t1/.

Bravo, G., Massa, H., Rose'Meyer, R., Chess-Williams, R., McDermott, C., & Sellers, D. J. (2017). Effect of short-term androgen deficiency on bladder contractility and urothelial mediator release. *Naunyn-Schmiedeberg's Archives of Pharmacology, 390*(5), 547–556. https://doi.org/10.1007/s00210-017-1355-6

Brink, P. R. (1998). Gap junctions in vascular smooth muscle. *Acta Physiologica Scandinavica, 164*(4), 349–356. https://doi.org/10.1046/j.1365-201X.1998.00439.x

Celayir, S. (2003). Effects of different sex hormones on male rabbit urodynamics: An experimental study. *Hormone Research, 60*(5), 215–220. https://doi.org/10.1159/000074034

Chen, J., Zhou, Y. X., Yu, Y. L., & Shen, Z. J. (2008). Effects of sex hormones on bladder function and structure: Experiment with ovariectomized female rats. *National Medical Journal of China, 88*(26), 1851–1854.

Cheng, C.-L., & de Groat, W. C. (2016). Effect of orchiectomy and testosterone replacement on lower urinary tract function in anesthetized rats. *American Journal of Physiology-Renal Physiology, 311*(5), F864–F870. https://doi.org/10.1152/ajprenal.00016.2016

Cody, J. D., Jacobs, M. L., Richardson, K., Moehrer, B., & Hextall, A. (2012). Oestrogen therapy for urinary incontinence in post-menopausal women. *Cochrane Database of Systematic Reviews, 10*, CD001405. https://doi.org/10.1002/14651858.CD001405.pub3

Coit, V. A., Gibson, I. F., Evans, N. P., & Dowell, F. J. (2008). Neutering affects urinary bladder function by different mechanisms in male and female dogs. *European Journal of Pharmacology, 584*(1), 153–158. https://doi.org/10.1016/j.ejphar.2008.02.037

Comeglio, P., Morelli, A., Cellai, I., Vignozzi, L., Sarchielli, E., Filippi, S., Maneschi, E., Corcetto, F., Corno, C., Gacci, M., & Vannelli, G. B. (2014). Opposite effects of tamoxifen on metabolic syndrome-induced bladder and prostate alterations: A role for GPR30/GPER? *The Prostate, 74*(1), 10–28. https://doi.org/10.1002/pros.22723

Constantinou, C. E., & Omata, S. (1996). Analysis of the relative biomechanical effects of alpha 1 and alpha 2 antagonists in modifying the compliance of the prostate and micturition parameters of the hormonally manipulated male rat. *Neurourology and Urodynamics, 15*(1), 85–101. https://doi.org/10.1002/(SICI)1520-6777(1996)15:1<85::AID-NAU9>3.0.CO;2-B

Dail, W. G., Evan, A. P., & Eason, H. R. (1975). The major ganglion in the pelvic plexus of the male rat: A histochemical and ultrastructural study. *Cell and Tissue Research, 159*(1), 49–62. https://doi.org/10.1007/BF00231994

Diep, N., & Constantinou, C. E. (1999). Age dependent response to exogenous estrogen on micturition, contractility and cholinergic receptors of the rat bladder. *Life Sciences, 64*(23), PL279–PL289. https://doi.org/10.1016/S0024-3205(99)00168-X

Dixon, J. S., & Gosling, J. A. (1990). Ultrastructure of smooth muscle cells in the urinary system. In P. M. Motta (Ed.), *Electron microscopy in Biology and medicineUltrastructure of smooth muscle* (pp. 153–169). Boston, MA: Springer US.

Dodds, E. C., & Lawson, W. (1936). Synthetic strogenic agents without the phenanthrene nucleus. *Nature*. https://www.nature.com/articles/137996a0.

Dohkita, S., Nishimoto, T., & Morita, T. (1987). Noradrenaline content in canine lower urinary tract tissue. *Tohoku Journal of Experimental Medicine, 151*(1), 117–119. https://doi.org/10.1620/tjem.151.117

Durrer, S., Maerkel, K., Schlumpf, M., & Lichtensteiger, W. (2005). Estrogen target gene regulation and coactivator expression in rat uterus after developmental exposure to the ultraviolet filter 4-methylbenzylidene camphor. *Endocrinology, 146*(5), 2130–2139. https://doi.org/10.1210/en.2004-1272

Egan, K. B., Miner, M. M., Suh, M., McVary, K., Ni, X., Roehrborn, C. G., Wittert, G., Wong, D. G., & Rosen, R. C. (2015). Do baseline estrogen and testosterone affect lower urinary tract symptoms (LUTS) prior to or after pharmacologic treatment with tadalafil? *Andrology, 3*(6), 1165–1172. https://doi.org/10.1111/andr.12114

Eika, B., Salling, L. N., Christensen, L. L., Andersen, A., Laurberg, S., & Danielsen, C. C. (1990). Long-term observation of the detrusor smooth muscle in rats. Its relationship to ovariectomy and estrogen treatment. *Urological Research, 18*(6), 439–442. https://doi.org/10.1007/BF00297382

Fejes, I., Koloszár, S., Závaczki, Z., Daru, J., Szöllösi, J., & Pál, A. (2006). Effect of body weight on testosterone/estradiol ratio in oligozoospermic patients. *Archives of Andrology, 52*(2), 97–102. https://doi.org/10.1080/01485010500315479

Felsen, D. (1998). Studies on gender differences in beta-adrenergic receptors in rabbit bladder. *The Journal of Urology, 159*(2), 346. https://doi.org/10.1016/S0022-5347(01)63915-6

Fernandes, V. S., Barahona, M. V., Recio, P., Martínez-Sáenz, A., Ribeiro, A. S. F., Contreras, C., Martínez, A. C., Bustamante, S., Carballido, J., García-Sacristán, A., Prieto, D., & Hernández, M. (2012). Mechanisms involved in testosterone-induced relaxation to the pig urinary bladder neck. *Steroids, 77*(5), 394–402. https://doi.org/10.1016/j.steroids.2011.12.020

Filippi, S., Morelli, A., Sandner, P., Fibbi, B., Mancina, R., Marini, M., Gacci, M., Vignozzi, L., Vannelli, G. B., Carini, M., & Forti, G. (2007). Characterization and functional role of androgen-dependent PDE5 activity in the bladder. *Endocrinology, 148*(3), 1019–1029. https://doi.org/10.1210/en.2006-1079

Flamm, J., Kiesswetter, H., & Englisch, M. (1979). An urodynamic study of patients with benign prostatic hypertrophy treated conservatively with phytotherapy or testosterone (author's transl). *Wiener Klinische Wochenschrift, 91*(18), 622–627.

Flum, A. S., Firmiss, P. R., Bowen, D. K., Kukulka, N., Santos, G. B. D., Dettman, R. W., & Gong, E. M. (2017). Testosterone modifies alterations to detrusor muscle after partial bladder outlet obstruction in juvenile mice. *Frontiers in Pediatrics, 5*. https://doi.org/10.3389/fped.2017.00132

Fraga, R. D., Dambros, M., Miyaoka, R., Riccetto, C. L. Z., & Palma, P. C. R. (2007). Role of 17 β-estradiol on type IV collagen fibers volumetric density in the basement membrane of bladder wall. *International Urogynecology Journal*. https://www.springermedizin.de/role-of-17-%CE%B2-estradiol-on-type-iv-collagen-fibers-volumetric-den/8144228.

Garcia-Segura, L. M., Azcoitia, I., & DonCarlos, L. L. (2001). Neuroprotection by estradiol. *Progress in Neurobiology, 63*(1), 29–60. https://doi.org/10.1016/s0301-0082(00)00025-3

Gevaert, T., Vriens, J., Segal, A., Everaerts, W., Roskams, T., Talavera, K., Owsianik, G., Liedtke, W., Daelemans, D., Dewachter, I., & Van Leuven, F. (2007). Deletion of the transient receptor potential cation channel TRPV4 impairs murine bladder voiding. *The Journal of Clinical Investigation, 117*(11), 3453–3462. https://doi.org/10.1172/JCI31766

Girard, B. M., Merrill, L., Malley, S., & Vizzard, M. A. (2013). Increased TRPV4 expression in urinary bladder and lumbosacral dorsal root ganglia in mice with chronic overexpression of NGF in urothelium. *Journal of Molecular Neuroscience, 51*(2), 602–614. https://doi.org/10.1007/s12031-013-0033-5

Gosling, J. A., Dixon, J. S., & Jen, P. Y. (1999). The distribution of noradrenergic nerves in the human lower urinary tract. A review. *European Urology, 36*(Suppl. 1), 23–30. https://doi.org/10.1159/000052314

Greenwood, D., Coggeshall, R. E., & Hulsebosch, C. E. (1985). Sexual dimorphism in the numbers of neurons in the pelvic ganglia of adult rats. *Brain Research, 340*(1), 160–162. https://doi.org/10.1016/0006-8993(85)90787-5

Gu, J., Yuan, T., Ni, N., Ma, Y., Shen, Z., Yu, X., Shi, R., Tian, Y., Zhou, W., & Zhang, J. (2019). Urinary concentration of personal care products and polycystic ovary syndrome: A case-control study. *Environmental Research, 168*, 48–53. https://doi.org/10.1016/j.envres.2018.09.014

Haider, A., Gooren, L. J. G., Padungtod, P., & Saad, F. (2010). A safety study of administration of parenteral testosterone undecanoate to elderly men over minimally 24 months. *Andrologia, 42*(6), 349–355. https://doi.org/10.1111/j.1439-0272.2009.01016.x

Haider, K. S., Haider, A., Doros, G., & Traish, A. (2018). Long-term testosterone therapy improves urinary and sexual function, and quality of life in men with hypogonadism: Results from a propensity matched subgroup of a controlled registry study. *The Journal of Urology, 199*(1), 257–265. https://doi.org/10.1016/j.juro.2017.07.039

Hall, R., Andrews, P. L. R., & Hoyle, C. H. V. (2002). Effects of testosterone on neuromuscular transmission in rat isolated urinary bladder. *European Journal of Pharmacology, 449*(3), 301–309. https://doi.org/10.1016/S0014-2999(02)02047-2

Hanus, M., & Matousková, M. (1993). Antiestrogens (tamoxifen) in the alternative therapy of benign prostatic hyperplasia. *Rozhledy v Chirurgii: Mesicnik Ceskoslovenske Chirurgicke Spolecnosti, 72*(7), 316–318.

Hayashi, N., Bella, A. J., Wang, G., Lin, G., Deng, D. Y., Nunes, L., & Lue, T. F. (2007). Effect of extended-term estrogen on voiding in a postpartum ovariectomized rat model. *Canadian Urological Association Journal, 1*(3), 256–263.

Hendrix, S. L. (2005). Effects of estrogen with and without progestin on urinary incontinence. *JAMA, 293*(8), 935. https://doi.org/10.1001/jama.293.8.935

Hirai, K., & Tsuda, H. (2009). Estrogen and urinary incontinence. *International Journal of Urology, 16*(1), 45–48. https://doi.org/10.1111/j.1442-2042.2008.02164.x

Hristov, K. L., Parajuli, S. P., Provence, A., & Petkov, G. V. (2016). Testosterone decreases urinary bladder smooth muscle excitability via novel signaling mechanism involving direct activation of the BK channels. *American Journal of Physiology-Renal Physiology, 311*(6), F1253–F1259. https://doi.org/10.1152/ajprenal.00238.2016

Hristov, K. L., Parajuli, S. P., Provence, A., Rovner, E. S., & Petkov, G. V. (2017). Nongenomic modulation of the large conductance voltage- and Ca^{2+}-activated K^+ channels by estrogen: A novel regulatory mechanism in human detrusor smooth muscle. *Physiological Reports, 5*(14). https://doi.org/10.14814/phy2.13351

Hughes, F. M., Sexton, S. J., Jin, H., Govada, V., & Todd Purves, J. (2017). Bladder fibrosis during outlet obstruction is triggered through the NLRP3 inflammasome and the production of IL-1β. *American Journal of Physiology-Renal Physiology, 313*(3), F603–F610. https://doi.org/10.1152/ajprenal.00128.2017

Janjua, N. R., Mogensen, B., Andersson, A.-M., Petersen, J. H., Henriksen, M., Skakkebaek, N. E., & Wulf, H. C. (2004). Systemic absorption of the sunscreens benzophenone-3, octyl-methoxycinnamate, and 3-(4-methyl-benzylidene) camphor after whole-body topical application and reproductive hormone levels in humans. *Journal of Investigative Dermatology, 123*(1), 57–61. https://doi.org/10.1111/j.0022-202X.2004.22725.x

Ji, G., Barsotti, R. J., Feldman, M. E., & Kotlikoff, M. I. (2002a). Stretch-induced calcium release in smooth muscle. *The Journal of General Physiology, 119*(6), 533–544. https://doi.org/10.1085/jgp.20028514

Ji, G., Barsotti, R. J., Feldman, M. E., & Kotlikoff, M. I. (2002b). Stretch-induced calcium release in smooth muscle. *The Journal of General Physiology, 119*(6), 533–543. https://doi.org/10.1085/jgp.20028514

Jung, C., Fandos, C., Lorenzo, I. M., Plata, C., Fernandes, J., Gené, G. G., Vázquez, E., & Valverde, M. A. (2009). The progesterone receptor regulates the expression of TRPV4 channel. *Pflugers Archiv: European Journal of Physiology, 459*(1), 105–113. https://doi.org/10.1007/s00424-009-0706-7

Kakeyama, M., Sone, H., Miyabara, Y., & Tohyama, C. (2003). Perinatal exposure to 2,3,7,8-tetrachlorodibenzo-p-dioxin alters activity-dependent expression of BDNF MRNA in the neocortex and male rat sexual behavior in adulthood. *Neurotoxicology, 24*(2), 207–217. https://doi.org/10.1016/S0161-813X(02)00214-0

Kallner, H. K., Elmér, C., Andersson, K.-E., & Altman, D. (2016). Hormonal influence on the effect of mirabegron treatment for overactive bladder. *Menopause, 23*(12), 1303–1306. https://doi.org/10.1097/GME.0000000000000708

Kane, D. D., Kerns, J. M., Lin, D. L., & Damaser, M. S. (2004). Early structural effects of oestrogen on pudendal nerve regeneration in the rat. *BJU International, 93*(6), 870–878. https://doi.org/10.1111/j.1464-410X.2003.04792.x

Kaplan, S. A., O'Neill, E., Lowe, R., Hanson, M., & Meehan, A. G. (2013). Prevalence of low testosterone in aging men with benign prostatic hyperplasia: Data from the Proscar Long-term Efficacy and Safety Study (PLESS). *The Aging Male, 16*(2), 48–51. https://doi.org/10.3109/13685538.2013.773421

Karazindiyanoğlu, S., & Cayan, S. (2008). The effect of testosterone therapy on lower urinary tract symptoms/bladder and sexual functions in men with symptomatic late-onset hypogonadism. *The Aging Male, 11*(3), 146–149. https://doi.org/10.1080/13685530802290438

Karl-Erik, A., & Anders, A. (2004). Urinary bladder contraction and relaxation: Physiology and pathophysiology. *Physiological Reviews*, 935–986. https://doi.org/10.1152/physrev.00038.2003

Kathrins, M., Doersch, K., Nimeh, T., Canto, A., Niederberger, C., & Seftel, A. (2016). The relationship between testosterone-replacement therapy and lower urinary tract symptoms: A systematic review. *Urology, 88*, 22–32. https://doi.org/10.1016/j.urology.2015.11.006

Kaufman, J. M., & Vermeulen, A. (2005). The decline of androgen levels in elderly men and its clinical and therapeutic implications. *Endocrine Reviews, 26*(6), 833–876. https://doi.org/10.1210/er.2004-0013

Keane, P. F., Timoney, A. G., Kiely, E., Williams, G., & Stamp, G. (1988). Response of the benign hypertrophied prostate to treatment with an LHRH analogue. *British Journal of Urology, 62*(2), 163–165. https://doi.org/10.1111/j.1464-410x.1988.tb04299.x

Keast, J. R., & Saunders, R. J. (1998). Testosterone has potent, selective effects on the morphology of pelvic autonomic neurons which control the bladder, lower bowel and internal reproductive organs of the male rat. *Neuroscience, 85*(2), 543–556. https://doi.org/10.1016/s0306-4522(97)00631-3

Keil, K. P., Abler, L. L., Altmann, H. M., Wang, Z., Wang, P., Ricke, W. A., Bjorling, D. E., & Vezina, C. M. (2015). Impact of a folic acid—enriched diet on urinary tract function in mice treated with testosterone and estradiol. *American Journal of Physiology-Renal Physiology, 308*(12), F1431–F1443. https://doi.org/10.1152/ajprenal.00674.2014

Kim, S. K., GyuKo, I., Park, H. J., Chung, J.-H., Cho, K. B., Kwon, O. Y., Park, K. H., Ahn, Y. S., Park, C. G., & Kim, Y. O. (2018). Effects of panax ginseng on the nerve growth factor expression in testosterone induced benign prostatic hyperplasia. *Saudi Journal of Biological Sciences, 25*(1), 66–70. https://doi.org/10.1016/j.sjbs.2016.07.005

Kim, D., Kyung, M. B., & Chancellor. (2006). Is estrogen for urinary incontinence good or bad? *Reviews in Urology, 8*(2), 91–92.

Kobayashi, H., Yoshiyama, M., Zakoji, H., Takeda, M., & Araki, I. (2009). Sex differences in the expression profile of acid-sensing ion channels in the mouse urinary bladder: A possible involvement in irritative bladder symptoms. *BJU International, 104*(11), 1746–1751. https://doi.org/10.1111/j.1464-410X.2009.08658.x

Kranse, R., & van Mastrigt, R. (2003). Weak correlation between bladder outlet obstruction and probability to void to completion. *Urology, 62*(4), 667–671. https://doi.org/10.1016/s0090-4295(03)00575-2

Krause, M., Klit, A., Blomberg Jensen, M., Søeborg, T., Frederiksen, H., Schlumpf, M., Lichtensteiger, W., Skakkebaek, N. E., & Drzewiecki, K. T. (2012). Sunscreens: Are they beneficial for health? An overview of endocrine disrupting properties of UV-filters. *International Journal of Andrology, 35*(3), 424–436. https://doi.org/10.1111/j.1365-2605.2012.01280.x

Lee, J. G., Wein, A. J., & Levin, R. M. (1994). Comparative pharmacology of the male and female rabbit bladder neck and urethra: Involvement of nitric oxide. *Pharmacology, 48*(4), 250–259. https://doi.org/10.1159/000139187

Legendre, G., Fritel, X., Ringa, V., Lesavre, M., & Fernandez, H. (2012). Urinary incontinence and menopause. *Progrès en Urologie: Journal De l'Association Francaise D'urologie Et De La Societe Francaise D'urologie, 22*(11), 615–621. https://doi.org/10.1016/j.purol.2012.08.267

Lepor, H. (2005). Pathophysiology of lower urinary tract symptoms in the aging male population. *Reviews in Urology, 7*(Suppl. 7), S3–S11.

Li, N., Ding, H., He, X., Li, Z., & Liu, Y. (2017). Expression and function of the small-conductance Ca^{2+}-activated K^+ channel is decreased in urinary bladder smooth muscle cells from female Guinea pig with partial bladder outlet obstruction. *International Urology and Nephrology, 49*(7), 1147–1155. https://doi.org/10.1007/s11255-017-1592-0

Liang, C.-C., Lee, T.-H., & Chang, S.-D. (2013). Effects of sex hormones on cell proliferation and apoptosis in the urinary bladder muscle of ovariectomized rat. *Taiwanese Journal of Obstetrics & Gynecology, 52*(3), 335–340. https://doi.org/10.1016/j.tjog.2012.11.003

Lin, W.-Y., Rehfuss, A., Whitbeck, C., Juan, Y.-S., Chichester, P., Mannikarottu, A., & Levin, R. M. (2007). Effect of letrozole on urinary bladder function in the female rabbit. *BJU International, 100*(6), 1391–1395. https://doi.org/10.1111/j.1464-410X.2007.07149.x

Lüthje, P., Brauner, H., Ramos, N. L., Ovregaard, A., Gläser, R., Hirschberg, A. L., Aspenström, P., & Brauner, A. (2013). Estrogen supports urothelial defense mechanisms. *Science Translational Medicine, 5*(190), 190ra80. https://doi.org/10.1126/scitranslmed.3005574

Madeiro, A., Girão, M., Sartori, M., Acquaroli, R., Baracat, E., & Rodrigues De Lima, G. (2002). Effects of the association of androgen/estrogen on the bladder and urethra of castrated rats. *Clinical & Experimental Obstetrics & Gynecology, 29*(2), 117–120.

Mahapokai, W., Xue, Y., van Garderen, E., van Sluijs, F. J., Mol, J. A., & Schalken, J. A. (2000). Cell kinetics and differentiation after hormonal-induced prostatic hyperplasia in the dog. *The Prostate, 44*(1), 40–48. https://doi.org/10.1002/1097-0045(20000615)44:1<40::aid-pros6>3.0.co;2-j

Mäkinen, J. I., Pitkänen, Y. A., Salmi, T. A., Grönroos, M., Rinne, R., & Paakkari, I. (1995). Transdermal estrogen for female stress urinary incontinence in postmenopause. *Maturitas, 22*(3), 233–238. https://doi.org/10.1016/0378-5122(95)00944-g

Mangera, A., Osman, N. I., & Chapple, C. R. (2013). Anatomy of the lower urinary tract. *Surgery, 31*(7), 319–325. https://doi.org/10.1016/j.mpsur.2013.04.013

McMurry, S. T., Lochmiller, R. L., Chandra, S. A., & Qualls, C. W. (1995). Sensitivity of selected immunological, hematological, and reproductive parameters in the cotton rat (*Sigmodon hispidus*) to subchronic lead exposure. *Journal of Wildlife Diseases, 31*(2), 193–204. https://doi.org/10.7589/0090-3558-31.2.193

Medicine, Institute of, Board on Health Sciences Policy, and Committee on Understanding the Biology of Sex and Gender Differences. (2001). *Exploring the biological contributions to human health: Does sex matter?* National Academies Press.

Mills, A. C., & Sengelaub, D. R. (1993). Sexually dimorphic neuron number in lumbosacral dorsal root ganglia of the rat: Development and steroid regulation. *Journal of Neurobiology, 24*(11), 1543–1553. https://doi.org/10.1002/neu.480241108

Morita, T., Masuda, H., Tosaka, A., Ishizaka, K., Tsujii, T., & Kondo, S. (1998). Sex differences in function and distribution of beta-adrenoceptors in rabbit urinary bladder. *The Journal of Urology, 159*(2), 555–558. https://doi.org/10.1016/s0022-5347(01)63982-x

Nakamura, N., Vijay, V., Desai, V. G., Hansen, D. K., Han, T., Chang, C.-W., Chen, Y.-C., Harrouk, W., McIntyre, B., Foster, P. M., & Fuscoe, J. C. (2018). Transcript profiling in the testes and prostates of postnatal day 30 Sprague-Dawley rats exposed prenatally and lactationally to 2-hydroxy-4-methoxybenzophenone. *Reproductive Toxicology, 82*, 111–123. https://doi.org/10.1016/j.reprotox.2018.10.001

Nappi, R. E., & Davis, S. R. (2012). The use of hormone therapy for the maintenance of urogynecological and sexual health post WHI. *Climacteric: The Journal of the International Menopause Society, 15*(3), 267–274. https://doi.org/10.3109/13697137.2012.657589

Nauman, C. H., & Schaum, J. L. (1987). Human exposure estimation for 2,3,7,8-TCDD. *Chemosphere, 16*(8–9), 1851–1856. https://doi.org/10.1016/0045-6535(87)90176-7

Nicholson, T. M., Moses, M. A., Uchtmann, K. S., Keil, K. P., Bjorling, D. E., Vezina, C. M., Wood, R. W., & Ricke, W. A. (2015). Estrogen receptor-α is a key mediator and therapeutic target for bladder complications of benign prostatic hyperplasia. *The Journal of Urology, 193*(2), 722–729. https://doi.org/10.1016/j.juro.2014.08.093

Nicholson, T. M., Nguyen, J. L., Leverson, G. E., Taylor, J. A., Vom Saal, F. S., Wood, R. W., & Ricke, W. A. (2018). Endocrine disruptor bisphenol A is implicated in urinary voiding dysfunction in male mice. *American Journal of Physiology-Renal Physiology, 315*(5), F1208–F1216. https://doi.org/10.1152/ajprenal.00582.2017

Nicholson, T. M., Ricke, E. A., Marker, P. C., Miano, J. M., Mayer, R. D., Timms, B. G., vom Saal, F. S., Wood, R. W., & Ricke, W. A. (2012). Testosterone and 17β-estradiol induce glandular prostatic growth, bladder outlet obstruction, and voiding dysfunction in male mice. *Endocrinology, 153*(11), 5556–5565. https://doi.org/10.1210/en.2012-1522

Nitti, V. W. (2001). The prevalence of urinary incontinence. *Reviews in Urology, 3*(Suppl. 1), S2–S6.

Pan, Z., Yang, H., & Reinach, P. S. (2011). Transient receptor potential (TRP) gene superfamily encoding cation channels. *Human Genomics, 5*(2), 108–116. https://doi.org/10.1186/1479-7364-5-2-108

Pohóczky, K., Kun, J., Szalontai, B., Szőke, É., Sághy, É., Payrits, M., Kajtár, B., Kovács, K., Környei, J. L., Garai, J., Garami, A., Perkecz, A., Czeglédi, L., & Helyes, Z. (2016). Estrogen-dependent up-regulation of TRPA1 and TRPV1 receptor proteins in the rat endometrium. *Journal of Molecular Endocrinology, 56*(2), 135–149. https://doi.org/10.1530/JME-15-0184

Preslock, J. P. (1980a). A review of in vitro testicular steroidogenesis in rodents, monkeys and humans. *Journal of Steroid Biochemistry, 13*(8), 965–975. https://doi.org/10.1016/0022-4731(80)90172-7

Preslock, J. P. (1980b). A review of in vitro testicular steroidogenesis in rodents, monkeys and humans. *Journal of Steroid Biochemistry, 13*(8), 90172–90177. https://doi.org/10.1016/0022-4731(80

Ratz, P. H., McCammon, K. A., Altstatt, D., Blackmore, P. F., Shenfeld, O. Z., & Schlossberg, S. M. (1999). Differential effects of sex hormones and phytoestrogens on peak and steady state contractions in isolated rabbit detrusor. *The Journal of Urology, 162*(5), 1821–1828. https://doi.org/10.1016/S0022-5347(05)68245-6

Ricke, W. A., Lee, C. W., Clapper, T. R., Schneider, A. J., Moore, R. W., Keil, K. P., Abler, L. L., Wynder, J. L., Alvarado, A. L., Beaubrun, I., Vo, J., Bauman, T. M., Ricke, E. A., Peterson, R. E., & Vezina, C. M. (2016). In utero and lactational TCDD exposure increases susceptibility to lower urinary tract dysfunction in adulthood. *Toxicological Sciences, 150*(2), 429–440. https://doi.org/10.1093/toxsci/kfw009

Roehrborn, C. G. (2008). Pathology of benign prostatic hyperplasia. *International Journal of Impotence Research, 20*(Suppl. 3), S11–S18. https://doi.org/10.1038/ijir.2008.55

Roehrborn, C. G., Spann, M. E., Myers, S. L., Serviss, C. R., Hu, L., & Jin, Y. (2015). Estrogen receptor beta agonist LY500307 fails to improve symptoms in men with enlarged prostate secondary to benign prostatic hypertrophy. *Prostate Cancer and Prostatic Diseases, 18*(1), 43–48. https://doi.org/10.1038/pcan.2014.43

Rüegg, J. C. (2012). *Calcium in muscle contraction: Cellular and molecular physiology*. Springer Science & Business Media.

Ruetten, H., Wegner, K. A., Kennedy, C. L., Turco, A., Zhang, H. L., Wang, P., Sandhu, J., Sandhu, S., Morkrid, J., Wang, Z., Macoska, J., Peterson, R. E., Bjorling, D. E., Ricke, W. A., Marker, P. C., & Vezina, C. M. (2020). Impact of sex, androgens, and prostate size on C57BL/6J mouse urinary physiology: Urethral histology. *American Journal of Physiology-Renal Physiology, 318*(3), F617–F627. https://doi.org/10.1152/ajprenal.00540.2019

Ruetten, H., Wegner, K. A., Zhang, H. L., Wang, P., Sandhu, J., Sandhu, S., Mueller, B., Wang, Z., Macoska, J., Peterson, R. E., & Bjorling, D. E. (2019). Impact of sex, androgens, and prostate size on C57BL/6J mouse urinary physiology: Functional assessment. *American Journal of Physiology-Renal Physiology, 317*(4), F996–F1009. https://doi.org/10.1152/ajprenal.00270.2019

Saad, F., Gooren, L., Haider, A., & Yassin, A. (2007). An exploratory study of the effects of 12 month administration of the novel long-acting testosterone undecanoate on measures of sexual function and the metabolic syndrome. *Archives of Andrology, 53*(6), 353–357. https://doi.org/10.1080/01485010701730880

Sacco, F., Rigon, G., Carbone, A., & Sacchini, D. (1990). Transvaginal estrogen therapy in urinary stress incontinence. *Minerva Ginecologica, 42*(12), 539–544.

Schulze, H., & Barrack, E. R. (1987). Immunocytochemical localization of estrogen receptors in the normal male and female canine urinary tract and prostate. *Endocrinology, 121*(5), 1773–1783. https://doi.org/10.1210/endo-121-5-1773

Schweikert, H. U., & Tunn, U. W. (1987). Effects of the aromatase inhibitor testolactone on human benign prostatic hyperplasia. *Steroids, 50*(1−3), 191−200. https://doi.org/10.1016/0039-128x(83)90071-5

Shabsigh, A., Kang, Y., Shabsign, R., Gonzalez, M., Liberson, G., Fisch, H., & Goluboff, E. (2005). Clomiphene citrate effects on testosterone/estrogen ratio in male hypogonadism. *The Journal of Sexual Medicine, 2*(5), 716−721. https://doi.org/10.1111/j.1743-6109.2005.00075.x

Shermadou, E. S., Rahman, S., & Leslie, S. W. (2020). Anatomy, abdomen and pelvis, bladder. In *StatPearls*. Treasure Island (FL): StatPearls Publishing. http://www.ncbi.nlm.nih.gov/books/NBK531465/.

Shortliffe, L. M. D., Ye, Y., Behr, B., & Wang, B. (2014). Testosterone changes bladder and kidney structure in juvenile male rats. *The Journal of Urology, 191*(6), 1913−1919. https://doi.org/10.1016/j.juro.2014.01.012

Sigala, S., Peroni, A., Mirabella, G., Fornari, S., Palazzolo, F., Pezzotti, G., Simeone, C., Cunico, S. C., & Spano, P. (2004). Alpha1 adrenoceptor subtypes in human urinary bladder: Sex and regional comparison. *Life Sciences, 76*(4), 417−427. https://doi.org/10.1016/j.lfs.2004.09.008

Stanton, M. C., Christopher Austin, J., Delaney, D. P., Gosfield, A., Marx, J. O., Zderic, S. A., Chacko, S., & Moreland, R. S. (2006). Partial bladder outlet obstruction selectively abolishes protein kinase C induced contraction of rabbit detrusor smooth muscle. *The Journal of Urology, 176*(6 Pt 1), 2716−2721. https://doi.org/10.1016/j.juro.2006.07.137

Stewart, W. F., Van Rooyen, J. B., Cundiff, G. W., Abrams, P., Herzog, A. R., Corey, R., Hunt, T. L., & Wein, A. J. (2003). Prevalence and burden of overactive bladder in the United States. *World Journal of Urology, 20*(6), 327−336. https://doi.org/10.1007/s00345-002-0301-4

Takyu, S. (1993). Effects of testosterone on the autonomic receptor-mediated function in lower urinary tract from male rabbits. *Nihon Hinyokika Gakkai Zasshi. The Japanese Journal of Urology, 84*(2), 330−338. https://doi.org/10.5980/jpnjurol1989.84.330

Tanidir, Y., Ercan, F., & Tarcan, T. (2011). Exogenous testosterone and estrogen affect bladder tissue contractility and histomorphology differently in rat ovariectomy model. *The Journal of Sexual Medicine, 8*(6), 1626−1637. https://doi.org/10.1111/j.1743-6109.2011.02227.x

Tatemichi, S., Akiyama, K., Kobayashi, M., Yamazaki, Y., Yokoyama, O., & Uruno, T. (2006). A selective α1A-adrenoceptor antagonist inhibits detrusor overactivity in a rat model of benign prostatic hyperplasia. *The Journal of Urology*. https://www.auajournals.org/doi/abs/10.1016/j.juro.2006.04.029.

Taylor, J. A., Jones, M. B., Besch-Williford, C. L., Berendzen, A. F., Ricke, W. A., & Vom Saal, F. S. (2020). Interactive effects of perinatal BPA or DES and adult testosterone and estradiol exposure on adult urethral obstruction and bladder, kidney, and prostate pathology in male mice. *International Journal of Molecular Sciences, 21*(11). https://doi.org/10.3390/ijms21113902

Tek, M., Balli, E., Cimen, B., Efesoy, O., Oğuz, I., & Cayan, S. (2010). The effect of testosterone replacement therapy on bladder functions and histology in orchiectomized mature male rats. *Urology, 75*(4), 886−890. https://doi.org/10.1016/j.urology.2009.08.016

Thornton, L. A., Burchell, R. K., Burton, S. E., Lopez-Villalobos, N., Pereira, D., MacEwan, I., Fang, C., Hatmodjo, A. C., Nelson, M. A., Grinberg, A., & Velathanthiri, N. (2018). The effect of urine concentration and PH on the growth of *Escherichia coli* in canine urine in vitro. *Journal of Veterinary Internal Medicine, 32*(2), 752−756. https://doi.org/10.1111/jvim.15045

Townsend, M. K., Curhan, G. C., Resnick, N. M., & Grodstein, F. (2009). Postmenopausal hormone therapy and incident urinary incontinence in middle-aged women. *American Journal of Obstetrics and Gynecology, 200*(1), 86.e1−86.e5. https://doi.org/10.1016/j.ajog.2008.08.009

Turco, A. E., Thomas, S., Crawford, L. K., Tang, W., Peterson, R. E., Li, L., Ricke, W. A., & Vezina, C. M. (2020). In utero and lactational 2,3,7,8-tetrachlorodibenzo-p-dioxin (TCDD) exposure exacerbates urinary dysfunction in hormone-treated C57BL/6J mice through a non-malignant mechanism involving proteomic changes in the prostate that differ from those elicited by testosterone and estradiol. *American Journal of Clinical and Experimental Urology, 8*(1), 59−72.

Valeri, A., Brain, K. L., Young, J. S., Sgaragli, G., & Pessina, F. (2009). Effects of 17β-oestradiol on rat detrusor smooth muscle contractility. *Experimental Physiology, 94*(7), 834−846. https://doi.org/10.1113/expphysiol.2009.047118

Valeri, A., Fiorenzani, P., Rossi, R., Aloisi, A. M., Valoti, M., & Pessina, F. (2012). The soy phytoestrogens genistein and daidzein as neuroprotective agents against anoxia-glucopenia and reperfusion damage in rat urinary bladder. *Pharmacological Research, 66*(4), 309−316. https://doi.org/10.1016/j.phrs.2012.06.007

Varella, L. R. D., da Silva, R. B., de Oliveira, M. C. E., Melo, P. H. A., de Oliveira Maranhão, T. M., & Micussi, M. T. A. B. C. (2016). Assessment of lower urinary tract symptoms in different stages of menopause. *Journal of Physical Therapy Science, 28*(11), 3116−3121. https://doi.org/10.1589/jpts.28.3116

Veronesi, M. C., Rota, A., Battocchio, M., Faustini, M., & Mollo, A. (2009). Spaying-related urinary incontinence and oestrogen therapy in the bitch. *Acta Veterinaria Hungarica, 57*(1), 171−182. https://doi.org/10.1556/AVet.57.2009.1.17

Vignozzi, L., Filippi, S., Morelli, A., Comeglio, P., Cellai, I., Sarchielli, E., Maneschi, E., Gacci, M., Vannelli, G. B., & Maggi, M. (2012). Testosterone/estradiol ratio regulates NO-induced bladder relaxation and responsiveness to PDE5 inhibitors. *The Journal of Sexual Medicine, 9*(12), 3028−3040. https://doi.org/10.1111/j.1743-6109.2012.02946.x

Walsh, P. C., & Wilson, J. D. (1976). The induction of prostatic hypertrophy in the dog with androstanediol. *The Journal of Clinical Investigation, 57*(4), 1093−1097. https://doi.org/10.1172/JCI108353

Walz, T., Häner, M., Wu, X. R., Henn, C., Engel, A., Sun, T. T., & Aebi, U. (1995). Towards the molecular architecture of the asymmetric unit membrane of the mammalian urinary bladder epithelium: A closed 'twisted ribbon' structure. *Journal of Molecular Biology, 248*(5), 887−900. https://doi.org/10.1006/jmbi.1995.0269

Wang, J.-Y., Liao, L., Liu, M., Sumarsono, B., & Cong, M. (2018). Epidemiology of lower urinary tract symptoms in a cross-sectional, population-based study. *Medicine, 97*(34). https://doi.org/10.1097/MD.0000000000011554

Wang, C., Symington, J. W., Ma, E., Cao, B., & Mysorekar, I. U. (2013). Estrogenic modulation of uropathogenic *Escherichia coli* infection pathogenesis in a murine menopause model. *Infection and Immunity, 81*(3), 733−739. https://doi.org/10.1128/IAI.01234-12

Winter, M. L., Bosland, M. C., Wade, D. R., Falvo, R. E., Nagamani, M., & Liehr, J. G. (1995). Induction of benign prostatic hyperplasia in intact dogs by near-physiological levels of 5 alpha-dihydrotestosterone and 17 beta-estradiol. *The Prostate, 26*(6), 325–333. https://doi.org/10.1002/pros.2990260608

Wu, X., Guo, X., Wang, H., Zhou, S., Li, L., Chen, X., Wang, G., Liu, J., Ge, H.-S., & Ge, R.-S. (2017). A brief exposure to cadmium impairs Leydig cell regeneration in the adult rat testis. *Scientific Reports, 7*(1), 6337. https://doi.org/10.1038/s41598-017-06870-0

Xu, S., Cheng, Y., Keast, J. R., & Osborne, P. B. (2008). 17β-estradiol activates estrogen receptor β-signalling and inhibits transient receptor potential vanilloid receptor 1 activation by capsaicin in adult rat nociceptor neurons. *Endocrinology, 149*(11), 5540–5548. https://doi.org/10.1210/en.2008-0278

Yamaguchi, K., Kobayashi, M., Kato, T., & Akita, K. (2011). Origins and distribution of nerves to the female urinary bladder: New anatomical findings in the sex differences. *Clinical Anatomy, 24*(7), 880–885. https://doi.org/10.1002/ca.21186

Yang, Y., Ozawa, H., Lu, H., Yuri, K., Hayashi, S., Nihonyanagi, K., & Kawata, M. (1998). Immunocytochemical analysis of sex differences in calcitonin gene-related peptide in the rat dorsal root ganglion, with special reference to estrogen and its receptor. *Brain Research, 791*(1), 35–42. https://doi.org/10.1016/S0006-8993(98)00021-3

Yassin, D.-J., Doros, G., Hammerer, P. G., & Yassin, A. A. (2014). Long-term testosterone treatment in elderly men with hypogonadism and erectile dysfunction reduces obesity parameters and improves metabolic syndrome and health-related quality of Life. *The Journal of Sexual Medicine, 11*(6), 1567–1576. https://doi.org/10.1111/jsm.12523

Yassin, A. A., El-Sakka, A. I., Saad, F., & Gooren, L. J. G. (2008). Lower urinary-tract symptoms and testosterone in elderly men. *World Journal of Urology, 26*(4), 359–364. https://doi.org/10.1007/s00345-008-0284-x

Yasui, M., Kawahara, T., Takamoto, D., Izumi, K., Uemura, H., & Miyamoto, H. (2019). Distribution of androgen receptor expression in the urinary bladder. *International Journal of Urology, 26*(2), 305–306. https://doi.org/10.1111/iju.13841

Yiloren, T., Feriha, E., & Tufan, T. (2011). Exogenous testosterone and estrogen affect bladder tissue contractility and histomorphology differently in rat ovariectomy model. *The Journal of Sexual Medicine*, 1626–1637. https://doi.org/10.1111/j.1743-6109.2011.02227.x

Zhou, Y., Xiao, X.-Q., Chen, L.-F., Yang, R., Shi, J.-D., Du, X.-L., Klocker, H., Park, I., Lee, C., & Zhang, J. (2009). Proliferation and phenotypic changes of stromal cells in response to varying estrogen/androgen levels in castrated rats. *Asian Journal of Andrology, 11*(4), 451–459. https://doi.org/10.1038/aja.2009.28

Further reading

Artero-Morales, Maite, González-Rodríguez, Sara, & Ferrer-Montiel, Antonio (2018). TRP channels as potential targets for sex-related differences in migraine pain. *Frontiers in Molecular Biosciences, 5*. https://doi.org/10.3389/fmolb.2018.00073

Chapter 8

Transcriptomic identification of cell types in the lower urinary tract

Douglas Strand
University of Texas Southwestern Medical Center, Department of Urology, Dallas, TX, United States

The cell is the fundamental unit of life. Historically, cell type was defined by morphology, location, and immunopositivity for 1 or 2 antibodies. With the revolution of single-cell RNA sequencing (scRNA-seq), cell type is now defined by an objective, complete molecular profile. It is true that cells can be reprogrammed and achieve alternate differentiation, which suggests that cell "type" is illusory and only a spectrum of cell "states" exist. But these transitions are typically achieved artificially or in advanced disease. An ontological model of normal cell types based on molecular signature and anatomical location is more pragmatic for understanding the function of a cell within its environment (Diehl et al., 2016; Mungall et al., 2012) (Fig. 8.1).

Tools for the objective transcriptomic identification of cell types are rapidly advancing. This chapter is intended to provide a brief overview of currently available technologies as well as several protocols and analysis tools for identification of cell types in the lower urinary tract. By categorizing the identity and location of cell types in the lower urinary tract, we will gain new insight into the control of organ function by the nervous system. The ultimate goal of consortiums such as the Human BioMolecular Atlas Project (hubmapconsortium.com) and the Human Cell Atlas (humancellatlas.org) (Rozenblatt-Rosen et al., 2021) is to "create the next generation of molecular analysis technologies and computational tools, enabling the generation of foundational 3D tissue maps and construction of an atlas of the function and relationships among cells in the human body." Once a broad characterization of the molecular identity of each cell type is completed with transcriptomics, more detailed analyses of the categorical function of cellular subpopulations can be determined with multiomics approaches. Detailed protocols and code for bioinformatics analysis are made available in an effort to facilitate rapid setup in any laboratory. Some of these approaches require advanced expertise and hardware. Furthermore, access to fresh tissues can be difficult to coordinate for single-cell analysis given the sometimes lengthy processing times. The pipelines detailed here will provide a clear-eyed perspective of the hurdles ahead, but it should also be noted that many (low cost) discoveries can be made using bioinformatics tools on data already produced and many new technologies that bypass the need for fresh tissue are available such as single nucleus profiling of frozen tissue.

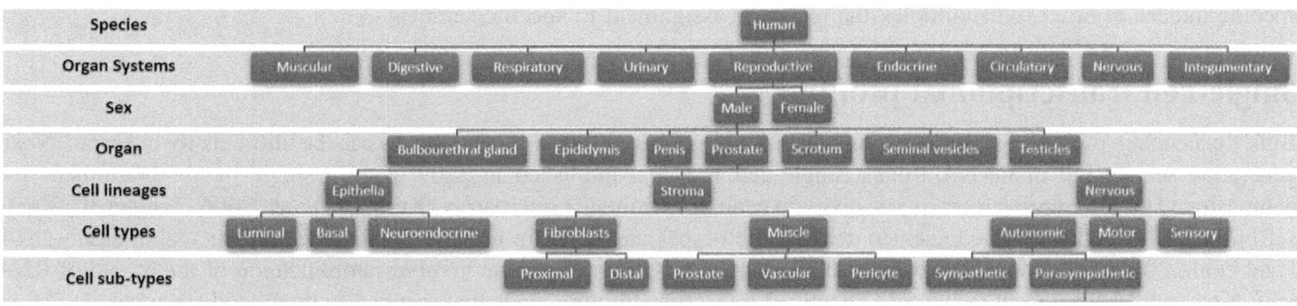

FIGURE 8.1 Example of the ontological categorization of human prostate cell types.

Neuro-Urology Research. https://doi.org/10.1016/B978-0-12-822455-7.00002-7

Bulk transcriptional profiling

To understand the molecular differences between two samples, that is, "differentially expressed genes" (DEGs), RNA sequencing was developed in the mid-2000s. This technology supplanted microarrays, which used hybridization to capture and count known transcripts. Instead, RNA sequencing captures each nucleotide of adapter ligated transcripts, and then aligns demultiplexed transcripts with the appropriate genome for identification and counting. A standard workflow starts with RNA extraction and purification, cDNA synthesis, and adapter ligation, which creates a library of transcripts with sample-specific barcodes that can be identified and counted after short-read sequencing. Long read sequencing is suited for more complex analyses such as RNA isoform identification, which is typically unnecessary for cell identification.

Example of bulk RNA sequencing protocol

- Extract mRNA with a method appropriate for amount of expected yield.
  1. High input (>100 ng): Follow method described in Hou et al. (2015).
  2. Low input (10–100 ng): Follow method described in Takara SMARTer Stranded Total RNA-Seq Kit (Cat #634861) with RiboGone to remove ribosomal RNA (Cat #634847).
- Quantify RNA concentration.
- Synthesize first-strand cDNA.
- Purify first-strand cDNA with solid-phase reversible immobilization (SPRI) AMPure beads.
- Amplify RNA-seq library by polymerase chain reaction (PCR).
- Purify RNA-seq library.
- Quantify DNA concentration for Illumina sequencing.

A Qubit fluorometer can be used to determine the concentration of RNA prior to library preparation. Total RNA is depleted of ribosomal RNA and fragmented before cDNA synthesis. cDNAs are then A-tailed and sample-specific, indexed adapters are ligated. The addition of polyAs to the blunt end of cDNA prepares them for ligation to the sequencing adapters, which contain polyTs and can be indexed to allow for the sequencing of multiple samples at a time. Indexed cDNA are PCR amplified and purified with beads. Before sequencing, the quality (Bioanalyzer) and quantity (Qubit) must once again be assessed so that a normalized concentration of each indexed sample can be sequenced. Sequencing is typically performed on Illumina platforms with options based on the length and number of sequencing reads required. Just a few years ago, a minimum of 1–2 mg of starting material was required, but now low input kits make it possible to sequence as little as 250 pg of RNA.

The pool of single-stranded DNA is loaded onto a flow cell that includes reservoirs of fluorescently labeled nucleotides. After each cycle of sequencing on an Illumina sequencer, the growing double-stranded DNA molecule is imaged to determine which of the four fluorescent nucleotides was added. Short-read sequencers can be programmed to sequence custom lengths. The vast majority of RNA transcripts can be aligned with 75 base pair length. Typically, a depth of 15–20 million reads per sample is suggested but depends on the experiment. The number of reads per sample will depend on the type of sequencer and the amount of input. The reads are spread across the entire transcriptome and the read depth required will depend on whether lowly expressed genes of interest are adequately sampled. Paired end sequencing of 150 base pairs certainly provides more information on longer variants, but single-end sequencing with 75 base pair reads is sufficient for alignment in many cases. A lane of sequencing on a NextSeq 550 in high output mode provides ~400 million reads, which means that ~25 bulk libraries can be sequenced together. This, of course, requires the use of 25 distinct, sample/library-specific indices in order to demultiplex the reads for assignment to specific samples.

Single-cell transcriptional profiling

Bulk sequencing pieces of tissue that contain a confounding number of cell types can be difficult to interpret. Next-generation sequencing of mRNA from a single cell was first published in 2009 using a single mouse blastomere and demonstrated a 75% improvement in sensitivity in gene detection over microarray (Kurimoto et al., 2006; Tang et al., 2009). scRNA-seq has produced unprecedented detail on cell types and states in normal and diseased organs (Jaitin et al., 2014; Patel et al., 2014; Treutlein et al., 2014; Villani et al., 2017). This technique involves amplification of the 3′ end of RNA transcripts, while subsequent techniques quantified transcripts through reads that mapped to the 5′ end (Islam et al., 2011). Alternatively, SMART-seq is a technique that is lower throughput (predominantly due to its high cost per cell and the need to sort an individual cell into an individual well), but can reliably read the full length of transcripts (Ramskold et al., 2012).

This approach offers the advantage of being able to identify alternative transcript isoforms and single-nucleotide polymorphisms, but this is generally unnecessary for the identification of cell type.

Each cell in a simple cellular system such as an 8-cell blastomere can be distinguished by relatively high transcriptional differences. However, most complex cellular systems such as an organ require at minimum 20–30 cells of an individual type for proper clustering and identification of subpopulations. Amplification of picograms of RNA from an individual cell in a microplate well (SMART-seq) is time consuming and expensive (Fan et al., 2015; Han et al., 2018; Ramskold et al., 2012). Droplet-based sequencing with advanced microfluidics approaches such as inDrop (Klein et al., 2015) and Drop-seq (Macosko et al., 2015) dramatically increased the number of cells that could be profiled by encapsulating thousands of individual cells in a nanoliter-sized oil droplet that acts as a tiny PCR chamber for incorporating unique molecular identifiers (UMIs) into each transcript. While these systems were laborious and technically challenging to set up in a lab, in 2012, commercially available platforms such as the Fluidigm C1 made it possible to profile 96 cells at a time. Because of the demand for higher throughput systems, the 10X Genomics Chromium Controller and the Bio-Rad ddSEQ became commercially available in 2016. These platforms dramatically improved the number of cells that could be profiled with a single cartridge (80,000 and 1,600 cells per cartridge, respectively). However, concerns with high doublet rates (two cells in a single droplet) and poor encapsulation efficiency (as low as 4% of input cells) with lab-based setups like Drop-seq and commercial platforms such as the Bio-Rad ddSEQ have allowed the 10x Genomics platform to dominate (with up to 65% capture efficiency). This means that precious and small specimens can be reliably profiled without losing the majority of cells during the encapsulation process. Two other commercial droplet-based platforms are the Takara ICell8, which is lower throughput than the 10x Genomics Chromium Controller, but offers the ability to perform full-length RNA sequencing using SMART-seq2 chemistry (Picelli et al., 2014) and the Mission Bio Tapestri, which unlike the 10x Genomics platform offers a multiomic DNA (targeted) and protein profiling capability, which is more useful for clinical oncology applications where the search is for rare cell types. Stuart and Satija provide a comprehensive review of uni- and multimodal single-cell technologies (Stuart & Satija, 2019).

The transcriptomic analysis of digested cell suspensions has its limitations. The digestion protocol itself leads to artificial stress signatures that can mask the true identity of the cell (van den Brink et al., 2017). Typically, fresh tissues are minced and digested in a certain type of enzyme for an optimized time period of just a few minutes up to a few hours. Most of these enzymes are active at 37°C. Extended periods of time at this high temperature can cause cellular stress. Cold proteases are enzymatically active at 6°C and can reduce digestion-induced cellular stress, but are not suitable for all types of tissue and are significantly more expensive (Adam et al., 2017). A direct comparison of heat-mediated enzymatic digestion at 37°C versus digestion with a psychrophilic protease at 6°C of the mouse kidney revealed that cold protease digestion reduces stress artifact, but fails to capture certain cell populations at expected proportions (Denisenko et al., 2020). It was also demonstrated that cryopreservation prior to scRNA-seq resulted in high stress and 30% fewer high-quality cells versus performing scRNA-seq on freshly isolated cells, with cryopreservation negatively affecting epithelia in particular (Denisenko et al., 2020). If using enzymatic digestion at 37°C, the investigator must optimize the type and concentration of enzyme used and the incubation time necessary to achieve release of the desired cell types without inducing so much stress that the cells have to be removed from the analysis in silico (van den Brink et al., 2017). A standardized protocol for tissue digestion is shown below and more detailed protocols are available (Goldstein et al., 2011).

Example of tissue digestion protocol

1. Cut and mince tissue in plate with cold phosphate-buffered saline (PBS) + 0.1% bovine serum albumin (BSA). Transfer into a tube with the capacity to hold no more than 0.1 g of minced tissue per ml digestion solution.
2. Spin down at 1800 rpm at 4°C.
3. Aspirate supernatant and wash in cold Hank's balanced salt solution (HBSS).
4. Spin down and resuspend in collagenase digestion solution in HBSS.

Reagents: Collagenase I (Gibco 17100-017), ROCK inhibitor Y-27632 (StemRD Y-001), DNase I (Roche 10104159001) 125 U/mL

5. Seal tube caps with parafilm. Incubate on rotating platform for 15 min to 4 h.
6. Spin down and resuspend in PBS to wash.
7. Spin down and resuspend in TrypLE (ThermoFisher 12605010). Incubate for 5 min in 37°C water bath.
8. Add Dulbecco's modified Eagle's medium (DMEM) + 10% fetal bovine serum (FBS) to quench trypsin.
9. Spin down and resuspend in DMEM + 10% FBS.

10. Break up small chunks by aspirating 2–3 times each through 18, 20, and 22 gauge needles attached to a 10 mL syringe barrel.
11. Filter through a 100 μm cell strainer.
12. Spin down and resuspend in 1x RBC Lysis Buffer (diluted from 10x stock in cell culture grade H_2O). Incubate for 5 min on ice.
13. Add PBS and resuspend to wash.
14. Spin down and resuspend in DMEM +10% FBS.
15. Filter through a 40 μm cell strainer.
16. Use hemocytometer to count.
17. At this stage, cells can be frozen down, used for single-cell RNA-seq or other techniques.
18. *To freeze down cells*: Spin down and resuspend pellet in 10% dimethyl sulfoxide (DMSO) + 90% FBS. Transfer to cryovials and place into cryocontainer and freeze at −80°C overnight. Transfer to liquid nitrogen storage.

Because of the limitations of encapsulation and digestion, droplet-based transcriptomic profiling of single cells is limited by cell size and shape. The 10x Chromium Controller cannot encapsulate a cell larger than ~40 um inside a nanoliter-sized oil droplet. This means that relatively large cell types such as muscle can be underrepresented. Furthermore, enzymatic digestion cannot capture the body of particular cell types, such as neurons with long processes, without damaging the cell membrane and spilling the internal RNA. With the artifact created by cell size and shape, and to circumvent enzymatic digestion, an alternative technique for profiling the RNA contained within the nucleus can be used on organs such as the heart and brain (Cui et al., 2020; Grindberg et al., 2013; Lacar et al., 2016). Nuclei can be isolated for high-throughput droplet-based platforms or for SMART-seq analysis of full length RNA on a limited number of cells (Lacar et al., 2016). Alternatively, anatomically dissected areas of interest can be embedded in optimal cutting temperature (OCT) and cut into 40-micron-thick sections and enzymatically digested for a short incubation time to release nuclei (Hu et al., 2017; Lake et al., 2019). While the sequencing of nuclear RNA is required for certain cell types, it should be noted that the diversity of RNA transcripts is generally reduced compared to profiling RNA from whole cells. A direct comparison of single-cell and single-nuclei datasets from matched tissue shows that immune cells are underrepresented in snRNA-seq data; alternatively, certain rare stromal cell types are only detectable with snRNA-seq because of the harsh conditions of enzymatic digestion (Denisenko et al., 2020; Wu et al., 2019). In summary, the choice of hot versus cold enzyme, single cell versus single nuclei, and Drop-seq versus SMART-seq protocols is highly influenced by desired cell types to be profiled, availability of fresh tissue, cost, labor, and desired type of RNA to be profiled.

Overview of single-cell RNA sequencing protocol

Whether RNA transcripts have been captured through whole cell or nuclear preparation, cDNA libraries must be produced for sequencing. Library production can be costly and time consuming. Most core laboratories offer this as a service with variable turnaround times and a high cost per library due to the amount of labor. If the plan is to profile more than a few samples, it is more efficient to perform library production in the lab. The most important consideration is the amount of time needed to produce each library. A single technician can reliably produce a small batch of libraries over the course of 2 days. As noted above, the process of tissue digestion and the cleanup of the single-cell suspension can take from 1 to 5 h depending on the tissue type. Once a single-cell or single-nucleus suspension is ready for encapsulation in massively parallel droplet-based protocols or isolated in individual wells in plate-based protocols, the next steps are relatively common between platforms with more detailed protocols available from manufacturers.

The first step in massively parallel droplet protocols is the encapsulation of individual cells with a unique poly(dT) primer that will incorporate an Illumina sequencing primer, a barcode, and a UMI into each transcript sequenced in that cell. With commercial platforms, the encapsulation can finish within minutes and is followed by a 55 min reverse transcription step to produce barcoded, full-length cDNA from poly-adenylated mRNA. The first-strand cDNA can be stored for up to a week before library construction.

In the second step, encapsulated cells are vortexed and broken, and silane magnetic beads are used to purify first-strand cDNA for cleanup and cDNA amplification. After several steps of magnetic size selection and cleanup, the pool of cDNA must be quantified. Cleanup and cDNA amplification take about 2 h and can be paused at this point for several weeks.

The final step of library construction takes 6–8 h and uses fragmentation, end repair, and A-tailing for the incorporation of a sample index and the P5, P7, and read 2 primer sequences needed for Illumina sequencers. The libraries produced are compatible with a variety of Illumina sequencers, the selection of which depends on the number of reads required. Because of the exhaustive number of steps and time needed for library construction, many will choose to store samples until enough are accumulated to perform the steps on several samples at once. The accumulation of samples also has the advantage of cost savings on sequencing.

Overview of bioinformatics

The analysis of uni- or multimodal single-cell data is a rapidly evolving field and has been covered extensively by many experts who provide tutorials and guidance for the new user (Amezquita et al., 2020; Andrews et al., 2021; Stuart & Satija, 2019; Stuart et al., 2019). Once the reads are demultiplexed from the sequencer, an expression matrix is created, which represents the number of transcripts counted in each cell and will be used for clustering and visualization. Unlike SMART-seq, the shallow sequencing depth of massively parallel droplet-based scRNA-seq can result in a zero count for a high percentage of the total genes. This high "dropout" rate can confound downstream analyses and therefore several quality control steps are required.

Before filter steps can be applied to the data, a gene-cell expression matrix must be created from the FASTQ files converted from the raw base call (BCL) files from Illumina sequencers. For libraries created with the 10x Genomics platform, the *cellranger mkfastq* pipeline can be downloaded for demultiplexing BCL files into FASTQ files. The *cellranger count* pipeline then takes FASTQ files and uses STAR (Dobin et al., 2013) to align reads to the appropriate genome(s) of interest so that UMIs associated with a specific gene can be counted and associated with a specific cell barcode. A tutorial for using the Cell Ranger analysis pipeline (Zheng et al., 2017) from 10x Genomics can be found at https://support.10xgenomics.com/single-cell-gene-expression/software/pipelines/latest/what-is-cell-ranger. Other Bioconductor software packages in R that can be run on command line to create an expression matrix are also available (Amezquita et al., 2020).

The first quality control step is to filter out barcodes that are not associated with a cell ("ambient" RNA) as well as those barcodes that are from a "multiplet" (two or more cells in one droplet). The 10x Genomics Chromium Controller has a ~65% efficiency of encapsulating a cell in an oil droplet, which means that many of the droplets in the final pool are empty. Ambient RNA can be inserted into empty droplets due to cell breakdown during digestion that has not been properly cleaned from the cell suspension. Additionally, the percentage of multiplets increases proportionally with the number of cells loaded. For each dataset, a minimum and a maximum threshold of UMIs is set to establish that a barcode is associated with a cell and is not due to ambient RNA or multiplets. The danger of setting this lower UMI threshold too high is that healthy cell types that have relatively low RNA levels (such as leukocytes) can be inappropriately removed from the data. The danger of setting the higher UMI threshold too low is that cells with naturally high RNA content can be inappropriately removed.

After read mapping and gene expression quantification, uneven sequencing depths (Hafemeister & Satija, 2019) and batch effects (Stuart et al., 2019) between samples need to be normalized. With the high number of genes displaying zero values due to dropouts, imputing the missing values based on mathematics-driven theory (van Dijk et al., 2018) or external datasets (Wang et al., 2019) can improve the quality of the data, although validation is essential since these are not real values.

The next set of quality control steps are highly dependent on the type of tissue and experimental questions. Only the highest variable genes are used for clustering (Macosko et al., 2015), so if multiple different cell types are stressed or proliferating, they may cluster based on gene expression signatures related to those cell states rather than clustering based on the identity of the cell types. Regressing out known stress (van den Brink et al., 2017; Henry et al., 2018) and cell cycle (Tirosh et al., 2016) signatures can unmask the true molecular identity of the cell without removing the entire cell from the dataset. Additional filtering steps prior to dimensionality reduction involve the removal of barcodes associated with damaged or dying cells. A high proportion of genes detected from the mitochondrial genome and a low expression of genes per cell are indicators of damaged or dying cells (Ilicic et al., 2016). It is important to note that some cell types such as muscle have a naturally higher proportion of mitochondrial genes relative to other cell types, which must be taken into account when applying the threshold. Deciding whether to set the threshold at 5% or 20% is highly dependent on the tissue type being studied.

In most droplet-based scRNA-seq datasets, from <1000 to >4000 genes per cell are detected with each gene representing a dimension in space. To visualize the structure of the data in two-dimensional space, a process called dimensionality reduction mathematically selects the most informative genes. DEGs between the clusters can be calculated from normalized data, using the MAST or similar methods (Finak et al., 2015), implemented in Seurat.

Validation of cell type identification

Once the filtered set of cells are clustered, it must be determined whether the cluster, or subsets within the cluster, represents real cell types. The simplest approach is to annotate the clusters hierarchically based on known major lineages (see Fig. 8.1). Epithelia, endothelia, leukocytes, fibroblasts, and muscle lineages share common gene expression profiles across most tissues (Muhl et al., 2020). Accordingly, a well-annotated reference dataset can be used to annotate the major lineages of clustered data. The tool SingleR (Aran et al., 2019) can be used to assign broad cell type lineage identities. Once these identities are transferred, heterogeneity within each major lineage can be examined by subsetting and reclustering.

The reclustering of major lineages can reveal substructures that relate to cellular subtypes or states. At this stage, the analysis of the data becomes more experimental. By increasing the mathematical stringency, or resolution, of nearest neighbor embedding, more and more divisions (subclusters) can be produced. It is imperative to test whether the DEGs within these smaller subclusters actually represent discrete cell types, cell subtypes, or different cell states. Validation of a DEG as an "anchor gene" that defines a cell type regardless of development or disease stage (Thiagarajan et al., 2011) is essential. Standard immunohistochemistry (IHC) with high-quality antibodies or in situ hybridization (ISH) with RNA probes should be used to determine whether the gene identifies an anatomically and/or morphologically distinct cell type.

Spatial transcriptomics

The data generated by single-cell sequencing must be validated in situ due to potential artifacts created by tissue dissociation or analysis in silico. While this is a powerful technology for objectively assessing cellular heterogeneity, a major limitation of droplet-based sequencing is the inability to spatially localize each profiled cell. New spatial transcriptomics technologies are emerging as a solution. The 10X Genomics Visium platform contains a 6.5 × 6.5 mm capture area overlaid with grid of gene expression spots that each contains uniquely barcoded oligonucleotides for labeling transcripts captured from the 1–10 cells within each location. The spatially naïve transcriptome data can then be clustered in the same way as droplet-based sequencing data for the molecular identification of cell types. Cluster data are then linked back to the grid to visualize gene expression in situ (www.10xgenomics.com/products/spatial-gene-expression). A slightly different approach to spatial profiling is the nanoString GeoMX Digital Spatial Profiler (DSP) (www.nanostring.com/products/geomx-digital-spatial-profiler/geomx-dsp-overview/). Instead of profiling a grid of locations across an entire piece of tissue, the DSP allows the user to select multiple regions of interest (ROI) under a microscope. In addition, users can use IHC or RNAscope with immunofluorescence to segment up to three spectral channels plus a nuclear stain within each ROI. A pool of >20,000 RNA oligos with photocleavable UMIs is hybridized to the tissue. A system of millions of micromirrors are used to release photocleavable RNA identifiers that are counted on a next-generation sequencer. These counts can be correlated with cluster-based DEGs from scRNA-seq data to determine whether the probe of interest is labeling the cell type(s) of interest.

Conclusion

Each platform has its own advantages, but it must be stressed that scRNA-seq data must be validated with other modalities such as ISH or IHC. Once each cell type is identified in the lower urinary tract, models can be created for perturbing the physiological function to solve the cellular pathogenesis of lower urinary tract diseases.

References

Adam, M., Potter, A. S., & Potter, S. S. (2017). Psychrophilic proteases dramatically reduce single-cell RNA-seq artifacts: A molecular atlas of kidney development. *Development, 144*(19), 3625–3632.

Amezquita, R. A., Lun, A. T. L., Becht, E., Carey, V. J., Carpp, L. N., Geistlinger, L., Marini, F., Rue-Albrecht, K., Risso, D., Soneson, C., Waldron, L., Pages, H., Smith, M. L., Huber, W., Morgan, M., Gottardo, R., & Hicks, S. C. (2020). Orchestrating single-cell analysis with Bioconductor. *Nature Methods, 17*(2), 137–145.

Andrews, T. S., Kiselev, V. Y., McCarthy, D., & Hemberg, M. (2021). Tutorial: Guidelines for the computational analysis of single-cell RNA sequencing data. *Nature Protocols, 16*(1), 1–9.

Aran, D., Looney, A. P., Liu, L., Wu, E., Fong, V., Hsu, A., Chak, S., Naikawadi, R. P., Wolters, P. J., Abate, A. R., Butte, A. J., & Bhattacharya, M. (2019). Reference-based analysis of lung single-cell sequencing reveals a transitional profibrotic macrophage. *Nature Immunology, 20*(2), 163–172.

Cui, M., Wang, Z., Chen, K., Shah, A. M., Duan, L., Sanchez-Ortiz, E., Li, H., Xu, L., Liu, N., Bassel-Duby, R., & Olson, E. N. (2020). Dynamic transcriptional responses to injury of regenerative and non-regenerative cardiomyocytes revealed by single-nucleus RNA sequencing. *Developmental Cell, 53*(1), 102–116 e8.

Denisenko, E., Guo, B. B., Jones, M., Hou, R., de Kock, L., Lassmann, T., Poppe, D., Clement, O., Simmons, R. K., Lister, R., & Forrest, A. R. R. (2020). Systematic assessment of tissue dissociation and storage biases in single-cell and single-nucleus RNA-seq workflows. *Genome Biology, 21*(1), 130.

Diehl, A. D., Meehan, T. F., Bradford, Y. M., Brush, M. H., Dahdul, W. M., Dougall, D. S., He, Y., Osumi-Sutherland, D., Ruttenberg, A., Sarntivijai, S., Van Slyke, C. E., Vasilevsky, N. A., Haendel, M. A., Blake, J. A., & Mungall, C. J. (2016). The cell ontology 2016: Enhanced content, modularization, and ontology interoperability. *Journal of Biomedical Semantics, 7*(1), 44.

Dobin, A., Davis, C. A., Schlesinger, F., Drenkow, J., Zaleski, C., Sonali, J., Batut, P., Chaisson, M., & Gingeras, T. R. (2013). STAR: Ultrafast universal RNA-seq aligner. *Bioinformatics, 29*(1), 15–21.

Fan, H. C., Fu, G. K., & Fodor, S. P. (2015). Expression profiling. Combinatorial labeling of single cells for gene expression cytometry. *Science, 347*(6222), 1258367.

Finak, G., McDavid, A., Yajima, M., Deng, J., Gersuk, V., Shalek, A. K., Slichter, C. K., Miller, H. W., McElrath, M. J., Prlic, M., Linsley, P. S., & Gottardo, R. (2015). MAST: A flexible statistical framework for assessing transcriptional changes and characterizing heterogeneity in single-cell RNA sequencing data. *Genome Biology, 16*, 278.

Goldstein, A. S., Drake, J. M., Burnes, D. L., Finley, D. S., Zhang, H., Reiter, R. E., Huang, J., & Witte, O. N. (2011). Purification and direct transformation of epithelial progenitor cells from primary human prostate. *Nature Protocols, 6*(5), 656–667.

Grindberg, R. V., Yee-Greenbaum, J. L., McConnell, M. J., Novotny, M., O'Shaughnessy, A. L., Lambert, G. M., Arauzo-Bravo, M. J., Lee, J., Fishman, M., Robbins, G. E., Lin, X., Venepally, P., Badger, J. H., Galbraith, D. W., Gage, F. H., & Lasken, R. S. (2013). RNA-sequencing from single nuclei. *Proceedings of the National Academy of Sciences of the United States of America, 110*(49), 19802–19807.

Hafemeister, C., & Satija, R. (2019). Normalization and variance stabilization of single-cell RNA-seq data using regularized negative binomial regression. *Genome Biology, 20*(1), 296.

Han, X., Wang, R., Zhou, Y., Fei, L., Sun, H., Lai, S., Saadatpour, A., Zhou, Z, Chen, H., Ye, F., Huang, D., Xu, Y., Huang, W., Jiang, M., Jiang, X., Mao, J., Chen, Y., Lu, C., Xie, J., et al. (2018). Mapping the mouse cell atlas by microwell-seq. *Cell, 172*(5), 1091–1107 e17.

Henry, G. H., Malewska, A., Joseph, D. B., Malladi, V. S., Lee, J., Torrealba, J., Mauck, R. J., Gahan, J. C., Raj, G. V., Roehrborn, C. G., Hon, G. C., MacConmara, M. P., Reese, J. C., Hutchinson, R. C., Vezina, C. M., & Strand, D. W. (2018). A cellular anatomy of the normal adult human prostate and prostatic urethra. *Cell Reports, 25*(12), 3530–3542 e5.

Hou, Z., Jiang, P., Swanson, S. A., Elwell, A. L., Nguyen, B. K. S., Bolin, J. M., Stewart, R., & Thomson, J. A. (2015). A cost-effective RNA sequencing protocol for large-scale gene expression studies. *Scientific Reports, 5*, 9570.

Hu, P., Fabyanic, E., Kwon, D. Y., Tang, S., Zhou, Z., & Wu, H. (2017). Dissecting cell-type composition and activity-dependent transcriptional state in mammalian brains by massively parallel single-nucleus RNA-seq. *Molecular Cell, 68*(5), 1006–1015 e7.

Ilicic, T., Kim, J. K., Kolodziejczyk, A. A., Bagger, F. O., McCarthy, D. J., Marioni, J. C., & Teichmann, S. A. (2016). Classification of low quality cells from single-cell RNA-seq data. *Genome Biology, 17*, 29.

Islam, S., Kjallquist, U., Moliner, A., Zajac, P., Fan, J. B., Lonnerberg, P., & Linnarsson, S. (2011). Characterization of the single-cell transcriptional landscape by highly multiplex RNA-seq. *Genome Research, 21*(7), 1160–1167.

Jaitin, D. A., Kenigsberg, E., Keren-Shaul, H., Elefant, N., Paul, F., Zaretsky, I., Mildner, A., Cohen, N., Jung, S., Tanay, A., & Amit, I. (2014). Massively parallel single-cell RNA-seq for marker-free decomposition of tissues into cell types. *Science, 343*(6172), 776–779.

Klein, A. M., Mazutis, L., Akartuna, I., Tallapragada, N., Veres, A., Li, V., Peshkin, L., Weitz, D. A., & Kirschner, M. W. (2015). Droplet barcoding for single-cell transcriptomics applied to embryonic stem cells. *Cell, 161*(5), 1187–1201.

Kurimoto, K., Yabuta, Y., Ohinata, Y., Ono, Y., Uno, K. D., Yamada, R. G., Ueda, H. R., & Saitou, M. (2006). An improved single-cell cDNA amplification method for efficient high-density oligonucleotide microarray analysis. *Nucleic Acids Research, 34*(5), e42.

Lacar, B., Linker, S. B., Jaeger, B. N., Krishnaswami, S. R., Barron, J. J., Kelder, M. J. E., Parylak, S. L., Paquola, A. C. M., Venepally, P., Novotny, M., O'Connor, C., Fitzpatrick, C., Erwin, J. A., Hsu, J. Y., Husband, D., McConnell, M. J., Lasken, R., & Gage, F. H. (2016). Nuclear RNA-seq of single neurons reveals molecular signatures of activation. *Nature Communications, 7*, 11022.

Lake, B. B., Chen, S., Hoshi, M., Plongthongkum, N., Salamon, D., Knoten, A., Vijayan, A., Venkatesh, R., Kim, E. H., Gao, D., Gaut, J., Zhang, K., & Jain, S. (2019). A single-nucleus RNA-sequencing pipeline to decipher the molecular anatomy and pathophysiology of human kidneys. *Nature Communications, 10*(1), 2832.

Macosko, E. Z., Basu, A., Satija, R., Nemesh, J., Shekhar, K., Goldman, M., Tirosh, I., Bialas, A. R., Kamitaki, N., Martersteck, E. M., Trombetta, J. J., Weitz, D. A., Sanes, J. R., Shalek, A. K., Regev, A., & McCarroll, S. A. (2015). Highly parallel genome-wide expression profiling of individual cells using nanoliter droplets. *Cell, 161*(5), 1202–1214.

Muhl, L., Genove, G., Leptidis, S., Liu, J., He, L., Mocci, G., Sun, Y., Gustafsson, S., Buyandelger, B., Chivukula, I. V., Segerstolpe, A., Raschperger, E., Hansson, E. M., Bjorkegren, J. L. M., Peng, X., Vanlandewijck, M., Lendahl, U., & Betsholtz, C. (2020). Single-cell analysis uncovers fibroblast heterogeneity and criteria for fibroblast and mural cell identification and discrimination. *Nature Communications, 11*(1), 3953.

Mungall, C. J., Torniai, C., Gkoutos, G. V., Lewis, S. E., & Haendel, M. A. (2012). Uberon, an integrative multi-species anatomy ontology. *Genome Biology, 13*(1), R5.

Patel, A. P., Tirosh, I., Trombetta, J. J., Shalek, A. K., Gillespie, S. M., Wakimoto, H., Cahill, D. P., Nahed, B. V., Curry, W. T., Martuza, R. L., Louis, D. N., Rozenblatt-Rosen, O., Suva, M. L., Regev, A., & Bernstein, B. E. (2014). Single-cell RNA-seq highlights intratumoral heterogeneity in primary glioblastoma. *Science, 344*(6190), 1396–1401.

Picelli, S., Faridani, O. R., Bjorklund, A. K., Winberg, G., Sagasser, S., & Sandberg, R. (2014). Full-length RNA-seq from single cells using Smart-seq2. *Nature Protocols, 9*(1), 171–181.

Ramskold, D., Luo, S., Wang, Y., Li, R., Deng, Q., Faridani, O. R., Daniels, G. A., Khrebtukova, I., Loring, J. F., Laurent, L. C., Schroth, G. P., & Sandberg, R. (2012). Full-length mRNA-Seq from single-cell levels of RNA and individual circulating tumor cells. *Nature Biotechnology, 30*(8), 777–782.

Rozenblatt-Rosen, O., Shin, J. W., Rood, J. E., Hupalowska, A., Regev, A., & Heyn, H. (2021). Human Cell Atlas Standards and Technology Working Group. Building a high-quality human cell atlas. *Nature Biotechnology, 39*(2), 149–153.

Stuart, T., Butler, A., Hoffman, P., Hafemeister, C., Papalexi, E., Mauck, W. M., 3rd, Hao, Y., Stoeckius, M., Smibert, P., & Satija, R. (2019). Comprehensive integration of single-cell data. *Cell, 177*(7), 1888–1902 e21.

Stuart, T., & Satija, R. (2019). Integrative single-cell analysis. *Nature Reviews Genetics, 20*(5), 257–272.

Tang, F., Barbacioru, C., Wang, Y., Nordman, E., Lee, C., Xu, N., Wang, X., Bodeau, J., Tuch, B. B., Siddiqui, A., Lao, K., & Surani, M. A. (2009). mRNA-Seq whole-transcriptome analysis of a single cell. *Nature Methods, 6*(5), 377–382.

Thiagarajan, R. D., Georgas, K. M., Rumballe, B. A., Lesieur, E., Chiu, H. S., Taylor, D., Tang, D. T. P., Grimmond, S. M., & Little, M. H. (2011). Identification of anchor genes during kidney development defines ontological relationships, molecular subcompartments and regulatory pathways. *PLoS One, 6*(2), e17286.

Tirosh, I., Izar, B., Prakadan, S. M., Wadsworth, M. H., 2nd, Treacy, D., Trombetta, J. J., Rotem, A., Rodman, C., Lian, C., Murphy, G., Fallahi-Sichani, M., Dutton-Regester, K., Lin, J., Cohen, O., Shah, P., Lu, D., Genshaft, A. S., Hughes, T. K., Ziegler, C. G. K., et al. (2016). Dissecting the multicellular ecosystem of metastatic melanoma by single-cell RNA-seq. *Science, 352*(6282), 189–196.

Treutlein, B., Brownfield, D. G., Wu, A. R., Neff, N. F., Mantalas, G. L., Espinoza, F. H., Desai, T. J., Krasnow, M. A., & Quake, S. R. (2014). Reconstructing lineage hierarchies of the distal lung epithelium using single-cell RNA-seq. *Nature, 509*(7500), 371–375.

van den Brink, S. C., Sage, F., Vertesy, A., Spanjaard, B., Peterson-Maduro, J., Baron, C. S., Robin, C., & van Oudenaarden, A. (2017). Single-cell sequencing reveals dissociation-induced gene expression in tissue subpopulations. *Nature Methods, 14*(10), 935–936.

van Dijk, D., Sharma, R., Nainys, J., Yim, K., Kathail, P., Carr, A. J., Burdziak, C., Moon, K. R., Chaffer, C. L., Pattabiraman, D., Bierie, B., Mazutis, L, Wolf, G., Krishnaswamy, S., & Pe'er, D. (2018). Recovering gene interactions from single-cell data using data diffusion. *Cell, 174*(3), 716–729 e27.

Villani, A. C., Satija, R., Reynolds, G., Sarkizova, S., Shekhar, K., Fletcher, J., Griesbeck, M., Butler, A., Zheng, S., Lazo, S., Jardine, L., Dixon, D., Stephenson, E., Nilsson, E., Grundberg, I., McDonald, D., Filby, A., Li, W., De Jager, P. L., et al. (2017). Single-cell RNA-seq reveals new types of human blood dendritic cells, monocytes, and progenitors. *Science, 356*(6335).

Wang, J., Agarwal, D., Huang, M., Hu, G., Zhou, Z., Ye, C., & Zhang, N. R. (2019). Data denoising with transfer learning in single-cell transcriptomics. *Nature Methods, 16*(9), 875–878.

Wu, H., Kirita, Y., Donnelly, E. L., & Humphreys, B. D. (2019). Advantages of single-nucleus over single-cell RNA sequencing of adult kidney: Rare cell types and novel cell states revealed in fibrosis. *Journal of the American Society of Nephrology, 30*(1), 23–32.

Zheng, G. X., Terry, J. M., Belgrader, P., Ryvkin, P., Bent, Z. W., Wilson, R., Ziraldo, S. B., Wheeler, T. D., McDermott, G. P., Zhu, J., Gregory, M. T., Shuga, J., Montesclaros, L., Underwood, J. G., Masquelier, D. A., Nishimura, S. Y., Schnall-Levin, M., Wyatt, P. W., ... Hindson, C. M., et al. (2017). Massively parallel digital transcriptional profiling of single cells. *Nature Communications, 8*, 14049.

Section III

Neurobiological tools applied to neuro-urology research

Chapter 9

Exploring urinary bladder neural circuitry through calcium imaging

William F. Jackson and Nathan R. Tykocki
Department of Pharmacology and Toxicology, Michigan State University, East Lansing, MI, United States

Traditional means of recording neuronal function

In some ways, the recording of electrical activity in neurons has changed very little since the time of Goldman, Hodgkin, and Katz's seminal work to understand the ionic components governing membrane potential (Fatt & Katz, 1953; Goldman, 1943; Hodgkin & Huxley, 1952). Sharp electrodes, coupled to low-noise, high-impedance amplifiers, and noise filtering/shielding apparatuses still serve as the backbone of electrophysiological measurements of neuronal function. While this chapter is devoted primarily to the Ca^{2+} imaging of neural function, we have included a brief overview of these techniques because they can often be used in conjunction with state-of-the-art imaging to provide never-before-seen detail of signaling circuits in the nervous system. For more information, we direct the reader to an excellent resource for neuroscience-related laboratory protocols through Cold Spring Harbor Laboratory (Sever, 2020).

Single-cell recordings from neurons, either with sharp electrodes or whole-cell patch, is a gold-standard technique for measuring and manipulating the membrane potential of a single neuron and for interrogating the various membrane currents responsible for cellular excitability. These techniques rely primarily on the use of glass electrodes or patch pipettes to measure minute changes in potential or current across cell membrane. At rest, most cell membranes have an extremely high input resistance (500 MΩ−30 GΩ). Therefore, small changes in membrane potential can be resolved reliably in response to extremely small changes in current (Hille, 1985). These two techniques have one major difference: sharp-electrode recording requires the insertion of a sharp glass electrode (usually containing a high concentration of salt, such as 3M KCl) through the cell membrane into the neuronal cell, while whole-cell patch recording affixes a heat-polished pipette to the outside of the cell and perforates the membrane physically or chemically to allow access to the cell cytoplasm (Bretag, 2017). The use of both techniques has persisted even though comparisons between sharp and patch electrode recordings uncovered stark differences in the scientific conclusions drawn when using each technique to make similar measurements (Li et al., 2004). An alternative to these two approaches is to use a suction electrode, wherein a cut end of a neuron or nerve bundle is pulled into the end of a large-bore glass or plastic pipette to create a seal (Florey & Kriebel, 1966). This tool is extremely useful for recording phenomena, such as afferent nerve activity from an intact organ, without the need for oil or grease to electrically isolate the preparation from the saline bath solution. This technique has proven useful in understanding afferent outflow in the whole urinary bladder and gut and is still often employed in these tissues (Heppner et al., 2016; Nullens et al., 2016). However, single fibers can be difficult to seal and discriminating individual unit firing patterns from multiple nerves at once requires reliance on specialized software algorithms.

While whole-cell patch recordings, sharp electrode recordings, and suction electrode recordings share common hardware in terms of performing the techniques, they also share another commonality: they are extremely difficult to learn and to perform repeatedly and reliably. Due to the change in resistance at the electrode tip as it bends, even the smallest movement of a sharp electrode will result in artifactual recordings difficult to discern from real results. Likewise, any disruption of the seal between the patch pipette and the membrane will markedly skew the experimental results (Standen et al., 1987).

Nonetheless, technical advances have been made that allow for in vivo recording of nerve function using whole-cell patch configurations. Protocols are now readily available that detail how to train and habituate animals in such a way as to allow patch recordings from within the brain of awake mice (albeit restrained and/or head-fixed) (Lee & Lee, 2017).

Neuro-Urology Research. https://doi.org/10.1016/B978-0-12-822455-7.00005-2

This does allow for unprecedented access into the neural circuitry of the brain; however, it is unclear of such a technique could work in studies elsewhere in the body where fixation and stabilization of the target organ are more complicated and postural training of mice more difficult. Thus, while these techniques have the advantage of resolution as high as single channel opening and individual action potential conduction, the nuance and complexity of the technique do not easily adapt to measuring complex circuits and interactions without extensive hardware and controls.

So, what can be done to measure neural activity in entire circuits if traditional techniques are found lacking? The answer lies not in reimagining how to measure electrical signals, but rather in reevaluating the signals we are trying to measure.

Overview of calcium signaling as it applies to neuronal function

Nearly all cellular processes—from life to death, metabolism to differentiation, trafficking to exocytosis—depend, in part, on a single second messenger: calcium (Bootman et al., 2001). This ubiquitous divalent cation has long been used in imaging studies to explore the functions of muscle, where calcium influx and intracellular calcium release represent the major driving forces behind muscle cell contraction and proliferation (Bozler, 1969; Cannell et al., 1995; Tsien & Tsien, 1990). But what about neurons? Given that action potentials (and thus neuronal signaling) are dependent on the influx of sodium and not calcium, is studying calcium signaling worthwhile for understanding neuronal function? In short, the answer is yes; however, the entire story is not nearly that simple.

Ca^{2+} signals versus Na^{+} signals: what does each say about a neuron?

Since the original observations of Hodgkin and Huxley in the 1940s and 1950s, the voltage-dependent movement of Na^{+} into nerve axons is recognized as the quintessential basis of neuronal signaling and function (Hodgkin & Huxley, 1947, 1952). As they (and others) continued to unravel the ionic conductances of membrane potential in the decades that followed, another ion had a clear and present function in all aspects of neuronal function: Ca^{2+} (Hodgkin & Keynes, 1957). Unlike Na^{+}—where the rapid opening and inactivation of voltage-gated Na^{+} channels drives a propagating and unidirectional depolarizing current along the length of the axon—Ca^{2+} has a distinct role in local, discrete cellular processes well beyond signal conduction. The pattern, location, and mechanisms behind Ca^{2+} signals in neurons drive cell maturation, cell death, long-term changes in gene expression, and (arguably most importantly) the release of neurotransmitter from presynaptic nerve terminals (Augustine et al., 2003).

The relationship between Na^{+} and Ca^{2+} in neurons is not unlike a computer network: Na^{+} currents form the "wire" that ensures the transmission of high-fidelity information along the axon, while Ca^{2+} is the "computer" that translates these signals into meaningful information to be sent and received from node to node. We can thus extract an immense amount of information about neuronal signaling by studying localized Ca^{2+} signals within both presynaptic terminals and cell bodies. This is especially important in the lower urinary tract due to the close apposition of efferent and afferent fibers within the urinary bladder, which continues through the pelvic ganglia and only begins to separate within the spinal cord and dorsal roots (de Groat & Yoshimura, 2009).

From local to global: types of neuronal Ca^{2+} signals

Like most cells, Ca^{2+} signals in neurons are a spatial and temporal composite of extracellular Ca^{2+} influx and intracellular Ca^{2+} stores release. A complete exploration of the tools employed to probe these mechanisms could fill an entire book; we instead direct the reader to several outstanding reviews describe the specific mechanisms responsible for functional and dysfunctional Ca^{2+} dynamics in neurons in detail (Augustine et al., 2003; Berridge, 1998; Brini et al., 2014). For the purposes of our discussion here, we will focus on two main concepts: (1) rapid, localized Ca^{2+} signals for investigating single neurons; and (2) propagating global Ca^{2+} signals for investigating interconnected neural circuitry.

Localized Ca^{2+} events in neurons are generally described to be the events initiating and controlling presynaptic neurotransmitter release. Though there is evidence that neurotransmitter release can be regulated through the release of Ca^{2+} from intracellular stores (Liang et al., 2002), these events mostly rely on the opening of voltage-gated Ca^{2+} channels in response to Na^{+} influx-mediated membrane depolarization (Augustine et al., 2003). These events rarely spread beyond the synapse and appear to drive the rate of release of neurotransmitter through opening of P/Q-type Ca^{2+} channels (Arai & Jonas, 2014). Postsynaptically, activation of local Ca^{2+} events can drive Ca^{2+}-driven action potentials or propagating Ca^{2+} "waves" down the length of the axon (Augustine et al., 2003). Thus, both pre- and postsynaptic neuronal excitability can be measured and interpreted through the study of local spatiotemporal Ca^{2+} signals as well as the resultant propagating Ca^{2+} signals in downstream neural networks. It is important to consider these distinctions, though, as some of the techniques discussed later in this chapter lack the resolution to determine the spatial characteristics of neuronal Ca^{2+} signals and could lead to the improper interpretation that a true "global" increase in intracellular Ca^{2+} occurs instead.

Using imaging to measure neuronal function

Traditional recording techniques have amazingly high sensitivity in their ability to resolve single-channel openings and subthreshold changes in membrane potential, but these results are—for the most part—two-dimensional. The resultant traces tell us a lot about the change in voltage or current versus time but tell us nearly nothing about what is happening elsewhere in the cell at the same moment—or for that matter, immediately adjacent to the patch pipette or electrode itself. While possible to measure conduction along or between neurons with multiple electrodes, we still only gain very high-resolution temporal data with an extremely low resolution spatially. To combat this limitation, we now turn to high-resolution, high-speed imaging techniques that employ fluorescent dyes or proteins, which respond to changes in intracellular ionic concentration. The most versatile and often-used fluorophores have a quasi-linear relationship between their emission intensity and the concentration of free Ca^{2+} within the cell, allowing for the spatial and temporal measurement of nanomolar to micromolar concentrations of Ca^{2+} reliably and consistently. Thus, imaging changes in neuronal Ca^{2+} allows us to see not only changes in neural excitability but also to see how these changes propagate along and between neurons within a cellular network.

In this section, we will introduce the two types of tools used to measure intracellular Ca^{2+} in cells: chemical calcium indicators (e.g., FURA-2 or FLUO-4) and genetically encoded Ca^{+2} indicators (GECIs, e.g., GCaMPs or RCaMPs). Each has its strengths and limitations, but these two tools revolutionized the way in which we can visualize cellular function individually and in the whole tissue. Also included are the pros and cons of dyes and GECIs to help the reader make an informed choice as to the best tool for their application.

Chemical calcium indicators

There are six key features of ideal chemical Ca^{2+} indicators:

1. Can be easily loaded into cells.
2. Distribute equally throughout the cell so that local and global Ca^{2+} signaling events can be recorded.
3. Chemically inert (nontoxic and nonpharmacologic)
4. Reversibly bind/interact with Ca^{2+} over the physiological range of $[Ca^{2+}]_{in}$ (50 nM−10 μM).
5. Accurately report $[Ca^{2+}]_{in}$ over the physiological range
6. Do not alter local Ca^{2+} concentration or Ca^{2+} buffering.

The extent to which chemical Ca^{2+} indicators meet some but not all of these features will be presented subsequently. A weakness of all of the chemical Ca^{2+} indicators is that they must be introduced, in some way, into the cytosol of the cells from which $[Ca^{2+}]_{in}$ is to be measured. The most common means to do this is through the use of acetoxymethyl (-AM) ester forms of these dyes, which are cell-permeant in their -AM ester form but which after cleavage of the -AM ester from the indicator by ubiquitous intracellular esterases become charged and are better retained in cytosol of cells. The major concern with this approach for dyes like FURA-2 (which are fluorescent and Ca^{2+} insensitive in their -AM form) is incomplete deesterification. To overcome this issue, the AM form of the dye used for loading the cells must be removed and sufficient time allowed for complete deesterification (usually at least 30 min wash period at 37°C). A second concern about -AM ester loading is unequal distribution of dyes in subcellular compartments (Thomas et al., 2000). This is mostly a problem with intensiometric indicators but may also skew $[Ca^{2+}]_{in}$ measurements with any indicator. A major limitation of -AM ester loading of chemical Ca^{2+} indicators is that this means of indicator loading does not allow for cell-specific loading. This is problematic in isolated tissues and in vivo applications because all cells will load with the indicator, not just nerves, and Ca^{2+} signals from other cells (like smooth muscle in bladder, for example) may dominate. To circumvent this problem, cells may be loaded with the impermeant K^{+}-salts of chemical Ca^{2+} indicators using either microinjection with sharp microelectrodes or through patch-pipettes (Takahashi et al., 1999). The limitation of this approach is that only a few cells can be loaded at one time limiting the study of cell population properties. Isolated peripheral nerves have been successfully loaded with Ca^{2+} indicators by using dyes-coupled to large molecular weight dextrans and soaking the cut end of the nerve in solutions containing the indicator-dextran solution (Fontaine et al., 2017; Gover et al., 2003). Axonal transport then picks up and moves the indicator-dextrans along the length of the nerve allowing Ca^{2+} measurements in these nerves (Fontaine et al., 2017) and at end-organs (Jackson et al., 2001). Electroporation also has been used to load chemical indicators into dorsal root ganglia, for example (Chen & Huang, 2017).

The last property, that the Ca^{2+} indicator does not alter local Ca^{2+} concentration or affect Ca^{2+} buffering or signaling, is a consideration for all Ca^{2+} indicators. Excessive loading with chemical Ca^{2+} indicators or expression of genetically encoded Ca^{2+} indicators (GECIs) can excessively buffer/dampen Ca^{2+} signals and alter Ca^{2+} signaling. Thus, in new systems, it is optimal to always compare responses of cells/tissues/organisms loaded with Ca^{2+} indicators to unloaded controls.

There are two main classes of chemical Ca^{2+} indicators: Ratiometric indicators, such as FURA-2 (Grynkiewicz et al., 1985), INDO-1 (Grynkiewicz et al., 1985), and FURA-red (Chen & Huang, 2017; Whitaker et al., 2010) and intensiometric dyes such as FLUO-4 (Gee et al., 2000) and its relatives (Thomas et al., 2000; Whitaker et al., 2010). Many of the GECIs also fall into the latter class as intensiometric indicators.

Ratiometric calcium indicators

Ratiometric Ca^{2+} indicators, as their name implies, use the ratio of emitted fluorescence either from two excitation wavelengths as with FURA-2 or the ratio of fluorescence intensity at two emission wavelengths as with INDO-1 (Grynkiewicz et al., 1985). The major benefit of this approach is that it eliminates artifacts in the estimation of Ca^{2+} levels based on the fluorescence signal due to differences in the effective path length for the fluorescent indicator (i.e., cell thickness at the point of measurement), differences in the concentration of the indicator within and between cells (affected by loading, local solubility and the presence of organelles as well as photobleaching), and variation in the excitation illumination intensity (which can be significant for some light sources).

By far and away, FURA-2 is the most commonly used ratiometric Ca^{2+} indicator. This indicator displays a Ca^{2+}-dependent shift in the excitation spectrum (Fig. 9.1) such that the ratio of the intensity of emitted light at 510 nm with

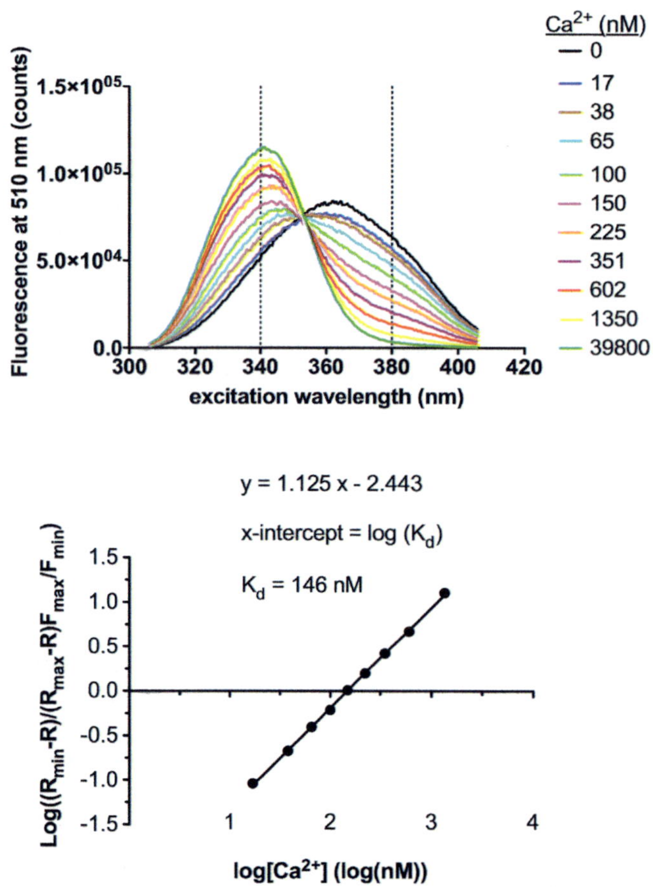

FIGURE 9.1 FURA-2AM ratiometric calibration. *(Top)* The intensity of emitted 510 nm light from 10 mM solutions of FURA-2 pentapotassium salt equilibrated with different Ca^{2+} concentrations with excitation wavelengths scanned from 300 to 410 nm. As can be seen, there is a leftward shift in the excitation maximum as $[Ca^{2+}]$ increases. *(Bottom)* The linear relationship between fluorescence ratio and Ca^{2+} concentration allows for accurate estimation of changes in intracellular $[Ca^{2+}]$ over ~3 orders of magnitude.

340 nm excitation (F_{340}) divided by the 510 nm emission with 380 nm excitation (F_{380}) increases as Ca^{2+} increases. In contrast to the fluorescence emission monitored at a single wavelength, the F_{340}/F_{380} ratio is relatively immune to differences in cell/tissue thickness, the concentration of FURA-2 from which the signal is generated, and changes in illumination intensity. Formula 9.1 is commonly used to measure $[Ca^{2+}]_{in}$ using FURA-2:

$$[Ca^{2+}]_{in} = K_d\left(\frac{(R - R_{min})}{(R_{max} - R)}\right) \times \left(\frac{F_{max}}{F_{min}}\right) \tag{9.1}$$

where:

K_d = dissociation constant of FURA-2 (measured from calibration experiments, see Fig. 9.1).
$R = (F_{340} - Background_{340})/(F_{380} - Background_{380})$, where F is the fluorescence intensity of emitted light at 510 nm.
R_{min} and R_{max} = minimum and maximum R-values obtained with 0 nM Ca^{2+} and 39 mM Ca^{2+}, respectively.
F_{max} and F_{min} = the background-corrected fluorescence emission maximum (in 0 nM Ca^{2+}) and minimum (in 39 mM Ca^{2+}), respectively, with 380 nm excitation recorded at 510 nM.

F_{min} and F_{max} may be estimated at the end of experiments by exposing cells/tissues to the Ca^{2+} ionophore, ionomycin, in Ca^{2+} containing solution to obtain the F_{min} value, and then equilibrating the cells/tissue with solutions containing 0 Ca^{2+} and 1–10 mM EGTA to obtain the F_{max} value.

The key measurements required for quantification of $[Ca^{2+}]_{in}$ with FURA-2 are an accurate estimate of the FURA-2 K_d in the system you are studying, accurate measurement of the background fluorescence elicited by 340 and 380 nm illumination, R_{min}, R_{max}, F_{min} and F_{max}.

The measurement of the FURA-2 K_d is very system and condition dependent with in vitro values being threefold lower than values estimated in cells due to differences in ionic strength and protein binding (Petr & Wurster, 1997). Because the K_d for FURA-2 is so dependent on the precise local chemistry, the exact value of this important variable is difficult to estimate in cells and tissues. For this reason, most investigators simply report background-corrected fluorescence ratios as an index of changes in $[Ca^{2+}]_{in}$. The problem of the Ca^{2+} K_d being dependent on the local chemical environment is not unique to FURA-2, but has been shown to occur with all known fluorescent Ca^{2+} indicators (*see* Johnson, 2010).

The accurate measurement of the background fluorescence of your system when illuminated by the appropriate excitation wavelength(s) (340 and 380 nm for FURA-2 for example) is essential for the use of any fluorescent indicator. This is particularly important when the minimum fluorescence at 380 nm is near the background, because F_{min} may approach zero, particularly when signal noise is present and can make calculation of ratio values indeterminant (due to division by small values close to zero). To this end, it is recommended that sufficient FURA-2 be loaded into your system such that F_{min} is about 4X the background fluorescence.

The measurement of R_{min}, R_{max}, F_{min}, and F_{max} also can be difficult, particularly when used in complex tissues and organs. Flooding cells with Ca^{2+} using ionomycin, for example, to measure F_{min} and R_{max} can lead to rapid extrusion of dye and activation of transporters as cells try and reduce $[Ca^{2+}]_{in}$, such that true F_{min} and R_{max} values may never be obtained. Obtaining accurate F_{max} and R_{min} values are also challenging because all sources of $[Ca^{2+}]_{in}$ including endoplasmic reticulum, mitochondria, etc. must be Ca^{2+} depleted which can take substantial time (on the order of hours) to reach true zero values. Because of these difficulties, most investigators rely on in vitro calibrations using FURA-2 solutions with known $[Ca^{2+}]$ to estimate R_{min}, R_{max}, F_{min}, and F_{max} for their imaging system. Despite the issues with measurement of the precise K_d and R_{min}, R_{max}, F_{min}, and F_{max}, FURA-2 has two significant strengths: the ability to quantitatively measure $[Ca^{2+}]_{in}$ between about 20 nM and 2 mM (given the caveats about the measurement of K_d, etc.) and the ratiometric nature of this indicator making its use relatively unaffected by non-homogeneous loading, changes in excitation intensity, dye extrusion, photobleaching, movement artifacts, etc.

Despite these strengths, there are several weaknesses of FURA-2 as a calcium indicator. First, the relatively high affinity of FURA-2 for Ca^{2+} means that the signal from this dye saturates at $[Ca^{2+}]$ above about 2 mM, so it cannot be used to measure large increase in $[Ca^{2+}]$in. There are low-affinity forms of FURA-2, such as FURA-4F (K_d = 770 nM, working range of 100 nM–10 mM), and FURA FF (K_d = 5.5 mM, working range of 500 nM–55 mM), that extend the range of $[Ca^{2+}]_{in}$ that can be measured. Second, FURA-2 has a slow off-rate for Ca^{2+} making it unsuitable for the measurement of rapid $[Ca^{2+}]_{in}$ events such as Ca^{2+} sparks and Ca^{2+} sparklets. This is further compounded by the dual excitation required for its use, such that the maximum sampling frequency is only about 1–5 images or ratios per s. Third, the UV light required to excite FURA-2 can be phototoxic, limiting exposure time to 10–30 s per site. Multiphoton excitation of FURA-2 has been reported that reduces the light-induced phototoxicity, but only for single wavelength excitation, eliminating ratiometric recording (Wokosin et al., 2004). Autofluorescence of tissue, meaning a high background, can also

TABLE 9.1 Pros and cons of ratiometric chemical calcium indicators excited by UV light.

Pros:	Cons:
Eliminate or reduce errors/artifacts due to unequal indicator loading, indicator photo-bleaching, differences in cell/tissue thickness, uneven illumination, etc.	UV excitation for Fura-2 can be phototoxic, limiting the duration of experiments and the intensity of excitation.
Can be calibrated to yield reasonable and accurate estimates of intracellular Ca^{2+} concentration.	Fura-2 has slow Ca^{2+}-binding-kinetics limiting use to slow Ca^{2+} events
	Dual wavelength excitation and Fura-2 kinetics limits image sampling frequency to 1–5 images/s.
	UV excitation for Fura-2 is not compatible with current commercially available confocal microscopes.
	Potential for Ca^{2+} buffering if cell/tissues are overloaded with indicator.

be a significant problem with UV-excitation of this class of Ca^{2+} indicators. To obviate the issues associated with the use of UV excitation, FURA-Red™ (excitation 457 and 488 nm; emission 660 nm) may be used (Lohr, 2003; Thomas et al., 2000; Whitaker et al., 2010). However, this ratiometric indicator has a low quantum yield and its fluorescence intensity decreases as $[Ca^{2+}]$ increases (Lohr, 2003; Thomas et al., 2000; Whitaker et al., 2010). A list of pros and cons for ratiometric calcium indicators is listed in Table 9.1.

Intensiometric calcium indicators

Chemical Ca^{2+} indicators, such as FLUO-4 (Gee et al., 2000) or its relatives (Minta et al., 1989), and GECIs such as the GCaMPs (Nakai et al., 2001), are examples of intensiometric Ca^{2+} indicators. As their name implies, intensiometric Ca^{2+} indicators increase their fluorescence emission intensity with an increase in Ca^{2+}. The main advantages of FLUO-4 and related chemical Ca^{2+} indicators are that they are excited within the visible range of wavelengths (excitation = 494 nm, emission = 516 nm for FLUO-4, for example) and display large ($\sim 100\times$) enhancement of fluorescence when bound to Ca^{2+} making them ideal for imaging on conventional confocal and multiphoton microscopes. The Ca^{2+} $K_d = 345$ nm for FLUO-4, allowing $[Ca^{2+}]_{in}$ measurements from about 35 nM to 3.5 mM. As with the ratiometric indicator FURA-2, there are lower affinity analogs of FLUO-4 available to allow for the detection of higher Ca^{2+} concentrations (Gee et al., 2000). In addition, FLUO-4, for example, has very fast kinetics and is good for the detection of rapid events like Ca^{2+} sparks and Ca^{2+} sparklets.

The main weakness of intensiometric Ca^{2+} indicators is the difficulty in conversion of the fluorescence output to an actual Ca^{2+} concentration within a cell. Formula 9.2 describes the relationship of the fluorescence of an intensiometric indicator to the $[Ca^{2+}]_{in}$:

$$[Ca^{2+}]_{in} = K_d \frac{(F - F_{min})}{(F_{max} - F)} \tag{9.2}$$

where

K_d = dissociation constant.
F = fluorescence emission − Background.
F_{max} and F_{min} = fluorescence emission maximum (at 39 mM Ca^{2+}) and minimum (at 0 nM Ca^{2+}), respectively.

As with FURA-2, the K_d for Ca^{2+} of FLUO-4 and other intensiometric indicators depends on their local environment, and exact values in cells are difficult to measure and may not be constant within a cell. While the values of F_{min} and F_{max} can be determined by using ionomycin or the Ca^{2+} ionophore A23187 to permeabilize cells to Ca^{2+} in calibration experiments, it must be recognized that the fluorescence emission intensity depends heavily on the concentration of the indicator within a cell, so that changes in the local concentration of dye due to photobleaching, transport of the indicator out of the cell, etc. will strongly impact such calibrations and limit the ability to exactly correlate the fluorescence emission intensity with the exact $[Ca^{2+}]_{in}$ (Thomas et al., 2000).

To get around this problem, a pseudoratio (Minta et al., 1989) approach can be used, as shown in Formula 9.3:

$$[Ca^{2+}]_{in} = \frac{K_d(F/F_o)}{\left(\frac{K_d}{[Ca^{2+}]_{Rest}}\right) + \left(1 - \frac{F}{F_o}\right)} \tag{9.3}$$

where

K_d and F are as defined in Formula 9.2.
F_o = fluorescence emission intensity at rest − background (at some time point before a Ca^{2+} transient starts)
$[Ca^{2+}]_{Rest}$ = the known $[Ca^{2+}]_{in}$ under resting conditions.

With this approach, the basal (resting) level of Ca^{2+} must be first measured, using FURA-2, for example. In practice, most investigators simply report the ratio F/F_o as an index of changes in intracellular Ca^{2+} from baseline. Approaches have been developed using a second indicator, such as FURA-Red along with FLUO-4 so that true ratio imaging can be achieved with intensiometric indicators (Lohr, 2003). The problem with this approach is that differences in dye loading, compartmentalization, bleach rate, and extrusion rate between the two dyes can be significant and make calibration very difficult. The pros and cons of intensiometric calcium indicators are listed in Table 9.2.

Genetically encoded Ca^{2+} indicators (GECIs)

A general weakness of the chemical Ca^{2+} indicators discussed thus far is that they must be loaded into cells in some way, which makes their use in whole isolated tissues and in vivo challenging because cell-specific loading can only be achieved by invasive approaches such as the use of patch pipettes or sharp microelectrodes. Genetically encoded Ca^{2+} indicators can get around this limitation by allowing cell-specific expression of proteins that serve as FRET or intensiometric Ca^{2+} indicators (Inoue, 2020; Mank & Griesbeck, 2008).

The first GECI was the coelenterate protein, aequorin, which displays Ca^{2+}-sensitive luminescence but requires the chemical cofactor coelenterazine, which is consumed during the process (Brini, 2008). More recently, GECIs have been developed based on fluorescent proteins, such as green fluorescent protein (GFP), coupled to the Ca^{2+} binding motifs of Calmodulin to form Ca^{2+}-sensitive fluorescent indicators. These have either been FRET-based indicators, such as the Chameleon family of GECIs (Inoue, 2020; Mank & Griesbeck, 2008) using Förster resonance energy transfer (FRET) between two fluorescent proteins, or single fluorescent proteins such as the GCaMPs based on GFP, that are the most popular GECIs used to date. The remainder of this review will focus on this latter class of GECIs (Inoue, 2020; Nakai et al., 2001).

GCaMP was engineered by attaching the M13 fragment of myosin-light chain kinase, which is the calmodulin interacting domain, to the N-terminus of circularly permuted enhanced GFP (cpEGFP). The Ca^{2+} binding protein, calmodulin (CaM), was then attached to the C-terminus of cpEGFP. The binding of Ca^{2+} to CaM results in CaM binding to M13 and a resultant conformational change in cpEGFP, increasing the fluorescence emission intensity of the cpEGFP (Nakai et al., 2001). Subsequent site-directed mutagenesis of GCaMP was performed to improve functional expression, sensitivity, dynamic range, and kinetics (Ca^{2+} on and off rates) resulting in the development of the GCaMP3 (Sun et al., 2013; Tian et al., 2009), GCaMP5 (Akerboom et al., 2012), GCaMP6 (Chen et al., 2013) and GCaMP8 (Ohkura et al., 2012) series of GECIs (Table 9.3). For example, three versions of GCaMP6 were developed with slow (GCaMP6s),

TABLE 9.2 Pros and cons of intensiometric chemical calcium indicators excited by visible light.

Pros:	Cons:
Excited by visible light and compatible with commercial wide-field, confocal, and 2-photon microscopes.	Emitted fluorescence is affected by unequal indicator loading, indicator photo-bleaching, differences in cell/tissue thickness, uneven illumination, etc.
Many versions are available such that $[Ca^{2+}]$ can be monitored over a wide range.	Difficult to calibrate, such that only relative changes in $[Ca^{2+}]$ can be measured.
Many have fast kinetics and can be used to measure fast events like Ca^{2+} sparklets and Ca^{2+} sparks.	Unequal loading, intracellular partitioning, and photo-bleaching make ratiometric measurements with two indicators difficult.
Many are very bright and can be detected by current sCMOS cameras.	Potential for Ca^{2+} buffering if cells/tissues are overloaded with indicator.
By using two indicators (e.g., Fluo4 and Fura-red), ratiometric measurements can be made.	

TABLE 9.3 Biophysical characteristics of GCaMPs.

Name	Excitation (nm)	ε (M^{-1} cm^{-1})	Emission (nm)	ϕ	Brightness ($\varepsilon \times \phi$)	K_d (nM)	Hill coefficient	F_{max}/F_{min}	τ_{On} (ms)	τ_{Off} (ms)	References
GCaMP1	487	1400	510	0.05	70	235	3.3	4.5	2.5	188	Nakai et al. (2001)
GCaMP1.6	488	3800	509	0.79	3002	146	3.8	4.9			Ohkura et al. (2005)
GCaMP2	487	19,000	508	0.93	17,670	146	3.8	4.9	14.29	75.19	Tallini et al. (2006)
GCaMP3	497	34,700	519	0.65	22,555	250	2.5	12		150	Chen et al. (2013) and Sun et al. (2013)
Fast-GCaMP-RS06	497	31,000	519	0.36	11,160	330	2.4	6.7		34	Sun et al. (2013)
GCaMP5g	497	44,900	519	0.63	28,287	413	2.4	34.4		154	Chen et al. (2013) and Sun et al. (2013)
GCaMP6s	497	68,000	519	0.61	41,480	144	2.9	63.2		893	Chen et al. (2013)
GCaMP6m	497	38,200	519	0.61	23,302	167	2.96	38.1		485	Chen et al. (2013)
GCaMP6f	497	61,900	519	0.59	36,521	375	2.27	51.8		254	Chen et al. (2013)
GCaMP8	498		515			200		37.5			Ohkura et al. (2012)

TABLE 9.4 Pros and cons of genetically encoded calcium indicators.

Pros:	Cons:
Through the use of appropriate promoters or target sequences can be expressed in a cell- and even organelle-specific manner.	Most are intensiometric so that differences in expression level, differences in cell/tissue thickness, uneven illumination, uneven photo-bleaching will affect the emitted fluorescence.
Available in forms that are excited by visible or near-infrared light and so are compatible with conventional wide-field, confocal, and 2-photon microscopes.	Many GECIs have slow kinetics limiting their ability to detect Ca^{2+}-signals associated with single action potentials.
Long-term recording of Ca^{2+} signals throughout the lifespan of an organism.	Potential for Ca^{2+} buffering or interference with intrinsic local Ca^{2+} binding proteins like calmodulin if overexpressed.

medium (GCaMP6m), and fast (GCaMP6f) kinetics (Chen et al., 2013). The rise time for Ca^{2+} signals associated with one action potential in mouse pyramidal cells is 179 ms for GCaMP6s versus 80 ms for GCaMP6m versus 45 ms for GCaMP6f, while the ½-time for decay of the Ca^{2+} signal from one action potential is ordered similarly at 550 ms for GCaMP6s versus 270 ms for GCaMP6m versus 142 ms for GCaMP6f. The Ca^{2+} sensitivity of this series occurred in reverse order with GCaMP6s having the highest sensitivity ($K_d = 144$ nM) versus GCaMP6m ($K_d = 167$ nM) versus GCaMP6f ($K_d = 375$ nM).

Longer wavelength GECIs also have been developed including the RCaMPs (excitation = 575 nm, emission = 594 nm) and R-GECO1 (excitation = 564 nm, emission = 598 nm), based on the fluorescent protein mRuby (Akerboom et al., 2013; Dana et al., 2016) and a FRET-based near-infrared GECI, iGECI (excitation 605 nm, emission 725 nm) (Shemetov et al., 2020). Newer GECIs like the XCaMP series appear to offer additional color choices, improved Ca^{2+} sensitivity, and kinetics (Inoue, 2020; Inoue et al., 2019). While the GECIs discussed thus far are inherently intensiometric indicators, coexpression of a red fluorescent protein, such as mCherry, or use of a fusion protein allows for ratiometric Ca^{2+} imaging (Cho et al., 2017). However, as with the use of two chemical Ca^{2+} indicators, differences in bleach rate or expression levels can make the quantitation of these ratioed signals challenging.

The major strength of GECIs is that their expression in cells can be driven by cell-specific promoters allowing cell-specific Ca^{2+} responses to be measured over the life-time of a cell in an organism (Lin & Schnitzer, 2016). Viruses or ballistic delivery of GECIs also can be used to transduce specific cells in whole organisms (Anderson et al., 2018, 2019; Kettunen et al., 2002). Alternatively, transgenic animals can be developed to express GECIs in a cell-specific manner as has been done for neurons in the CNS (Chen et al., 2012), the heart (Tallini et al., 2006), and vascular cells (Tallini et al., 2007), for example. These approaches eliminate the damage associated with sharp electrode or patch pipet loading of indicators and allow for longer-term monitoring of Ca^{2+} signals. The pros and cons of GECIs are listed in Table 9.4.

Exploring urinary bladder neural circuitry through calcium imaging

Now that we understand the tools in our proverbial toolbox, our focus now turns to implementing these techniques in the lower urinary tract. The urinary bladder has some distinct anatomical and physiological advantages that favor the implementation of imaging techniques in exploring its function. As a hollow organ, the bladder wall is thin enough to enable full-thickness image stacks relatively rapidly. The hollow nature of the urinary bladder also allows the use of epifluorescent microscopy with less interference of out-of-focus light above and below the plane of interest. Also, given its placement within the abdominal cavity, the bladder can be imaged in vivo with minimal disruption to its function through an abdominal window. Unfortunately, these advantages are also met with several key disadvantages that complicate the use of many state-of-the-art imaging tools. These include the spherical nature of the full bladder, which makes it difficult to maintain a focal plane using confocal microscopy or high magnification. The bladder wall area expands in all three dimensions simultaneously as it fills, complicating imaging of specific areas of interest over time during in vivo or ex vivo bladder filling. Constant micromotions and transient contractions in vivo and ex vivo often require the use of pharmacological tools to inhibit smooth muscle contractility to maintain focal plane, which may in turn alter neuronal activation (especially for sensory nerves).

In the following sections, we will discuss the implementation of each of these imaging modalities in the urinary bladder and provide helpful information and resources to aid in the successful use of these tools in the future. Each section will contain a summary of the pros and cons of each technique, as to aid the reader in deciding for themselves as to which might best suit their particular questions of interest. Remember—no matter what technique you choose to use to image calcium signals, some rules cannot be broken! Before delving into specific microscopy paradigms, we ask you to read, consider (or perhaps even memorize) the following *Commandments of Calcium Imaging* found in Table 9.5.

TABLE 9.5 Commandments of calcium imaging.

Commandment	Explanation
1. Load low, load slow, and load cool.	It is better to load with a lower concentration of dye for more time than a higher concentration of dye for a short time. Likewise, it is better to load at a lower temperature for a longer time than at a higher temperature for a shorter time. Higher temperatures often lead to compartmentalization of dyes or non-specific loading of other cells, so proceed with caution above room temperature.
2. No signal does not mean no dye.	A lack of signal can just as easily be overloading as it can be caused by underloading because dyes can act as a calcium chelator.
3. A watched sample never glows … or survives.	Do not image the sample until ample time has elapsed for the dye to deesterify and the tissue to reach physiological temperature. Doing so often leads to tissue death.
4. Nice pictures in the dark make bad pictures in the light.	You are collecting data—not filming a movie. Use the full dynamic range of your detector, such that little to no signal is below the detection limit and little to no signal is above. That way, the maximum amount of information is collected with regard to changes in fluorescence.
5. Image as slowly as necessary.	Higher frame rates may catch more spatial/temporal data, but it comes at a price: these rates typically require much higher light intensity due to the shortened exposure time. Once you understand the properties of the calcium events being explored, image no faster and no brighter than is needed to record the image.
6. Lasers burn things.	Always be mindful of laser intensity and volume of the bath surrounding the sample. Lasers (especially in confocal systems) can generate an immense amount of heat during imaging and can completely destroy a sample. This is especially true in multiphoton systems where high-power IR lasers are used.
7. When in doubt, move around.	Calcium dye loading is unlikely to be completely uniform. One lousy spot could be immediately adjacent to the perfect spot. Do not be afraid to look around your sample, just do so with care and minimal exposure.
8. Always treat your tissue like a Vampire.	Whether using GECIs or calcium dyes, treat your tissue as if light is deadly. Sunlight, fluorescent light bulbs, and halogen dissecting microscope lights produce more than enough broad-spectrum light to completely destroy a tissue before it even makes it to the microscope.
9. Higher magnification does not mean better resolution.	Resolution has nothing to do with the magnification, per se. The minimum distance one can resolve between two objects is $\lambda/2(NA)$, where λ is the wavelength of light and NA is the numerical aperture of the objective lens. We may think higher zoom gives more detail since more pixels of the camera are used to gather information on a smaller area, but in reality, it doesn't. Resolution is diffraction limited.
10. If you can see it through the eye pieces, you've probably killed it.	Use the minimum excitation intensity and the maximum detector gain (within a reasonable signal to noise ratio) to image fluorescence in live cells and tissues.
11. It's okay to break Commandments 1 and 2.	Dye loading is extremely finicky and will take a lot of trial and error as you get accustomed to the systems and tissues you are using. You may find high dye concentrations at 35°C give the best signal; by all means, use that instead of any suggestions given here or in the literature.

Epifluorescent/intravital microscopy

Epifluorescence, or "wide-field" fluorescence imaging, is what comes to mind for many of us when we think of microscopy. The basic features of an epifluorescence microscope are outlined in Fig. 9.2. Briefly, broad-spectrum white light is passed through an excitation filter (e.g., 480/30 bandpass filter for GCaMP) and bounced off of a dichroic mirror (e.g., 505DC for GCaMP) on its way through the objective lens. The emitted reflected light (~520 nm) then passes back up the objective, through the dichroic mirror and then through an emission filter (535/40 bandpass) before making its way to the camera or photometer (Webb & Brown, 2013). For many other target organs, the idea of illuminating the entire sample with high-intensity light means certain death and destruction of live tissues—not to mention extensive out-of-focus light obscuring the signal of interest. In the urinary bladder, however, the thin nature of the bladder wall, the hollow internal space, and the ability to mount a bladder *en face* lends it to exquisite calcium imaging using these simple epifluorescent techniques.

As shown in Fig. 9.3, traditional epifluorescence imaging is capable of remarkable images of neuronal networks within the bladder wall (under the right circumstances). Individual *Trpv1-tdTomato*-positive nerve fibers can be resolved throughout the bladder wall, and in close apposition to blood vessels running through the lamina propria. Although these images do not allow us to resolve where individual nerve fibers are with regard to their depth within the tissue, images like these are extremely useful for tracing the neural connections as they meander throughout the bladder wall. Ca^{2+} signals that propagate from one nerve fiber to another can be resolved in this manner. The only real limitations are in intensity and duration: if the calcium indicator is not robustly expressed or sufficiently deesterified within the cell, the signal of interest is unlikely to be significantly greater than the background fluorescence caused by illumination of the whole tissue. Similarly, imaging over long periods of time can be problematic due to photobleaching of such a wide area of tissue (though this is more apparent with AM dyes than with GECIs). Still, with regard to imaging calcium signals in nerves, epifluorescent imaging is a solid and feasible place to begin before investing heavily in the time and money required to implement more complex microscopic techniques. It also has a distinct advantage over both laser scanning confocal microscopy and multiphoton microscopy: speed. Since the entire focal plane is illuminated all the time, the entire field of view can be imaged at high resolution (i.e., full camera chip size) at rates of 30–60 fps or even faster if enough signal is present. Thus, kinetic changes in calcium signals can be recorded over large areas of a neuron or neural network in a way that can be analyzed with high precision and accuracy both spatially and temporally. So, while you might bleach your tissue to the point of complete loss of signal, you will get an immense amount of data as you do it.

One variant of epifluorescent microscopy we have used with great success is imaging with a stereomicroscope designed for intravital microscopy. These systems have a wide and incremental range of zoom capabilities that do not rely on

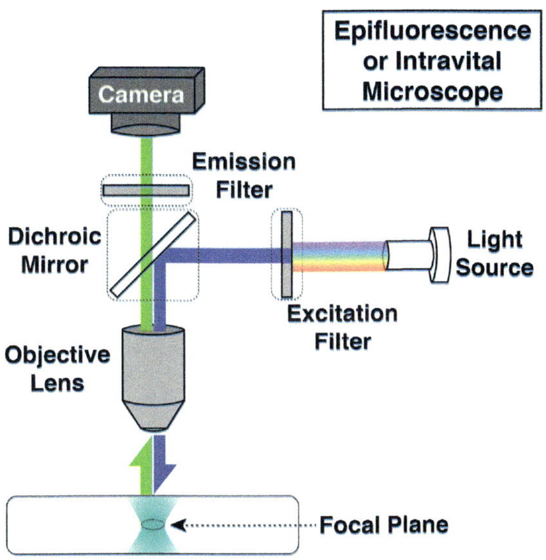

FIGURE 9.2 Epifluorescence microscope design. White light (from an LED, halogen, or mercury lamp) is first passed through an excitation filter and bounced off of a dichroic mirror before being sent through the objective to the sample. The emitted light from the fluorophore is then reflected back up the objective and through the dichroic mirror before passing through a final emission filter en route to the detector or camera. This design results in illuminating much of the tissue above and below the focal plane of interest with out-of-focus light, but it also allows for very rapid imaging of large areas of the bladder wall.

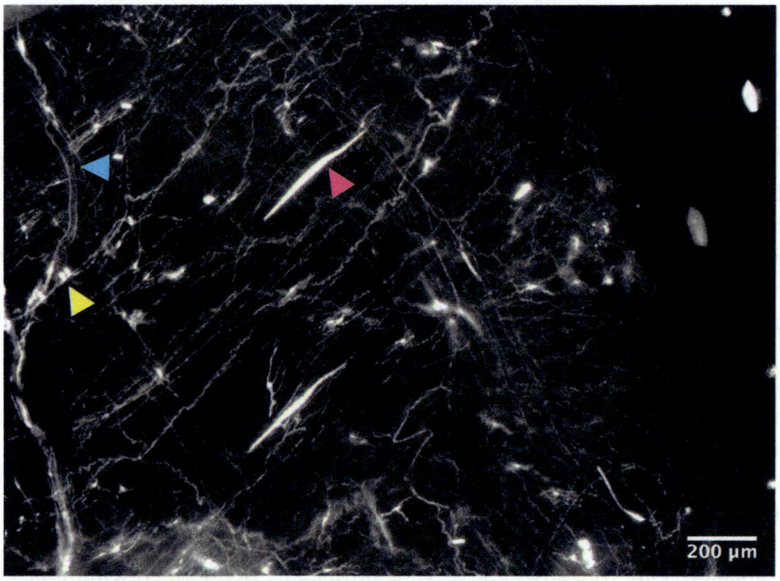

FIGURE 9.3 Image of Trpv1-tdTomato mouse bladder using epifluorescent microscopy. Tissue was mounted en face on Sylgard blocks before being imaged at 4× magnification using an inverted Nikon microscope. Images were captured using an Andor Clara interline CCD camera. While individual nerve fibers are easy to discern, expression of tdTomato in nonnerve cells can be seen in vascular smooth muscle (*yellow arrow*) and possibly interstitial cells (*magenta arrow*). Perivascular sensory nerves are also clearly visible (*cyan arrow*).

switching objective lenses (albeit at the sacrifice of numerical aperture). Intravital stereomicroscopes also have long working distance objectives and a thick focal plane, which allows one to image an in vivo urinary bladder while maintaining an entire section of the spherical surface in focus (see Fig. 9.4). Thus, while widely underutilized in favor of far more complex systems able to image at much higher magnification, epifluorescence and intravital microscopes offer distinct advantages and capabilities for the study of neuronal network calcium signaling ex vivo and in vivo in ways other techniques cannot. The pros and cons of epifluorescent calcium imaging are listed in Table 9.6.

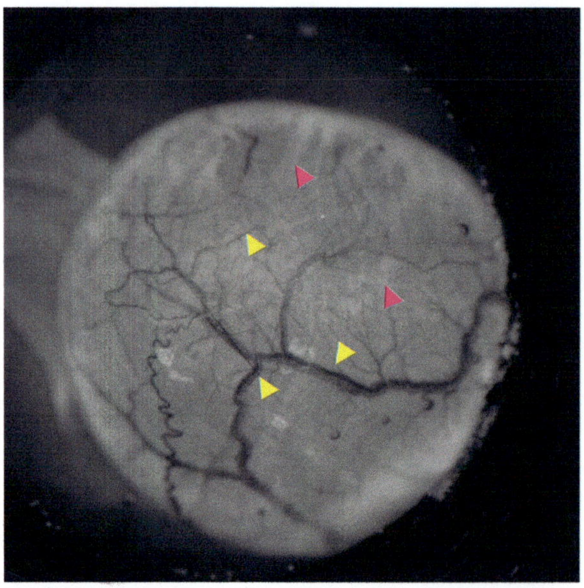

FIGURE 9.4 In vivo imaging of Acta2-GCaMP5 fluorescence in an anesthetized mouse. Images were taken on an Olympus intravital microscope system using a specially designed and fabricated abdominal window imaging device and an Andor Zyla 4.2Plus scMOS digital camera. Even in this fairly wide-field image, the use of the full imaging chip (2048 × 2048 pixels) allows individual arteries, arterioles, and veins to be resolved (*yellow arrows*) as well as increases in intracellular calcium within urinary bladder smooth muscle cells while filling. Though not showing calcium signals in nerves specifically, this does show the utility and versatility of the technique as applied to other cells within the bladder wall.

TABLE 9.6 Pros and cons of epifluorescent calcium imaging.

Pros:	Cons:
Easily accessible to many laboratories using existing microscopy systems, without requiring major capital investment.	Out-of-focus light and tissue autofluorescence can lead to rapid photobleaching and tissue damage as well as skew data interpretation.
Broad field of view.	At best, the maximum resolution in the *z* plane is 3 times lower than the resolution in the *xy* plane (i.e., 200 nm *xy* resolution leads to >600 nm *z* resolution).
Easily stitch together multiple images to create a "big picture" view of the neural network in the urinary bladder.	While z-stacks are theoretically possible, it requires extensive controls and expensive, time-consuming deconvolution software.
Can be used for ratiometric imaging using Fura2-AM dyes or GCaMP fusion proteins.	Requires special photometry equipment in addition to a camera for many ratiometric applications.

Confocal imaging

The major problem associated with epifluorescence microscopes for calcium imaging is the presence of out-of-focus light. The excess illumination can speed photobleaching of the sample and also obscure the dynamic changes in calcium dye fluorescence that are of interest. Confocal microscopy was specifically invented to deal with the problem of out-of-focus light and has quickly become the gold-standard calcium imaging system. As shown in Fig. 9.5, confocal microscopes differ very little from conventional epifluorescence microscopes except for the addition of two pinholes: one immediately adjacent to the light source and another proximal to the detector. These pinholes block out-of-focus light from illuminating the sample above or below the focal plane and also prevent out-of-focus light from reaching the detector (Stockert & Blázquez-Castro, 2017, pp. 159–218). The result is an image of stark clarity from a very narrow focal plane from within the sample.

Confocal microscopes are broken down further into two main subtypes (each with advantages and disadvantages): laser scanning confocal microscopes and spinning disc confocal microscopes. Laser scanning microscopes do just as the name implies: an acousto-optic-controlled or galvo-controlled mirror rapidly moves the focal point of the laser along the sample

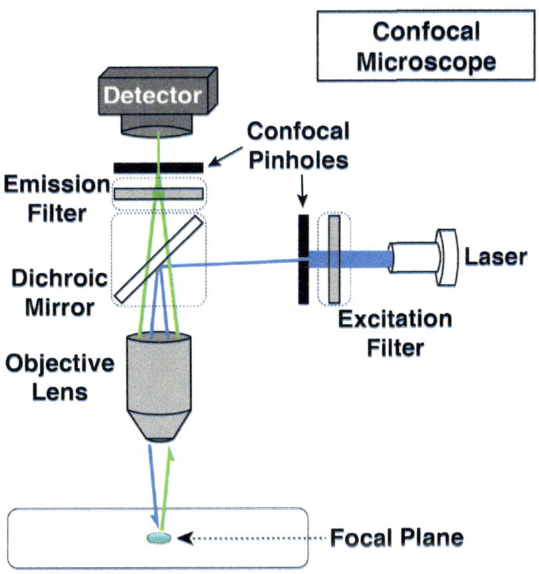

FIGURE 9.5 Confocal microscope design. Confocal microscopes utilize two pinholes to limit out-of-focus light from reaching the sample and the detector. The result is the illumination of only a very small spot in a narrow plane of the sample, which can be altered by adjusting the size of the pinhole. In laser scanning confocal microscopes, additional mirrors (not shown) are used to "scan" the point of illumination along the x and y axes of the sample to generate a two-dimensional image. Alternatively, spinning disc confocal microscopes instead use a disc with multiple pinholes in different locations to move the point of excitation rapidly along the sample. Unlike conventional microscopy, confocal microscopy usually employs high-powered lasers as the light source.

in discrete rows. Emission from each point is then recorded by the detector, a single point at a time, and reconstituted into a single image (Stockert & Blázquez-Castro, 2017, pp. 159–218). These systems allow for a wide array of flexibility in terms of laser light, excitation/emission spectra, pinhole size, and speed. These systems are thus the workhorse for multichannel immunofluorescence because each fluorophore can be imaged sequentially using a single detector. Unfortunately, most of these systems acquire images too slowly to be useful for imaging dynamic changes in fluorescence in live tissues due to the point-by-point nature of excitation. One should not stereotype this apparent lack of speed to all laser scanning confocal systems, however. Some of the first measurements of calcium sparks in human cerebral vascular smooth muscle cells were acquired at frame rates of 60–120 Hz using an early laser scanning confocal system produced by Noran Instruments (Wellman et al., 2002). Frame rates like this were reached by limiting the number of pixels being illuminated to 128 × 128 pixels or 128 × 256 pixels, as opposed to the maximal 512 × 512 (Wellman et al., 2002). Thus, laser scanning confocal microscopes can achieve the high sampling rates needed for calcium imaging in live tissue, but only at the expense of *xy* resolution.

The second subtype of confocal microscope is the spinning disc confocal microscope, which employs a series of pinholes in a rapidly spinning disc ("Nipkow" disc) to limit the location of illumination. The effect is similar to that of laser scanning confocal microscopes but with much faster acquisition rates that require much less excitation energy. Thus, spinning disc confocal microscopes can reduce the adverse effects of high-intensity light as well as allow for faster imaging of a larger area, since the speed of image collection does not depend on how quickly the illuminated point can be moved along the sample. The early Nipkow disc design was further adapted in the 1990s to include a second disc, which contained microlenses in each pinhole to focus more light through the open pinhole and increase illumination (Tanaami et al., 2002). These Yokogawa Spinning Disc Confocal units are now the dominant workhorse for live-cell and live-tissue Ca^{2+} imaging due to their high speed and excellent resolution as compared to laser scanning confocal systems (Wang et al., 2005). However, these systems also have shortcomings as compared to laser-scanning systems. First, the size of each pinhole cannot be altered. While some systems have additional discs for different *z* resolutions (e.g., 25 μm pinholes and 50 μm pinholes), the size cannot be further optimized on a cell or tissue-dependent basis. Also, simultaneous measurements of more than one wavelength of light require the use of multiple EMCCD or scMOS cameras, which can increase the cost of the system by $60,000 USD or more. So, in short: if all that will be done on the system is Ca^{2+} imaging, accept no substitute for a spinning disc confocal system. If money is tight and the microscope needs capabilities beyond imaging live tissue Ca^{2+} signals, then a laser-scanning confocal system may be more appropriate. The pros and cons of confocal microscopes (as a whole) can be found in Table 9.7.

Two examples of confocal images taken from the urinary bladder and associated tissues are presented in Fig. 9.6. On the left (Fig. 9.6A), a confocal image of FLUO-4AM fluorescence from a 120 μm-thick bladder wall transverse section shows loading of dye in nerves as well as urothelial cells and smooth muscle cells. On the right (Fig. 9.6B), dissociated dorsal root ganglia (DRG) cell bodies, also loaded with FLUO-4AM dye, show an increase in fluorescence after exposure to capsaicin.

Multiphoton imaging

Another strategy to reduce out of focus light in fluorescence microscopy is to use two- or multiphoton excitation (Gopal et al., 2021; Tauer, 2002; Xu et al., 1996). The idea is to excite the fluorophore with very rapid (femtosecond) pulses of

TABLE 9.7 Pros and cons of confocal calcium imaging.

Pros:	Cons:
Minimal out-of-focus light and high *xyz* resolution allows for spatiotemporal calcium measurements throughout a cell.	Expensive. Even the simplest confocal systems requires special training/maintenance.
Multiple fluorophores can be used simultaneously.	High-intensity lasers can easily damage samples, and each fluorophore requires its own laser.
Using a spinning disc system can achieve frame rates higher than 120 fps.	Deeper measurements in thick tissues are not possible without significant tissue damage.
	Samples must be flat and immobile; any movement or curvature of the sample severely limits the field of illumination and disrupts the focal plane.

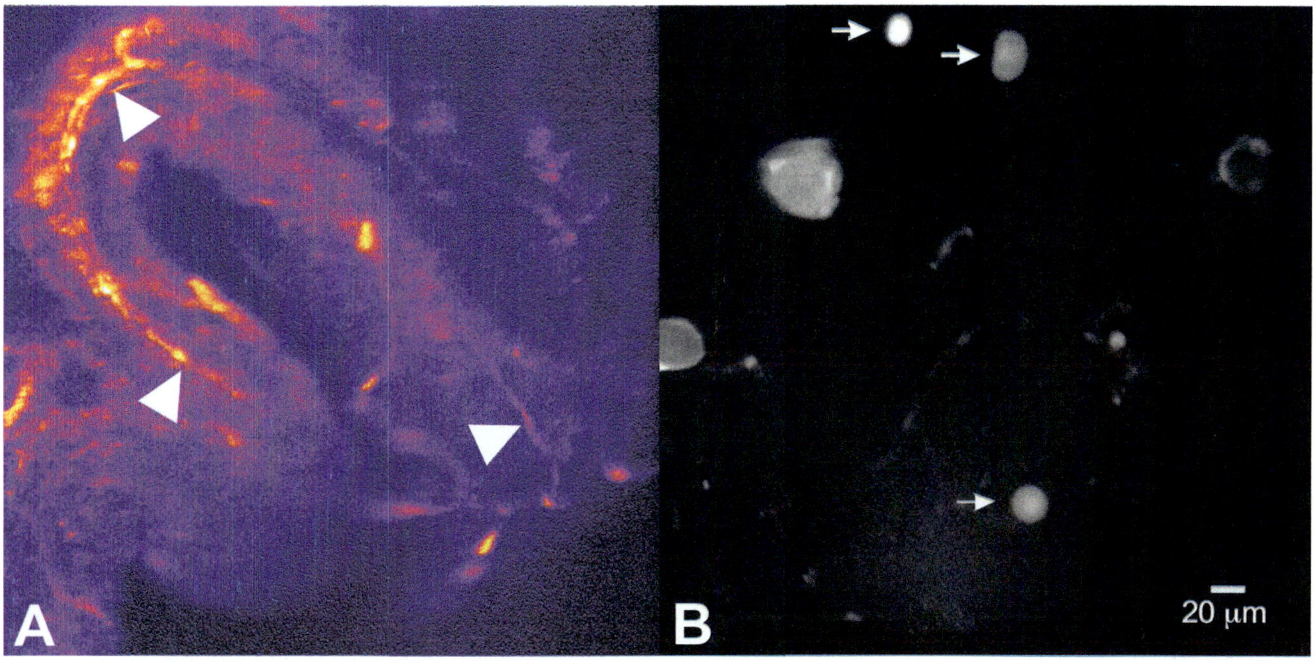

FIGURE 9.6 Images of urinary bladder and dorsal root ganglia using confocal microscopy. (A) Bladders were sliced in 120 μm-thick sections using a vibratome. Tissue was then loaded with 1 μM FLUO-4AM dye before imaging using a Noran Oz laser-scanning confocal microscope (Prairie Technologies, USA) on an inverted Nikon microscope. *White arrows* indicate loading of dye in nerves as well as urothelial cells and smooth muscle cells. (B) Dissociated dorsal root ganglia DRG cell bodies, loaded with FLUO-4AM dye (5 μM), and after exposure to capsaicin. *White arrows* indicate DRG cell bodies that responded to capsaicin with an increase in fluorescence intensity. Images were collected using an Andor Revolution spinning disc confocal microscope (Oxford Instruments, UK) coupled to a Nikon upright microscope.

infrared light (approximately 2X the excitation wavelength of the fluorophore) with a minute interpulse interval. The two or more of the long wavelength photons are absorbed by the fluorophore in sequence, raising the energy level of the electrons in the shell of the fluorophore molecule sufficiently to emit photons at the normal emission wavelength. Because only fluorophore molecules at the focal point of the microscope objective (in an estimated femtoliter volume) absorb sufficient energy to fluoresce, out of focus fluorescence in the planes above and below the focal plane is eliminated (Tauer, 2002; Xu et al., 1996). As with a laser scanning confocal system, the excitation light is scanned over the sample, line-by-line. All of the light emitted by the fluorophores in each small volume of sample is collected by a sensitive photomultiplier tube and the image reconstructed as with a laser scanning confocal microscope (Fig. 9.7).

The major advantages of multiphoton microscopy are: (1) the ability to excite fluorophores normally excited by UV or visible light with much longer wavelength light ($\sim 2\times$ the 1-photon excitation wavelength) reducing or eliminating photo-induced toxicity associated with shorter wavelength excitation; (2) much less photobleaching above and below the focal plane than occurs with other methods (Tauer, 2002); (3) the ability to excite chemicals within a femtoliter volume at the focal point of the objective allowing precise control of photo-release of caged compounds or localized photobleaching (Tauer, 2002); (4) the ability often to excite multiple fluorophores using a single IR laser line (Xu et al., 1996); and (5) the ability to image deeper within tissues (up to 1 mm) than can be accomplished with shorter wavelength illumination (Gopal et al., 2021). These advantages make multiphoton imaging of fluorescent probes and proteins in thick (200 mm) intact tissues and organs (in vivo or ex vivo) the method of choice. Because phototoxicity is reduced, longer imaging sequences can often be captured.

Weaknesses of multiphoton imaging include the high cost of tunable (700–1000 nm) high peak power femtosecond IR lasers (>$150K) making system costs higher than for laser scanning confocal microscopes; photobleaching and potential photodamage at the site of illumination (Tauer, 2002); IR-induced heating of samples that can easily boil solutions under a coverslip and literally cook your sample (personal experience and (Tauer, 2002)); relatively slow image acquisition times (10–15, 512 × 512 pixel images per second on commercial systems) compared to current laser scanning and spinning disk confocal systems due to low signal intensity; and lower resolution than confocal systems due to the long excitation wavelengths. These pros and cons are summarized in Table 9.8.

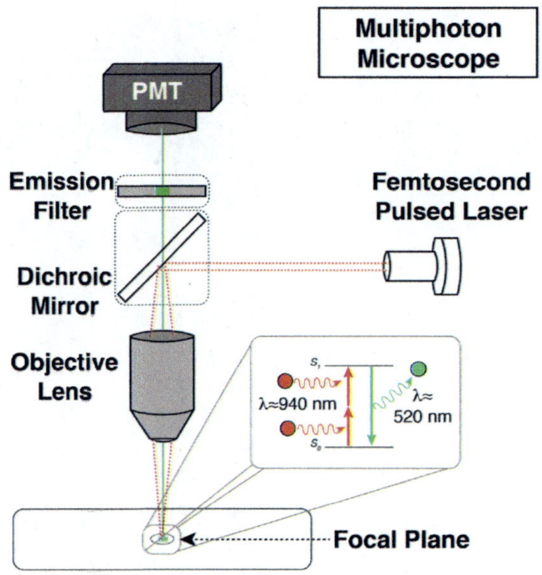

FIGURE 9.7 Multiphoton microscope design. Multiphoton microscopes utilize a femtosecond pulsed laser to deliver pulses of long-wavelength light, which arrive at the sample in sequence. These pulses have a near-additive effect, exciting the fluorophore as if a shorter wavelength of light had been used. The result is only excitation of the fluorophore at the focal point without the need for pinholes as used in confocal systems. Multiphoton microscopes are also ideal for deep-tissue imaging, since the low energy, long wavelength light can penetrate much deeper and does not illuminate above or below the plane of interest.

TABLE 9.8 Pros and cons of multiphoton calcium imaging.

Pros:	Cons:
Use of longer (700–1000 nm) excitation wavelengths reduce phototoxicity compared to UV excitation.	High cost of tunable, femtosecond IR lasers.
Reduced photo bleaching and photodamage above and below the objective focal plane.	Photobleaching and photodamage in the focal plane of the objective
The ability to excite multiple fluorophores with a single IR laser line.	IR-induced sample heating and subsequent thermal damage.
Ability to excite and release caged-compounds or photo bleach fluorophores in femtoliter volumes.	Image acquisition speed is limited to 10–15 images/s (at 512 × 512 pixels per image).
Lower scattering of IR excitation allows deep-tissue imaging.	Because of the long wavelengths used for imaging, resolution is less than can be achieved with shorter wavelength methods.

An example of the power of multiphoton microscopy as a tool in the lower urinary tract is shown in Fig. 9.8. A mouse urinary bladder was first incubated with the nuclear labeling dye Sytox Green, and the cell viability dye calcein-AM before being cannulated through the urethra and pressurized to 25 mmHg. Then, *z*-stack images were acquired across the bladder wall for both the green dyes and for collagen using second harmonic generation (shown in magenta). Images were capable of being collected across the entire bladder wall—even considering the curvature of the organ—that clearly showed both the cellular structures and the extracellular collagen matrix. While not being employed to highlight changes in intracellular calcium, these images do show the power of the technique to take deep images otherwise impossible to acquire with this level of detail in whole tissue using other microscopic methods.

Fiber photometry

In fiber photometry, an optical filament is implanted within specific brain regions of a living animal (usually as discussed above) to deliver and collect light. The resultant measurements of changes in global calcium fluorescence intensity at the

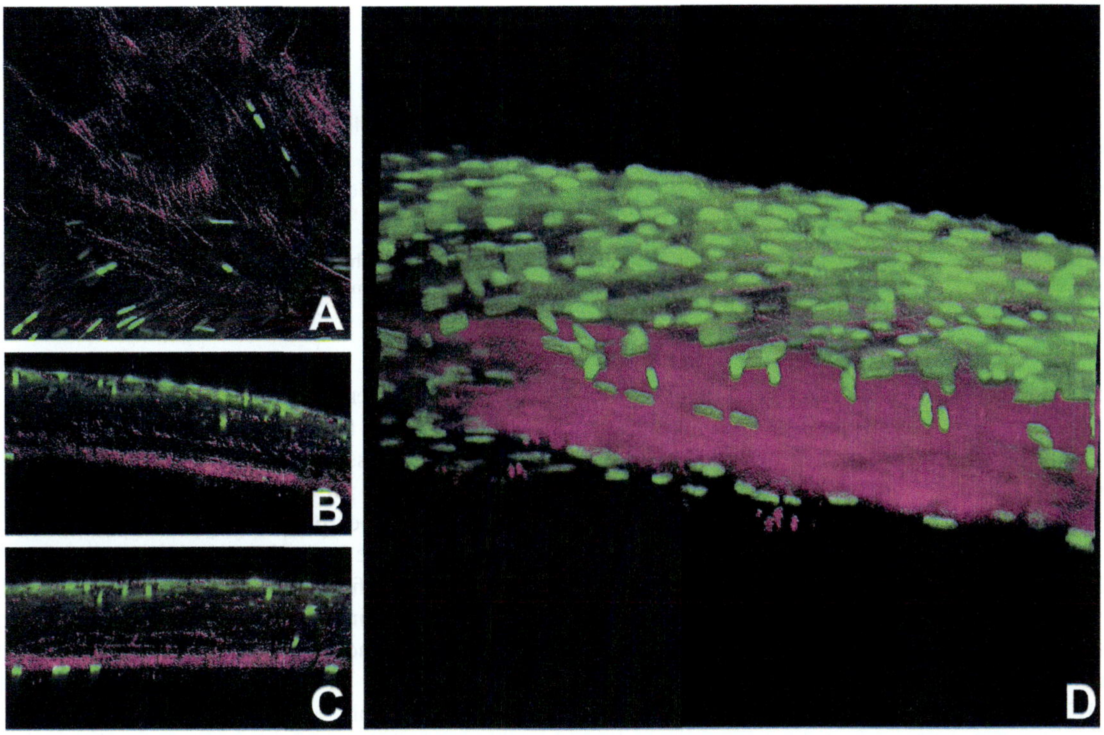

FIGURE 9.8 3D reconstruction of the mouse urinary bladder wall using multiphoton microscopy. Intact, cannulated, and pressurized mouse urinary bladder was incubated with Sytox green to label cell nuclei and calcein to label cell cytoplasm prior to imaging (*green*). Second harmonic generation was used to image collagen (*magenta*). Z-stacks were acquired using a Zeiss 880 LSM NLO microscope system. Stacks were then analyzed using ImageJ software (NIH, USA) to show views from an example z plane (A), x plane (B), and y plane (C). An orthogonal view of the entire bladder wall is shown in (D).

area of the fiber probe can be recorded at rates far greater than any other imaging technology and can be used to see exactly how cells in a given brain region respond to external stimuli (Resendez & Stuber, 2015). Even though basic fiber photometry is the only microscopy-based system discussed herein that does not create a traditional image of Ca^{2+} dynamics, this technique still offers never-before-seen insights into how neural excitability changes within the living brain. As shown in Fig. 9.9, fiber photometry uses many of the same basic optical tools as any other microscope: a laser or LED

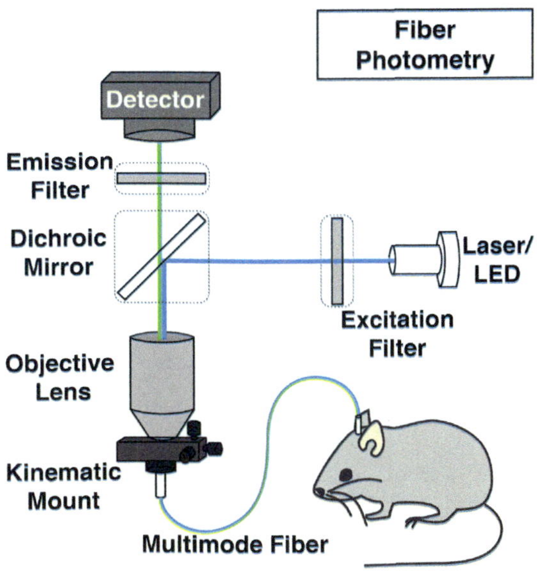

FIGURE 9.9 Fiber photometry design. In fiber photometry, excitation light is passed through a fiber optic filament implanted within the brain of a live mouse. The resulting emitted light then passes back up the fiber to the detector, where its intensity is measured. These systems can employ multiple fibers and lasers, allowing for several fluorophores to be imaged in a single living animal and from deep within the brain at very high acquisition rates.

TABLE 9.9 Pros and cons of fiber photometry calcium imaging.

Pros:	Cons:
Can be used to measure excitability in specific brain regions in living animals.	No images are collected from the results.
Higher recording rates than other microscope-based systems.	Requires very delicate surgery to accurately place fiber optics within the brain.
Allows measurements from deep brain areas otherwise inaccessible by microscopic imaging.	Usually requires the use of GECI animals to achieve cellular loading of calcium indicator.

light source, excitation and emission filters, dichroic mirrors, and objective lenses. The only difference is that instead of being focused on a tissue or sample, the light is instead focused on an optical fiber that is implanted within the brain of a living animal. These fibers can also be designed to allow independent transmission of different wavelengths of light at the same time, essentially allowing for ratiometric-like recordings of changes in GECI fluorescence within a population of neurons within the brain (Martianova et al., 2019). However, traditional fiber photometry can only record changes in fluorescence intensity and cannot generate a traditional image from within the brain. Modified fiber photometric systems can employ a graded-index (GRIN) lens, implanted similarly to the optical fiber, to generate images. Though larger in diameter than the traditional fiber (and thus creating more tissue damage during implantation), the information collected from these probes more closely resembles an image of discrete cells within the brain and allows for the determination of excitability within each cell in a network and not just the integrated activity of a whole population of cells (Girven & Sparta, 2017). The pros and cons of fiber photometry are summarized in Table 9.9.

An example of fiber photometry placement can be found in Fig. 9.10. This histological image shows the placement of the photometry fiber within the pons micturition center to record changes in Ca^{2+} in neurons projecting from the

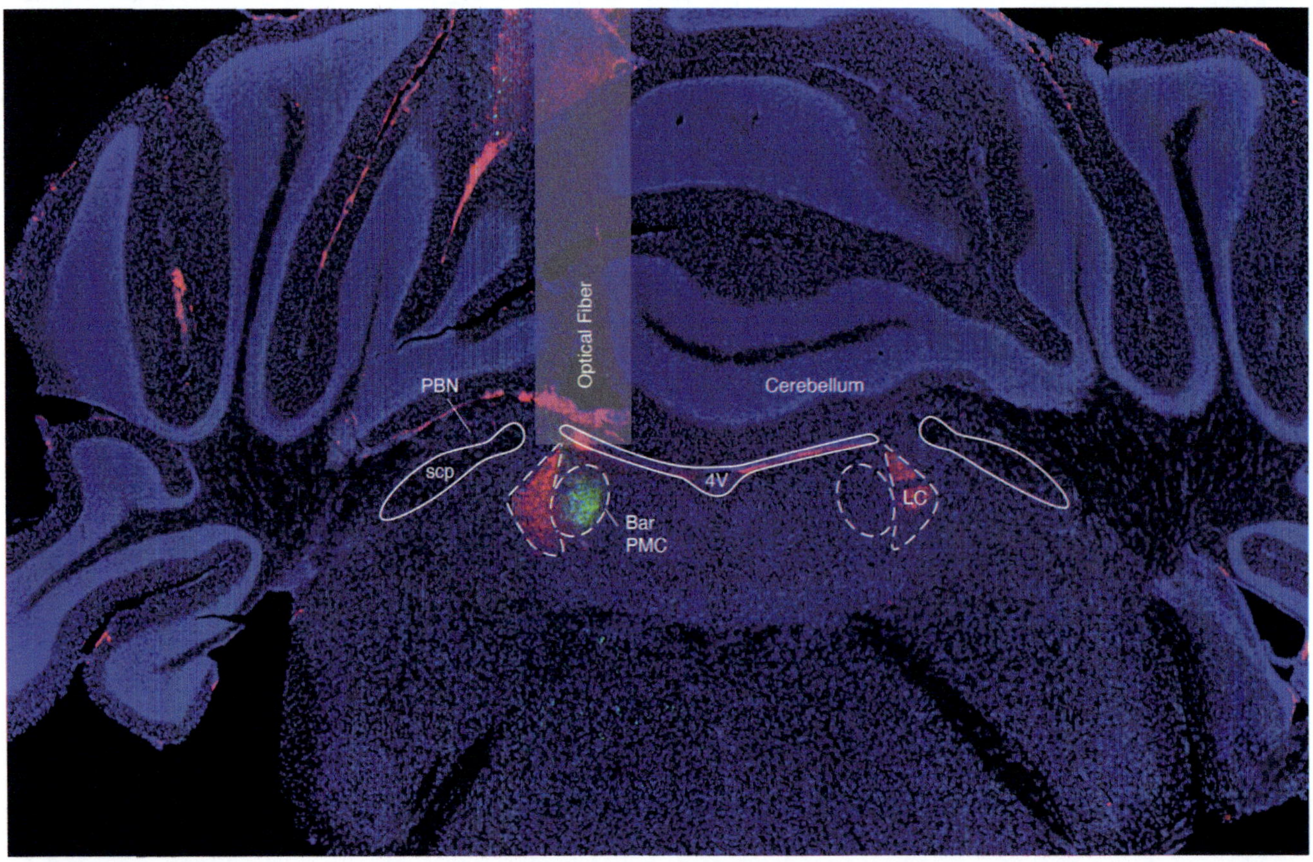

FIGURE 9.10 In vivo fiber photometry of inhibitory periaqueductal gray axon terminals projecting to pons micturition center. *4V*, 4th ventricle; *bar*, Barrington's nucleus; *LC*, locus coeruleus; *PAG*, periaqueductal gray; *PBN*, parabrachial nucleus; *PMC*, pontine micturition center. Histological image of mouse coronal brain section with DIO.GCaMP6s expressing vlPAG$^{Vgat\text{->}Bar}$ axon terminals (*green*), optical fiber placement, and tyrosine hydroxylase-labeled neurons (*red*) at the level of the pontine micturition center. *From Verstegen Lab. (2021). Beth Israel Deaconess Medical Center.*

periaqueductal gray. Since only these neurons express the GECI GCaMP6s, the signals recorded from these mice will only show excitability within this population of neurons without interference from other cells. Without fiber photometry, regions this deep within the brain would be impossible to record using optical techniques.

Beyond calcium: new frontiers to measure neuronal function

As the implementation of calcium imaging techniques has grown, so has the innovation in fluorescent dyes and protein sensors. New tools for measuring intracellular concentrations of monovalent cations are beginning to open the door to new ways of imaging neuronal function and dysfunction. In this section, we will explore some of the newest technologies for measuring fluxes of other ions in live cells and tissues and discuss their applicability to the study of neuronal connectivity in the lower urinary tract.

Membrane potential dyes and genetically encoded voltage indicators

As mentioned above, changes in calcium are often measured as a surrogate for neuronal firing; however, changes in calcium can also occur for reasons completely unrelated to action potentials traveling along an axon. Thus, tools to measure other ionic fluxes could be extremely useful in terms of understanding the nature of neural excitation. Membrane potential dyes are one such tool, as depolarization of neurons results in a very large change in voltage. Unfortunately, most membrane potential dyes are either too slow in their kinetics to reliably measure action potentials [e.g., DiBAC4(3); Epps et al., 1994] or too low a signal/noise ratio to affectively be used at high imaging rates [e.g., di-3-ANEPPDHQ; Fisher et al., 2008]. Newer membrane potential dyes have since been developed that provide a higher signal/noise ratio and faster, intensiometric kinetics capable of tracking action potentials. The FLIPR membrane potential (FMP) dye, commonly used in high-throughout screening, was recently discovered to accumulate in the cytosol of isolated neurons and to function well as a rapid voltage sensor (Fairless et al., 2013). Another such dye, FluoVolt (ThermoFisher Inc. USA), has shown promise in measuring electrical activity in intact mouse hearts (Salerno et al., 2020). This dye also has the distinct advantage of having the same excitation and emission wavelengths as FLUO-4 calcium imaging dyes, thus nearly all of the calcium imaging systems discussed above could be rapidly adapted to measure membrane potential changes instead.

Over the last 20 years, genetically encoded voltage indicators (GEVIs) were developed to try to match the response times and signal/noise ratio of the voltage-sensitive dyes with the selectivity of expression seen with the GECIs outlined above. Successful versions of these probes, such as SPARC and ArcLight, link eGFP to rat $Na_V1.4$ channel or a to transmembrane phosphatase isolated from the sea squirt *Ciona intestinalis* (*Ci*VSD), respectively (Kannan et al., 2019). Thus, SPARC and ArcLight take advantage of the voltage sensing domains of these two proteins to intensiometrically measure changes in voltage in much the same way GCaMPs utilized calmodulin to force a transformational change upon calcium binding. While still in development, GEVIs have been successfully used to measure in vivo neuronal excitability and represent a potential new avenue for investigating neuronal function in a more targeted manner.

pH sensitive indicators

Another useful measure of neuronal function is pH, specifically with regard to synaptic vesicle tracking. Since the pH of vesicles is extremely acidic as compared to the surrounding cytoplasm, pH-sensitive dyes have been extremely useful in tracking exocytosis and endocytosis (Aziz et al., 2013). The problem is that most of the dye conjugates used for tracking vesicle pH are membrane-impermeant, thus requiring endocytic uptake of the dye or introduction to the cytosol through a patch pipette. The tracking of neuronal vesicle pH using the genetically encoded pH sensor pHluorin compensates for the dye uptake issues but relies on the assumption that all synaptic vesicles have and use the same complement of proteins. The recent creation of a pH-sensitive organic cypHer5E dye, coupled with phospholipids, allows for the best of both worlds: it does not rely on genetic modification to insert the dye and it also allows for the repeated tracking of exocytosis and endocytosis at the presynaptic terminal (Kahms & Klingauf, 2018). The major downside of this new dye conjugate is its excitation wavelength is 640 nm, which requires the addition of a HeNe laser to the traditional argon-ion laser used for most intensiometric dyes. It requires different excitation and emission filters than most calcium indicators. Nonetheless, these dyes offer huge potential to measure synaptic release and recycling directly.

Genetically encoded K^+ indicators

Just as measuring Na^+ is necessary to record neuron depolarization, measuring intracellular K^+ is equally necessary when measuring the kinetics and mechanisms governing repolarization and maintenance of resting membrane potential. Until recently, intracellular K^+ was measured almost exclusively with electrodes that provided little to no data regarding spatiotemporal changes in K^+ (Frant & Ross, 1970). The K^+-sensitive dyes that have been developed suffer from either a distinct lack of specificity for K^+ over Na^+ or difficulty crossing the cell membrane, thus severely limiting their applicability to neuronal assays (Shen et al., 2019; Zhou et al., 2011). Now, new FRET-based genetically encoded K^+ indicators (GEKIs) and eGFP-based GEKIs are becoming promising new probes for measuring intracellular K^+ dynamics. As discussed earlier in this chapter, FRET-based probes rely on the transfer of energy between two proteins only when in close apposition to one another to ultimately cause fluorescence emission from the fluorophore. The FRET-based K^+ sensor KIRIN1 shows excellent selectivity for K^+ versus Na^+, making it ideal from the standpoint of specificity (Shen et al., 2019). But, FRET is complex to accurately measure as it relies heavily on both the efficiency of energy transfer from one fluorophore to the other and their alignment with one another (Piston & Kremers, 2007). To address this, the eGFP-based K^+ indicator GINKO1 was also developed for intensiometric measurements of K^+ dynamics at similar wavelengths to GCaMP proteins (Shen et al., 2019). Thus, accurate measurements of intracellular K^+ dynamics are on the horizon using the same systems already optimized for measuring Ca^{2+}.

Next-generation microelectrode arrays

As discussed above, traditional microelectrode-based techniques have three rather important drawbacks. First, they could only record for very short periods of time (seconds to minutes; occasionally hours in certain circumstances) (Soffe & Roberts, 1982). Second, a very limited number of neurons could be recorded simultaneously without interference between electrodes and within spatial constraints (Marx, 2014). Third, both sharp electrodes and patch pipets disrupted the cell membrane, which can lead to nonphysiological changes to the cell's electrical and functional properties (Hooper et al., 2015; Li et al., 2004; Standen et al., 1987). The advent of implantable microelectrode arrays over a decade ago meant that scientists were no longer limited in the duration of recording or by the number of electrodes one could cram in a cranial window. Animals could be fitted with an array that allowed the measurement of neuronal field potentials from multiple cells at once over a long period of time. This meant neural function could now be measured over weeks to months, as opposed to seconds to minutes with electrode-based recordings. The only drawback was resolution: while able to measure field potentials generated by action potentials, subthreshold potentials, and individual action potential characteristics remained below the detection limits (Spira & Hai, 2013).

With the continued development of nanotechnology and wireless communication, newer types of arrays and sensors are now available that gain the resolution needed to see single action potential firings in individual neurons from the brain of relatively free-moving animals (Spira & Hai, 2013). As the development of new communication modalities and better power management and delivery systems are developed, these techniques have the potential to enable the lifelong recording of neuronal activity within and outside of the brain. Also, these tools couple well with many of the optogenetic techniques used to manipulate cells and tissues in the periphery and may allow for a high-resolution mapping of neural excitation in response to peripheral manipulation (Luan et al., 2020). However, the intricacies of the surgical techniques required to implant these devices and the costs associated with them still place them out of reach for many scientists looking to study complex neural network interactions.

Summary

When it comes to calcium imaging, pictures are indeed worth much more than a thousand words—they offer an unprecedented view of the spatial and temporal characteristics of cellular excitability. For this reason, immensely important innovations in microscope and camera technology now allow us to see not just how changes in calcium drive cellular function but how it drives interactions between cells and tissues. Though each system may have its drawbacks and strengths, each can be used to explore how the neural networks of the brain and body interact to drive information to and from the lower urinary tract.

References

Akerboom, J., Carreras Calderon, N., Tian, L., Wabnig, S., Prigge, M., Tolo, J., Gordus, A., Orger, M. B., Severi, K. E., Macklin, J. J., Patel, R., Pulver, S. R., Wardill, T. J., Fischer, E., Schuler, C., Chen, T. W., Sarkisyan, K. S., Marvin, J. S., Bargmann, C. I., … Looger, L. L. (2013). Genetically encoded calcium indicators for multi-color neural activity imaging and combination with optogenetics. *Frontiers in Molecular Neuroscience, 6*(2), 2. https://doi.org/10.3389/fnmol.2013.00002

Akerboom, J., Chen, T. W., Wardill, T. J., Tian, L., Marvin, J. S., Mutlu, S., Calderon, N. C., Esposti, F., Borghuis, B. G., Sun, X. R., Gordus, A., Orger, M. B., Portugues, R., Engert, F., Macklin, J. J., Filosa, A., Aggarwal, A., Kerr, R. A., Takagi, R., … Looger, L. L. (2012). Optimization of a GCaMP calcium indicator for neural activity imaging. *Journal of Neuroscience, 32*(40), 13819–13840. https://doi.org/10.1523/JNEUROSCI.2601-12.2012

Anderson, H. E., Fontaine, A. K., Caldwell, J. H., & Weir, R. F. (2018). Imaging of electrical activity in small diameter fibers of the murine peripheral nerve with virally-delivered GCaMP6f. *Scientific Reports, 8*(1), 3219. https://doi.org/10.1038/s41598-018-21528-1

Anderson, H. E., Schaller, K. L., Caldwell, J. H., & Weir, R. F. F. (2019). Intravascular injections of adenoassociated viral vector serotypes rh10 and PHP.B transduce murine sciatic nerve axons. *Neuroscience Letters, 706*, 51–55. https://doi.org/10.1016/j.neulet.2019.05.010

Arai, I., & Jonas, P. (2014). Nanodomain coupling explains Ca^{2+} independence of transmitter release time course at a fast central synapse. *Elife, 3*. https://doi.org/10.7554/eLife.04057

Augustine, G. J., Santamaria, F., & Tanaka, K. (2003). Local calcium signaling in neurons. *Neuron, 40*(2), 331–346. https://doi.org/10.1016/s0896-6273(03)00639-1

Aziz, M., Yang, W. L., & Wang, P. (2013). Measurement of phagocytic engulfment of apoptotic cells by macrophages using pHrodo succinimidyl ester. *Current Protocols in Immunology*, 14–31. https://doi.org/10.1002/0471142735.im1431s100 (Chapter 14).

Berridge, M. J. (1998). Neuronal calcium signaling. *Neuron, 21*(1), 13–26. https://doi.org/10.1016/s0896-6273(00)80510-3

Bootman, M. D., Collins, T. J., Peppiatt, C. M., Prothero, L. S., MacKenzie, L., De Smet, P., Travers, M., Tovey, S. C., Seo, J. T., Berridge, M. J., Ciccolini, F., & Lipp, P. (2001). Calcium signalling—an overview. *Seminars in Cell & Developmental Biology, 12*(1), 3–10. https://doi.org/10.1006/scdb.2000.0211

Bozler, E. (1969). Role of calcium in initiation of activity of smooth muscle. *American Journal of Physiology, 216*(3), 671–674. https://doi.org/10.1152/ajplegacy.1969.216.3.671

Bretag, A. H. (2017). The glass micropipette electrode: A history of its inventors and users to 1950. *The Journal of General Physiology, 149*(4), 417–430. https://doi.org/10.1085/jgp.201611634

Brini, M. (2008). Calcium-sensitive photoproteins. *Methods, 46*(3), 160–166. https://doi.org/10.1016/j.ymeth.2008.09.011

Brini, M., Cali, T., Ottolini, D., & Carafoli, E. (2014). Neuronal calcium signaling: Function and dysfunction. *Cellular and Molecular Life Sciences, 71*(15), 2787–2814. https://doi.org/10.1007/s00018-013-1550-7

Cannell, M. B., Cheng, H., & Lederer, W. J. (1995). The control of calcium release in heart muscle. *Science, 268*(5213), 1045–1049. https://doi.org/10.1126/science.7754384

Chen, Q., Cichon, J., Wang, W., Qiu, L., Lee, S.-J., Campbell, N., DeStefino, N., Goard, M., Fu, Z., Yasuda, R., Looger, L., Arenkiel, B., Gan, W.-B., & Feng, G. (2012). Imaging neural activity using Thy1-GCaMP transgenic mice. *Neuron, 76*(2), 297–308.

Chen, Y., & Huang, L. M. (2017). A simple and fast method to image calcium activity of neurons from intact dorsal root ganglia using fluorescent chemical Ca^{2+} indicators. *Molecular Pain, 13*. https://doi.org/10.1177/1744806917748051, 1744806917748051.

Chen, T. W., Wardill, T. J., Sun, Y., Pulver, S. R., Renninger, S. L., Baohan, A., Schreiter, E. R., Kerr, R. A., Orger, M. B., Jayaraman, V., Looger, L. L., Svoboda, K., & Kim, D. S. (2013). Ultrasensitive fluorescent proteins for imaging neuronal activity. *Nature, 499*(7458), 295–300. https://doi.org/10.1038/nature12354

Cho, J. H., Swanson, C. J., Chen, J., Li, A., Lippert, L. G., Boye, S. E., Rose, K., Sivaramakrishnan, S., Chuong, C. M., & Chow, R. H. (2017). The GCaMP-R family of genetically encoded ratiometric calcium indicators. *ACS Chemical Biology, 12*(4), 1066–1074. https://doi.org/10.1021/acschembio.6b00883

Dana, H., Mohar, B., Sun, Y., Narayan, S., Gordus, A., Hasseman, J. P., Tsegaye, G., Holt, G. T., Hu, A., Walpita, D., Patel, R., Macklin, J. J., Bargmann, C. I., Ahrens, M. B., Schreiter, E. R., Jayaraman, V., Looger, L. L., Svoboda, K., & Kim, D. S. (2016). Sensitive red protein calcium indicators for imaging neural activity. *Elife, 5*, e12727. https://doi.org/10.7554/eLife.12727

Epps, D. E., Wolfe, M. L., & Groppi, V. (1994). Characterization of the steady-state and dynamic fluorescence properties of the potential-sensitive dye bis-(1,3-dibutylbarbituric acid)trimethine oxonol (Dibac4(3)) in model systems and cells. *Chemistry and Physics of Lipids, 69*(2), 137–150. https://doi.org/10.1016/0009-3084(94)90035-3

Fairless, R., Beck, A., Kravchenko, M., Williams, S. K., Wissenbach, U., Diem, R., & Cavalie, A. (2013). Membrane potential measurements of isolated neurons using a voltage-sensitive dye. *PLoS One, 8*(3), e58260. https://doi.org/10.1371/journal.pone.0058260

Fatt, P., & Katz, B. (1953). The electrical properties of crustacean muscle fibres. *The Journal of Physiology, 120*(1–2), 171–204. https://doi.org/10.1113/jphysiol.1953.sp004884

Fisher, J. A., Barchi, J. R., Welle, C. G., Kim, G. H., Kosterin, P., Obaid, A. L., Yodh, A. G., Contreras, D., & Salzberg, B. M. (2008). Two-photon excitation of potentiometric probes enables optical recording of action potentials from mammalian nerve terminals in situ. *Journal of Neurophysiology, 99*(3), 1545–1553. https://doi.org/10.1152/jn.00929.2007

Florey, E., & Kriebel, M. E. (1966). A new suction-electrode system. *Comparative Biochemistry & Physiology, 18*(1), 175–178. https://doi.org/10.1016/0010-406x(66)90342-2

Fontaine, A. K., Gibson, E. A., Caldwell, J. H., & Weir, R. F. (2017). Optical read-out of neural activity in mammalian peripheral axons: Calcium signaling at nodes of Ranvier. *Scientific Reports, 7*(1), 4744. https://doi.org/10.1038/s41598-017-03541-y

Frant, M. S., & Ross, J. W., Jr. (1970). Potassium ion specific electrode with high selectivity for potassium over sodium. *Science, 167*(3920), 987–988. https://doi.org/10.1126/science.167.3920.987

Gee, K. R., Brown, K. A., Chen, W. N., Bishop-Stewart, J., Gray, D., & Johnson, I. (2000). Chemical and physiological characterization of fluo-4 Ca^{2+}-indicator dyes. *Cell Calcium, 27*(2), 97–106. https://doi.org/10.1054/ceca.1999.0095

Girven, K. S., & Sparta, D. R. (2017). Probing deep brain circuitry: New advances in in vivo calcium measurement strategies. *ACS Chemical Neuroscience, 8*(2), 243–251. https://doi.org/10.1021/acschemneuro.6b00307

Goldman, D. E. (1943). Potential, impedance, and rectification in membranes. *The Journal of General Physiology, 27*(1), 37–60. https://doi.org/10.1085/jgp.27.1.37

Gopal, A. A., Kazarine, A., Dubach, J. M., & Wiseman, P. W. (2021). Recent advances in nonlinear microscopy: Deep insights and polarized revelations. *The International Journal of Biochemistry & Cell Biology, 130*, 105896. https://doi.org/10.1016/j.biocel.2020.105896

Gover, T. D., Kao, J. P., & Weinreich, D. (2003). Calcium signaling in single peripheral sensory nerve terminals. *Journal of Neuroscience, 23*(12), 4793–4797. https://doi.org/10.1523/jneurosci.23-12-04793.2003

de Groat, W. C., & Yoshimura, N. (2009). Afferent nerve regulation of bladder function in health and disease. *Handbook of Experimental Pharmacology, 194*, 91–138. https://doi.org/10.1007/978-3-540-79090-7_4

Grynkiewicz, G., Poenie, M., & Tsien, R. Y. (1985). A new generation of Ca^{2+} indicators with greatly improved fluorescence properties. *Journal of Biological Chemistry, 260*, 3440–3450.

Heppner, T. J., Tykocki, N. R., Hill-Eubanks, D., & Nelson, M. T. (2016). Transient contractions of urinary bladder smooth muscle are drivers of afferent nerve activity during filling. *The Journal of General Physiology, 147*(4), 323–335. https://doi.org/10.1085/jgp.201511550

Hille, B. (1985). Ionic channels of excitable membranes. In *Journal of neuroscience research* (3rd ed., pp. 1–22). Sinauer Associates.

Hodgkin, A. L., & Huxley, A. F. (1947). Potassium leakage from an active nerve fibre. *The Journal of Physiology, 106*(3), 341–367. https://doi.org/10.1113/jphysiol.1947.sp004216

Hodgkin, A. L., & Huxley, A. F. (1952). A quantitative description of membrane current and its application to conduction and excitation in nerve. *The Journal of physiology, 117*(4), 500–544. https://doi.org/10.1113/jphysiol.1952.sp004764

Hodgkin, A. L., & Keynes, R. D. (1957). Movements of labelled calcium in squid giant axons. *The Journal of physiology, 138*(2), 253–281. https://doi.org/10.1113/jphysiol.1957.sp005850

Hooper, S. L., Thuma, J. B., Guschlbauer, C., Schmidt, J., & Buschges, A. (2015). Cell dialysis by sharp electrodes can cause nonphysiological changes in neuron properties. *Journal of Neurophysiology, 114*(2), 1255–1271. https://doi.org/10.1152/jn.01010.2014

Inoue, M. (2020). Genetically encoded calcium indicators to probe complex brain circuit dynamics in vivo. *Neuroscience Research*. https://doi.org/10.1016/j.neures.2020.05.013

Inoue, M., Takeuchi, A., Manita, S., Horigane, S. I., Sakamoto, M., Kawakami, R., Yamaguchi, K., Otomo, K., Yokoyama, H., Kim, R., Yokoyama, T., Takemoto-Kimura, S., Abe, M., Okamura, M., Kondo, Y., Quirin, S., Ramakrishnan, C., Imamura, T., Sakimura, K., ... Bito, H. (2019). Rational engineering of XCaMPs, a multicolor GECI suite for in vivo imaging of complex brain circuit dynamics. *Cell, 177*(5), 1346–1360 e24. https://doi.org/10.1016/j.cell.2019.04.007

Jackson, V. M., Trout, S. J., Brain, K. L., & Cunnane, T. C. (2001). Characterization of action potential-evoked calcium transients in mouse postganglionic sympathetic axon bundles. *The Journal of Physiology, 537*(Pt 1), 3–16. https://doi.org/10.1111/j.1469-7793.2001.0003k.x

Johnson, I. D. (2010). *The Molecular Probes handbook. A guide to fluorescent probes and labeling technologies* (11th, pp. 1–1060). Life Technologies Corporation. https://www.thermofisher.com/us/en/home/references/molecular-probes-the-handbook.html. (Accessed 10 January 2021).

Kahms, M., & Klingauf, J. (2018). Novel pH-sensitive lipid based exo-endocytosis tracers reveal fast intermixing of synaptic vesicle pools. *Frontiers in Cellular Neuroscience, 12*, 18. https://doi.org/10.3389/fncel.2018.00018

Kannan, M., Vasan, G., & Pieribone, V. A. (2019). Optimizing strategies for developing genetically encoded voltage indicators. *Frontiers in Cellular Neuroscience, 13*, 53. https://doi.org/10.3389/fncel.2019.00053

Kettunen, P., Demas, J., Lohmann, C., Kasthuri, N., Gong, Y., Wong, R. O., & Gan, W. B. (2002). Imaging calcium dynamics in the nervous system by means of ballistic delivery of indicators. *Journal of Neuroscience Methods, 119*(1), 37–43. https://doi.org/10.1016/s0165-0270(02)00154-1

Lee, D., & Lee, A. K. (2017). In vivo patch-clamp recording in awake head-fixed rodents. *Cold Spring Harbour Protocols, 2017*(4). https://doi.org/10.1101/pdb.prot095802. pdb-prot095802.

Li, W. C., Soffe, S. R., & Roberts, A. (2004). A direct comparison of whole cell patch and sharp electrodes by simultaneous recording from single spinal neurons in frog tadpoles. *Journal of Neurophysiology, 92*(1), 380–386. https://doi.org/10.1152/jn.01238.2003

Liang, Y., Yuan, L. L., Johnston, D., & Gray, R. (2002). Calcium signaling at single mossy fiber presynaptic terminals in the rat hippocampus. *Journal of Neurophysiology, 87*(2), 1132–1137. https://doi.org/10.1152/jn.00661.2001

Lin, M. Z., & Schnitzer, M. J. (2016). Genetically encoded indicators of neuronal activity. *Nature Neuroscience, 19*(9), 1142–1153. https://doi.org/10.1038/nn.4359

Lohr, C. (2003). Monitoring neuronal calcium signalling using a new method for ratiometric confocal calcium imaging. *Cell Calcium, 34*(3), 295–303. https://doi.org/10.1016/s0143-4160(03)00105-2

Luan, L., Robinson, J. T., Aazhang, B., Chi, T., Yang, K., Li, X., Rathore, H., Singer, A., Yellapantula, S., Fan, Y., Yu, Z., & Xie, C. (2020). Recent advances in electrical neural interface engineering: minimal invasiveness, longevity, and scalability. *Neuron, 108*(2), 302–321. https://doi.org/10.1016/j.neuron.2020.10.011

Mank, M., & Griesbeck, O. (2008). Genetically encoded calcium indicators. *Chemical Reviews, 108*(5), 1550–1564. https://doi.org/10.1021/cr078213v

Martianova, E., Aronson, S., & Proulx, C. D. (2019). Multi-fiber photometry to record neural activity in freely-moving animals. *Journal of Visualized Experiments, 152*. https://doi.org/10.3791/60278

Marx, V. (2014). Neurobiology: Rethinking the electrode. *Nature Methods, 11*(11), 1099–1103. https://doi.org/10.1038/nmeth.3149

Minta, A., Kao, J. P., & Tsien, R. Y. (1989). Fluorescent indicators for cytosolic calcium based on rhodamine and fluorescein chromophores. *Journal of Biological Chemistry, 264*(14), 8171–8178.

Nakai, J., Ohkura, M., & Imoto, K. (2001). A high signal-to-noise Ca(2+) probe composed of a single green fluorescent protein. *Nature Biotechnology, 19*(2), 137–141. https://doi.org/10.1038/84397

Nullens, S., Deiteren, A., Jiang, W., Keating, C., Ceuleers, H., Francque, S., Grundy, D., De Man, J. G., & De Winter, B. Y. (2016). In vitro recording of mesenteric afferent nerve activity in mouse jejunal and colonic segments. *Journal of Visualized Experiments, 116*. https://doi.org/10.3791/54576

Ohkura, M., Matsuzaki, M., Kasai, H., Imoto, K., & Nakai, J. (2005). Genetically encoded bright Ca2+ probe applicable for dynamic Ca2+ imaging of dendritic spines. *Analytical Chemistry, 77*(18), 5861–5869. https://doi.org/10.1021/ac0506837

Ohkura, M., Sasaki, T., Sadakari, J., Gengyo-Ando, K., Kagawa-Nagamura, Y., Kobayashi, C., Ikegaya, Y., & Nakai, J. (2012). Genetically encoded green fluorescent Ca^{2+} indicators with improved detectability for neuronal Ca^{2+} signals. *PLoS One, 7*(12), e51286. https://doi.org/10.1371/journal.pone.0051286

Petr, M. J., & Wurster, R. D. (1997). Determination of in situ dissociation constant for Fura-2 and quantitation of background fluorescence in astrocyte cell line U373-MG. *Cell Calcium, 21*(3), 233–240. https://doi.org/10.1016/s0143-4160(97)90047-6

Piston, D. W., & Kremers, G. J. (2007). Fluorescent protein FRET: The good, the bad and the ugly. *Trends in Biochemical Sciences, 32*(9), 407–414. https://doi.org/10.1016/j.tibs.2007.08.003

Resendez, S. L., & Stuber, G. D. (2015). In vivo calcium imaging to illuminate neurocircuit activity dynamics underlying naturalistic behavior. *Neuropsychopharmacology, 40*(1), 238–239. https://doi.org/10.1038/npp.2014.206

Salerno, S., Garten, K., Smith, G. L., Stolen, T., & Kelly, A. (2020). Two-photon excitation of FluoVolt allows improved interrogation of transmural electrophysiological function in the intact mouse heart. *Progress in Biophysics and Molecular Biology, 154*, 11–20. https://doi.org/10.1016/j.pbiomolbio.2019.08.007

Sever, R. (2020). *Cold Spring Harbor protocols*. Cold Spring Harbor Laboratory Press. http://cshprotocols.cshlp.org.

Shemetov, A. A., Monakhov, M. V., Zhang, Q., Canton-Josh, J. E., Kumar, M., Chen, M., Matlashov, M. E., Li, X., Yang, W., Nie, L., Shcherbakova, D. M., Kozorovitskiy, Y., Yao, J., Ji, N., & Verkhusha, V. V. (2020). A near-infrared genetically encoded calcium indicator for in vivo imaging. *Nature Biotechnology*. https://doi.org/10.1038/s41587-020-0710-1

Shen, Y., Wu, S. Y., Rancic, V., Aggarwal, A., Qian, Y., Miyashita, S. I., Ballanyi, K., Campbell, R. E., & Dong, M. (2019). Genetically encoded fluorescent indicators for imaging intracellular potassium ion concentration. *Communications Biology, 2*, 18. https://doi.org/10.1038/s42003-018-0269-2

Soffe, S. R., & Roberts, A. (1982). Tonic and phasic synaptic input to spinal cord motoneurons during fictive locomotion in frog embryos. *Journal of Neurophysiology, 48*(6), 1279–1288. https://doi.org/10.1152/jn.1982.48.6.1279

Spira, M. E., & Hai, A. (2013). Multi-electrode array technologies for neuroscience and cardiology. *Nature Nanotechnology, 8*(2), 83–94. https://doi.org/10.1038/nnano.2012.265

Standen, N. B., Gray, P. T. A., Whitaker, M. J., & Biologists, C. of (1987). *Microelectrode techniques: The Plymouth workshop handbook*. Company of Biologists. https://books.google.com/books?id=dtqyAAAAIAAJ.

Stockert, J. C., & Blázquez-Castro, A. (2017). *Fluorescence microscopy in life sciences*. Bentham Science Publishers Ltd. http://search.ebscohost.com/login.aspx?direct=true&db=e000xna&AN=1684109&site=ehost-live.

Sun, X. R., Badura, A., Pacheco, D. A., Lynch, L. A., Schneider, E. R., Taylor, M. P., Hogue, I. B., Enquist, L. W., Murthy, M., & Wang, S. S. (2013). Fast GCaMPs for improved tracking of neuronal activity. *Nature Communications, 4*(1), 2170. https://doi.org/10.1038/ncomms3170

Takahashi, A., Camacho, P., Lechleiter, J. D., & Herman, B. (1999). Measurement of intracellular calcium. *Physiological Reviews, 79*(4), 1089–1125. https://journals.physiology.org/doi/pdf/10.1152/physrev.1999.79.4.1089.

Tallini, Y. N., Brekke, J. F., Shui, B., Doran, R., Hwang, S. M., Nakai, J., Salama, G., Segal, S. S., & Kotlikoff, M. I. (2007). Propagated endothelial Ca^{2+} waves and arteriolar dilation in vivo: Measurements in Cx40BAC GCaMP2 transgenic mice. *Circulation Research, 101*(12), 1300–1309. https://doi.org/10.1161/CIRCRESAHA.107.149484

Tallini, Y. N., Ohkura, M., Choi, B. R., Ji, G., Imoto, K., Doran, R., Lee, J., Plan, P., Wilson, J., Xin, H. B., Sanbe, A., Gulick, J., Mathai, J., Robbins, J., Salama, G., Nakai, J., & Kotlikoff, M. I. (2006). Imaging cellular signals in the heart in vivo: Cardiac expression of the high-signal Ca^{2+} indicator GCaMP2. *Proceedings of the National Academy of Sciences of the United States of America, 103*(12), 4753–4758. https://doi.org/10.1073/pnas.0509378103

Tanaami, T., Otsuki, S., Tomosada, N., Kosugi, Y., Shimizu, M., & Ishida, H. (2002). High-speed 1-frame/ms scanning confocal microscope with a microlens and Nipkow disks. *Applied Optics, 41*(22), 4704–4708. https://doi.org/10.1364/ao.41.004704

Tauer, U. (2002). Advantages and risks of multiphoton microscopy in physiology. *Experimental Physiology, 87*(6), 709–714. https://doi.org/10.1113/eph8702464

Thomas, D., Tovey, S. C., Collins, T. J., Bootman, M. D., Berridge, M. J., & Lipp, P. (2000). A comparison of fluorescent Ca^{2+} indicator properties and their use in measuring elementary and global Ca^{2+} signals. *Cell Calcium, 28*(4), 213–223. https://doi.org/10.1054/ceca.2000.0152

Tian, L., Hires, S. A., Mao, T., Huber, D., Chiappe, M. E., Chalasani, S. H., Petreanu, L., Akerboom, J., McKinney, S. A., Schreiter, E. R., Bargmann, C. I., Jayaraman, V., Svoboda, K., & Looger, L. L. (2009). Imaging neural activity in worms, flies and mice with improved GCaMP calcium indicators. *Nature Methods, 6*(12), 875–881. https://doi.org/10.1038/nmeth.1398

Tsien, R. W., & Tsien, R. Y. (1990). Calcium channels, stores, and oscillations. *Annual Review of Cell Biology, 6*, 715–760. https://doi.org/10.1146/annurev.cb.06.110190.003435

Wang, E., Babbey, C. M., & Dunn, K. W. (2005). Performance comparison between the high-speed Yokogawa spinning disc confocal system and single-point scanning confocal systems. *Journal of Microscopy, 218*(Pt 2), 148–159. https://doi.org/10.1111/j.1365-2818.2005.01473.x

Webb, D. J., & Brown, C. M. (2013). Epi-fluorescence microscopy. *Methods in Molecular Biology, 931*, 29–59. https://doi.org/10.1007/978-1-62703-056-4_2

Wellman, G. C., Nathan, D. J., Saundry, C. M., Perez, G., Bonev, A. D., Penar, P. L., Tranmer, B. I., & Nelson, M. T. (2002). Ca^{2+} sparks and their function in human cerebral arteries. *Stroke, 33*(3), 802–808.

Whitaker, M., Kao, J. P. Y., Li, G., & Auston, D. A. (2010). Chapter 5—practical aspects of measuring intracellular calcium signals with fluorescent indicators. In , *99. Methods in Cell Biology* (pp. 113–152). Academic Press.

Wokosin, D. L., Loughrey, C. M., & Smith, G. L. (2004). Characterization of a range of fura dyes with two-photon excitation. *Biophysical Journal, 86*(3), 1726–1738. https://doi.org/10.1016/S0006-3495(04)74241-1

Xu, C., Zipfel, W., Shear, J. B., Williams, R. M., & Webb, W. W. (1996). Multiphoton fluorescence excitation: New spectral windows for biological nonlinear microscopy. *Proceedings of the National Academy of Sciences of the United States of America, 93*(20), 10763–10768. https://doi.org/10.1073/pnas.93.20.10763

Zhou, X., Su, F., Tian, Y., Youngbull, C., Johnson, R. H., & Meldrum, D. R. (2011). A new highly selective fluorescent K^+ sensor. *Journal of the American Chemical Society, 133*(46), 18530–18533. https://doi.org/10.1021/ja207345s

Chapter 10

The periaqueductal gray and control of bladder function

Takeya Kitta
Department of Renal and Urologic Surgery, Asahikawa Medical University, Asahikawa, Hokkaido, Japan

Introduction

The periaqueductal gray (PAG) is one of the most preserved evolutionary components of the brain. After the discovery of a role for the PAG in modulating nociception (Basbaum & Fields, 1978), many researchers believe that the PAG only controls pain perception. However, pain control greatly depends on the state of the individual. The goal of the brain is survival of the individual and hence of the species. In this context, nociception control is but one of the many other systems that attempt to achieve that goal. For example, individuals fighting for their life should not be bothered by perceptions of pain. In a situation like this, pain or nociception is inhibited by the PAG because all energy has to go to the motor systems that allow the individual to survive. These motor systems produce the fight-or-flight response (Bandler et al., 1991). However, in safe situations, the PAG plays a role in a variety of functions including autonomic and emotional control (Mouton & Holstege, 1994), vocalization, respiration, as well as micturition control. Due to its axial location, it is well situated at the crossroads of ascending sensory information and inputs from higher centers that modulate these processes. **Neural activity is high in the PAG during micturition as detected by immunohistochemistry** (Zare et al., 2019). Many type of neurotransmitters were revealed in this activity. As a result of the development and progress of functional brain imaging, more insight has been gained into the specific brain regions involved in the micturition reflex. In this chapter, we provide an overview of the current state of research on the PAG region and its neurons, the complex circuitry within the PAG, functional brain imaging of micturition reflex, and how this area acts as a "switch" for the micturition reflex.

Anatomy and neural network of the PAG

PAG columns

The PAG is the gray matter structure surrounding the cerebral aqueduct within the midbrain. It is shaped like a celery stalk, with the regions **in the midline ventral to the aqueduct and which is** arranged into separate well-differentiated nuclei from other structure. The PAG is composed of functionally separate columns (Bandler & Shipley, 1994; Parvizi et al., 2000) (Fig. 10.1). Based on neuroanatomical connectivity patterns and cytoarchitectural features, the PAG has been subdivided into the following four longitudinal columns **spanning the rostral to caudal extent**: the dorsomedial PAG (dmPAG), dorsolateral PAG (dlPAG), lateral PAG (lPAG), and ventrolateral PAG (vlPAG) (Keay & Bandler, 2002).

Functional classification of PAG columns

PAG columns are functionally divided into two groups with opposite autonomic functions: the vlPAG, which has parasympathetic functions, and the dlPAG and lPAG, which have sympathetic functions (Benarroch, 2012). All PAG columns have ipsilateral and contralateral connections and all PAG columns are interconnected (Jansen et al., 1998), and the PAG consists of distinct columns that receive selective inputs from a lot of regions. These intrinsic network activation or inhibition of one column will affect the output neurons of the other columns as a part of this intrinsic neuromodulatory circuit (Katz & Frost, 1996). The vlPAG receives direct and indirect afferents from the lumbosacral cord (Ding et al., 1997) and

Neuro-Urology Research. https://doi.org/10.1016/B978-0-12-822455-7.00010-6

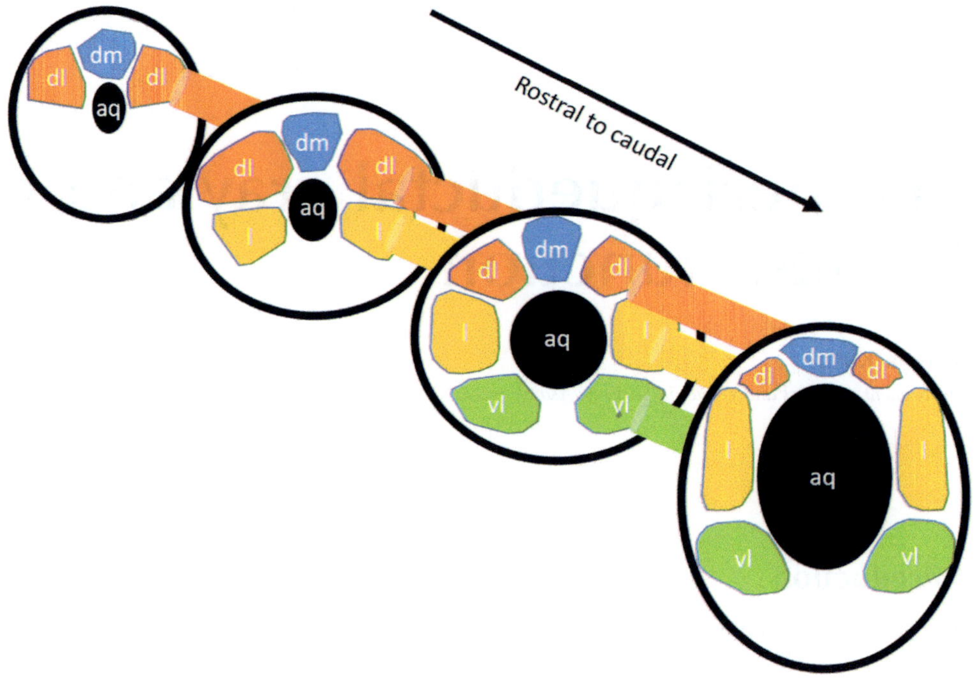

FIGURE 10.1 Schematic illustration of the periaqueductal gray (PAG) columns. *aq*, aqueduct of midbrain; *dl*, dorsolateral; *dm*, dorsomedial; *l*, lateral; *PAG*, periaqueductal gray; *vl*, ventrolateral. Schematic illustration of the dorsomedial, dorsolateral, lateral, and ventrolateral neuronal columns. A coronal section through the rostral to caudal part of the PAG showing columnar segmentations. Two pairs of columns with major functional contribution in micturition.

has direct connections with the Pontine micturition center (PMC), which controls micturition by its projections to sacral parasympathetic segments. Various studies have mentioned the vlPAG as the column associated with bladder function (Matsuura et al., 1998; Mitsui et al., 2003; Taniguchi et al., 2002**). Noto et al. showed that stimulation of the pelvic nerve afferents induced evoked potentials in the dorsal PAG and that the optimal site for inducing bladder contractions was in the ventral PAG. This suggests that afferent and efferent pathways are distinct in the PAG. Based on their results, they suggested that it is unlikely that the PAG is only a relay station and raised the possibility that the dorsal part of the PAG may have modulatory effects on micturition** (Noto et al., 1991). This has been confirmed by methods probing both ascending and descending tracts (Loewy et al., 1979). **Loewy et al.** were using the autoradiographic transport technique and horseradish peroxidase method combined with monoamine oxidase staining technique to explore the exact origin, course, and termination of the micturition reflex pathway. **And columns of dPAG and vlPAG play a role as follows.** Noxious stimuli at the level of the bladder have been shown to increase c-Fos expression in the vlPAG in rats (Mitsui et al., 2003). Electrical or chemical stimulation of the vlPAG in rats resulted in contraction of the bladder or increased micturition frequency (Stone et al., 2015). Likewise, the injection of inhibitory mediators into the vlPAG has been shown to attenuate bladder contractions and external urethral sphincter electromyographic activity in rats (Matsuura et al., 2000).

PAG projections to the pontine micturition center

The gold standard of neural network research remains anterograde and retrograde tract tracing in animals (Aggleton et al., 1980; Mantyh, 1982). Last decade, diffusion-weighted imaging has been continuously improved to probe random microscopic motion of water protons on a per pixel basis (Moseley et al., 1990). **Diffusion-weighted imaging has become an important modality in the diagnostic workup in the central nervous system (CNS). Diffusion tensor imaging is a novel imaging technique that can reveal unique information of white matter microstructures within the CNS** (Meoded & Huisman, 2019) **. This methodology can confirm brain pathways in humans** (Sillery et al., 2005; Pereira et al., 2010). The extensive connections of the PAG in humans have been studied using diffusion tensor imaging (Table 10.1) (Sillery et al., 2005). PAG fibers project to the lateral pontine tegmentum, and particularly strong projections have been observed to Barrington's nucleus (Barrington, 1927; Verstegen et al., 2019), also known as the M-region (Holstege et al., 1986), the pelvic organ−stimulating center (Holstege, 2014), and the PMC in the dorsal pons. In turn, the cells in the PMC project to sacral parasympathetic neurons (Gert Holstege & Kuypers, 1982). In this chapter, this region is described as the "PMC."

TABLE 10.1 PAG connections based on diffusion tensor imaging.
PAG projections to many regions.Diffusion tensor imaging is a novel imaging technique that can reveal unique information of white matter microstructures within the central nervous system (Meoded & Huisman, 2019). This methodology can confirm brain pathways in humans (Sillery et al., 2005; Pereira et al., 2010).

Cerebellum	Dorsomedial prefrontal cortex
Medulla	Ventromedial prefrontal cortex
Thalamus	Amygdala
Precentral gyrus	Hypothalamus
Postcentral gyrus	Subcaudate white matter
Anterior cingulate cortex	Accumbens
Ventrolateral prefrontal cortex	Internal capsule

The PAG is a "gate" for the activity of PMC neurons

The role of the PMC in the control of micturition and urinary bladder function has been well documented (Griffiths & Fowler, 2013; Matsuura et al., 2000; Noto et al., 1989). The axons of PMC neurons project directly to the sacral spinal cord (Holstege, 2005), where the pelvic organs, including the ureter, urinary bladder, urinary sphincter, vagina, rectum, anal sphincter, prostate, and penis, are innervated by projection of downward motoneurons (Banrezes et al., 2002; Dorofeeva et al., 2009; Papka et al., 1995); hence, PMC has been termed the pelvic organ–stimulating center (Holstege, 2014). A complete **synergistic** micturition response is initiated upon activation of the PMC through the induction of detrusor contractions and the relaxation of the striated muscles of the pelvic floor. In addition, the **PMC has reciprocal connections with the vlPAG** (Kuipers et al., 2006). Additional evidence in regard to the importance of the vlPAG in micturition control is presented as follows. A previous study reported that electrical or chemical stimulation of the vlPAG by electrodes or D,L-homocysteic acid increased the frequency of micturition in rats (Stone et al., 2015; Taniguchi et al., 2002). By contrast, stereotaxic injection of cobalt chloride ($CoCl_2$), an inhibitory mediator, into the caudal vlPAG was found to attenuate bladder contractions and the electromyographic activity of the external urethral sphincter reversibly in rats (Matsuura et al., 1998). The injection of $CoCl_2$ or other inhibitory or stimulatory agents, such as L-glutamate, into the vlPAG has been found to lead to voiding suppression or stimulation, respectively, in rats (Matsuura et al., 2000). Taking these studies into account, the PAG is therefore considered to lie at the crossroads between ascending sensory information and inputs from higher centers modulating these processes, thereby serving as a coordination center (Blok et al., 1995; Fowler & Griffiths, 2010; Holstege et al., 1985; Holstege, 1987; Hopkins & Holstege, 1978; Krout & Loewy, 2000). Subsequently, it may relay these signals to the other PAG columns secondarily. Higher brain structures decide whether micturition should take place based on visual, auditory, olfactory, and somatosensory information. The major forebrain input to the PAG is provided by the prefrontal cortex, in which the medial wall, anterior cingulate gyrus, and posterior orbitofrontal and anterior insular cortices target the dlPAG, lPAG, vlPAG, and dmPAG, and the vlPAG, respectively. Regarding inputs, those from the amygdala originate in the central nucleus and ventrolateral portion of the basal nucleus, while those from the hypothalamus originate in many different nuclei and lateral hypothalamic areas. These areas also contribute to the decision regarding whether the PAG should activate the PMC (**final go sign of micturition reflex**) (Holstege et al., 1985; Holstege, 1987; Hopkins & Holstege, 1978). These reports are corresponding with human functional magnetic resonance imaging (fMRI) study during micturition (Griffiths & Fowler, 2013). Few studies report on projection targets of the vlPAG. Axonal projections of vlPAG neurons that project to the thalamus and the insula will finally reach the medial prefrontal cortex.

Moreover, the PAG receives a bundle of noradrenergic and adrenergic fibers that originate in the ventrolateral (A1 and C1 groups) and dorsomedial (A2 and C3 groups) medulla (Herbert & Saper, 1992). Nociceptive information is sent to the contralateral PAG by neurons of lamina I within the superficial dorsal horn and caudal trigeminal nucleus, and these projections target the lPAG and vlPAG (Mouton et al., 2001). From our study and related reports, **vlPAG neurons play the central role in micturition control.**

Study using c-Fos expression levels in the PAG (activity of bladder control)

In our previous study (Mitsui et al., 2003), we used immunohistochemical detection of the c-Fos protein, which is a transcription factor expressed after neuronal activation, in the CNS of rats after urinary bladder stimulation to identify sensory processing after receiving the sensory bladder—related signals. A variety of stimulation protocols of the lower urinary tract have been reported to cause changes in central pathways such as vlPAG, PMC, and spinal cord (Birder & De Groat, 1992; Mitsui et al., 2003). In addition, the vlPAG has been found to be responsible for sensorimotor communication with the bladder, and **c-Fos reactivity is** significantly higher in the vlPAG of rats exposed to noxious stimuli (acetic acid) compared with a saline-infused group (Mitsui et al., 2003). The stimulation of different types of afferent fibers may explain this difference in columnar activation. For example, in terms of acetic acid infusion, the response to noxious stimuli is primarily by unmyelinated C-fibers (Aizawa et al., 2018). It should be kept in mind that c-Fos expression is a marker of neuronal activity, but not of an efferent result (such as a muscle contraction) or the effect on a distant neuronal target such as the PMC, which is controlled by the PAG. As with the activation of excitatory neurons, that of inhibitory neurons (-GABAergic interneurons) causes distinctive actions at the distant side; therefore, the type of neuron that was activated cannot be differentiated. This issue has recently been investigated by Zare et al., who studied the coexpression of c-Fos and neuronal markers in the PAG (Zare et al., 2018). To gain a better understanding of these mechanisms (what kind of neurotransmitter play distinct action, etc.), the exact functional cell populations in the PAG must be determined (given later in this chapter).

Functional cell populations in the PAG

The PAG contains a wide variety of neurons, including glutamatergic (Siegfried & Nunes de Souza, 1989), dopaminergic (DA) neurons (Lu et al., 2006), g-aminobutyric acid (GABA) (Huo et al., 2005), serotoninergic (Clements et al., 1985), m-opioid (Fields, 2004), neurokinin-1 (Gregg & Siegel, 2003), neuronal nitric oxide synthase (Chiavegatto et al., 1998), and other neurotransmitters, all of which have been confirmed by immunohistochemically study.

Glutamatergic control of micturition

Glutamate acts as an excitatory neurotransmitter. In studies in the cat brain, 30% of PAG neurons have been reported to be positively labeled for glutamate (Barbaresi et al., 1997). Zare et al. highlighted the role of glutamatergic cells in the bladder—PAG circuitry and noted that glutamatergic cells in the vlPAG control other important functions such as analgesia and freezing behavior (Samineni et al., 2017). As glutamatergic receptors in the vlPAG of the rat are activated following electrical bladder stimulation, they are involved in sensorimotor processing by relaying peripheral sensory signals to the micturition motor control centers. In addition, to facilitate the micturition reflex, **glutamatergic neurons** in the vlPAG are thought to stimulate the PMC when receiving suprathreshold sensory signals from a full bladder and may project to higher cortical regions for further analysis and decision-making.

Dopaminergic control of micturition

Histological studies have indicated that the PAG is composed of two different types of DA neurons: small and large(Flores et al., 2004; Hasue & Shammah-Lagnado, 2002). A pharmacological and behavioral study reported that PAG neurons expressing D1 (but not D2) DA receptors are involved in the modulation of opiate-induced analgesia and furthermore that D1 DA receptors in the PAG play a functional role in nociceptive responses (Flores et al., 2004). At least, D1 DA receptor could inhibit nociceptive responses. And in that study, the D1 receptor agonist SKF-38,393 was shown not to alter the micturition reflex, whereas this reflex was facilitated by the D1 DA receptor antagonist. Those findings indicate the presence of tonic DA regulation of the micturition reflex through the D1 DA receptors. In other words, D1 DA but not D2 DA receptors seem to have an essential role in the DA mechanism involved in regulating the nociceptive signaling part of the micturition reflex, at least in the PAG.

GABAergic control of micturition

The PAG is composed of a rich population of GABAergic interneurons (Griffiths & Lovick, 2005; Reichling & Basbaum, 1990), which exert a tonic inhibitory influence on output neurons (Behbehani et al., 1990; Brack & Lovick, 2007; Ogawa et al., 1994). A previous study reported that microinjection of a GABA agonist into the PAG suppressed the voiding reflex during the storage phase, whereas microinjection of a GABA antagonist increased voiding activity (Stone et al., 2011). To our knowledge, the effect of deep brain stimulation (DBS) of the PAG on GABA levels has not been investigated; however, DBS of the subthalamic nucleus has been shown to generate inhibitory postsynaptic potentials in subthalamic nucleus neurons through the activation of GABAergic afferents (Lee et al., 2004). Thus, DBS of the PAG may induce local

GABA release from the axon terminals attributing to inhibited micturition. A number of different behavioral responses controlled by the PAG, including defensive behavior, flight, antinociception, and tonic immobility, have been found to be subject to ongoing inhibitory GABAergic control (Monassi et al., 1999; Morgan & Clayton, 2005). The primary inhibitory GABAergic input to the PMC relevant to micturition is projected from the dlPAG (Numata et al., 2008). **Numata et al.** speculated that dlPAG includes the micturition-suppressing region that seems to have neural connections with the PMC of the cat. From their electrical and pharmacological approach, GABA is assumed to be one of the neurotransmitters that are involved in the PMC inhibition from the micturition-suppressing region in the PAG (Numata et al., 2008). Furthermore, in our previous study in conscious rats, we found an association between voiding and a decrease in the concentration of extracellular GABA in the PAG (Kitta et al., 2008), which suggests that transmission in the micturition reflex pathway is inhibited by ongoing GABAergic activity between voids.

Serotonergic (5-hydroxytryptamine) control of micturition

There are 14 serotoninergic (5-hydroxytryptamine [5-HT]) receptor subtypes in the CNS that have been classified into seven receptor families (5-HT 1–7) by their characteristics (structural, functional, and pharmacological) (Barnes & Sharp, 1999). Although few studies have investigated the micturition-related pathways involving 5-HT cells in the PAG, the serotoninergic vlPAG is known to be an important regulatory mechanism for inhibiting ejaculation in rats and may contribute to the selective serotonin receptor inhibitor–induced inhibition of ejaculation. However, the regulation of the autonomic function of the genital organs differs somewhat from that of the bladder (Normandin & Murphy, 2011). **Normandin and Murphy** focused on the serotonergic regulation of the nucleus paragigantocellularis (a primary source of inhibition of ejaculation in male rats) and vlPAG. Their data that serotonergic lesions of the vlPAG facilitated sexual behavior provide evidence of the functional serotonergic connectivity between the vlPAG and paragigantocellularis.

In vivo microdialysis study

A variety of experimental approaches have been taken to gain a better understanding of the supraspinal control of the lower urinary tract. Animal studies using a variety of techniques, including tract tracing, electrical stimulation, lesioning, and neural recording, have been instrumental for identifying key subcortical regions for bladder control. The PAG plays a major role in sending bladder-filling information to higher centers and receives afferent inputs from a number of higher brain regions, including the prefrontal cortex; therefore, it is a major control center in terms of the facilitation or inhibition of the PMC, depending on the safety of danger for voiding. Thus, the role of the PAG appears to be more important under an arousal than under an anesthetized status. In our previous study, we first implemented this protocol under an anesthetized condition. We found that under this condition, not only bladder functional parameters but also changes in neurotransmitters were inhibited. As neural activity in the PAG was affected under the anesthetized condition (Table 10.2), we decided to extend this protocol to freely moving rats. We found that the glutamate and 5-HT neurons in the PAG participated in the modulation of both nociception and the micturition reflex differently. Furthermore, 5-HT levels in the PAG appeared to reflect the nociceptive stimuli (Kitta et al., 2016).

TABLE 10.2 Extracellular levels of glutamate and serotonin in the ventrolateral PAG. *Extracellular levels of glutamate and serotonin in the ventrolateral PAG after saline (micturition reflex) or acetic acid infusion (nociceptive reaction) into the bladder in free moving rats (Kitta et al., 2016).*

	Glutamate	Serotonin
Micturition reflex (saline infusion into the bladder)		
Under anesthesia	–	–
Awake	+	–
Nociceptive reaction (acetic acid infusion into the bladder)		
Under anesthesia	+	–
Awake	++	+

–, no increase; +,++, increase.

Electrical stimulation (deep brain stimulation) of the PAG in humans

Several studies regarding the electrical stimulation of the PAG have been carried out in anesthetized animals. Neurons in the vlPAG are known to react to bladder stimulation by changing their firing rate and modulating the micturition reflex (Matsuura et al., 2000). DBS of the vlPAG has been shown to reduce the severity of central poststroke pain, neuropathic pain, and cephalalgia (Gray et al., 2014). Stimulation of the PAG has been reported to increase dramatically the maximum cystometric capacity, which is the volume at which patients ask for a saline infusion to be discontinued; however, it does not affect the volumes at which voiding is desired (Green et al., 2012). Therefore, by regulating the micturition reflex, DBS of the vlPAG could be beneficial in patients requiring pain control. **For now, there are many obstacles to select DBS for patients with overactive bladder symptoms (OAB). However, for poststroke patients with pain and OAB and few treatment options available, DBS may work beneficially** (Owen et al., 2006).

A central switch for micturition

Normally, micturition is under volitional control in humans, and voiding is appropriate only under specific circumstances. Even when having a full bladder, humans can suppress voiding until the relevant social conditions are met, for relatively long periods of time. At the first suitable opportunity, humans can empty a partially full bladder to maximize the interval before the next void. If necessary, humans can also halt the micturition process midstream. During the storage phase, bladder afferent activity in the pelvic nerve increases gradually before being transmitted to the brain via the sacral spinal cord to relay information regarding the extent of bladder filling (Fowler et al., 2008). In the reflex pathway, the PMC is inactive when the volume is below the threshold level for triggering micturition; however, **some neurotransmitters activate and some neurotransmitters deactivate in PAG neurons, total integrated signal might not be above the threshold during storage phase of urine**, and when the bladder volume reaches the micturition threshold, the activation of the **micturition reflex "switch" of PAG activates the PMC, which sends excitatory and inhibitory signals to the lower urinary tract via the spinal cord to induce a sustained bladder contraction (excite pelvic nerve) and the reciprocal relaxation of the external urethral sphincter (inhibit pudendal nerve), respectively, leading to the release of urine**.

Human functional magnetic resonance imaging study on the PAG in bladder control as part of functional brain imaging

As a result of the development and progress of functional brain imaging (Table 10.3), more insights have been gained into the specific brain regions involved in the micturition reflex. Functional imaging can noninvasively visualize the activation of specific brain regions in response to various stimuli, taking sensory afferent and motor efferent functioning into account. At present, fMRI is widely used to determine the brain regions activated in different phases of micturition. Functional imaging studies have helped to elucidate the different connections to the PAG and to detect defects in some structural or functional pathologies associated with PAG or upstream (or other) sites in the brain. A recent meta-analysis summarized the results of numerous studies using activation likelihood estimation (Arya et al., 2017). From 14 neuroimaging studies included our study (Matsuura et al., 2002), Arya et al. reported 89 regions for activation likelihood estimation analysis.

TABLE 10.3 Comparison of functional brain imaging modalities.
Comparison of spatial and temporal sensitivity between various functional brain imaging modalities (Kitta et al., 2015).

Modality	Spatial resolution	Temporal resolution	Advantages	Disadvantages
SPECT	10 mm	>60s	Low cost available	Invasive, limited resolution
PET	5 mm	45s (using $H_2^{15}O$)	Sensitive, good resolution, metabolic studies, receptor mapping	Invasive, very expensive
fMRI	3 mm	5s	Excellent resolution, noninvasive	Expensive, limited to activation studies
fNIRS	3 cm	<10s	High temporal resolution, noninvasive, not requiring severe limitation of motion	Very expensive, limited resolution for deep structures

The meta-analysis revealed significant activation in multiple brain regions including thalamus, right insula, cerebellum, and brainstem (right PAG) is commonly activated during bladder filling in healthy adults across different studies (Arya et al., 2017). Their meta-analysis confirmed that the PAG is the interface or the "central relay station" between the afferent and the efferent arm of micturition in many previous studies using different imaging modalities such as MRI, positron emission tomography (PET), and single-photon emission computed tomography (SPECT).

A new imaging modality, near-infrared spectroscopy (NIRS), started to be used in the early 1990s. Oxygenated and deoxygenated hemoglobin have differential absorption spectra; thus, functional NIRS (fNIRS) can been used to assess changing concentrations of hemoglobin to a depth of 1 cm beneath the skull (Matsumoto et al., 2009). Evaluating functional structural relationships in large brain structures, particularly in the cortical regions, is considered relatively simple. However, given the resolution of functional imaging, which is in the millimeter range, challenges remain for subcortical and brainstem structures. Therefore, it is important to be able to evaluate brain activation in smaller brain structures robustly, reproducibly, and sensitively. Previous studies have typically used a slice thickness of <7 mm, whereas current standard neuroimaging protocols typically have a resolution of around 2−5 mm, which is only sufficient for covering a few voxels of the PAG. This limited resolution makes it difficult to identify unique activation regions and introduces partial volume effects, which degrades the quantitative accuracy of the image. In future research, the use of more modern modalities (e.g., 9 T MRI) can be expected to provide more precise information regarding the connections between specific parts of the brain and individual PAG columns.

Issues in functional brain imaging of micturition reflex

Evaluating the voiding phase of the micturition cycle during brain neuroimaging requires the individual to urinate while lying down in the scanner, which is difficult. This is even more challenging during brain MRI scans where the head is secured to a helmet-like head coil and the individual is requested to initiate and complete the task of voiding while trying to remain as still as possible because motion artifacts (even those 4 mm in any direction) interfere with the quality of the data acquired. Therefore, the majority of functional neuroimaging studies conducted over the past 20 years, especially those using MRI, have focused on the storage phase. Recent modalities such as fNIRS may help solve this issue.

Working model of lower urinary tract control

A meta-analysis of data regarding micturition in humans accumulated during the first decade of brain imaging allowed Fowler and Griffiths to propose a working model of lower urinary tract control via higher centers (Fowler & Griffiths, 2010) (Fig. 10.2). In that study, they hypothesized that during the storage phase, the information on bladder status carried by spinal afferents is relayed directly to the midbrain PAG. In addition, this activates the hypothalamus, thalamus, dorsal anterior cingulate cortex, right insula, and lateral prefrontal cortex. The decision regarding whether to void is made in the prefrontal cortex. If the decision is to "go," the medial prefrontal cortex tonically inhibits transmission through the PAG and disinhibits the downstream PMC neurons, which allow voiding to proceed. Concurrently, the hypothalamus provides a **"safe" (to urinate) signal to the PAG.**

In a more recent comprehensive review of forebrain influences, Griffiths and Fowler extended their original model and proposed the existence of a backup continence mechanism that can bypass the brainstem switch to promote continence in cases of extreme urgency by activating a pathway from the anterior cingulate region to the PAG and pontine storage center (the existence of which has not yet been confirmed in humans) (Griffiths & Fowler, 2013) (Fig. 10.3).

PAG and recent research with lower urinary tract function

Activation of the PAG during the storage phase

Experiments are typically designed so that the brain can be scanned while the bladder is being filled passively by intravesical saline infusion. Marked brain activation was observed in the pons, midbrain PAG, anterior insula, putamen, thalamus, and anterior cingulate gyrus during maximum urine storage (bladder distention). In a previous study (Matsuura et al., 2002), the increased activation of the PAG observed when the participants had a full bladder, but not when they were subjected to intravesical ice water stimulation, suggests that the activation of the PAG is specific to bladder distention, not to other bladder sensations (Matsuura et al., 2002). In addition, an fMRI study involving healthy females observed that cold (pain) sensation was processed differently than was bladder distension at the supraspinal level (Mehnert et al., 2011). The results of those studies suggest that bladder mechanosensory and nociceptive signals have separate pathways.

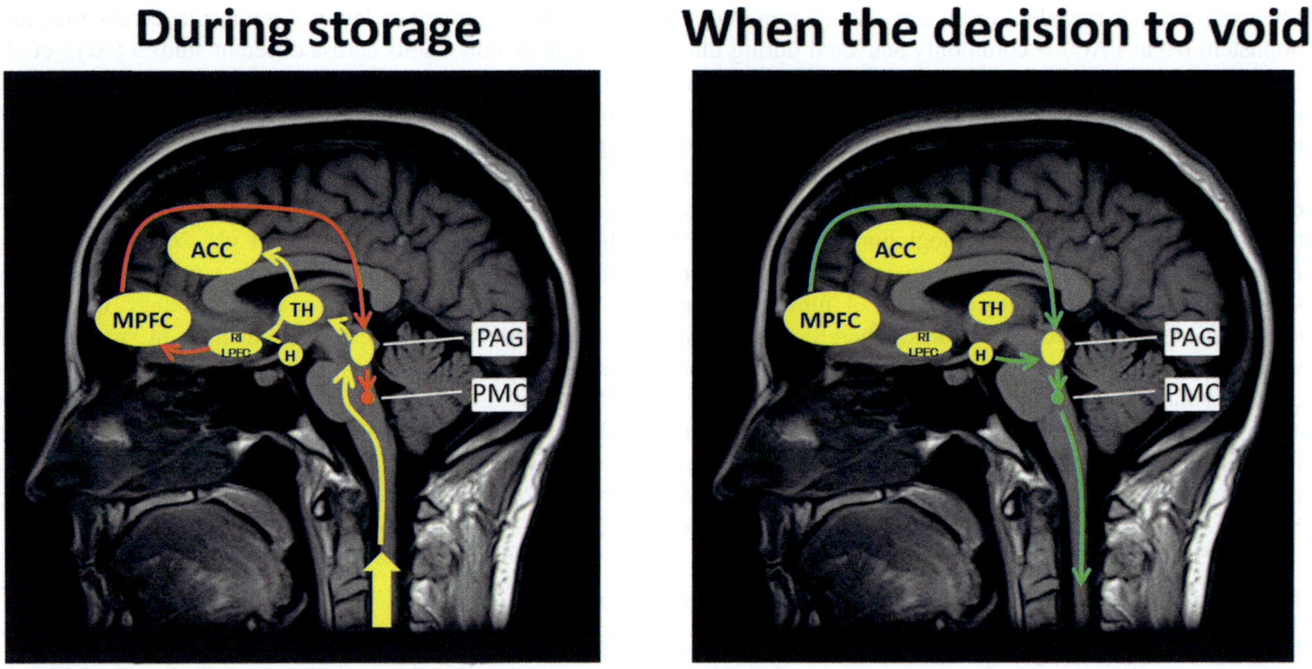

FIGURE 10.2 Preliminary working model of lower urinary tract control via higher brain centers (Fowler & Griffiths, 2010). The figure shows a signal ascending to the midbrain periaqueductal gray (PAG) during the storage phase; signals are relayed to the dorsal anterior cingulate cortex (ACC), the right insula (RI), and the lateral prefrontal cortex (LPFC) via the hypothalamus (H) and thalamus (TH). During the storage phase, these signals pass to the medial prefrontal cortex (MPFC), where the decision regarding whether to void is made. During this phase, the decision not to void is maintained by chronic inhibition of the PAG via a long pathway from the MPFC; consequently, the Pontine micturition center (PMC) is also suppressed, and no voiding occurs. When the decision to void is made, the MPFC relaxes its inhibition of the PAG and the H provides a "safe" (we can urinate here) signal; consequently, the PAG excites the PMC, which in turn sends descending motor output to the sacral spinal cord, which ultimately relaxes the urethral sphincter and contracts the detrusor, leading to voiding. Voiding is continued to completion by continuing afferent input, probably to the PAG. *Modified from Fig. 7 in Fowler, C.J., & Griffiths, D. J. (2010). A decade of functional brain imaging applied to bladder control.* Neurourology and Urodynamics, *29(1), 49–55. https://doi.org/10.1002/nau.20740.*

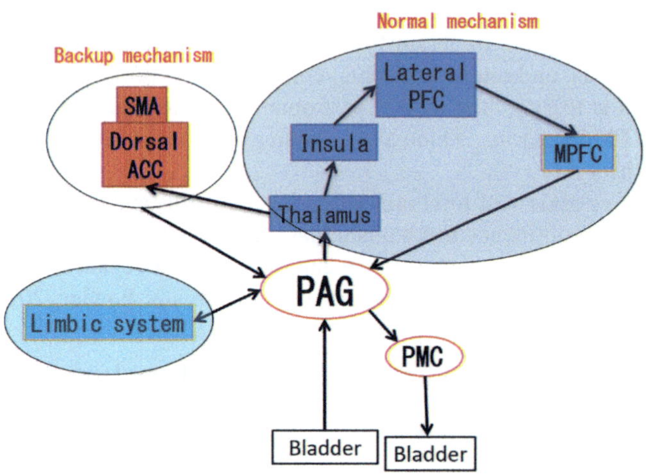

FIGURE 10.3 Working model of forebrain control of micturition, showing neural circuits mediating two possible continence mechanisms (Griffiths & Fowler, 2013). This shows neural circuits mediating two possible continence mechanisms (Griffiths & Fowler, 2013).

PAG activation during the voiding phase

The PAG plays a critical role in the initial moments of bladder emptying. During "attempted micturition," the PAG is consistently active (Kuhtz-Buschbeck et al., 2009), causing individuals to direct their attention to sensations arising from the bladder and urethra, which increases their desire to void. The initiation of voiding induces significant activity in not only the PAG but also the cortical regions, as shown on fMRI, whereas actual micturition is associated with significantly less activity (Kuhtz-Buschbeck et al., 2009).Unsuccessful micturition occurs as a result of the inefficient activation of the PAG and PMC during the initiation phase, which may be secondary to the inadequate antecedent activation of the frontoparietal and cingulate cortices; these are involved in decision-making in regard to micturition (Michels et al., 2015). The same results have been reported in studies utilizing positron emission tomography (Takao et al., 2008). A study on healthy women using fMRI reported that voiding imitation or interruption through the relaxation or contraction of the pelvic floor muscles could induce activation patterns in the PAG, the cortical regions, and other micturition control centers (Seseke et al., 2006). These findings highlight the role of the PAG in the relay and integration of information from different areas, including the spinal cord and corticolimbic system, which are essential in the physiology of micturition, and as a trigger for making the decision to void.

Function of the PAG in patients with lower urinary tract dysfunction

The results of clinical studies have suggested the role of the PAG in lower urinary tract dysfunction. Bladder filling showed **no change in PAG activity between healthy controls and patients with detrusor overactivity** (Griffiths et al., 2005). A subsequent study **found only minimal responses or PAG activity between the controls and patients with detrusor overactivity** (Griffiths et al., 2007), and the results of follow-up analyses using the same data revealed rostral insula and anterior cingulate activity related to the PAG, during bladder filling in healthy controls (Tadic et al., 2008) in addition to a slight effect of age on PAG connectivity (Griffiths et al., 2009). In addition, in women with Fowler's syndrome (a urethral sphincter abnormality), it has been reported that PAG activity normalizes when lower urinary function is improved by neuromodulation (Kavia et al., 2010). **Detrusor overactivity is associated with PAG activity in patients with Parkinson's disease (PD)** (Kitta et al., 2006) and PAG activity has been reported to influence activity in the thalamus and insula only during DBS of the subthalamic nucleus, which suggests improved sensory gating of bladder afferents. Another study involving patients with PD found that PAG activity is enhanced while the bladder is full, compared to while the bladder is empty (Herzog et al., 2008). Furthermore, DBS of the subthalamic nucleus has been shown to increase PAG activity in patients with PD (Herzog et al., 2008), which would indirectly affect the activity of other cortical regions related to micturition, ultimately restoring afferent bladder information processing. Herzog et al. tested PD patients with DBS switched on or off during cystometry (infusion saline into bladder), while changes in regional cerebral blood flow were measured by PET. DBS switched on led to improved afferent urinary bladder information processing.

Animal studies and human research as two wheels of the same cart

Although animal models have traditionally been used to study the neural control of micturition, it should be kept in mind that animals differ from humans in numerous ways in regard to the voiding phase. Considering all the social, emotional, and mechanical criteria for voiding, major differences exist between voluntary voiding in humans and the more instinctive bladder emptying in animals. Given these differences in brain control over voiding, the conscious and complex human intention to void, or to withhold from voiding, emphasizes the need for research in functional neuroimaging in humans, as opposed to relying exclusively on animal data. Functional brain imaging in healthy and clinical populations has confirmed results in the animal literature suggesting the involvement of the PAG in the homeostatic regulation of salient functions, including pain, anxiety, and autonomic function. More basic research needs to be carried out to elucidate the underlying pathological mechanisms and develop new diagnostic and therapeutic strategies, thereby making animal models indispensable (Sartori et al., 2020). Translational research brings innovations from bench to bedside, whereas reverse translation works backward from experiences in daily clinical practice to the bench; both are highly relevant for pushing the limits.

DBS could be a possible strategy for treating lower urinary tract dysfunction and has already been performed for many types of diseases, such as PD, essential tremor, and dystonia. In addition, treatment for PD of DBS has been shown to be useful for lower urinary tract function. However, DBS is still an invasive modality. The use of noninvasive brain-centered magnetic stimulation in the treatment of fecal incontinence has recently been reported (Paris et al., 2013). In this study, transcranial magnetic stimulation (TMS) of the motor cortex was carried out using a magnetic stimulator connected to a 110-mm-diameter double-cone coil. When the coil was positioned over the vertex of the cranium, monophasic currents were induced along the interhemispheric fissure, which bilaterally stimulated the region of the motor cortex innervating the pelvic musculature. This result suggests that TMS may be a potential therapeutic tool for lower urinary tract dysfunction. Pelvic muscle stimulation (contraction) could be control the OAB (Bo et al., 2020). However, such promising devices cannot be considered as appropriate therapeutic interventions until voiding and storage disorders are more clearly understood.

Conclusions and future directions

In conclusion, the PAG is an attractive target to invoke treatment strategies to restore the normal function of the urinary bladder in urological disorders. This research could be expected to aid future studies investigating treatment options for patients with lower urinary tract dysfunction, such as OAB and interstitial cystitis/painful bladder syndrome. In future research, the use of new imaging modalities (such as high-Tesla MRI or functional brain imaging) can be expected to provide more precise information regarding the connections between specific parts of the brain and individual PAG columns. Although there is no current way to direct access to brain regions certainly, precise information ensures the availability of safe and effective treatments for the patient with lower urinary tract dysfunction.

References

Aggleton, J. P., Burton, M. J., & Passingham, R. E. (1980). Cortical and subcortical afferents to the amygdala of the rhesus monkey (*Macaca mulatta*). *Brain Research, 190*(2), 347–368. https://doi.org/10.1016/0006-8993(80)90279-6

Aizawa, N., Fujimori, Y., Kobayashi, J.i., Nakanishi, O., Hirasawa, H., Kume, H., Homma, Y., & Igawa, Y. (2018). KPR-2579, a novel TRPM8 antagonist, inhibits acetic acid-induced bladder afferent hyperactivity in rats. *Neurourology and Urodynamics, 37*(5), 1633–1640. https://doi.org/10.1002/nau.23532

Arya, N. G., Weissbart, S. J., Xu, S., & Rao, H. (2017). Brain activation in response to bladder filling in healthy adults: An activation likelihood estimation meta-analysis of neuroimaging studies. *Neurourology and Urodynamics, 36*(4), 960–965. https://doi.org/10.1002/nau.23058

Bandler, R., Carrive, P., & Zhang, S. P. (1991). Chapter 13 Integration of somatic and autonomic reactions within the midbrain periaqueductal grey: Viscerotopic, somatotopic and functional organization. *Progress in Brain Research, 87*(C), 269–305. https://doi.org/10.1016/S0079-6123(08)63056-3

Bandler, R., & Shipley, M. T. (1994). Columnar organization in the midbrain periaqueductal gray: Modules for emotional expression? *Trends in Neurosciences, 17*(9), 379–389.

Banrezes, B., Andrey, P., Maschino, E., Schirar, A., Peytevin, J., Rampin, O., & Maurin, Y. (2002). Spatial segregation within the sacral parasympathetic nucleus of neurons innervating the bladder or the penis of the rat as revealed by three-dimensional reconstruction. *Neuroscience, 115*(1), 97–109. https://doi.org/10.1016/S0306-4522(02)00405-0

Barbaresi, P., Gazzanelli, G., & Malatesta, M. (1997). Glutamate-positive neurons and terminals in the cat periaqueductal gray matter (PAG): A light and electron microscopic immunocytochemical study. *The Journal of Comparative Neurology, 383*(3), 381–396.

Barnes, N. M., & Sharp, T. (1999). A review of central 5-HT receptors and their function. *Neuropharmacology, 38*(8), 1083–1152.

Barrington, F. J. F. (1927). Affections of micturition resulting from lesions of the nervous system. *Journal of the Royal Society of Medicine, 20*(5), 722–727. https://doi.org/10.1177/0035915727020005125

Basbaum, A. I., & Fields, H. L. (1978). Endogenous pain control mechanisms: Review and hypothesis. *Annals of Neurology, 4*(5), 451–462. https://doi.org/10.1002/ana.410040511

Behbehani, M. M., Jiang, M., Chandler, S. D., & Ennis, M. (1990). The effect of GABA and its antagonists on midbrain periaqueductal gray neurons in the rat. *Pain, 40*(2), 195–204. https://doi.org/10.1016/0304-3959(90)90070-T

Benarroch, E. E. (2012). Periaqueductal gray: An interface for behavioral control. *Neurology, 78*(3), 210–217. https://doi.org/10.1212/WNL.0b013e31823fcdee

Birder, L. A., & De Groat, W. C. (1992). Increased c-fos expression in spinal neurons after irritation of the lower urinary tract in the rat. *Journal of Neuroscience, 12*(12), 4878–4889. https://doi.org/10.1523/jneurosci.12-12-04878.1992

Blok, B. F. M., De Weerd, H., & Holstege, G. (1995). Ultrastructural evidence for a paucity of projections from the lumbosacral cord to the pontine micturition center or M-region in the cat: A new concept for the organization of the micturition reflex with the periaqueductal gray as central relay. *Journal of Comparative Neurology, 359*(2), 300–309. https://doi.org/10.1002/cne.903590208

Bo, K., Fernandes, A. C. N. L., Duarte, T. B., Brito, L. G. O., & Ferreira, C. H. J. (2020). Is pelvic floor muscle training effective for symptoms of overactive bladder in women? A systematic review. *Physiotherapy, 106*, 65–76. https://doi.org/10.1016/j.physio.2019.08.011

Brack, K. E., & Lovick, T. A. (2007). Neuronal excitability in the periaqueductal grey matter during the estrous cycle in female Wistar rats. *Neuroscience, 144*(1), 325–335. https://doi.org/10.1016/j.neuroscience.2006.08.058

Chiavegatto, S., Scavone, C., & Canteras, N. S. (1998). Nitric oxide synthase activity in the dorsal periaqueductal gray of rats expressing innate fear responses. *NeuroReport, 9*(4), 571–576. https://doi.org/10.1097/00001756-199803090-00002

Clements, J. R., Beitz, A. J., Fletcher, T. F., & Mullett, M. A. (1985). Immunocytochemical localization of serotonin in the rat periaqueductal gray: A quantitative light and electron microscopic study. *Journal of Comparative Neurology, 236*(1), 60–70. https://doi.org/10.1002/cne.902360106

Ding, Y. Q., Zheng, H. X., Gong, L. W., Lu, Y., Zhao, H., & Qin, B. Z. (1997). Direct projections from the lumbosacral spinal cord to Barrington's nucleus in the rat: A special reference to micturition reflex. *Journal of Comparative Neurology, 389*(1), 149–160. https://doi.org/10.1002/(SICI)1096-9861(19971208)389:1<149::AID-CNE11>3.0.CO;2-G

Dorofeeva, A. A., Panteleev, S. S., & Makarov, F. N. (2009). Involvement of the sacral parasympathetic nucleus in the innervation of the descending colon and rectum in cats. *Neuroscience and Behavioral Physiology, 39*(2), 207–210. https://doi.org/10.1007/s11055-009-9104-z

Fields, H. (2004). State-dependent opioid control of pain. *Nature Reviews Neuroscience, 5*(7), 565–575. https://doi.org/10.1038/nrn1431

Flores, J. A., El Banoua, F., Galán-Rodríguez, B., & Fernandez-Espejo, E. (2004). Opiate anti-nociception is attenuated following lesion of large dopamine neurons of the periaqueductal grey: Critical role for D1 (not D 2) dopamine receptors. *Pain, 110*(1–2), 205–214. https://doi.org/10.1016/j.pain.2004.03.036

Fowler, C. J., & Griffiths, D. J. (2010). A decade of functional brain imaging applied to bladder control. *Neurourology and Urodynamics, 29*(1), 49–55. https://doi.org/10.1002/nau.20740

Fowler, Clare J., Griffiths, D., & de Groat, W. C. (2008). The neural control of micturition. *Nature Reviews Neuroscience, 9*(6), 453–466. https://doi.org/10.1038/nrn2401

Gray, A. M., Pounds-Cornish, E., Eccles, F. J. R., Aziz, T. Z., Green, A. L., & Scott, R. B. (2014). Deep brain stimulation as a treatment for neuropathic pain: A longitudinal study addressing neuropsychological outcomes. *The Journal of Pain, 15*(3), 283–292. https://doi.org/10.1016/j.jpain.2013.11.003

Green, A. L., Stone, E., Sitsapesan, H., Turney, B. W., Coote, J. H., Aziz, T. Z., Hyam, J. A., & Lovick, T. A. (2012). Switching off micturition using deep brain stimulation at midbrain sites. *Annals of Neurology, 72*(1), 144–147. https://doi.org/10.1002/ana.23571

Gregg, T. R., & Siegel, A. (2003). Differential effects of NK1 receptors in the midbrain periaqueductal gray upon defensive rage and predatory attack in the cat. *Brain Research, 994*(1), 55–66. https://doi.org/10.1016/j.brainres.2003.09.024

Griffiths, D., Derbyshire, S., Stenger, A., & Resnick, N. (2005). Brain control of normal and overactive bladder. *The Journal of Urology, 174*(5), 1862–1867. https://doi.org/10.1097/01.ju.0000177450.34451.97

Griffiths, D. J., & Fowler, C. J. (2013). The micturition switch and its forebrain influences. *Acta Physiologica, 207*(1), 93–109. https://doi.org/10.1111/apha.12019

Griffiths, J. L., & Lovick, T. A. (2005). GABAergic neurones in the rat periaqueductal grey matter express α4, β1 and δ GABAA receptor subunits: Plasticity of expression during the estrous cycle. *Neuroscience, 136*(2), 457–466. https://doi.org/10.1016/j.neuroscience.2005.08.013

Griffiths, D., Tadic, S. D., Schaefer, W., & Resnick, N. M. (2007). Cerebral control of the bladder in normal and urge-incontinent women. *NeuroImage, 37*(1), 1–7. https://doi.org/10.1016/j.neuroimage.2007.04.061

Griffiths, D. J., Tadic, S. D., Schaefer, W., & Resnick, N. M. (2009). Cerebral control of the lower urinary tract: How age-related changes might predispose to urge incontinence. *NeuroImage, 47*(3), 981–986. https://doi.org/10.1016/j.neuroimage.2009.04.087

Hasue, R. H., & Shammah-Lagnado, S. J. (2002). Origin of the dopaminergic innervation of the central extended amygdala and accumbens shell: A combined retrograde tracing and immunohistochemical study in the rat. *Journal of Comparative Neurology, 454*(1), 15–33. https://doi.org/10.1002/cne.10420

Herbert, H., & Saper, C. B. (1992). Organization of medullary adrenergic and noradrenergic projections to the periaqueductal gray matter in the rat. *Journal of Comparative Neurology, 315*(1), 34–52. https://doi.org/10.1002/cne.903150104

Herzog, J., Weiss, P. H., Assmus, A., Wefer, B., Seif, C., Braun, P. M., Pinsker, M. O., Herzog, H., Volkmann, J., Deuschl, G., & Fink, G. R. (2008). Improved sensory gating of urinary bladder afferents in Parkinson's disease following subthalamic stimulation. *Brain : Journal of Neurology, 131*(Pt 1), 132–145.

Holstege, G. (1987). Some anatomical observations on the projections from the hypothalamus to brainstem and spinal cord: An HRP and autoradiographic tracing study in the cat. *Journal of Comparative Neurology, 260*(1), 98–126. https://doi.org/10.1002/cne.902600109

Holstege, G. (2005). Micturition and the soul. *Journal of Comparative Neurology, 493*(1), 15–20. https://doi.org/10.1002/cne.20785

Holstege, G. (2014). The periaqueductal gray controls brainstem emotional motor systems including respiration. In *Progress in brain research* (Vol 209, pp. 379–405). Elsevier B.V. https://doi.org/10.1016/B978-0-444-63274-6.00020-5

Holstege, G., Griffiths, D., De Wall, H., & Dalm, E. (1986). Anatomical and physiological observations on suprapinal control of bladder and urethral sphincter muscles in the cat. *Journal of Comparative Neurology, 250*(4), 449–461. https://doi.org/10.1002/cne.902500404

Holstege, G., & Kuypers, H. G. (1982). The anatomy of brain stem pathways to the spinal cord in cat. A labeled amino acid tracing study. *Progress in Brain Research, 57*, 145. https://doi.org/10.1016/S0079-6123(08)64128-X

Holstege, G., Meiners, L., & Tan, K. (1985). Projections of the bed nucleus of the stria terminalis to the mesencephalon, pons, and medulla oblongata in the cat. *Experimental Brain Research, 58*(2), 379–391. https://doi.org/10.1007/BF00235319

Hopkins, D. A., & Holstege, G. (1978). Amygdaloid projections to the mesencephalon, pons and medulla oblongata in the cat. *Experimental Brain Research, 32*(4), 529–547. https://doi.org/10.1007/BF00239551

Huo, F. Q., Wang, J., Li, Y. Q., Chen, T., Han, F., & Tang, J. S. (2005). GABAergic neurons express μ-opioid receptors in the ventrolateral orbital cortex of the rat. *Neuroscience Letters, 382*(3), 265–268. https://doi.org/10.1016/j.neulet.2005.03.070

Jansen, A. S., Farkas, E., Mac Sams, J., & Loewy, A. D. (1998). Local connections between the columns of the periaqueductal gray matter: A case for intrinsic neuromodulation. *Brain Research, 784*(1–2), 329–336.

Katz, P. S., & Frost, W. N. (1996). Intrinsic neuromodulation: Altering neuronal circuits from within. *Trends in Neurosciences, 19*(2), 54–61.

Kavia, R., DasGupta, R., Critchley, H., Fowler, C., & Griffiths, D. (2010). A functional magnetic resonance imaging study of the effect of sacral neuromodulation on brain responses in women with Fowler's syndrome. *BJU International, 105*(3), 366–372. https://doi.org/10.1111/j.1464-410X.2009.08819.x

Keay, K. A., & Bandler, R. (2002). Distinct central representations of inescapable and escapable pain: Observations and speculation. *Experimental Physiology, 87*(2), 275–279. https://doi.org/10.1113/eph8702355

Kitta, T., Kakizaki, H., Furuno, T., Moriya, K., Tanaka, H., Shiga, T., Tamaki, N., Yabe, I., Sasaki, H., & Nonomura, K. (2006). Brain activation during detrusor overactivity in patients with Parkinson's disease: A positron emission tomography study. *The Journal of Urology, 175*(3), 994–998. https://doi.org/10.1016/S0022-5347(05)00324-1

Kitta, T., Matsumoto, M., Tanaka, H., Mitsui, T., Yoshioka, M., & Nonomura, K. (2008). GABAergic mechanism mediated via D1 receptors in the rat periaqueductal gray participates in the micturition reflex: An in vivo microdialysis study. *European Journal of Neuroscience, 27*(12), 3216–3225. https://doi.org/10.1111/j.1460-9568.2008.06276.x

Kitta, T., Mitsui, T., Kanno, Y., Chiba, H., Moriya, K., & Shinohara, N. (2015). Brain-bladder control network: the unsolved 21st century urological mystery. *International Journal of Urology, 22*(4), 342–348. https://doi.org/10.1111/iju.12721

Kitta, T., Mitsui, T., Kanno, Y., Chiba, H., Moriya, K., Yoshioka, M., & Shinohara, N. (2016). Differences in neurotransmitter systems of ventrolateral periaqueductal gray between the micturition reflex and nociceptive regulation: An in vivo microdialysis study. *International Journal of Urology, 23*(7), 593–598. https://doi.org/10.1111/iju.13096

Krout, K. E., & Loewy, A. D. (2000). Periaqueductal gray matter projections to midline and intralaminar thalamic nuclei of the rat. *The Journal of Comparative Neurology, 424*(1), 111–141.

Kuhtz-Buschbeck, J. P., Gilster, R., van der Horst, C., Hamann, M., Wolff, S., & Jansen, O. (2009). Control of bladder sensations: An fMRI study of brain activity and effective connectivity. *NeuroImage, 47*(1), 18–27. https://doi.org/10.1016/j.neuroimage.2009.04.020

Kuipers, R., Mouton, L. J., & Holstege, G. (2006). Afferent projections to the pontine micturition center in the cat. *The Journal of Comparative Neurology, 494*(1), 36–53.

Lee, K. H., Chang, S. Y., Roberts, D. W., & Kim, U. (2004). Neurotransmitter release from high-frequency stimulation of the subthalamic nucleus. *Journal of Neurosurgery, 101*(3), 511–517. https://doi.org/10.3171/jns.2004.101.3.0511

Loewy, A. D., Saper, C. B., & Baker, R. P. (1979). Descending projections from the pontine micturition center. *Brain Research, 172*(3), 533–538.

Lu, J., Jhou, T. C., & Saper, C. B. (2006). Identification of wake-active dopaminergic neurons in the ventral periaqueductal gray matter. *Journal of Neuroscience, 26*(1), 193–202. https://doi.org/10.1523/JNEUROSCI.2244-05.2006

Mantyh, P. W. (1982). Forebrain projections to the periaqueductral gray in the monkey, with observations in the cat and rat. *Journal of Comparative Neurology, 206*(2), 146–158. https://doi.org/10.1002/cne.902060205

Matsumoto, S., Ishikawa, A., Kume, H., Takeuchi, T., & Homma, Y. (2009). Near Infrared Spectroscopy study of the central nervous activity during artificial changes in bladder sensation in men. *International Journal of Urology, 16*(9), 760–764. https://doi.org/10.1111/j.1442-2042.2009.02358.x

Matsuura, S., Allen, G. V., & Downie, J. W. (1998). Volume-evoked micturition reflex is mediated by the ventrolateral periaqueductal gray in anesthetized rats. *American Journal of Physiology - Regulatory, Integrative and Comparative Physiology, 275*(6), R2049–R2055.

Matsuura, S., Downie, J. W., & Allen, G. V. (2000). Micturition evoked by glutamate microinjection in the ventrolateral periaqueductal gray is mediated through Barrington's nucleus in the rat. *Neuroscience, 101*(4), 1053–1061.

Matsuura, S., Kakizaki, H., Mitsui, T., Shiga, T., Tamaki, N., & Koyanagi, T. (2002). Human brain region response to distention or cold stimulation of the bladder: A positron emission tomography study. *The Journal of Urology, 168*(5), 2035–2039. https://doi.org/10.1016/S0022-5347(05)64290-5

Mehnert, U., Michels, L., Zempleni, M. Z., Schurch, B., & Kollias, S. (2011). The supraspinal neural correlate of bladder cold sensation-An fMRI study. *Human Brain Mapping, 32*(6), 835–845. https://doi.org/10.1002/hbm.21070

Meoded, A., & Huisman, T. A. G. M. (2019). Diffusion tensor imaging of brain malformations: Exploring the internal architecture. *Neuroimaging Clinics of North America, 29*(3), 423–434. https://doi.org/10.1016/j.nic.2019.03.004

Michels, L., Blok, B. F. M., Gregorini, F., Kurz, M., Schurch, B., Kessler, T. M., Kollias, S., & Mehnert, U. (2015). Supraspinal control of urine storage and micturition in men-an fMRI study. *Cerebral Cortex, 25*(10), 3369–3380. https://doi.org/10.1093/cercor/bhu140

Mitsui, T., Kakizaki, H., Matsuura, S., Tanaka, H., Yoshioka, M., & Koyanagi, T. (2003). Chemical bladder irritation provokes c-fos expression in the midbrain periaqueductal gray matter of the rat. *Brain Research, 967*(1–2), 81–88. https://doi.org/10.1016/S0006-8993(02)04226-9

Monassi, C. R., Ramos Andrade Leite-Panissi, C., & Menescal-de-Oliveira, L. (1999). Ventrolateral periaqueductal gray matter and the control of tonic immobility. *Brain Research Bulletin, 50*(3), 201–208. https://doi.org/10.1016/S0361-9230(99)00192-6

Morgan, M. M., & Clayton, C. C. (2005). Defensive behaviors evoked from the ventrolateral periaqueductal gray of the rat: Comparison of opioid and GABA disinhibition. *Behavioural Brain Research, 164*(1), 61–66. https://doi.org/10.1016/j.bbr.2005.05.009

Moseley, M. E., Kucharczyk, J., Mintorovitch, J., Cohen, Y., Kurhanewicz, J., Derugin, N., Asgari, H., & Norman, D. (1990). Diffusion-weighted MR imaging of acute stroke: Correlation with T2-weighted and magnetic susceptibility-enhanced MR imaging in cats. *AJNR. American Journal of Neuroradiology, 11*(3), 423–429.

Mouton, L. J., & Holstege, G. (1994). The periaqueductal gray in the cat projects to lamina VIII and the medial part of lamina VII throughout the length of the spinal cord. *Experimental Brain Research, 101*(2), 253–264. https://doi.org/10.1007/BF00228745

Mouton, L. J., Klop, E., & Holstege, G. (2001). Lamina I-periaqueductal gray (PAG) projections represent only a limited part of the total spinal and caudal medullary input to the PAG in the cat. *Brain Research Bulletin, 54*(2), 167–174.

Normandin, J. J., & Murphy, A. Z. (2011). Serotonergic lesions of the periaqueductal gray, a primary source of serotonin to the nucleus paragigantocellularis, facilitate sexual behavior in male rats. *Pharmacology Biochemistry and Behavior, 98*(3), 369–375. https://doi.org/10.1016/j.pbb.2011.01.024

Noto, H., Roppolo, J. R., Steers, W. D., & de Groat, W. C. (1989). Excitatory and inhibitory influences on bladder activity elicited by electrical stimulation in the pontine micturition center in the rat. *Brain Research, 492*(1–2), 99–115. https://doi.org/10.1016/0006-8993(89)90893-7

Noto, H., Roppolo, J. R., Steers, W. D., & de Groat, W. C. (1991). Electrophysiological analysis of the ascending and descending components of the micturition reflex pathway in the rat. *Brain Research, 549*(1), 95–105. https://doi.org/10.1016/0006-8993(91)90604-T

Numata, A., Iwata, T., Iuchi, H., Taniguchi, N., Kita, M., Wada, N., Kato, Y., & Kakizaki, H. (2008). Micturition-suppressing region in the periaqueductal gray of the mesencephalon of the cat. *American Journal of Physiology - Regulatory, Integrative and Comparative Physiology, 294*(6), R1996–R2000. https://doi.org/10.1152/ajpregu.00393.2006

Ogawa, S., Kow, L. M., & Pfaff, D. W. (1994). In vitro electrophysiological characterization of midbrain periaqueductal gray neurons in female rats: Responses to GABA- and met-enkephalin-related agents. *Brain Research, 666*(2), 239–249. https://doi.org/10.1016/0006-8993(94)90778-1

Owen, S. L. F., Green, A. L., Stein, J. F., & Aziz, T. Z. (2006). Deep brain stimulation for the alleviation of post-stroke neuropathic pain. *Pain, 120*(1–2), 202–206.

Papka, R. E., McCurdy, J. R., Williams, S. J., Mayer, B., Marson, L., & Platt, K. B. (1995). Parasympathetic preganglionic neurons in the spinal cord involved in uterine innervation are cholinergic and nitric oxide-containing. *The Anatomical Record, 241*(4), 554–562. https://doi.org/10.1002/ar.1092410413

Paris, G., Chastan, N., Gourcerol, G., Verin, E., Menard, J. F., Michot, F., Weber, J., & Leroi, A. M. (2013). Evoked pressure curves from the external anal sphincter following transcranial magnetic stimulation in healthy volunteers and patients with faecal incontinence. *Colorectal Disease, 15*(12), e732–e740. https://doi.org/10.1111/codi.12386

Parvizi, J., Van Hoesen, G. W., & Damasio, A. (2000). Selective pathological changes of the periaqueductal gray matter in Alzheimer's disease. *Annals of Neurology, 48*(3), 344–353. https://doi.org/10.1002/1531-8249(200009)48:3<344::AID-ANA9>3.0.CO;2-S

Pereira, E. A. C., Lu, G., Wang, S., Schweder, P. M., Hyam, J. A., Stein, J. F., Paterson, D. J., Aziz, T. Z., & Green, A. L. (2010). Ventral periaqueductal grey stimulation alters heart rate variability in humans with chronic pain. *Experimental Neurology, 223*(2), 574–581. https://doi.org/10.1016/j.expneurol.2010.02.004

Reichling, D. B., & Basbaum, A. I. (1990). Contribution of brainstem GABAergic circuitry to descending antinociceptive controls: I. GABA-immunoreactive projection neurons in the periaqueductal gray and nucleus raphe magnus. *Journal of Comparative Neurology, 302*(2), 370–377. https://doi.org/10.1002/cne.903020213

Samineni, V. K., Grajales-Reyes, J. G., Copits, B. A., O'Brien, D. E., Trigg, S. L., Gomez, A. M., Bruchas, M. R., & Gereau, R. W. (2017). Divergent modulation of nociception by glutamatergic and GABAergic neuronal subpopulations in the periaqueductal gray. *eNeuro, 4*(2). https://doi.org/10.1523/ENEURO.0129-16.2017

Sartori, A. M., Kessler, T. M., & Schwab, M. E. (2020). Methods for assessing lower urinary tract function in animal models. *European Urology Focus*. https://doi.org/10.1016/j.euf.2019.12.004

Seseke, S., Baudewig, J., Kallenberg, K., Ringert, R. H., Seseke, F., & Dechent, P. (2006). Voluntary pelvic floor muscle control-an fMRI study. *NeuroImage, 31*(4), 1399–1407. https://doi.org/10.1016/j.neuroimage.2006.02.012

Siegfried, B., & Nunes de Souza, R. L. (1989). NMDA receptor blockade in the periaqueductal grey prevents stress-induced analgesia in attacked mice. *European Journal of Pharmacology, 168*(2), 239–242. https://doi.org/10.1016/0014-2999(89)90570-0

Sillery, E., Bittar, R. G., Robson, M. D., Behrens, T. E. J., Stein, J., Aziz, T. Z., & Johansen-Berg, H. (2005). Connectivity of the human periventricular-periaqueductal gray region. *Journal of Neurosurgery, 103*(6), 1030–1034.

Stone, E., Coote, J. H., Allard, J., & Lovick, T. A. (2011). GABAergic control of micturition within the periaqueductal grey matter of the male rat. *Journal of Physiology, 589*(8), 2065–2078. https://doi.org/10.1113/jphysiol.2010.202614

Stone, E., Coote, J. H., & Lovick, T. A. (2015). Effect of electrical vs. chemical deep brain stimulation at midbrain sites on micturition in anaesthetized rats. *Acta Physiologica, 214*(1), 135–145. https://doi.org/10.1111/apha.12491

Tadic, S. D., Griffiths, D., Schaefer, W., & Resnick, N. M. (2008). Abnormal connections in the supraspinal bladder control network in women with urge urinary incontinence. *NeuroImage, 39*(4), 1647–1653. https://doi.org/10.1016/j.neuroimage.2007.10.059

Takao, T., Tsujimura, A., Miyagawa, Y., Kiuchi, H., Ueda, T., Hirai, T., Komori, K., Takada, S., Nonomura, N., Osaki, Y., Enomoto, K., Hatazawa, J., & Okuyama, A. (2008). Brain responses during the first desire to void: A positron emission tomography study. *International Journal of Urology: Official Journal of the Japanese Urological Association, 15*(8), 724–728. https://doi.org/10.1111/j.1442-2042.2008.02076.x

Taniguchi, N., Miyata, M., Yachiku, S., Kaneko, S., Yamaguchi, S., & Numata, A. (2002). A study of micturition inducing sites in the periaqueductal gray of the mesencephalon. *The Journal of Urology, 168*(4 I), 1626–1631. https://doi.org/10.1097/00005392-200210010-00096

Verstegen, A. M. J., Klymko, N., Zhu, L., Mathai, J. C., Kobayashi, R., Venner, A., Ross, R. A., VanderHorst, V. G., Arrigoni, E., Geerling, J. C., & Zeidel, M. L. (2019). Non-crh glutamatergic neurons in Barrington's nucleus control micturition via glutamatergic afferents from the midbrain and hypothalamus. *Current Biology: CB, 29*(17), 2775–2789.e7. https://doi.org/10.1016/j.cub.2019.07.009

Zare, A., Jahanshahi, A., Meriaux, C., Steinbusch, H. W. M., & van Koeveringe, G. A. (2018). Glutamatergic cells in the periaqueductal gray matter mediate sensory inputs after bladder stimulation in freely moving rats. *International Journal of Urology, 25*(6), 621–626. https://doi.org/10.1111/iju.13562

Zare, A., Jahanshahi, A., Rahnama'i, M. S., Schipper, S., & van Koeveringe, G. A. (2019). The role of the periaqueductal gray matter in lower urinary tract function. *Molecular Neurobiology, 56*(2), 920–934. https://doi.org/10.1007/s12035-018-1131-8

Section IV

Research directions and research opportunities

Chapter 11

Impact of spinal neuromodulation on spinal neural networks controlling lower urinary tract function

Parag Gad[1,2] and Evgeniy Kreydin[1,2,3]

[1]*SpineX Inc, Los Angeles, CA, United States;* [2]*Rancho Research Institute, Rancho Los Amigos National Rehabilitation Center, Downey, CA, United States;* [3]*Institute of Urology, Keck School of Medicine of University of Southern California, Los Angeles, CA, United States*

Neurophysiology of the lower urinary tract

The lower urinary tract (LUT) is responsible for two mechanically simple functions: storage of urine without incontinence and complete emptying of urine at volitionally determined times. To achieve these functions, the LUT has evolved to consist primarily of two components—the detrusor (the smooth muscle reservoir which expands at a low pressure to collect urine produced by the kidneys and contract periodically in order to empty) and the outlet, consisting of the bladder neck (a funnel-like continuation of the detrusor) and the external urethral sphincter (EUS), a distinct skeletal muscle surrounding the urethra. Together, the bladder neck and the EUS contract during urinary storage to promote continence and relax during urinary voiding to promote complete emptying. In a healthy state, the detrusor and the outlet work in synchrony (synergistically) to make storage and emptying maximally efficient. Despite seemingly simple function, the innervation of the LUT is complex, involving each component of the central nervous system (CNS) and both the autonomic and somatic components of the peripheral nervous system. While the reasons for this complexity are speculative, one can appreciate the multifactorial decisions the nervous system must make in order to ensure proper LUT function in humans. In addition to receiving and integrating information about the physiological state of the LUT (e.g., bladder fullness, bladder pain), humans (and many other mammals) must also assess whether it is safe and socially appropriate to continue urinary storage or to initiate voiding. Sensory information reflecting the physiological state of the LUT primarily flows along the parasympathetic pelvic nerve and somatosensory pudendal nerve to the spinal cord and is eventually thought to reach a brainstem region known as the periaqueductal gray (PAG). This area, also known to play an important role in pain perception, may serve as a relay point sending and receiving information between the LUT and higher brain regions, such as the cortex. Although brain regions involved in bladder control are still being discovered, areas such as the insula, the cingulate cortex, and the motor cortex are some that are consistently found to be activated during urinary storage and emptying in functional brain imaging studies (Griffiths et al., 2005, 2015; Khavari et al., 2017; Krhut et al., 2012; Michels et al., 2015). Once the decision to void is made, neural firing in the PAG increases and reaches a certain threshold, which subsequently results in the activation of the Pontine micturition center (PMC, Barrington's nucleus), also located in the brainstem. The PMC then sends an excitatory signal to the parasympathetic nucleus (the origin of the pelvic nerve) causing detrusor contraction and an inhibitory signal to Onuf's nucleus (the origin of the pudendal nerve) causing EUS relaxation and leading to urinary voiding. When the bladder is emptied, urinary storage begins again and the cycle repeats. While this description is thought to be accurate, it is unquestionably simplified. The micturition cycle is strongly modulated by the sympathetic component of the autonomic nervous system, by multiple storage and voiding reflexes, and by mechanisms in the spinal cord and the brain that are yet to be discovered. Thus, it is not surprising that even small perturbations in the neural control of the LUT can have significant consequences for its function, let alone severe injuries, such as spinal cord injury (SCI), stroke, or multiple sclerosis.

Neuro-Urology Research. https://doi.org/10.1016/B978-0-12-822455-7.00001-5

The role of the spinal cord in LUT control

The spinal cord is more than a simple highway for transmission of afferent and efferent information between the brain and the end organ. Anatomically, the spinal cord is arranged into the exterior facing white matter and the internal butterfly-shaped gray matter that surrounds the central canal. The white matter is composed primarily of myelinated axons optimized for fast signal transduction along the CNS. It houses the various neural tracts that carry sensory and motor information between different neural centers. The gray matter, on the other hand, contains the spinal command centers that are influenced by higher brain regions, but can also generate end-organ activity autonomously. Several types of such command centers exist, but central pattern generators (CPGs) are the best studied. CPGs are defined as spinal cord networks that can generate periodic rhythmic activity even in the absence of periodic sensory input (while afferent input is often required to activate or "gate" a CPG, it does not need to be periodic or rhythmic) (Steuer & Guertin, 2019). Spinal and brainstem CPGs are thought to be responsible for the stereotyped motor behavior associated with actions such as swallowing, chewing, and respiration (brainstem and cervical CPGs) and ejaculation, defecation, and micturition (sacral CPGs).

As mentioned above, there are two sacral neural networks tasked with LUT control: the sacral parasympathetic nucleus (SPN), whose firing induces rhythmic contractions of the detrusor muscle, and Onuf's nucleus, which tonically contracts the EUS, and whose tonic (human and other species) or phasic (rats and other species) inhibition allows efficient voiding during a detrusor contraction. Together, the SPN and Onuf's nucleus likely form the basis for the sacral neural network that represents the micturition CPG. Evidence of linkage between these effector nuclei and afferent inputs to form a closed loop network is abundant. For example, dendrites arising from the SPN make connections not only with bulbospinal axons that carry descending inputs from the brainstem (PMC and PAG) but also directly to afferent projections that enter the spinal cord from the LUT via the pelvic plexus and spinal interneurons that are known to exist at the sacral level (Nadelhaft & Vera, 1995; Sugaya et al., 1997; Vizzard et al., 1995). While poorly understood and not well-defined anatomically, these spinal interneurons likely constitute the key regulatory and modulating influence that enables autonomous and efficient functioning of the micturition CPG. Pseudorabies virus neural labeling in rats identified sacral interneurons as receiving afferent input from the bladder. In addition, bladder and urethral stimulation has been shown to generate a transcription response in sacral interneurons located near the dorsal commissure, superficial dorsal horn, and the SPN (Birder & de Groat, 1993; Birder et al., 1999). Again, while some of these interneurons form networks and send projections to higher centers, others make direct connections to the key LUT nuclei of the sacral spinal cord (Beckel & Holstege, 2011). Few studies have examined the cellular characteristics of the putative CPG interneurons. However, at least one group has demonstrated that they may have pacemaker-like properties further adding to the evidence that an autonomous sacral CPG tasked with LUT control may exist (Derjean et al., 2005).

SCI, in particular, has a markedly negative impact on LUT function. A total of 1.3M patients live with chronic SCI worldwide and the incidence of new cases is approximately 17,500 per year (Singh et al., 2014). In addition to the medical risks of renal injury, recurrent urinary tract infections, and urinary incontinence, *neurogenic lower urinary tract dysfunction (NLUTD) that develops after an SCI takes a devastating toll on quality of life*: a 2004 survey of 681 individuals with SCI found that regaining bowel and bladder function ranks second on the list of medical priorities for paraplegics—ahead of walking (fourth), and second only to sexual function (Anderson, 2004). Even more striking, roughly 54% of hospitalized patients, in a recently published survey, identified the loss of bladder and bowel control as a prospect they considered worse than death, compared to just 3% saying that same of being confined to a wheelchair (Rubin et al., 2016). These findings are not surprising when one considers the impact of LUT on quality of life (Creasey & Craggs, 2012; Stiens et al., 1997). After SCI, the connection between the CNS and the LUT is interrupted, leading to impairment in both bladder and EUS activity. Almost every aspect of the urinary cycle goes awry after SCI: (1) due to the loss of inhibitory influence from the brain and brainstem structures, the bladder contracts spontaneously (detrusor overactivity, DO), leading to incontinence, (2) conscious sensation of bladder fullness is lost and the patient is forced to empty the bladder on schedule, rather than spontaneously, and (3) coordination between the detrusor and the EUS is lost, resulting in detrusor sphincter dyssynergia: detrusor contraction against a closed sphincter and consequent incomplete bladder emptying. In addition, as the bladder contracts against a closed sphincter, it generates high pressures, which can lead to loss of bladder compliance and reflux of urine to the kidneys causing kidney injury and even failure.

While health effects of NLUTD can be severe, its effects on quality of life are equally important. As mentioned above, many patients experience incontinence and depend on continence supplies to avoid soiling their clothes. Similarly, most patients with SCI rely on catheters to empty the bladder and must always carry spare catheters and collection bags. In addition, locating a wheelchair-accessible bathroom and catheterizing in a wheelchair (especially for female patients) can be time consuming, inconvenient, and cumbersome. When one considers how central normal LUT function is to daily living, one can appreciate how NLUTD can lead to embarrassment, avoidance of social activities, and, ultimately, significant mental health dysfunction. The balance between urinary storage and emptying at appropriate times is not achieved by any current therapy for LUT dysfunction after SCI. Thus, the holy grail for any novel treatment would be to maximize urinary storage without compromising emptying, while concurrently restoring LUT sensation and enabling low-pressure, efficient voiding. Because several distinct tasks are effected by the LUT, correction of overall LUT function remains an elusive goal in the SCI population.

If an autonomous sacral micturition CPG exists, why does SCI lead to severe LUT dysfunction?

As discussed above, there is abundant anatomic and functional evidence that the sacral cord houses a neural network that not only modulates afferent input from the bladder and urethra and efferent signaling from the brainstem but is also capable of autonomously coordinating the activity of the LUT. Indeed, persistent LUT activity after an SCI demonstrates that detrusor and EUS contraction can be mediated in a reflexogenic fashion by the sacral spinal cord. However, despite its preservation, the neural activity that remains after an SCI is insufficient and uncoordinated, in most cases, to promote (1) efficient storage and emptying of urine; (2) coordination between the detrusor and the outlet; and (3) rapid switching between and maximal activation of either urinary storage or emptying (de Groat & Wickens, 2013). The inability to achieve "normal" LUT activity after a suprasacral and anatomically complete SCI would, thus, argue against the existence of an autonomous sacral CPG or neural network that is able to regulate LUT activity on its own.

However, other examples from both animal models and humans point to the contrary. In human and animal neonates, there is probably minimal reliance on brainstem and cerebral influences to allow efficient, coordinated bladder emptying (de Groat, 2002). Although human neonates are incontinent, they do not develop recurrent UTIs or expose their upper tracts to high-pressure voiding seen in patients with SCI. Likewise, some animal species depend on the sacral spinal cord reflexes to allow efficient voiding during the neonatal period (as evidenced by the reliance on the perineal-to-bladder reflex to empty the bladder for several weeks after birth in some species) (de Groat et al., 1998). A significant realignment of spinal and supraspinal circuitry occurs as the CNS matures with the consequent loss of some sacral reflexes and the development of bulbospinal mechanisms that eventually control LUT function in the adult (Araki & de Groat, 1997).

Unfortunately, an SCI that interrupts the bulbospinal pathway does not lead to a simple regression of sacral circuitry to a neonatal state. Indeed, it appears that the significant neural reorganization that occurs is highly pathological and eventually leads to severe LUT dysfunction associated with chronic SCI (de Groat & Yoshimura, 2006; de Groat & Yoshimura, 2012; Vizzard, 2006). Reversal or prevention of this pathological neural reorganization may be the key to allowing a more "benign" neonatal phenotype to predominate. Several preclinical and clinical studies have demonstrated that early intervention after an SCI can have significant beneficial effect on long-term LUT function. Neural reorganization leading to transformation of neural networks controlling LUT function that occurs after an SCI appears to be driven at least in part by neurotrophic factors such as nerve growth factor (NGF) (Vizzard, 2006). Thus, administration of anti-NGF antibody to the spinal cords of rats with SCI was shown to diminish DSD and DO (Seki et al., 2002, 2004). In a mechanistic study examining sacral spinal control of the LUT, intrathecal administration of a serotonin receptor agonist resulted in increased bladder capacity in a feline model of SCI and more physiologic EUS contractions in a rodent model of SCI (Dolber et al., 2007; Espey & Downie, 1995). The authors hypothesized that a spinal cord transection results in deactivation of modulating sacral centers, leading to sequelae such as DSD, DO, and inefficient voiding, and that reactivation of these centers with serotonin improves LUT parameters (Dolber et al., 2007). In our own work examining locomotor activity, we found that serotonin analogues mediate a stronger response to spinal neuromodulation in lower extremity functional recovery in paralyzed patients (Gerasimenko et al., 2015a). Perhaps the best-known clinical example of the potential of early intervention to improve post-SCI LUT function is the study by Sievert et al., who implanted a commercially available sacral nerve neuromodulator (SNM) (InterStim, Medtronic, USA) within 5 months of injury (Sievert et al., 2010). Compared to controls who did not undergo implantation, patients experienced less incontinence, fewer infections, and decreased reliance on oral medications for both bowel and bladder control. Altogether, these results suggest that SCI not only interrupts communication between the brainstem and the sacral micturition circuits but also changes the behavior of the sacral circuitry proper. Prevention (as described in the above examples) or reversal (as demonstrated by our work and the work of others [Hubscher, Pikov, Lu, Havton] described below) of these changes may prove to be an effective means of preventing the deterioration of and restoring LUT function after SCI.

Current options to alter LUT function after SCI

Managing inability to empty urine: In order to address the inability to void, a common approach to bladder dysfunction is the use of a catheter to empty the bladder every few hours. While effective, if this is not done often enough (due to poor sensation, scheduling, or compliance), the patient may become incontinent between catheterizations and/or develop high bladder pressures leading to increased risk of kidney injury and renal insufficiency. In addition, even with scrupulous attention to maintaining a hygienic environment, UTIs are inevitable. Further, for subjects who are tetraplegic, self-catheterization is impossible; the patient is dependent on a care provider, an indwelling catheter, or major surgical procedure (such as a sphincterotomy or incontinent stoma) to allow the bladder to empty. Notably, there is no pharmacological therapy to improve or facilitate bladder contraction. Even if such therapy existed, it would lack the temporal and spatial specificity needed to induce restoration of natural micturition.

Managing inability to store urine: A number of drugs improve storage of urine and decrease DO. Cholinergic antagonists, such as oxybutynin, inhibit bladder contraction and thus decrease DO and increase bladder capacity. Unfortunately, anticholinergics rarely cure incontinence, do not address the need for catheterization, and have significant adverse effects, such as dry mouth and constipation. Intradetrusor botulinum toxin injection is another modality to reduce DO and improve capacity. While effective and well tolerated, botulinum toxin injection is an invasive procedure, does not address the need for catheterization either, and has to be repeated every 6–9 months to maintain efficacy. Finally, dorsal sacral rhizotomy is a surgical option to treat DO and improve bladder capacity. This is a morbid neurosurgical procedure that usually results in permanent loss of erectile function in males and pelvic floor dysfunction in females.

Neuromodulatory devices for LUT: A strategy sometimes used is selective stimulation of the relevant circuitry via peripheral roots and nerves. For example, the posterior/anterior stimulator also known as the Finetech–Brindley stimulator has been used since 1978. Despite good acceptance by patients, this is a major neurosurgical procedure that is usually accompanied by sacral rhizotomy, again leaving males with erectile dysfunction and females with a weakened pelvic floor (Van Kerrebroeck et al., 1993). Additionally, the Finetech–Brindley stimulator is dependent on timed voiding due to lack of sensation. In contrast to direct motor nerve stimulation to elicit a bladder contraction, there are two FDA-approved neuromodulation techniques (sacral neuromodulation [SNM] and percutaneous tibial nerve stimulation [PTNS]) that alter peripheral nerve excitability and thus modify the activity of the LUT. While effective in addressing the symptomatology of nonneurogenic overactive bladder, SNM and PTNS are not designed or yet approved to manage voiding dysfunction in the neurogenic population (Amend et al., 2011).

Novel neuromodulation techniques to improve LUT function after SCI

Historically, multiple techniques have been proposed to induce micturition after an SCI, including stimulation of the bladder wall and pelvic plexus (Walter et al., 2005a,b; Bresler et al., 2008), the pelvic nerve (Holmquist and Olin, 1968a, b), and/or the sacral nerve (Granger et al., 2013), resulting in contractions in the bladder with or without relaxation of the EUS. Some of these strategies may involve surgical severing of nerves, resulting in permanent loss of some motor and sensory function. Also, these prior strategies completely ***bypass the automaticity*** that is intrinsic to the sensorimotor neural circuitries present in the spinal cord (Edgerton et al., 2001, 2008). Some of the spinal neuromodulation technologies are discussed below.

Implantable neuromodulation techniques

Intraspinal stimulation

Targeted electrical microstimulation of the spinal cord has been previously explored for restoring volitional control in the animal model of experimental SCI. Intraspinal microstimulation array technology was developed by Pikov and team (Pikov et al., 2007, 2020) to evaluate its ability to provide more focused and reliable bladder control in the preclinical model. A mechanically robust intraspinal multisite silicon array was built using novel microfabrication processes to provide custom-designed tip geometry and 3D electrode distribution. These arrays were tested in various animal models without and with spinal cord injuries for a period of several months. The objective of these arrays was to target the dorsal gray commissure area in the S2 sacral cord that contains the neural networks responsible for the coordination between the detrusor and EUS. They observed increased bladder responsiveness to stimulation starting at 1 month postspinal transection, possibly due to supraspinal disinhibition of the spinal circuitry and/or hypertrophy and hyperexcitability of the spinal bladder afferents. Further, bladder voiding is coordinated via a supraspinal descending tract that terminates in the dorsal gray commissure at the S1–S2 spinal levels and inhibits the EUS motoneurons. This same region of the spinal cord is also innervated from primary bladder afferents. The authors reported that intraspinal electrode sites that produced activation of the bladder and suppression of the EUS tone were located in the same area and in adjacent dorsal columns. Taking this into account and considering the effects of direct stimulation in the dorsal columns that persisted even after spinal cord transection, the authors speculated that changes in bladder function occurred predominantly due to modulation of spinal activity rather than supraspinally, even in the animals with intact spinal cords.

Epidural spinal stimulation

Ongoing studies in the preclinical and clinical model have demonstrated encouraging functional outcomes of the use of epidural stimulation to enable neurologic rehabilitation. An initial group of human subjects with complete paralysis after an SCI were implanted with a spinal cord electrode array. These subjects showed the ability to generate standing with full weight-bearing and recovery of some voluntary control of movement in the lower extremity in the presence of epidural stimulation. All four subjects showed anecdotal evidence of improved bladder function, including the ability to sense a filled bladder as well as enable volitional bladder voiding (Angeli et al., 2014; Harkema et al., 2011). This improvement was attributed to the combination of weight-bearing standing and epidural stimulation, although the mechanisms involved were unknown. Along with an improvement in bladder function, the subjects also reported an improvement in cardiovascular function, thermoregulation, and sexual function. *These unexpected, but positive and encouraging results could be associated with the overlap in the neural circuits controlling both locomotor and bladder function.* It has recently been shown that LUT function can be improved in rats undergoing epidural stimulation in combination with pharmacological neuromodulation. Specifically, enabling of locomotor-related spinal neuronal circuits by epidural stimulation can influence neural networks controlling voiding with the onset of bladder emptying within seconds of the onset of the epidural stimulation (Gad et al., 2014). The use of epidural stimulation and a serotonergic agonist in an acute experimental setting, without locomotor training, recently also has been shown to improve micturition reflexes and voiding function in rats with a chronic midthoracic complete SCI. The 5-HT1A receptor agonist, 8-hydroxy-2-(di-n-propylamino)-tetralin (8-OH-DPAT), facilitated the relaxation of the EUS muscle during voiding when this treatment was combined with epidural stimulation over the spinal EUS controlling center in the L3 spinal cord segment (Abud et al., 2015). The use of 8-OH-DPAT, in the absence of epidural electrical stimulation, also may improve voiding efficiency, the amplitude of bladder pressure during voiding contractions, and the amplitude and duration of EUS EMG bursting during voiding in rats with an incomplete cauda equina injury (Chang & Havton, 2013). Combined, the human and animal studies to date suggest that epidural stimulation may enable improvement in autonomic functions, specifically the LUT. More recently Hubscher and team have explored the use of epidural spinal stimulation over the lumbosacral spinal cord to enable bladder voiding in patients with complete paralysis due to SCI. They have observed that site-specific stimulation can activate detrusor and EUS muscles to improve coordination and induce a voiding contraction. In addition, these modalities could be further enhanced with the use of more specific training paradigms including locomotor training in the presence of epidural spinal stimulation.

Noninvasive neuromodulation techniques

Over the course of the past several years, the following question has been raised: "Invasive spinal stimulation versus noninvasive spinal stimulation, which is better?" Though the above question has not been completely answered, several observations reflect important advantages, disadvantages, and similarities of noninvasive (Ievins & Moritz, 2017; Gerasimenko et al., 2015b; Grahn et al., 2017) versus invasive neuromodulation. Some of the advantages of noninvasive neuromodulation include (1) lack of the need for surgery or a neural implant; (2) markedly lower cost; (3) the ability to change the location of stimulation according to the patient and the bodily function of interest; and (4) the ability to deliver stimulation immediately and only when needed. Disadvantages of the noninvasive approach include (1) the necessity to don and doff the stimulation electrodes and (2) the inability of selecting a more focused neural network for stimulation using implanted electrodes. While theoretically valid, the latter advantage of epidural stimulation has not been practically demonstrated, nor has it been demonstrated to have an advantage. On the contrary, it appears that neuromodulation of a broader neural network can prove to be a significant advantage in improving motor and autonomic functions that involve large and complex groups of neurons and muscles needed for complex functions.

Noninvasive magnetic spinal stimulation

To overcome the need to implant electrodes over the spinal cord, several groups are exploring the use of noninvasive techniques including the use of magnetic stimulation over the lumbosacral spinal cord. Using traditional magnetic stimulators, Lu and team performed a proof-of-concept study in five SCI patients with complete injuries (Niu et al., 2018). They initially demonstrated a preferential activation between the urethra and the detrusor based on the frequency of stimulation. Specifically, low frequency (1 Hz) increased the activation in the detrusor while reducing the activity in the urethra. On the other hand, high frequency (30 Hz) increased activation of the urethra while inhibiting the detrusor. This reciprocal activation pattern at low frequency was used to train SCI patients to voluntarily void. They reported that all five patients were capable of voluntarily voiding after 16 weeks of therapy and also reported an increase in bladder capacity. However, the changes in sensation and any incontinence episodes were not reported.

Transcutaneous electrical spinal cord neuromodulation

Over the past decade, Edgerton and team have developed transcutaneous electrical spinal cord neuromodulation (TESCoN) as a novel modality to deliver sufficient electrical signal noninvasively to activate the neural structures of the spinal cord without significant cutaneous discomfort (Hofstoetter et al., 2013, 2015; Minassian et al., 2016; Gerasimenko et al., 2015b, 2015c, 2016). To test the safety and efficacy of delivering TESCoN to activate the spinal circuits that control detrusor and the EUS, the initial studies were done in adult nonhuman primates to neuromodulate spinal network to physiological states that facilitate the control of initiation and completion of voiding. In intact (non-SCI) animals, TESCoN at specific levels along the spinal cord induced contractions in the detrusor that enabled coordinated (detrusor contracting along with EUS relaxing), efficient voiding during urodynamic studies. Also, these studies were conducted in animals across a wide age range of 3–30 years (equivalent to 9–90 human years) (Gad et al., 2018a; Havton et al., 2019). No adverse events were recorded.

TESCoN was developed with the aim of restoring and improving functional capability of neural control of LUT function (storage and voiding) after severe neurological injury such as SCI (Gad et al., 2018b; Kreydin et al., 2020). This approach consists of using patient-specific neuromodulatory parameters that can directly and indirectly transform the neural networks controlling the LUT to newly formed functionally physiological states. Rather than relying on stimulation of peripheral nerves, TESCoN directly neuromodulates the physiological states of the spinal networks, which may provide autonomous monitoring of LUT states and allow control of urinary storage and voiding at the sacral spinal level. Unlike traditional approaches (such as functional electrical stimulation), which directly induce a specific motor response rather than neuromodulating the controlling spinal networks toward their normal states of automaticity, TESCoN delivers modest stimulation (at or below the motor threshold) to facilitate changes in the excitability of the spinal networks from a nonfunctional to a functional physiological state post paralysis. In addition to facilitating automaticity in the lumbosacral spinal cord, TESCoN appears to have a retrograde effect by enabling bidirectional communication between spinal and supraspinal centers rostral to the site of injury. These networks appear to allow sufficient functional supraspinal–spinal connectivity to enable restoration of critical levels of voluntary control even in individuals thought to have a complete injury.

The effect of TESCoN on the LUT was first noted in seven individuals with SCI at levels ranging from C3 to T11 and injury severity ranging from AIS A to AIS C (Gad et al., 2018b). Patients underwent urodynamic studies without and with TESCoN at 1 and 30 Hz. Acute data demonstrated that the LUT can be selectively activated based on the site and parameters of stimulation, that is, these experiments successfully demonstrated that TESCoN has the spatial (site-specific) and temporal (frequency and intensity of stimulation) resolution required to selectively activate different components of the LUT.

Thus, even though the stimulation occurs at the skin and not directly over the spinal cord, TESCoN has sufficient resolution (spatial and temporal) to generate unique responses. This fine-tuning ability is critical in defining effective stimulation protocols for voiding and storage. Acute TESCoN at 1Hz at T11–12 resulted in reduced DO, decreased detrusor sphincter dyssynergia, and improved voiding efficiency. TESCoN at T11–12 and L1–2 at 30 Hz resulted in a significant increase in bladder capacity (defined as the volume at which first DO occurred) and restoration of sensation of bladder fullness in all of the patients. In a recent chronic study of 13 patients with chronic urinary incontinence ($n = 5$ SCI, $n = 5$ stroke, $n = 3$ MS), 8 weeks of TESCoN therapy in clinic reduced DO and increased bladder capacity (in SCI from 142 ± 24 mL to 253 ± 10 mL) during clinical urodynamic study leading to reduced episodes of urinary incontinence and reduced frequency of urination. The key benefit of a prolonged course of stimulation was the persistence of LUT changes after TESCoN therapy was completed. This effect led to fewer incontinence episodes (68% reduction in SCI patients) and increased and persistent sensation of bladder fullness, allowing patients to catheterize/void when the bladder was full but before they experience an incontinence episode (Kreydin et al., 2020). In addition, patients (one SCI and two stroke patients) also reported improvement in bowel function and trunk stability while sitting. The mean Neurogenic Bladder Symptom Score (NBSS) (Welk et al., 2018) decreased from 32.3 ± 2.4 to 22.5 ± 3.1 ($P < .05$). No serious adverse events were reported (Kreydin et al., 2020). No patient reported any incidences of autonomic dysreflexia during TESCoN therapy sessions.

Spinal neuromodulation alters the responsiveness of spinal networks to bladder filling and emptying and increases the conscious awareness of these states. Together these findings suggest that parts of the brain and spinal cord responsible for conscious sensation of bladder fullness and automatically controlled spinal coordination of detrusor and sphincter contractions have been reengaged, reflecting a highly significant level of functional neural plasticity. It appears that TESCoN can transform nonfunctional spinal neuronal networks due to an SCI back to a functional state. With the return of conscious awareness of bladder fullness and voluntary of detrusor and sphincter, a critical level of functional connectivity between the brain and the spinal cord can be gradually reestablished after an SCI using a relatively inexpensive, safe, and noninvasive neuromodulatory approach.

A key advantage of spinal cord neuromodulation, either via spinal epidural or transcutaneous stimulation, is to engage rather than bypass this exquisite control system by directly stimulating the muscles. In addition to capitalizing on the intrinsic control of a wide range of muscle groups, as the supraspinal networks reorganize in concert with the spinal networks, spinal cord neuromodulation results in recruitment of motor units in a more normal order than occurs with direct muscle stimulation. This recruitment order of motor unit phenotypes provides a greater resistance to neuromuscular fatigue. In addition, the modulation of the physiological state of spinal interneurons plays a key role in enabling a continuum of the coordinated activation of more motor pools to enable and/or facilitate successful execution of a wide range of intended motor events.

New directions

The neural circuits responsible for LUT control can be divided into three anatomic and functional groups: (1) the spinal centers containing the SPN, Onuf's nucleus, and the sympathetic chain; (2) the brainstem centers that give rise to the bulbospinal tracts and are thought to coordinate urinary storage and voiding in the adult state; and (3) higher brain and cortical regions that likely integrate sensory information from groups 1 and 2 and provide a conscious awareness and conscious control of the LUT. In this chapter, we focused on the first group of neural circuits and explored how their activity can be harnessed to improve LUT function after SCI. However, many questions remain unanswered both for this group of neurons and those in the remaining two groups. As alluded earlier in the chapter, there is substantial evidence but no definitive proof that the sacral micturition circuitry represents an autonomous CPG capable of providing efficient control of urinary storage and emptying in the absence of neural input from the brainstem and the brain. While the idea that micturition may, at least at some developmental time points, be under CPG control is unconventional, the existence of independent rhythm-generating centers in the spinal cord for locomotion and other functions is much better studied and more accepted. In the case of locomotion, evidence for the existence of an autonomous spinal network dates back to the early 20th century when stepping patterns resembling natural walking were demonstrated in spinalized animal models after afferent, electrical, or pharmacological stimulation (Clarac, 2008). Over the last century, the locomotor CPG has been well characterized on an anatomic, cellular, and functional scale. Interneurons thought to provide the main modulating influence in this spinal network have been described using labeling studies, selective lesioning, and local pharmacologic interventions. Several of these have been shown to have rhythmogenic pacemaker properties (Steuer & Guertin, 2019). A number of potential functional models have been proposed and several groups, including ours, have taken advantage of locomotor CPG properties to modulate the activity of these centers to a new functional state and improve lower extremity function after SCI (Gerasimenko et al., 2015a). Although the sacral micturition circuitry has been partially characterized as described earlier in the chapter, many pathways remain to be discovered and described. Although some candidate neurons have been identified, they remain poorly characterized and only one group has attempted their physiological description (Derjean et al., 2005). Likewise, several neurotransmitters are thought to be "generally" involved in spinal control of the LUT and serotonergic agonists have been shown to have a potentially beneficial effect on LUT function in an SCI animal model. However, the biochemical, like the physiological, details of the sacral micturition CPG remain to be teased out. The challenge of recording activity remains a significant barrier to understanding the mechanisms of LUT-related neural transmission in the human spinal cord. The spinal cord is a small structure, making it difficult to image with standard functional techniques, such as functional magnetic resonance imaging (fMRI). However, novel techniques that allow functional imaging and characterization of the spinal cord are emerging. Ultrafast Doppler neurofunctional ultrasound measures real-time blood flow changes as low as 5% and miniaturization of Doppler ultrasound probes has made imaging of small anatomical structures such as the spinal cord possible (Dizeux et al., 2019). This technique appears to circumvent some of the important limitations of other functional imaging techniques with excellent temporal and spatial resolution, as well as an adjustable and variable field of view. Since changes in blood flow appear to correlate strongly with neural activity, ultrafast Doppler ultrasound represents a novel, unique, and potentially more effective approach to studying spinal cord neural centers and connectivity involved in micturition control. An important barrier to widespread utilization of ultrafast Doppler ultrasound is its limited ability to penetrate bony structures and the concordant requirement for a laminotomy or craniotomy (in the case of brain imaging) to obtain useful data. However, future refinements of signal-to-noise ratio parameters may make this requirement obsolete.

Supracollicular decerebration does not eliminate the urinary voiding or storage reflexes, yet transection of the spinal cord leads to the loss of normal LUT reflexes. This suggests that the brainstem, rather than the forebrain or the spinal cord, houses the neural networks responsible for normal storage and emptying, at least in the adult state. Barrington's nucleus is a pontine center that sends output directly to the sacral spinal cord and stimulates bladder emptying. In turn, Barrington's nucleus itself is richly innervated by afferent neurons arising in the PAG, the hypothalamus, and other higher brain structures. However, much like the spinal cord, the exact pathways regulating the brainstem micturition centers remain to be discovered. Novel imaging techniques, such as optogenetics, are now allowing a deeper understanding of the neural circuitry and the biochemical signal patterns that make up and control this region of the micturition pathway (Verstegen et al., 2019). Animal knockout models are also shedding light on neuronal subpopulations in Barrington's nucleus and their divergent roles in LUT control. In particular, functional and biochemical characterization of these subpopulations may lead not only to an improved mapping of the brainstem micturition pathways but also to the development of novel pharmacological and neuromodulatory interventions to improve LUT dysfunction in both neurogenic and idiopathic cases. Several studies have begun to characterize LUT-related electrophysiological behavior of brainstem and subcortical structures in humans, with an inventive approach of recording field potentials from electrodes implanted for deep brain stimulation (e.g., to improve motor function in patients with Parkinson's disease). Using this approach, the behavior and interaction of nuclei of the basal ganglia during urinary storage and voiding have been partially described, with direct relationships identified between the PAG and this subcortical brain region (Roy et al., 2018a, 2018b, 2019). As the field of deep brain stimulation broadens, becomes safer, and targets a more diverse assortment of brainstem and subcortical neural structures, future studies will focus on recording and characterizing field potentials in areas of the human CNS, previously thought to be inaccessible to electrophysiology.

Finally, higher brain regions including the cortex probably represent the least characterized regions with respect to LUT function. However, this designation is rapidly changing with the advent of noninvasive radiation-free functional imaging techniques. Several brain areas from the frontal lobe to the cerebellum have been implicated in LUT sensation and control, and a number of brain regions have been shown to exhibit differential activation in patients with various forms of LUT dysfunction (Griffiths et al., 2005, 2015; Khavari et al., 2017; Krhut et al., 2012; Michels et al., 2015). Not surprisingly, it appears that higher cortical regions are responsible for conscious bladder sensation and control and their activity, as measured (most commonly) by fMRI blood oxygen level–dependent (BOLD) response, is diminished in conditions that affect such conscious function. Although BOLD fMRI signal has revolutionized noninvasive functional brain imaging, it has important shortcomings that may be addressed by novel imaging modalities. In particular, BOLD fMRI has fairly poor temporal resolution and may not be able to detect neural events associated with LUT control on a less than 1–2 s timescale. Additionally, the neural activity measured by BOLD fMRI is indirect, as this modality detects changes in hemoglobin oxygenation, which appears to correlate with neural activity. Electrophysiological modalities, such as electroencephalography (EEG) and electrocorticography (ECoG), offer excellent temporal resolution and represent true neural activity. While suffering from poor spatial resolution (EEG) and being highly invasive (ECoG), electrophysiology may offer an important and useful complement to fMRI. Simultaneous EEG and fMRI recording remains a technical challenge both because of acquisition and data analysis. Novel imaging techniques such as magnetoencephalography (MEG), functional Doppler ultrasound, and photoacoustic imaging are also gaining prominence and are likely to contribute significantly to the understanding of cerebral control over the LUT. In addition, the application of novel fMRI sequences, such as diffusion tensor imaging (DTI), which permits visualization of white fiber tracts in the human brain, and arterial spin labeling (ASL), which detects absolute changes in cerebral blood flow, will also shed light on the cerebral circuitry of LUT control in health and disease.

Funding

The authors declare that this study received funding from Walkabout Foundation, Dana & Albert R. Broccoli Charitable Foundation, Nanette, and Burt Forester, including matching by PwC LLP and Roberta Wilson. The funder was not involved in the study design, collection, analysis, interpretation of data, the writing of this article, or the decision to submit it for publication.

Author disclosure statement

PG holds shareholder interest in SpineX Inc. and holds certain inventorship rights on intellectual property licensed by the Regents of the University of California to SpineX Inc.

References

Abud, E. M., et al. (2015). Spinal stimulation of the upper lumbar spinal cord modulates urethral sphincter activity in rats after spinal cord injury. *American Journal of Physiology - Renal Physiology, 308*(9), F1032–F1040.

Amend, B., et al. (2011). How does neuromodulation work. *Neurourology and Urodynamics, 30*(5), 762–765.

Anderson, K. D. (2004). Targeting recovery: Priorities of the spinal cord-injured population. *Journal of Neurotrauma, 21*(10), 1371–1383.

Angeli, C. A., et al. (2014). Altering spinal cord excitability enables voluntary movements after chronic complete paralysis in humans. *Brain, 137*(Pt 5), 1394–1409.

Araki, I., & de Groat, W. C. (1997). Developmental synaptic depression underlying reorganization of visceral reflex pathways in the spinal cord. *Journal of Neuroscience, 17*(21), 8402–8407.

Beckel, J. M., & Holstege, G. (2011). Neurophysiology of the lower urinary tract. *Handbook of Experimental Pharmacology*, (202), 149–169.

Birder, L. A., & de Groat, W. C. (1993). Induction of c-fos expression in spinal neurons by nociceptive and nonnociceptive stimulation of LUT. *American Journal of Physiology, 265*(2 Pt 2), R326–R333.

Birder, L. A., et al. (1999). Increased c-fos expression in spinal lumbosacral projection neurons and preganglionic neurons after irritation of the lower urinary tract in the rat. *Brain Research, 834*(1–2), 55–65.

Bresler, L., Walter, J. S., Jahoda, A., Wheeler, J. S., Turk, T., & Wurster, R. D. (2008). Effective methods of pelvic plexus nerve and bladder stimulation in anesthetized animal model. *Journal of Rehabilitation Research & Development, 45*(4).

Chang, H. H., & Havton, L. A. (2013). Serotonergic 5-HT(1A) receptor agonist (8-OH-DPAT) ameliorates impaired micturition reflexes in a chronic ventral root avulsion model of incomplete cauda equina/conus medullaris injury. *Experimental Neurology, 239*, 210–217.

Clarac, F. (2008). Some historical reflections on the neural control of locomotion. *Brain Research Reviews, 57*(1), 13–21.

Creasey, G. H., & Craggs, M. D. (2012). Functional electrical stimulation for bladder, bowel, and sexual function. *Handbook of Clinical Neurology, 109*, 247–257.

Derjean, D., et al. (2005). Plateau potentials and membrane oscillations in parasympathetic preganglionic neurones and intermediolateral neurones in the rat lumbosacral spinal cord. *Journal of Physiology, 563*(Pt 2), 583–596.

Dizeux, A., et al. (2019). Functional ultrasound imaging of the brain reveals propagation of task-related brain activity in behaving primates. *Nature Communications, 10*(1), 1400.

Dolber, P. C., et al. (2007). Activation of the external urethral sphincter central pattern generator by a 5-HT(1A) receptor agonist in rats with chronic spinal cord injury. *American Journal of Physiology - Regulatory, Integrative and Comparative Physiology, 292*(4), R1699–R1706.

Edgerton, V. R., Courtine, G., Gerasimenko, Y. P., Lavrov, I., Ichiyama, R. M., Fong, A. J., Cai, L. L., Otoshi, C. K., Tillakaratne, N. J., Burdick, J. W., & Roy, R. R. (2008). Training locomotor networks. *Brain research reviews, 57*(1), 241–254.

Edgerton, V. R., de Leon, R. D., Harkema, S. J., Hodgson, J. A., London, N., Reinkensmeyer, D. J., Roy, R. R., Talmadge, R. J., Tillakaratne, N. J., Timoszyk, W., & Tobin, A. (2001). Retraining the injured spinal cord. *The Journal of physiology, 533*(1), 15–22.

Espey, M. J., & Downie, J. W. (1995). Serotonergic modulation of cat bladder function before and after spinal transection. *European Journal of Pharmacology, 287*(2), 173–177.

Gad, P. N., et al. (2014). Initiation of bladder voiding with epidural stimulation in paralyzed, step trained rats. *PLoS One, 9*(9), e108184.

Gad, P. N., et al. (2018a). Noninvasive neurophysiological mapping of the lower urinary tract in adult and aging rhesus macaques. *Journal of Neurophysiology, 119*(4), 1521–1527.

Gad, P. N., et al. (2018b). Non-invasive neuromodulation of spinal cord restores lower urinary tract function after paralysis. *Frontiers in Neuroscience, 12*, 432.

Gerasimenko, Y., et al. (2015a). Transcutaneous electrical spinal-cord stimulation in humans. *Annals of Physical and Rehabilitation Medicine, 58*(4), 225–231.

Gerasimenko, Y. P., et al. (2015b). Noninvasive reactivation of motor descending control after paralysis. *Journal of Neurotrauma, 32*(24), 1968–1980.

Gerasimenko, Y., et al. (2015c). Initiation and modulation of locomotor circuitry output with multisite transcutaneous electrical stimulation of the spinal cord in noninjured humans. *Journal of Neurophysiology, 113*(3), 834–842.

Gerasimenko, Y., et al. (2016). Integration of sensory, spinal, and volitional descending inputs in regulation of human locomotion. *Journal of Neurophysiology, 116*(1), 98–105.

Grahn, P. J., et al. (2017). Enabling task-specific volitional motor functions via spinal cord neuromodulation in a human with paraplegia. *Mayo Clinic Proceedings, 92*(4), 544–554.

Granger, Nicolas, D., Chew, P., Fairhurst, J. W., Fawcett, S. P., Lacour, M., Craggs, C. A., Mosse, N., Donaldson, & Jeffery, N. D. (2013). Use of an implanted sacral nerve stimulator to restore urine voiding in chronically paraplegic dogs. *Journal of Veterinary Internal Medicine, 27*(1), 99–105.

Griffiths, D., et al. (2005). Brain control of normal and overactive bladder. *The Journal of Urology, 174*(5), 1862–1867.

Griffiths, D., et al. (2015). Brain mechanisms underlying urge incontinence and its response to pelvic floor muscle training. *The Journal of Urology, 194*(3), 708–715.

de Groat, W. C. (2002). Plasticity of bladder reflex pathways during postnatal development. *Physiology and Behavior, 77*(4–5), 689–692.

de Groat, W. C., et al. (1998). Developmental and injury induced plasticity in the micturition reflex pathway. *Behavioural Brain Research, 92*(2), 127–140.

de Groat, W. C., & Wickens, C. (2013). Organization of the neural switching circuitry underlying reflex micturition. *Acta Physiologica, 207*(1), 66–84.

de Groat, W. C., & Yoshimura, N. (2006). Mechanisms underlying the recovery of lower urinary tract function following spinal cord injury. *Progress in Brain Research, 152*, 59–84.

de Groat, W. C., & Yoshimura, N. (2012). Plasticity in reflex pathways to the lower urinary tract following spinal cord injury. *Experimental Neurology, 235*(1), 123–132.

Harkema, S., et al. (2011). Effect of epidural stimulation of the lumbosacral spinal cord on voluntary movement, standing, and assisted stepping after motor complete paraplegia: A case study. *Lancet, 377*(9781), 1938–1947.

Havton, L. A., et al. (2019). Noninvasive spinal neuromodulation to map and augment lower urinary tract function in rhesus macaques. *Experimental Neurology, 322*, 113033.

Hofstoetter, U. S., et al. (2013). Effects of transcutaneous spinal cord stimulation on voluntary locomotor activity in an incomplete spinal cord injured individual. *Biomedizinische Technik.*

Hofstoetter, U. S., et al. (2015). Augmentation of voluntary locomotor activity by transcutaneous spinal cord stimulation in motor-incomplete spinal cord-injured individuals. *Artificial Organs, 39*(10), E176—E186.

Holmquist, B., & Olin, T. (1968a). Electromicturition in male dogs at pelvic nerve stimulation: an urethrocystographic study. *Scandinavian Journal of Urology and Nephrology, 2*(2), 115—127.

Holmquist, B., & Olin, T. (1968b). Electromicturition in female dogs at pelvic nerve stimulation: An urethrocystographic study. *Scandinavian journal of urology and nephrology, 2*(2), 129—135.

Ievins, A., & Moritz, C. T. (2017). Therapeutic stimulation for restoration of function after spinal cord injury. *Physiology, 32*(5), 391—398.

Khavari, R., et al. (2017). Functional magnetic resonance imaging with concurrent urodynamic testing identifies brain structures involved in micturition cycle in patients with multiple sclerosis. *The Journal of Urology, 197*(2), 438—444.

Kreydin, E., et al. (2020). Transcutaneous electrical spinal cord neuromodulator (TESCoN) improves symptoms of overactive bladder. *Frontiers in Systems Neuroscience, 14*, 1.

Krhut, J., et al. (2012). A preliminary report on the use of functional magnetic resonance imaging with simultaneous urodynamics to record brain activity during micturition. *The Journal of Urology, 188*(2), 474—479.

Michels, L., et al. (2015). Supraspinal control of urine storage and micturition in men–an fMRI study. *Cerebral Cortex, 25*(10), 3369—3380.

Minassian, K., et al. (2016). Spinal rhythm generation by step-induced feedback and transcutaneous posterior root stimulation in complete spinal cord-injured individuals. *Neurorehabilitation and Neural Repair, 30*(3), 233—243.

Nadelhaft, I., & Vera, P. L. (1995). Central nervous system neurons infected by pseudorabies virus injected into the rat urinary bladder following unilateral transection of the pelvic nerve. *The Journal of Comparative Neurology, 359*(3), 443—456.

Niu, T., et al. (2018). A proof-of-concept study of transcutaneous magnetic spinal cord stimulation for neurogenic bladder. *Scientific Reports, 8*(1), 12549.

Pikov, V., Bullara, L., & McCreery, D. B. (2007). Intraspinal stimulation for bladder voiding in cats before and after chronic spinal cord injury. *Journal of Neural Engineering, 4*(4), 356—368.

Pikov, V., McCreery, D. B., & Han, M. (2020). Intraspinal stimulation with a silicon-based 3D chronic microelectrode array for bladder voiding in cats. *Journal of Neural Engineering, 17*, 065004.

Roy, H. A., Aziz, T. Z., & Green, A. L. (2019). Sensory thalamus and periaqueductal grey area local field potential signals during bladder filling. *Journal of Clinical Neuroscience, 68*, 342—343.

Roy, H. A., et al. (2018a). Beta oscillations and urinary voiding in Parkinson disease. *Neurology, 90*(17), e1530—e1534.

Roy, H. A., et al. (2018b). Effects of pedunculopontine nucleus stimulation on human bladder function. *Neurourology and Urodynamics, 37*(2), 726—734.

Rubin, E. B., Buehler, A. E., & Halpern, S. D. (2016). States worse than death among hospitalized patients with serious illnesses. *JAMA Internal Medicine, 176*(10), 1557—1559.

Seki, S., et al. (2002). Immunoneutralization of nerve growth factor in lumbosacral spinal cord reduces bladder hyperreflexia in spinal cord injured rats. *The Journal of Urology, 168*(5), 2269—2274.

Seki, S., et al. (2004). Suppression of detrusor-sphincter dyssynergia by immunoneutralization of nerve growth factor in lumbosacral spinal cord in spinal cord injured rats. *The Journal of Urology, 171*(1), 478—482.

Sievert, K. D., et al. (2010). Early sacral neuromodulation prevents urinary incontinence after complete spinal cord injury. *Annals of Neurology, 67*(1), 74—84.

Singh, A., et al. (2014). Global prevalence and incidence of traumatic spinal cord injury. *Clinical Epidemiology, 6*, 309—331.

Steuer, I., & Guertin, P. A. (2019). Central pattern generators in the brainstem and spinal cord: An overview of basic principles, similarities and differences. *Reviews in the Neurosciences, 30*(2), 107—164.

Stiens, S. A., Bergman, S. B., & Formal, C. S. (1997). Spinal cord injury rehabilitation. 4. Individual experience, personal adaptation, and social perspectives. *Archives of Physical Medicine and Rehabilitation, 78*(3 Suppl. 1), S65—S72.

Sugaya, K., et al. (1997). The central neural pathways involved in micturition in the neonatal rat as revealed by the injection of pseudorabies virus into the urinary bladder. *Neuroscience Letters, 223*(3), 197—200.

Van Kerrebroeck, P. E., Koldewijn, E. L., & Debruyne, F. M. (1993). Worldwide experience with the Finetech-Brindley sacral anterior root stimulator. *Neurourology and Urodynamics, 12*(5), 497—503.

Verstegen, A. M. J., et al. (2019). Non-Crh glutamatergic neurons in Barrington's nucleus control micturition via glutamatergic afferents from the midbrain and hypothalamus. *Current Biology, 29*(17), 2775—2789 e7.

Vizzard, M. A. (2006). Neurochemical plasticity and the role of neurotrophic factors in bladder reflex pathways after spinal cord injury. *Progress in Brain Research, 152*, 97—115.

Vizzard, M. A., et al. (1995). Transneuronal labeling of neurons in the adult rat brainstem and spinal cord after injection of pseudorabies virus into the urethra. *The Journal of Comparative Neurology, 355*(4), 629—640.

Walter, J. S., et al. (2005a). Bladder-wall and pelvic-plexus stimulation with model microstimulators: Preliminary observations. *Journal of Rehabilitation Research & Development, 42*(2).

Walter, J. S., Wheeler, J. S., Fitzgerald, M. P., McDonnell, A., & Wurster, R. D. (2005b). A spinal cord injured animal model of lower urinary tract function: observations using direct bladder and pelvic plexus stimulation with model microstimulators. *The Journal of Spinal Cord Medicine, 28*(3), 73—81.

Welk, B., et al. (2018). The Neurogenic Bladder Symptom Score (NBSS): A secondary assessment of its validity, reliability among people with a spinal cord injury. *Spinal Cord, 56*(3), 259—264.

Chapter 12

Neural control of continence

Danielle J. Gordon, Abdo E. Kabarriti and Jeffrey P. Weiss
Department of Urology, SUNY Downstate Health Sciences University, Brooklyn, NY, United States

Introduction

Micturition, or bladder filling and voiding, is a process that requires great neural coordination between the brain, spinal cord, and peripheral ganglia. The neural interplay between the forebrain, brainstem, and spinal pathways is integral in the control of smooth muscle of the bladder and striated muscle of the urethra. An exceedingly interesting aspect of neural control of voiding is the "switch-like" mechanism of voiding. The urinary bladder only has two modes: filling/storage and voiding/elimination. For this reason, unlike other autonomic pathways with a baseline continuous autonomic neural control, that is, cardiovascular or gastrointestinal systems, the lower urinary tract system relies on consistent coordinated and organized central nervous system pathways (de Groat et al., 2015).

Urine storage (bladder filling) and periodic elimination of urine (voiding) are dependent on the coordination of the urinary bladder (reservoir) and bladder outlet (bladder neck, urethra, urethral sphincter). The coordination between the bladder and bladder outlet is dependent upon an intricate neural system involving the brain and spinal cord (central nervous system) as well as peripheral ganglia (de Groat et al., 2015). In short, during bladder filling, the bladder outlet is closed and bladder smooth muscle is relaxed, mainly a result of neural control mechanisms of the spinal cord (Yoshimura et al., 2020). As the bladder reaches capacity, the pontine micturition center (PMC) is activated, which in turn results in activation of parasympathetic sacral (pelvic) nerves and subsequent micturition (Yoshimura et al., 2020). However, bladder filling and voiding is much more complex than the series of events as outlined above. The complex higher brain centers such as the cerebral cortex as well as various neurotransmitters (dopamine, serotonin, norepinephrine, GABA, opioids, acetylcholine) are also involved in regulation of micturition (Yoshimura et al., 2020).

This chapter will review central as well as peripheral nervous system pathways involved in micturition and then discuss neurological insults to these pathways (identified through either functional brain imaging or our knowledge of neurologic and molecular pathways) that result in clinical conditions causing incontinence. Great strides have been made in treatments that target these pathways and improve patients' lower urinary tract symptoms (LUTS). However, there remains tremendous potential for further research, as will be discussed throughout the chapter.

Central nervous nystem pathways

Brain

Neural control of voiding relies heavily on the coordination between various parts of the brain including medullary raphe nuclei, locus coeruleus, Barrington's nucleus (PMC), periaqueductal gray (PAG), hypothalamus, and the cerebral cortex. It is especially important to establish the function of the micturition center located in the pons and the PAG region, two parts of the brain that are integral to the control of continence and micturition.

Neurons in the PMC receive input from PAG and anterior and caudal hypothalamus and in turn send descending neuronal output to the parasympathetics in the spinal cord (Fowler et al., 2008). The PAG region works as a critical relay system. It controls input from the higher brain centers to the PMC. In particular, it receives inhibitory inputs from the prefrontal cortex, which then suppresses excitation to the PMC, inhibiting voiding. In contrast, when voiding is favored,

Neuro-Urology Research. https://doi.org/10.1016/B978-0-12-822455-7.00004-0

the PMC is able to be excited (Holstege, 2005). Furthermore, the PAG receives bladder afferents during filling and relays the information to the insula, which is in turn activated. Recent imaging studies using functional magnetic resonance imaging (fMRI) have shown that the insula is activated during bladder filling (Griffiths et al., 2005). In addition to having neural connections with the thalamus, prefrontal cortex, and insula, the PAG also interacts with other higher brain centers, most importantly the anterior cingulate gyrus (ACG).

As a part of the limbic/emotional system, the role of the ACG in the control of continence implies the involvement of emotion in micturition. The dorsal part of the ACG, which is activated during bladder filling, is also involved in regulation of heart rate in response to an emotional stimulus (Carmichael & Price, 1995). Using fMRI imaging, a study evaluated brain response to bladder filling in female patients with and without urge incontinence, thus determining which parts of the brain were instrumental in maintaining continence (Carmichael & Price, 1995). Interestingly, the study found that the female subjects with urge incontinence had strong ACG activation. This finding suggests the involvement in emotion in the abnormal sense of urgency associated with urge incontinence (Carmichael & Price, 1995). Activation of ACG in this study resulted in inhibition of the PMC, suggesting that the sensation of urgency led to recruitment of other pathways to try to prevent incontinence. Further research should be conducted on these other pathways, as it is likely the failure of these pathways as opposed to the activation of the ACG that directly results in urge incontinence.

The frontal lobes of the brain, particularly the prefrontal cortex, are the cognitive sites of planning and determination of voluntary behaviors, including micturition. The prefrontal cortex has connections with the PAG, hypothalamus, and ACG (Carmichael & Price, 1995). It is the connection with the PAG that likely results in suppression of micturition when undesired or socially inappropriate (Holstege, 2005). This connection is also of great importance in understanding etiologies of incontinence. Patients with poor bladder control likely have a prefrontal cortex that responds weakly to bladder filling, thus interfering with the tonic suppression of micturition usually supplied by the prefrontal cortex. One study using fMRI imaging found that among subjects with good bladder control, at larger bladder volumes, the activation of the orbitofrontal cortex in particular was strengthened. In contrast, among the subjects with poor bladder control, at larger bladder volumes, activation of the orbitofrontal cortex was significantly weaker (Griffiths et al., 2005).

Voiding is primarily controlled by reflex pathways in the brain. Activation of the PMC occurs only when deemed appropriate by the prefrontal cortex, which in turn leads to interruption of the tonic suppression of input to the PMC from the PAG. When the PMC is activated, descending pathways lead to urethral smooth muscle relaxation and bladder contraction (through activation of sacral parasympathetics). (Fowler et al., 2008).

Spinal cord

A fundamental understanding of the parts of the spinal cord integral in the control of micturition is necessary in order to elaborate on the more complex pathways of bladder filling and voiding, which involve both the brain and spinal cord.

Parasympathetic preganglionic neurons are located in the intermediolateral cell column of the spinal cord at level L6/S1 (sacral intermediate gray matter) and send dendrites to the dorsal commissure, lateral funiculus, and lateral dorsal horn of the spinal cord (Morgan et al., 1991). These axons originating in the sacral parasympathetic nucleus coalesce to form the pelvic (splanchnic) nerves. Sympathetic preganglionic innervation arises in the intermediolateral cell column (lateral horn) of T1-L2. The hypogastric nerve then carries axons from T10 to L2. The somatic motor neurons, which innervate the external urethral sphincter, originates from the ventral part (laminae IX) of the anterior horn of S2, S3, and S4, also referred to as Onuf's nucleus.

The spinal cord is also integral in receiving input from the bladder and urethra through afferent nerves. Bladder and urethral afferents project to the lateral dorsal horn and dorsal commissure, which, as described above, are also sites of parasympathetic preganglionic dendrites. This suggests that the lateral dorsal horn and dorsal commissure are sites of great coordination of bladder and sphincter activity (Fowler et al., 2008). Bladder filling, during which the detrusor muscle is relaxed and muscles of the urethral sphincter are activated/contracted, is primarily controlled by the spinal cord. Bladder afferents project through the pelvic nerves to the spinal cord, which in turn results in sympathetic outflow in the hypogastric nerve to the bladder and urethra (preventing contraction/micturition) and pudendal outflow to the external urethral sphincter (promoting contraction).

Link between brain and spinal cord

Control of the lower urinary tract system is not dependent on the solitary activity of the brain or spinal cord, but is instead dependent on the interconnections between the two. The PAG and PMC receive input from interneurons in the spinal cord. Additionally, the PMC receives input from the PAG and the hypothalamus. The PMC in turn sends output to the parasympathetic nucleus of the spinal cord (Blok et al., 1995). Furthermore, there are direct projections from the paraventricular nucleus of the hypothalamus to preganglionic motor neurons in the spinal cord and from the lateral pons to the sphincter motor nucleus of the spinal cord (Holstege et al., 1986; Holstege & Mouton, 2003). While most of these pathways were proven in the cat or other laboratory animals (and not human) due to availability of testing, similar pathways are presumed in the human.

Neurotransmitters

While not the focus of this last chapter, it is important to briefly discuss the neurotransmitters involved in the control of continence. Excitatory neurotransmitters include glutamic acid, tachykinins, pituitary adenylate cyclase–activating polypeptide, NO, and ATP. Glutamic acid is the main neurotransmitter involved in the spinal reflex pathways, which control the bladder and external urethral sphincter (Yoshiyama & de Groat, 2005). The main inhibitory neurotransmitters include GABA, glycine, and opioids and are instrumental in the tonic inhibition of the PMC. Specifically, GABA regulates the threshold at which micturition would occur, which in turn influences bladder capacity (Mallory et al., 1991).

The role of human brain imaging in understanding neural circuits of continence

Functional brain imaging, specifically positron emission tomography (PET) scans and fMRI, has been instrumental in revealing normal neural circuitry controlling continence as well as etiologies implicated in pathological conditions including overactive bladder and urge urinary incontinence. Understanding the sites of neural activation and deactivation in both normal and pathological processes allows for new research to target sites of abnormal neural activity, thus treating disorders of incontinence. Recently, fMRI has been used to a greater extent given its relative inexpensiveness and lack of radioactivity (de Groat et al., 2015).

One very important review article compiled results of multiple imaging studies (using both PET and fMRI) to determine locations of activation during bladder filling and emptying. During bladder filling, the insula, prefrontal and frontal cortexes, as well as ACG are frequently activated. During voiding, the PAG and PMC were found to be activated (Griffiths & Tadic, 2008).

These activations found on imaging findings support the neural control of continence. Afferent signals from the bladder synapse in the PAG and are relayed to the insula, where sensation of bladder fullness is interpreted. The insula then sends inhibitory signals to the PAG and PMC during bladder storage until a voluntary decision about voiding is made by the prefrontal cortex (Griffiths & Tadic, 2008).

Human brain imaging and urge incontinence

fMRI observations in patients with urge incontinence

Differences in neural activation on brain imaging among individuals with urge incontinence have been found as compared to individuals with normal micturition habits. At small bladder volumes, the limbic system (hippocampus, parahippocampal gyrus, amygdala, parts of the temporal lobe, medial orbitofrontal cortex, and parts of posterior cortex) of patients with urge incontinence showed deactivation (which was statistically significant) as compared to patients without urge incontinence. Whether this deactivation is indicative of a desire to suppress emotions associated with undesired incontinence or is involved in the mechanism contributing to loss of bladder control is unclear (Griffiths & Tadic, 2008). At larger bladder volumes, there is significantly greater activation in the ACG (part of limbic system) and frontoparietal regions in patients with urge incontinence (Kuhtz-Buschbeck et al., 2007). The frontoparietal region is involved in regulation of pelvic floor muscle activity. Thus, nonsurprisingly, as the bladder fills, patients with incontinence become emotionally aroused (likely secondary to concern about impending incontinence) and try to maintain bladder control by recruiting additional pelvic floor muscle control. While these differences in activation found on fMRI are interesting, they do not explain the neural origins of urge incontinence, but instead represent likely neural adaptations as a result of incontinence.

Urge incontinence can theoretically be secondary to either increased sensitivity or abnormal activity at the level of the bladder afferents themselves or impaired processing and supraspinal control of signals sent from the bladder afferents. The latter is much more likely because even if there is abnormal activity at the level of bladder afferents, with an intact neural circuit, the brain should be able to properly process and handle the afferent signals. Furthermore, the lack of increased activation of the PAG (which receives afferents) and insula (where sensation is processed) on fMRI among patients with urge incontinence supports that the incontinence is likely secondary to a neural dysfunction in the brain as opposed to change in peripheral firing.

Functional brain imaging has revealed multiple possible brain origins of urge incontinence including the prefrontal cortex, limbic system, basal ganglia, and PMC. In a study involving 12 patients, 6 of whom had poor bladder control and 6 with good bladder control, cerebral response to bladder filling (infusion) was reviewed with fMRI (Griffiths et al., 2005). Specifically, the cerebral regions studied included PAG, thalamus, insula, dorsal anterior cingulate, ventromedial cerebellum, orbitofrontal cortex, PMC, and preoptic hypothalamus. Among the subjects with good bladder control, as the bladder was filled, responses in most of the regions studied only increased slightly. However, there was a significant increase in the activation of the orbitofrontal cortex among these subjects as the bladder filled. Subjects with poor bladder control, on the other hand, had increased activation in most of the studied regions as compared to those with good control. Strikingly, the response of the orbitofrontal cortex as the bladder filled paled in comparison to their counterparts with good bladder control. As expected, various studies have found that the orbitofrontal cortex is integral in voluntary control of the bladder (Sakakibara et al., 1996). Thus, theoretically, activating the orbitofrontal cortex should result in improvement in incontinence. Furthermore, the exaggerated activation of other regions (especially the limbic system) among patients with poor bladder control is likely responsible for the symptoms of urgency and frequency.

Potential cerebral therapeutic targets for urge incontinence

Various treatments aimed to improve symptomatic urgency and incontinence have been studied to determine their influence at the cerebral level including sacral neuromodulation (clinically referred to as InterStim) and pelvic floor muscle training. By studying the cerebral effects of currently used efficacious treatments, we can develop more, and potentially better therapies with supraspinal targets.

Sacral neuromodulation is a long-term treatment for fecal and urinary incontinence. The exact mechanism of action of sacral neuromodulation has been a topic of debate for years. Research has elucidated that its effects are likely secondary to modulation of spinal cord reflexes and brain circuits through its influence on peripheral afferents (Elkelini et al., 2010). However, more recent research has supported supraspinal effects of sacral neuromodulation. The first research conducted on these cortical effects used EEG activity as an indicator of a supraspinal mechanism of action of sacral neuromodulation. This study evaluated the EEGs of 10 patients with the InterStim and found electrical changes in the postcentral gyrus, a sensory cortical region. Thus, this study supported a supraspinal site of modulation of sacral neuromodulators (Braun et al., 2002). A more recent study used PET scanning to measure changes in cerebral blood flow in different regions among with acute and chronic sacral neuromodulation. Specifically, this study found that subjects with acute sacral neuromodulation (recent implantation of neuromodulator), there were significant decreases in blood flow to the medial cerebellum and increases in blood flow in the right postcentral gyrus cortex, right insular cortex, and ventromedial orbitofrontal cortex (Blok et al., 2006). This is particularly interesting as a study discussed earlier, which used fMRI, found that patients with poor bladder control lacked proper activation of the orbitofrontal cortex with bladder filling (Griffiths et al., 2005). Thus, the efficacy of sacral neuromodulators may be partly due to their acute effects on the orbitofrontal cortex. However, this would be very challenging to prove as it would likely be impossible to isolate from their effects on other cortical regions.

Pelvic floor muscle training is the least invasive treatment for incontinence and is frequently used by clinicians, especially in elderly patients who are likely to experience greater complications of other medical and surgical therapies. Research has shown that pelvic floor muscle training does not only have effects locally on the pelvic floor (through strengthening of musculature) but also have cerebral effects (Griffiths et al., 2015). One particular study conducted among 65 women over the age of 60 with urge urinary incontinence used fMRI to evaluate brain reaction to bladder filling before and after pelvic floor muscle training. The study found that among subjects with improvement in incontinence secondary to pelvic floor muscle training, activation of the dorsal anterior cingulate cortex, insula, and nearby supplementary motor area was diminished after training. This correlates well with the other studies discussed above, which found that activation of the limbic system was associated with sensation of urgency and urinary frequency (Carmichael & Price, 1995). Furthermore, this study also found that responders to pelvic floor muscle training had deactivation of the prefrontal cortex.

This is a very interesting finding and potentially direction of future research. An fMRI study discussed above found that subjects with poor bladder control lacked adequate activation of the orbitofrontal cortex, and instead had exaggerated activation of other regions of the brain (Griffiths et al., 2005). Thus, this implied that activation of the orbitofrontal cortex, a specific region of the prefrontal cortex, would likely improve bladder control. However, the research study discussing pelvic floor muscle training found that urge incontinence improved among subjects and correlated with fMRI findings of deactivation in the prefrontal cortex (Griffiths et al., 2015). Future research focusing on the complexities of the prefrontal cortex, specifically as to whether activating the orbitofrontal cortex while deactivating the rest of the prefrontal cortex during bladder filling would decrease urgency and incontinence, would be especially interesting and contributory.

Imaging and Fowler's syndrome

Fowler's syndrome (FS) is a very interesting, relatively uncommon condition that can result in functional bladder outlet obstruction (BOO) in young women. It results from failure of urethral sphincter relaxation and often results in urinary retention. While not commonly a cause of incontinence (does not frequently cause overflow incontinence), research into the neurological abnormalities associated with the condition has highlighted the functions of various brain regions in maintaining continence. Women with FS often have absent bladder sensation. In essence, one can think of the etiology of FS as opposite of that of urge urinary incontinence. Theoretically, it would be interesting to consider whether activating regions of the brain activated in FS in patents with urge incontinence would lead to improvement in incontinence symptoms.

Urinary retention secondary to FS is thought to be due to increased afferent signals from an overactive urethra. These signals in turn inhibit bladder afferents at the level of the sacral spinal cord (Griffiths & Fowler, 2010). Interestingly, sacral neuromodulation is also a treatment for FS as it aids in artificially restoring bladder afferents. As stated above, it is very interesting and applicable to study brain imaging of FS patients as these patients are "procontinent." Among patients with FS, one study revealed that as the bladder was filled, there were multiple regions of deactivation including the insula (Griffiths et al., 2015). This is of great importance as the insula is responsible for processing the sensation of fullness. Furthermore, among patients with FS, bladder filling did not produce the normal activation of the PAG. With this knowledge, a potential mechanism of action of sacral neuromodulation in FS was proposed. Through generating artificial afferent signals, sacral neuromodulation increases PAG activity, which results in insula stimulation (responsible for bladder sensation), and thus resulting in voiding (Griffiths et al., 2015). It is very interesting that sacral neuromodulation serves as a treatment for both urge urinary incontinence as well as FS as etiologically these represent nearly opposite conditions. Given the research conducted, it can be theoretically hypothesized that decreasing neural activation of the PAG may lead to decreased stimulation of the insula, thus reducing feelings of urgency and potentially incontinent episodes. This is a potential area of future research and would require interdisciplinary cooperation between neuroscientists/neurologists and urologists to both target the correct brain region and measure the response.

Overview of peripheral innervation

The central nervous system plays an integral role in the coordination of micturition and has numerous targets with direct effects on LUTS. However, micturition would not be possible without the complex integration of the peripheral nervous system including the autonomic (parasympathetic and sympathetic) and somatic pathways (pudendal nerves). The sympathetic pathway originates from the thoracolumbar spinal cord, while parasympathetic and somatic pathways originate from the sacral spinal cord.

The sympathetic nervous system, responsible for the "fight or flight" response, suppresses micturition. Simply stated, it would not be particularly advantageous to be voiding while attempting a great escape. Postganglionic sympathetic neurons release norepinephrine, which in turn relaxes the detrusor muscle and contracts the bladder neck and urethra. This occurs through activation of beta-adrenergic inhibitory receptors in the detrusor muscle and alpha adrenergic excitatory receptors in the urethra and bladder neck (Fowler et al., 2008). The parasympathetic nervous system, on the other hand, releases cholinergic (acetylcholine) neurotransmitters, which bind to M_3 muscarinic receptors and trigger bladder smooth muscle (detrusor) contraction and subsequent urinary flow (Fowler et al., 2008). Noncholinergic neurotransmission also plays a role in the parasympathetic nervous system. For example, nitric oxide released by parasympathetic nerves inhibits urethral smooth muscle contraction, thus permitting urinary flow. Peripheral innervation of the lower urinary tract system is outlined in Fig. 12.1.

The somatic nervous system houses motor nerves that supply the external urethral sphincter. These nerves arise in S2-S4 (Onuf's nucleus) and travel through the pudendal nerves (Fowler et al., 2008). Axons that innervate pelvic floor musculature originate from a motor nucleus medial to Onuf's nucleus. Input from the bladder (sensation of fullness) is relayed to the spinal cord by pelvic and hypogastric nerves, while input from bladder neck and urethra is relayed by pudendal and hypogastric nerves. These nerves consist of myelinated and unmyelinated fibers, which respond to bladder filling and contraction and noxious stimuli, respectively (Jänig & Morrison, 1986). The cell bodies of these fibers are located in the dorsal root ganglia (DRG) and spinal cord level S2-S4 and T11-L2. These axons then interact directly with interneurons and neurons of the spinal tract, which in turn project directly to the higher coordination centers of the brain. Thus, micturition is impossible without coordination between the peripheral (autonomic and somatic) nervous system and central nervous system.

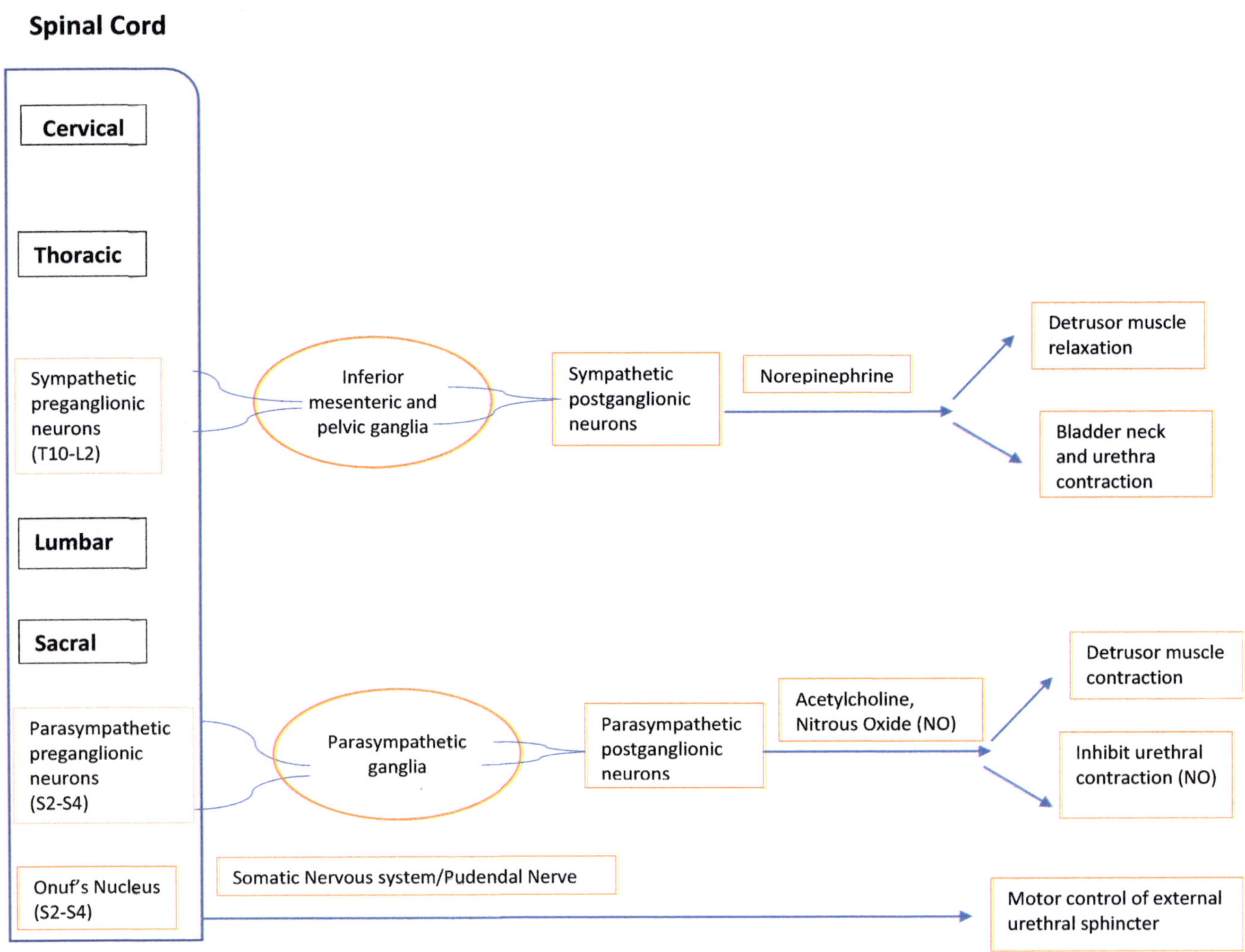

FIGURE 12.1 Peripheral innervation of the lower urinary tract system.

Disruptions in neural control of voiding and associated clinical findings

Studying deviations from normal neural control of voiding and the clinical manifestations of these deviations is necessary in order to determine potential therapeutic targets for the treatment of incontinence. Functional brain imaging, discussed above, has been helpful in determining differences in sites of neural activation among patients with normal micturition and those with incontinence.

Suprapontine lesions

Suprapontine lesions can be secondary to various etiologies including anterior cerebral lesions or Parkinson's disease, which results in degeneration of dopaminergic neurons. As a result of both of these conditions, there is a loss of the inhibitory control over the PMC. This in turn leads to a reduction in bladder capacity and detrusor overactivity (Fowler et al., 2008).

Parkinson's disease is associated with LUTS including urgency, frequency, and incontinence. As many as 50%–70% of individuals with Parkinson's disease experience LUTS (Araki et al., 2000). A study using PET imaging found that during bladder filling, there was a lack of activation in the pons and ACG among subjects with Parkinson's disease as compared to the healthy subjects (Kitta et al., 2006). Furthermore, other studies focused on the neural etiologies of LUTS and incontinence in Parkinson's disease have found that it is not only loss of dopaminergic inhibition of micturition but also the resulting overactivity of excitatory adenosine A_{2A} receptor mechanisms that result in the urinary symptoms (Kitta et al., 2012). Through studies in rats, it was determined that A_{2A} receptor pathways are downstream from dopamine D1 receptor pathways. Activation of the D1 receptor usually results in inhibition of A_{2A} activation, which in turn results in control of micturition (de Groat et al., 2015). In the case of Parkinson's disease, lack of activation of the D1 receptors (due to degeneration of dopaminergic neurons) as well as the subsequent lack of inhibition of A_{2A} receptor activation results in bladder overactivity (de Groat et al., 2015). Interestingly, treatment of Parkinson's disease with the widely used levodopa often does not help the urinary symptoms as it also activates dopamine D2 receptors. One study conducted among 87 subjects with Parkinson's disease found that stimulation of D2 receptors results in reduced bladder capacity and worsening of detrusor overactivity (Brusa et al., 2006). A new potential pharmacologic treatment targeting and antagonizing the A_{2A} receptors has been proposed as it would inhibit bladder overactivity caused by A_{2A} activation while leaving D2 receptors unaffected (de Groat et al., 2015).

About 20%−50% of patients with brain lesions (tumors) and cerebrovascular accidents (CVAs) have urinary symptoms such as frequency, urgency. and incontinence (de Groat et al., 2015). The etiologies thought to contribute to loss of bladder control among patients with cerebral lesions and CVAs include impaired sphincter control and lack of sensation of bladder filling and subsequent contraction (Griffiths, 2004). Based on research studies conducted on rats, NMDA glutamatergic activation is involved in urinary symptoms. This study in rats found that treatment with NMDA receptor antagonists prior to inducing a middle cerebral artery stroke reduced bladder overactivity (Griffiths, 2004). Furthermore, the rats with middle cerebral artery infarctions had upregulation of D2 receptor excitation. Thus, as long as the findings of these studies using rats can be extrapolated to humans, excitation of D2 and NMDA receptors, as well as disruption of the GABA inhibitory mechanism, can be targeted to improve urinary symptoms in patients with CVAs (Griffiths, 2004).

Bladder outlet obstruction

BOO is often thought of as a functional issue, resulting in hypertrophy of the detrusor muscle. This hypertrophy is accompanied by various neurological changes at the level of the detrusor muscle itself. These include increased sensitivity of receptors (muscarinic, purinergic) at the level of the muscle as well as changes in cell-to-cell communication in the muscle, specifically an upregulation of gap junction proteins and an increase in the number of interstitial cells (Boselli, Govoni, Condino, D'Agostino; Christ et al., 2003; Kim et al., 2011). Most of these neurologic changes at the level of the muscle are expected given the detrusor overactivity associated with detrusor hypertrophy secondary to outlet obstruction.

Interestingly, there are also changes in the central nervous system that accompany BOO. This proves that BOO is not only associated with functional changes at the level of the detrusor muscle but is also associated with more intricate changes in neural networks. Patients with BOO secondary to benign prostatic hyperplasia (BPH) have an upregulation of C-fiber afferent mechanisms, such as the spinal micturition reflex (Chai et al., 1998). A positive ice water test, defined as uninhibited bladder contraction with instillation of zero degrees saline at 50 cc/minute up to a total of 250 cc, triggers the spinal micturition reflex. This reflex involves C-fibers. One study conducted among 111 patients found that 71% of patients with BOO had a positive ice water test, whereas only 7% of nonobstructed patients had a positive test. This supports the concept of enhanced spinal micturition reflex secondary to upregulation of bladder afferents due to obstruction (Chai et al., 1998). This allows for the proposal of a future research direction potentially targeting the spinal micturition reflex (likely in combination with bladder outlet procedure) to decrease uninhibited bladder contractions secondary to prolonged obstruction.

Additionally, BOO also stimulates an increased expression of GAP-43 (growth-associated protein 43), a protein associated with neuronal growth and axonal regeneration. In this way, BOO operates similarly to spinal cord injury. This leads to the development of new spinal circuits after both BOO and spinal cord injury (Steers et al., 1990, 1991a). Relieving the obstruction often resulted in reversal of these neurological changes (Steers et al., 1991b). As clinicians, we often think about BOO as a functional issue and associate medical and surgical treatments with physical changes, that is, reduction in prostate size (5-alpha reductase inhibitors or transurethral resection of prostate) or relaxation of bladder neck smooth muscle (alpha-1 adrenergic receptor antagonists). However, these studies prove that successful treatment of BPH also influences the neurogenic aspects of detrusor overactivity associated with BOO. Thus, as stated above, combining bladder outlet reducing surgical interventions with other interventions targeting the spinal micturition reflex may completely treat LUTS secondary to the anatomical obstruction and the neurological changes that accompany it.

Interstitial cystitis and bladder pain syndrome

Interstitial cystitis (IC) and bladder pain syndrome (BPS) are chronic conditions for which significant research is required to fully understand the etiology and in turn treatment options. Unlike BOO, in which the etiology is mostly functional (although we now know there is also a neurologic component), a probable etiology of IC is hypersensitivity to both painful and nonpainful stimuli. Clinically, the most common complaints of patients with IC/BPS are urinary frequency and pain with bladder filling. This is believed to be caused by urothelial dysfunction and hypersensitivity secondary to chronic inflammation. Specifically, C-fiber afferent hyperactivity and mechanosensitivity are likely responsible for the clinical manifestations of IC/BPS (Yoshimura et al., 2014). The bladders of patients with IC/BPS showed edema, vasodilation, and proliferation of nerve cells fibers and mast cells on histological analysis (Theoharides et al., 2001). Furthermore, bladders with IC have a statistically greater amount of nerve fibers expressing substance P, which is a neurotransmitter of C-fiber afferents. C-fiber desensitization through intravesical application of capsaicin and resiniferatoxin (analog of capsaicin) has been reported successful in treating IC/BPS symptoms (Lazzeri et al., 1996). However, a prospective randomized control trial that studied intravesical instillation of resiniferatoxin among 163 IC patients found that there was no significant improvement in pain, urgency, frequency, nocturia, or average void volume during 12 weeks follow-up (Payne et al., 2005). Thus, laboratory findings have not necessarily correlated well with clinical application. Furthermore, patients with IC have been found to have elevated antiproliferative factor (APF), heparin-binding epidermal growth factor-like growth factor, and epidermal growth factor (Zhang et al., 2005). These proteins are often used as biomarkers of IC. Another potentially interesting research direction involves targeting these biomarkers to determine whether decreasing their density will result in improvement in symptoms. If successful, this would reflect that these proteins are not only biomarkers but are also contributory to the symptoms of IC/BPS.

While it is theorized that the etiology of IC/BPS is at least partly related to neurological hypersensitivity, it is negligent not to at least briefly mention presumed functional etiologies that may also contribute. These include bladder permeability and fibrosis. Denudation of the glycosaminoglycan layer of the bladder by protamine sulfate leads to increased bladder permeability. Some intravesical instillations, such as pentosan polysulfate, and other GAG replacement therapies are aimed at treating patients whose IC is secondary to increased bladder permeability (which can be determined with imaging modalities such as MRI) (Tyagi et al., 2018). Fibrosis and tissue scarring resulting from chronic tissue inflammation is also believed to be part of the pathology of IC/BPS (Tyagi et al., 2018).

The span of treatments currently utilized for IC/BPS and lack of reliable evidence available for these treatments underscore our lack of fundamental understanding of the etiology of IC/BPS. Because this chapter is aimed at understanding the neurological basis of syndromes of incontinence, treatments focused on neurological targets will be discussed. Cystoscopy with hydrodistension and neurostimulation are both used in the treatment of IC. Unsurprisingly, the true mechanism of action of these treatments is up for debate. Without complete understanding of the neurological etiologies of IC/BPS, the mechanisms of actions of treatments aimed at these neurological targets can only be theorized.

Cystoscopy with hydrodistension is a procedure during which under general anesthesia, the bladder is filled with normal saline until intravesical pressure reaches 80–100 cm of the water column and then stretched in that condition for 2 minutes, after which the saline is removed and repeat cystoscopy is conducted. The theoretical mechanism of action of the treatment involves ischemic necrosis of nerves in the bladder wall, which in turn results in decrease in urgency and the other urinary symptoms (Khullar et al., 2019). Additionally, hydrodistension has been associated with a decrease in the level of heparin-bound epidermal factor, which causes a decrease in APF, both biomarkers of IC. One study conducted among 71 female patients with IC/BPS found that hydrodistension for 2 min was effective in reducing pelvic pain and urgency/frequency (Al'-Shukri et al., 2018). However, there is a lack of both large-scale randomized control trials to determine the effectiveness of the treatment.

Neuromodulation is another treatment used for IC/BPS. It is a fourth line treatment that is only recommended if other treatments including behavioral and dietary changes, physical therapy techniques, multimodal pain management treatments, oral medications, intravesical instillations, and hydrodistension fail. Two prospective studies, albeit with small sample sizes (25 and 33 patients, respectively), found that sacral neuromodulation was associated with statistically significant improvement in frequency, pain, average voided volume, and maximum voided volume (Comiter, 2003; Whitmore et al., 2003). Additionally, one study found that percutaneous sacral third nerve root neurostimulation normalized the biomarkers of IC (Chai et al., 2000). While these studies are supportive of neuromodulation as a possible therapeutic agent for IC/BPS (albeit not a first line treatment modality), the exact mechanism of action is unknown. While sacral neuromodulation is proposed to activate or "reset" the somatic afferent inputs, thus modulating sensory processing and micturition reflex pathways in the spinal cord, this does not explain its mechanism of action in a chronic inflammatory condition such as IC/BPS (Mouritsen et al., 2010). To standardize neuromodulation as a treatment for IC/BPS, further research must be conducted to fully understand its mechanism of action.

Diabetes mellitus and detrusor underactivity

Diabetes mellitus is a condition frequently associated with urinary symptoms. Most commonly, it is associated with urinary retention secondary to decreased urinary sensation and reduced bladder capacity (Limberg et al., 2010). However, what is less common knowledge is the detrusor overactivity and LUTS that often precede urinary retention (diabetic cystopathy) in these patients. Initially diabetic patients often develop detrusor hypertrophy and overactivity secondary to polyuria (which is in turn secondary to glycosuria). This stage is then followed by diabetic cystopathy with detrusor underactivity (Daneshgari et al., 2009) The etiology of diabetic cystopathy is thought to be from demyelination of Schwann cells, which results in the impairment of nerve conduction (Mastri, 1980). Specifically, this demyelination results from changes in glucose metabolism, free radical formation, ischemia, and impaired transport in axons (Apfel, 1999). A study conducted among rats found that diabetic rats had a reduction in nitrous oxide—mediated relaxation of urethral smooth muscle and an increase in alpha-1 adrenoceptor—mediated contraction of the urethra, which contributes to impaired emptying (Torimoto et al., 2004).

Nerve growth factor (NGF) has been found to be associated with diabetic cystopathy. In rat studies, NGF levels have been found to be elevated in the initial stages of DM (with detrusor overactivity) and decreased in the advanced stages of DM (with diabetic cystopathy). In diabetic rats, studies have shown that NGF gene therapy (using HSV vectors) into the bladder prevents diabetic cystopathy (increased bladder capacity and retention) (Sasaki et al., 2004). While NGF gene therapy has been studied in humans with Alzheimer's disease, it has not yet been studied in patients with diabetic cystopathy. This is an avenue of research that should be pursued as, depending on the safety profile, it can potentially be used for prevention of diabetic cystopathy in humans.

Developmental changes in neural control of continence

Review of neurophysiology and pathology of bladder storage and voiding would not be complete without addressing neural development of the lower urinary tract. As discussed above, bladder filling and voiding involves complex coordination between the parasympathetic, sympathetic, and somatic nervous systems. At birth, all mammals lack the ability to voluntarily control micturition. As neural pathways develop, continence is obtained. Knowledge about the neural development of the lower urinary tract is not only critical in understanding the effects of congenital anomalies but also in understanding variations (mainly pathologic) in neural pathways and possible treatment targets. For example, pelvic surgery often results in damage to nerves resulting in incontinence and sexual dysfunction, and chronic cystitis is often accompanied by bladder hyperinnervation (Keast et al., 2015). Thus, understanding neural development of the urogenital system would likely aid us in understanding how to repair these neural connections to better treat these conditions.

Sensory and autonomic axons that innervate the bladder originate largely from neuronal populations within the DRG and pelvic ganglia. These axons enter the urinary bladder of the human by 13 weeks postconception, and their density gradually increases from that time (Keast et al., 2015). Research focusing on the factors that promote neuronal growth, survival, synaptic targeting, and signaling of these neurons in human tissue remains limited and is mostly based on rodent studies. In animal studies, retrograde labeling is used, which allows for separation of bladder sensory and autonomic neurons from the remainder of neurons located in the pelvic ganglia and DRG. This allows for them to be traced during the course of development. In living embryos, however, retrograde labeling cannot be performed and nerve tracing in vitro requires applying dyes to fixed bladder tissue (Keast et al., 2015). This is very tedious, as it requires examination of each target tissue within the bladder as well as multiple ganglion systems (Keast et al., 2015). However, it remains a great direction for future research.

Aging

Aging is often associated with LUTS symptoms among both males and females. A study conducted among 85 women found that as patients age, there is diminished detrusor contractility, bladder sensation, and intraurethral pressures (Pfisterer et al., 2006). Aging rats have been found to have decreased sensitivity of pelvic nerve afferents as well as a diminished bladder pressure during pelvic nerve stimulation compared to younger rats (Hotta et al., 1995). Additionally, in human bladders, one study found that with increasing age, human bladders showed a linear decline in the amount of acetylcholinesterase nerves (Gilpin et al., 1986). Thus, there is likely a decrease in parasympathetic nerve function with age, which also affects bladder function. Increasing acetylcholine levels may be a future research direction to decrease LUTS associated with aging.

There are also changes in the central nervous system that have been associated with LUTS that accompany aging. Studies conducted among rats have found decreased spinal cord sympathetic (serotonergic and noradrenergic) innervation in regions that involve bladder control including intermediolateral cell nucleus, sacral parasympathetic nucleus, dorsal gray commissure, and ventral horn (with Onuf's nucleus) (Ranson et al., 2003). Additionally, another study has found that preganglionic neurons in L1-L2, which project to the pelvic ganglion, have reductions in cell number and the length of the dendrites along with age (Santer et al., 2002). It is unlikely that treatments will be able to target these central nervous system age-related neurological changes that contribute to LUTS (at least in the foreseeable future). However, given that every person will experience aging, it is important to thoroughly understand the aging-associated pathogenesis of LUTS.

Spinal cord damage

This chapter would not be complete without addressing spinal cord injury and the types of neurogenic bladder dysfunction that result, as well as the various treatment modalities (current and future) aimed at improving the quality of life of these patients. More than 80% of patients with spinal cord injury develop some form of bladder dysfunction with the most common symptoms being urinary tract infection, damage to upper and lower urinary tract, and bladder/renal stones (Taweel & Seyam, 2015).

The type of bladder dysfunction, and thus treatment options, is dependent upon the timing and location of the spinal injury. Up to 3 months following a spinal cord injury, a patient can experience spinal shock, during which the bladder is unable to be activated by the parasympathetic nervous system due to interruption of the neural axis inferior to the pons. Patients lose all sensation of bladder filling during this period, and the micturition reflex is eliminated, resulting in urinary retention (Taweel & Seyam, 2015). Spinal shock should be managed with clean intermittent catheterization or indwelling urethral catheters as some degree of bladder function may return after the acute shock phase resolves.

Suprasacral lesions (above S1) result in loss of inhibition of the micturition reflex, resulting in detrusor overactivity and detrusor striated sphincter dyssynergia (Taweel & Seyam, 2015). Thus, bladder contractions are not coordinated with the sensation of bladder fullness or with sphincter relaxation, resulting in high voiding pressures, incomplete emptying, and incontinence. Eventually, if left untreated, these patients will develop renal and upper tract damage (McGuire & Savastano, 1983; Rudy et al., 1988). Sacral lesions, on the other hand, can have varying presentations. A complete sacral lesion results in lack of sensation of bladder filling and loss of the micturition reflex. In most patients, this results in a highly compliant but acontractile bladder with fixed, contracted sphincters that are unable to relax (Taweel & Seyam, 2015). However, there are patients with sacral lesions that have low bladder compliance, thought to be a result of changes in the sympathetic pathway (Abdel-Azim et al., 1991; Clarke & Thomas, 1981).

Conservative treatments of bladder dysfunction secondary to spinal cord injury are usually first line. These include clean intermittent catheterization or indwelling catheters (either urethral or suprapubic) in the case of urinary retention and detrusor overactivity associated with incomplete emptying, high voiding pressures, and low compliance. First-line treatment for neurogenic detrusor overactivity also involves anticholinergic medications, which improve incontinence, increase bladder capacity, reduce bladder pressure, and improve compliance.

A few of the less studied (and thus less utilized) surgical treatments involve targeting the neurological pathways (neuromodulation) themselves. These treatments including electrical stimulation of the bladder and nerve grafting/nerve transfer are especially applicable to this chapter and are an avenue of great further research potential. One study conducted on transcutaneous electrical bladder stimulation among 28 patients with neurogenic bladder dysfunction following spinal cord injury involved 30 sessions of electrical stimulation to the bladder, 5 sessions per week. Results were obtained using urodynamic testing immediately prior to treatment, immediately after, as well as 2 years posttreatment. The study found that there was a statistically significant increase in bladder capacity both immediately posttreatment (117 mL) and after 2 years (102 mL) as well as statistically significant decrease in postvoid residual at both time points (82 mL and 77 mL, respectively). Furthermore, there were significant decreases in voiding pressure at maximum flow with treatment (Radziszewski, 2013).

A systematic review of electrical stimulation treatments used for bladder dysfunction after spinal cord injury included studies with varying sites of electrode stimulation such as bladder wall, skin, peripheral nerves, sacral roots, and spinal cord (McGee et al., 2015). Bladder wall stimulation has not been successful as it results in activation of the sphincter as well due to activation of pelvic afferents in the bladder, thus causing detrusor sphincter dyssynergia (DSD) (Hald et al., 1967). Stimulation of the pelvic nerve in dogs, while resulting in bladder contraction, also caused DSD (Ju & Liao, 2016). Spinal cord stimulation, while successful in animal studies, is an invasive procedure that has not shown much benefit over sacral root simulation (McGee et al., 2015). Finally, transcutaneous electrical stimulation (often conducted over the dorsal nerve of the penis), as discussed previously, is successful in improving bladder dysfunction. However, given the location of the stimulation, its utility is somewhat limited.

Sacral anterior root, sacral nerve, percutaneous tibial nerve, and pudendal nerve stimulation have varying degrees of success among patients with spinal cord injuries. Sacral anterior root stimulation, while successful in producing bladder contractions and reducing residual volumes, urinary tract infections, and aiding in concurrent fecal incontinence, also requires transection of the sacral dorsal root (posterior rhizotomy). Posterior rhizotomy aids in increasing bladder capacity and compliance and improving voiding. However, it has long-term effects including the elimination of reflex erection, micturition, and pelvic sensation (Gaunt & Prochazka, 2006; Van Kerrebroeck et al., 1996). Sacral nerve stimulation, an effective and commonly used treatment for urge incontinence and urinary retention, has not been found effective in patients with complete spinal cord injuries (Schurch et al., 2003). However, among patients with incomplete spinal cord lesions, the effectiveness of sacral nerve stimulation is more promising. It has been shown to improve continence by increasing capacity and improve incontinence by increasing urinary frequency and voided volume. However, the study with the above promising results was only conducted among 24 subjects. A recent systematic review/meta-analysis of all eight research studies on sacral neuromodulation in spinal cord injury patients between 1971 and 2018 found an overall merged success rate of 75%, which was not statistically significant (Hu et al., 2019). Thus, larger scale randomized control trials are necessary to prove effectiveness of sacral nerve stimulation among incomplete spinal cord injury patients.

Percutaneous tibial and pudendal nerve stimulation also have varying degrees of success among spinal cord injury patients. Percutaneous tibial nerve stimulation (PTNS), less invasive as compared to sacral nerve stimulation as it does not require a surgical implantation, has had limited success in patients with neurological lesions. One study, conducted among 29 patients with bladder dysfunction secondary to various neurological lesions (15 patient with spinal cord injuries) found that about 50% had improvement in either capacity or voided volume (Amarenco et al., 2003). However, PTNS has been studied much more extensively for overactive bladder (with multiple randomized control trials) with great clinical improvement. Pudendal nerve stimulation, conducted through penile squeeze, inhibits the bladder and decreases bladder contractions. One study comparing sacral nerve to pudendal nerve stimulation for treatment of overactive bladder found pudendal nerve stimulation to produce a greater reduction in urinary symptoms (Peters et al., 2005). However, as with PTNS, studies involving pudendal nerve stimulation for the treatment of bladder dysfunction secondary to spinal cord injury are limited and are a great avenue of future research.

Nerve grafting/nerve transfer is much less studied as compared to neuromodulation. This technique, initially conducted among 12 paraplegic patients with neurogenic bladders and detrusor external sphincter dyssynergia in China, involves anastamosing the proximal end of the S1 ventral root and the distal end of S2 and/or S3 ventral roots. Thus, an Achilles tendon to bladder reflex was created (Lin et al., 2009). These 12 patients had an increase in bladder capacity, decrease in urinary residuals, no urinary tract infections, and improvement in renal function (Lin et al., 2009). In a follow-up research study in patients with low conus medullaris lesions of S2-S5, seven of nine patients regained satisfactory bladder function for 8–12 months (Lin & Hou, 2013). While the treatment has proven effective among these small cohorts, its widespread application is limited given its invasive nature. However, it is likely plausible to conduct larger-scale studies, especially with patients who have failed alternative, less-invasive treatment options.

Future research directions

There remains great potential for future research in treating LUTS secondary to pathological insults to the neural pathways that maintain continence. While we have a strong understanding of these neural pathways themselves, our understanding of ways to target these pathways as well as the mechanisms of action of some frequented treatments is still lacking. The avenues of future research include application of our findings in functional brain imaging in the treatment of urge incontinence and FS, targeting of specific biochemical receptors in the treatment of suprapontine lesions (Parkinson's disease, CVAs), further elicitation of the mechanism of action and application of neuromodulation in the treatment of IC/BPS and spinal cord injuries, nerve grafting/transfer for spinal cord injuries, and utilization of NGF in the treatment of diabetic cystopathy. These have all been addressed in their corresponding sections above.

Bladder tissue engineering and stem cell research are another two very exciting yet exceedingly complex other avenues of potential research. While the concept of bladder tissue engineering was developed in the 1950s, and it has been used experimentally in animals, clinical application in humans has been more challenging (Taweel & Seyam, 2015). Given the numerous functional requirements of the bladder (elasticity, contractility), as well as its anatomic complexity, it can be difficult to replicate through bladder engineering. Furthermore, mesenchymal stem cells have been proposed for bladder tissue engineering. However, there is a risk of tumor induction as well as urine toxicity with the usage of multipotent stem cells (Lam Van Ba et al., 2015) While with definite challenges to its application, bladder tissue engineering still remains an interesting and promising alternative to gastrointestinal techniques to bladder augmentation or replacement, which are often accompanied by various complications including electrolyte imbalances, vitamin deficiencies, stone formation, and cancer. Potential future research directions involving bladder tissue engineering include management of bladder cancer (muscle invasive), neurogenic bladder conditions that result in renal compromise, severe bladder injury due to radiation, refractory female incontinence, chronic pain syndromes, and congenital syndromes such as myelomeningocele and bladder exstrophy. While basic science has almost impeccably outlined the neural pathways of micturition, there remains great work to be done to apply this knowledge clinically for the treatment of neurological conditions resulting in LUTS.

References

Abdel-Azim, M., Sullivan, M., & Yalla, S. V. (1991). Disorders of bladder function in spinal cord disease. *Neurologic Clinics, 9*(3), 727–740.

Al'-Shukri, S. K., Kuz'min, I. V., Slesarevskaya, M. N., & Ignashov, Y. A. (2018). Bladder hydrodistension in treating patients with interstytial cystitis/bladder pain syndrome. *Urologiia, 1*, 26–29.

Amarenco, G., Ismael, S. S., Even-Schneider, A., Raibaut, P., Demaille-Wlodyka, S., Parratte, B., & Kerdraon, J. (2003). Urodynamic effect of acute transcutaneous posterior tibial nerve stimulation in overactive bladder. *The Journal of Urology, 169*(6), 2210–2215.

Apfel, S. C. (1999). Neurotrophic factors and diabetic peripheral neuropathy. *European Neurology, 41*(Suppl. 1), 27–34.

Araki, I., Kitahara, M., Oida, T., & Kuno, S. (2000). Voiding dysfunction and Parkinson's disease: Urodynamic abnormalities and urinary symptoms. *The Journal of Urology, 164*(5), 1640–1643.

Blok, B. F., De Weerd, H., & Holstege, G. (1995). Ultrastructural evidence for a paucity of projections from the lumbosacral cord to the pontine micturition center or M-region in the cat: A new concept for the organization of the micturition reflex with the periaqueductal gray as central relay. *The Journal of Comparative Neurology, 359*(2), 300–309.

Blok, B. F. M., Groen, J., Bosch, J. L. H. R., Veltman, D. J., & Lammertsma, A. A. (2006). Different brain effects during chronic and acute sacral neuromodulation in urge incontinent patients with implanted neurostimulators. *BJU International, 98*(6), 1238–1243.

Braun, P. M., Baezner, H., Seif, C., Boehler, G., Bross, S., Eschenfelder, C. C., Alken, P., Hennerici, M., & Juenemann, P. (2002). Alterations of cortical electrical activity in patients with sacral neuromodulator. *European Urology, 41*(5), 562–566. discussion 566-7.

Brusa, L., Petta, F., Pisani, A., Miano, R., Stanzione, P., Moschella, V., Galati, S., & Finazzi Agrò, E. (2006). Central acute D2 stimulation worsens bladder function in patients with mild Parkinson's disease. *The Journal of Urology, 175*(1), 202–206. discussion 206-7.

Carmichael, S. T., & Price, J. L. (1995). Limbic connections of the orbital and medial prefrontal cortex in macaque monkeys. *The Journal of Comparative Neurology, 363*(4), 615–641.

Chai, T. C., Gray, M. L., & Steers, W. D. (1998). The incidence of a positive ice water test in bladder outlet obstructed patients: Evidence for bladder neural plasticity. *The Journal of Urology, 160*(1), 34–38.

Chai, T. C., Zhang, C., Warren, J. W., & Keay, S. (2000). Percutaneous sacral third nerve root neurostimulation improves symptoms and normalizes urinary HB-EGF levels and antiproliferative activity in patients with interstitial cystitis. *Urology, 55*(5), 643–646.

Christ, G. J., Day, N. S., Day, M., Zhao, W., Persson, K., Pandita, R. K., & Andersson, K.-E. (2003). Increased connexin43-mediated intercellular communication in a rat model of bladder overactivity in vivo. *American Journal of Physiology - Regulatory, Integrative and Comparative Physiology, 284*(5), R1241–R1248.

Clarke, S. J., & Thomas, D. G. (1981). Characteristics of the urethral pressure profile in flaccid male paraplegics. *British Journal of Urology, 53*(2), 157–161.

Comiter, C. V. (2003). Sacral neuromodulation for the symptomatic treatment of refractory interstitial cystitis: A prospective study. *The Journal of Urology, 169*(4), 1369–1373.

Daneshgari, F., Liu, G., Birder, L., Hanna-Mitchell, A. T., & Chacko, S. (2009). Diabetic bladder dysfunction: Current translational knowledge. *The Journal of Urology, 182*(6 Suppl. l), S18–S26. https://doi.org/10.1016/j.juro.2009.08.070

Elkelini, M. S., Abuzgaya, A., & Hassouna, M. M. (2010). Mechanisms of action of sacral neuromodulation. *International Urogynecology Journal, 21*(Suppl. 2), S439–S446. https://doi.org/10.1007/s00192-010-1273-3

Fowler, C. J., Griffiths, D., & de Groat, W. C. (2008). The neural control of micturition. *Nature Reviews Neuroscience, 9*(6), 453–466. https://doi.org/10.1038/nrn2401

Gaunt, R. A., & Prochazka, A. (2006). Control of urinary bladder function with devices: Successes and failures. *Progress in Brain Research, 152*, 163–194.

Gilpin, S. A., Gilpin, C. J., Dixon, J. S., Gosling, J. A., & Kirby, R. S. (1986). The effect of age on the autonomic innervation of the urinary bladder. *British Journal of Urology, 58*(4), 378–381.

Griffiths, D. J. (2004). Cerebral control of bladder function. *Current Urology Reports, 5*(5), 348–352.

Griffiths, D., Clarkson, B., Tadic, S. D., & Resnick, N. M. (2015). Brain mechanisms underlying urge incontinence and its response to pelvic floor muscle training. *The Journal of Urology, 194*(3), 708–715. https://doi.org/10.1016/j.juro.2015.03.102

Griffiths, D., Derbyshire, S., Stenger, A., & Resnick, N. (2005). Brain control of normal and overactive bladder. *The Journal of Urology, 174*(5), 1862–1867.

Griffiths, D., & Fowler, C. (2010). Brain imaging in fowler's syndrome. *Current Bladder Dysfunction Reports, 5*, 114–118. https://doi.org/10.1007/s11884-010-0045-8

Griffiths, D., & Tadic, S. D. (2008). Bladder control, urgency, and urge incontinence: Evidence from functional brain imaging. *Neurourology and Urodynamics, 27*(6), 466–474.

de Groat, W. C., Griffiths, D., & Yoshimura, N. (2015). Neural control of the lower urinary tract. *Comprehensive Physiology, 5*(1).

Hald, T., Meier, W., Khalili, A., Agrawal, G., Benton, J. G., & Kantrowitz, A. (1967). Clinical experience with a radio-linked bladder stimulator. *The Journal of Urology, 97*(1), 73–78.

Holstege, G. (2005). Micturition and the soul. *The Journal of Comparative Neurology, 493*(1), 15–20.

Holstege, G., Griffiths, D., de Wall, H., & Dalm, E. (1986). Anatomical and physiological observations on supraspinal control of bladder and urethral sphincter muscles in the cat. *The Journal of Comparative Neurology, 250*(4), 449–461.

Holstege, G., & Mouton, L. J. (2003). Central nervous system control of micturition. *International Review of Neurobiology, 56*, 123–145.

Hotta, H., Morrison, J. F., Sato, A., & Uchida, S. (1995). The effects of aging on the rat bladder and its innervation. *The Japanese Journal of Physiology, 45*(5), 823–836.

Hu, M., Lai, S., Zhang, Y., Liu, M., & Wang, J. (2019). Sacral neuromodulation for lower urinary tract dysfunction in spinal cord injury: A systematic review and meta-analysis. *Urologia Internationalis, 103*(3), 337–343. https://doi.org/10.1159/000501529

Jänig, W., & Morrison, J. F. (1986). Functional properties of spinal visceral afferents supplying abdominal and pelvic organs, with special emphasis on visceral nociception. *Progress in Brain Research, 67*, 87–114.

Ju, Y.-H., & Liao, L.-M. (2016). Electrical stimulation of dog pudendal nerve regulates the excitatory pudendal-to-bladder reflex. *Neural Regeneration Research, 11*(4), 676–681. https://doi.org/10.4103/1673-5374.180757

Keast, J. R., Smith-Anttila, C. J. A., & Osborne, P. B. (2015). Developing a functional urinary bladder: A neuronal context. *Frontiers in Cell and Developmental Biology, 3*, 53. https://doi.org/10.3389/fcell.2015.00053

Khullar, V., Digesu, G. A., Veit-Rubin, N., Sahai, A., Rahnama'i, M. S., Tarcan, T., Chermansky, C., & Dmochowski, R. (2019). How can we improve the diagnosis and management of bladder pain syndrome? Part 2:ICI-RS 2018. *Neurourology and Urodynamics, 38*(Suppl. 5), S71–S81. https://doi.org/10.1002/nau.24245

Kim, S.-O., Oh, B. S., Chang, I.-Y., Song, S. H., Ahn, K., Hwang, E. C., Oh, K. J., Kwon, D., & Park, K. (2011). Distribution of interstitial cells of Cajal and expression of nitric oxide synthase after experimental bladder outlet obstruction in a rat model of bladder overactivity. *Neurourology and Urodynamics, 30*(8), 1639–1645. https://doi.org/10.1002/nau.21144

Kitta, T., Chancellor, M. B., de Groat, W. C., Kuno, S., Nonomura, K., & Yoshimura, N. (2012). Suppression of bladder overactivity by adenosine A2A receptor antagonist in a rat model of Parkinson disease. *The Journal of Urology, 187*(5), 1890–1897. https://doi.org/10.1016/j.juro.2011.12.062

Kitta, T., Kakizaki, H., Furuno, T., Moriya, K., Tanaka, H., Shiga, T., Tamaki, N., Yabe, I., Sasaki, H., & Nonomura, K. (2006). Brain activation during detrusor overactivity in patients with Parkinson's disease: A positron emission tomography study. *The Journal of Urology, 175*(3 Pt 1), 994–998.

Kuhtz-Buschbeck, J. P., van der Horst, C., Wolff, S., Filippow, N., Nabavi, A., Jansen, O., & Braun, P. M. (2007). Activation of the supplementary motor area (SMA) during voluntary pelvic floor muscle contractions–an fMRI study. *NeuroImage, 35*(2), 449–457.

Lam Van Ba, O., Aharony, S., Loutochin, O., & Corcos, J. (2015). Bladder tissue engineering: A literature review. *Advanced Drug Delivery Reviews, 82–83*, 31–37. https://doi.org/10.1016/j.addr.2014.11.013

Lazzeri, M., Beneforti, P., Benaim, G., Maggi, C. A., Lecci, A., & Turini, D. (1996). Intravesical capsaicin for treatment of severe bladder pain: A randomized placebo controlled study. *The Journal of Urology, 156*(3), 947–952.

Limberg, B. J., Andersson, K.-E., Aura Kullmann, F., Burmer, G., de Groat, W. C., & Rosenbaum, J. S. (2010). β-Adrenergic receptor subtype expression in myocyte and non-myocyte cells in human female bladder. *Cell and Tissue Research, 342*(2), 295–306. https://doi.org/10.1007/s00441-010-1053-x

Lin, H., & Hou, C. (2013). Transfer of normal S1 nerve root to reinnervate atonic bladder due to conus medullaris injury. *Muscle and Nerve, 47*(2), 241–245. https://doi.org/10.1002/mus.23507

Lin, H., Hou, C., Zhen, X., & Xu, Z. (2009). Clinical study of reconstructed bladder innervation below the level of spinal cord injury to produce urination by Achilles tendon-to-bladder reflex contractions. *Journal of Neurosurgery: Spine, 10*(5), 452–457. https://doi.org/10.3171/2009.1.SPINE08540

Mallory, B. S., Roppolo, J. R., & de Groat, W. C. (1991). Pharmacological modulation of the pontine micturition center. *Brain Research, 546*(2), 310–320.

Mastri, A. R. (1980). Neuropathology of diabetic neurogenic bladder. *Annals of Internal Medicine, 92*(2 Pt 2), 316–318.

McGee, M. J., Amundsen, C. L., & Grill, W. M. (2015). Electrical stimulation for the treatment of lower urinary tract dysfunction after spinal cord injury. *The Journal of Spinal Cord Medicine, 38*(2), 135–146. https://doi.org/10.1179/2045772314Y.0000000299

McGuire, E. J., & Savastano, J. A. (1983). Long-term followup of spinal cord injury patients managed by intermittent catheterization. *The Journal of Urology, 129*(4), 775–776.

Morgan, C. W., de Groat, W. C., Felkins, L. A., & Zhang, S. J. (1991). Axon collaterals indicate broad intraspinal role for sacral preganglionic neurons. *Proceedings of the National Academy of Sciences of the United States of America, 88*(15), 6888–6892.

Mouritsen, L., Kronschnabl, M., & Lose, G. (2010). Long-term results of vaginal repairs with and without xenograft reinforcement. *International Urogynecology Journal, 21*(4), 467–473. https://doi.org/10.1007/s00192-009-1061-0

Payne, C. K., Mosbaugh, P. G., Forrest, J. B., Evans, R. J., Whitmore, K. E., Antoci, J. P., Perez-Marrero, R., Jacoby, K., Diokno, A. C., O'Reilly, K. J., Griebling, T. L., Vasavada, S. P., Yu, A. S., & Frumkin, L. R. (2005). Intravesical resiniferatoxin for the treatment of interstitial cystitis: A randomized, double-blind, placebo controlled trial. *The Journal of Urology, 173*(5), 1590–1594.

Peters, K. M., Feber, K. M., & Bennett, R. C. (2005). Sacral versus pudendal nerve stimulation for voiding dysfunction: A prospective, single-blinded, randomized, crossover trial. *Neurourology and Urodynamics, 24*(7), 643–647.

Pfisterer, M. H.-D., Griffiths, D. J., Rosenberg, L., Schaefer, W., & Resnick, N. M. (2006). The impact of detrusor overactivity on bladder function in younger and older women. *The Journal of Urology, 175*(5), 1777–1783. discussion 1783.

Radziszewski, K. (2013). Outcomes of electrical stimulation of the neurogenic bladder: Results of a two-year follow-up study. *NeuroRehabilitation, 32*(4), 867–873. https://doi.org/10.3233/NRE-130911

Ranson, R. N., Dodds, A. L., Smith, M. J., Santer, R. M., & Watson, A. H. D. (2003). Age-associated changes in the monoaminergic innervation of rat lumbosacral spinal cord. *Brain Research, 972*(1–2), 149–158.

Rudy, D. C., Awad, S. A., & Downie, J. W. (1988). External sphincter dyssynergia: An abnormal continence reflex. *The Journal of Urology, 140*(1), 105–110.

Sakakibara, R., Hattori, T., Yasuda, K., & Yamanishi, T. (1996). Micturitional disturbance after acute hemispheric stroke: Analysis of the lesion site by CT and MRI. *Journal of the Neurological Sciences, 137*(1), 47–56.

Santer, R. M., Dering, M. A., Ranson, R. N., Waboso, H. N., & Watson, A. H. D. (2002). Differential susceptibility to ageing of rat preganglionic neurones projecting to the major pelvic ganglion and of their afferent inputs. *Autonomic Neuroscience: Basic & Clinical, 96*(1), 73–81.

Sasaki, K., Chancellor, M. B., Goins, W. F., Phelan, M. W., Glorioso, J. C., de Groat, W. C., & Yoshimura, N. (2004). Gene therapy using replication-defective herpes simplex virus vectors expressing nerve growth factor in a rat model of diabetic cystopathy. *Diabetes, 53*(10), 2723–2730.

Schurch, B., Reilly, I., Reitz, A., & Curt, A. (2003). Electrophysiological recordings during the peripheral nerve evaluation (PNE) test in complete spinal cord injury patients. *World Journal of Urology, 20*(6), 319–322.

Steers, W. D., Ciambotti, J., Erdman, S., & de Groat, W. C. (1990). Morphological plasticity in efferent pathways to the urinary bladder of the rat following urethral obstruction. *Journal of Neuroscience: The Official Journal of the Society for Neuroscience, 10*(6), 1943–1951.

Steers, W. D., Ciambotti, J., Etzel, B., Erdman, S., & de Groat, W. C. (1991). Alterations in afferent pathways from the urinary bladder of the rat in response to partial urethral obstruction. *The Journal of Comparative Neurology, 310*(3), 401–410.

Steers, W. D., Kolbeck, S., Creedon, D., & Tuttle, J. B. (1991). Nerve growth factor in the urinary bladder of the adult regulates neuronal form and function. *The Journal of Clinical Investigation, 88*(5), 1709–1715.

Taweel, W. A., & Seyam, R. (2015). Neurogenic bladder in spinal cord injury patients. *Research and Reports in Urology, 7*, 85–99. https://doi.org/10.2147/RRU.S29644

Theoharides, T. C., Kempuraj, D., & Sant, G. R. (2001). Mast cell involvement in interstitial cystitis: A review of human and experimental evidence. *Urology, 57*(6 Suppl. 1), 47–55.

Torimoto, K., Fraser, M. O., Hirao, Y., De Groat, W. C., Chancellor, M. B., & Yoshimura, N. (2004). Urethral dysfunction in diabetic rats. *The Journal of Urology, 171*(5), 1959–1964.

Tyagi, P., Moon, C.-H., Janicki, J., Kaufman, J., Chancellor, M., Yoshimura, N., & Chermansky, C. (2018). Recent advances in imaging and understanding interstitial cystitis. *F1000Research, 7*. https://doi.org/10.12688/f1000research.16096.1

Van Kerrebroeck, P. E., Koldewijn, E. L., Rosier, P. F., Wijkstra, H., & Debruyne, F. M. (1996). Results of the treatment of neurogenic bladder dysfunction in spinal cord injury by sacral posterior root rhizotomy and anterior sacral root stimulation. *The Journal of Urology, 155*(4), 1378–1381.

Whitmore, K. E., Payne, C. K., Diokno, A. C., & Lukban, J. C. (2003). Sacral neuromodulation in patients with interstitial cystitis: A multicenter clinical trial. *International Urogynecology Journal and Pelvic Floor Dysfunction, 14*(5), 305–308. discussion 308-9.

Yoshimura, N., Kitta, T., Kadekawa, K., Miyazato, M., & Shimizu, T. (2020). [Overview of pharmacological mechanisms controlling micturition in the central nervous system]. Nihon Yakurigaku Zasshi. *Folia Pharmacologica Japonica, 155*(1), 4–9. https://doi.org/10.1254/fpj.19107

Yoshimura, N., Oguchi, T., Yokoyama, H., Funahashi, Y., Yoshikawa, S., Sugino, Y., Kawamorita, N., Kashyap, M. P., Chancellor, M. B., Tyagi, P., & Ogawa, T. (2014). Bladder afferent hyperexcitability in bladder pain syndrome/interstitial cystitis. *International Journal of Urology: Official Journal of the Japanese Urological Association, 21*(Suppl. 1), 18–25. https://doi.org/10.1111/iju.12308

Yoshiyama, M., & de Groat, W. C. (2005). Supraspinal and spinal alpha-amino-3-hydroxy-5-methylisoxazole-4-propionic acid and N-methyl-D-aspartate glutamatergic control of the micturition reflex in the urethane-anesthetized rat. *Neuroscience, 132*(4), 1017–1026.

Zhang, C.-O., Li, Z.-L., & Kong, C.-Z. (2005). APF, HB-EGF, and EGF biomarkers in patients with ulcerative vs. non-ulcerative interstitial cystitis. *BMC Urology, 5*, 7.

Further reading

Boselli, C., Govoni, S., Condino, A. M., & D'Agostino, G. (n.d.). Bladder instability: A re-appraisal of classical experimental approaches and development of new therapeutic strategies. Journal of Autonomic Pharmacology, 21(5–6), 219–229.

Yokoyama, O., Yoshiyama, M., Namiki, M., & de Groat, W. C. (1999). Glutamatergic and dopaminergic contributions to rat bladder hyperactivity after cerebral artery occlusion. *American Journal of Physiology, 276*(4), R935–R942. https://doi.org/10.1152/ajpregu.1999.276.4.R935

Index

'*Note:* Page numbers followed by "f" indicate figures and "t" indicate tables.'

C

M

N

O

P

Q

R

S

T

U

V

W

9780128224557